HUMAN AGING

HUMAN AGING
From Cellular Mechanisms to Therapeutic Strategies

Edited by

CALOGERO CARUSO
Professor Emeritus, Department of Biomedicine, Neurosciences and
Advanced Diagnostics, University of Palermo, Palermo, Italy

GIUSEPPINA CANDORE
Associate Professor of General Pathology, Laboratory of
Immunopathology and Immunosenescence, Department of Biomedicine,
Neurosciences and Advanced Diagnostics, University of Palermo,
Palermo, Italy

ACADEMIC PRESS
An imprint of Elsevier

ELSEVIER

Academic Press is an imprint of Elsevier
125 London Wall, London EC2Y 5AS, United Kingdom
525 B Street, Suite 1650, San Diego, CA 92101, United States
50 Hampshire Street, 5th Floor, Cambridge, MA 02139, United States
The Boulevard, Langford Lane, Kidlington, Oxford OX5 1GB, United Kingdom

Notices
Knowledge and best practice in this field are constantly changing. As new research and experience broaden our understanding, changes in research methods, professional practices, or medical treatment may become necessary.

Practitioners and researchers must always rely on their own experience and knowledge in evaluating and using any information, methods, compounds, or experiments described herein. In using such information or methods they should be mindful of their own safety and the safety of others, including parties for whom they have a professional responsibility.

To the fullest extent of the law, neither the Publisher nor the authors, contributors, or editors, assume any liability for any injury and/or damage to persons or property as a matter of products liability, negligence or otherwise, or from any use or operation of any methods, products, instructions, or ideas contained in the material herein.

Library of Congress Cataloging-in-Publication Data
A catalog record for this book is available from the Library of Congress

British Library Cataloguing-in-Publication Data
A catalogue record for this book is available from the British Library

ISBN 978-0-12-822569-1

For information on all Academic Press publications
visit our website at https://www.elsevier.com/books-and-journals

Publisher: Wolff, Andre Gerhard
Editorial Project Manager: Ashdown, Megan
Production Project Manager: Raviraj, Selvaraj
Cover Designer: Hitchen, Miles

Typeset by SPi Global, India

Contents

Contributors xi
About the editors xv
Preface xvii

1. Aging and longevity: An evolutionary approach 1

Giuseppina Candore and Calogero Caruso

1.1 Introduction 1
1.2 Why does aging occur? 3
1.3 Mechanisms of aging 6
1.4 Causality and chance in aging and longevity 7
1.5 Conclusions and future perspectives 9
References 10

2. Demographic aspects of aging 13

Annalisa Busetta and Filippa Bono

2.1 Introduction 13
2.2 Understanding the process: Browsing around the demographic
transition theories 20
2.3 Aging inequalities 22
2.4 Conclusions and perspectives 31
References 33

3. Pathobiology of aging: An introduction to age-related diseases 35

Calogero Caruso, Giulia Accardi, Mattia Emanuela Ligotti, Sonya Vasto, and
Giuseppina Candore

3.1 Introduction 35
3.2 Complexity 37
3.3 Hallmarks of aging 41
3.4 Genomic instability 41
3.5 Epigenetic alteration 43
3.6 Deregulated nutrient sensing pathways 46
3.7 Loss of proteostasis 54
3.8 Mitochondrial dysfunction 55
3.9 Telomere attrition 60

3.10 Cellular senescence   60
3.11 Stem cell exhaustion   61
3.12 Altered intercellular communication   63
3.13 Cancer and aging   63
3.14 Conclusion and future perspectives   66
References   67

4. Cellular senescence and senescence-associated secretory phenotype (SASP) in aging process   **75**

Fabiola Olivieri, Antonio Domenico Procopio, and Maria Rita Rippo

4.1 Introduction   75
4.2 Signaling pathway stimulating the appearance of SASP   76
4.3 SASP components   78
4.4 MiRNA and extracellular vesicles as new regulators and components of SASP   79
4.5 SASP profile in different cell types   80
4.6 Cellular senescence, SASP, and aging   82
4.7 Conclusions and future perspectives   83
References   84

5. The role of inflammaging in the development of chronic diseases of older people   **89**

Jacek M. Witkowski, Ewa Bryl, and Tamas Fulop

5.1 Introduction   89
5.2 Basic mechanisms: Cellular senescence, inflammaging, molecular inflammation, and senoinflammation   89
5.3 Is chronic inflammatory state a common denominator of ARDs?   93
5.4 The case of COVID-19   97
5.5 Proposed interventions to prevent ARDs   98
5.6 Conclusion and future perspective   99
References   99

6. A new perspective on ROS in aging with an integrated view of the gut microbiota   **105**

Lu Wu and Ciriaco Carru

6.1 Introduction   105
6.2 The reactive oxygen species   105
6.3 The biological function of ROS   107
6.4 The oxidative stress theory of aging   107
6.5 ROS signaling in gut barrier, inflammation, and dysbiosis of gut microbiota in aging   108

6.6 Conclusion and future perspective   109
Acknowledgments   109
References   109

7. Aging of immune system   113

Marcello Pinti, Sara De Biasi, Lara Gibellini, Domenico Lo Tartaro, Anna De Gaetano,
Marco Mattioli, Lucia Fidanza, Milena Nasi, and Andrea Cossarizza

7.1 Introduction   113
7.2 Changes in the adaptive immunity   114
7.3 Changes in the innate immunity   116
7.4 Inflammaging   119
7.5 Conclusions and future perspectives   121
References   122

8. Vaccination in old age: Challenges and promises   129

Calogero Caruso, Anna Aiello, Graham Pawelec, and Mattia Emanuela Ligotti

8.1 Introduction   129
8.2 The state of the art   130
8.3 Challenges and promises   139
8.4 Conclusion and future perspectives   146
Note added in proof   147
Funding   148
References   148

9. Resilience signaling and hormesis in brain health and disease   155

Vittorio Calabrese, Angela Trovato, Maria Scuto, Maria Laura Ontario,
Mario Tomasello, Rosario Perrotta, and Edward Calabrese

9.1 Introduction   155
9.2 Regional specificity of brain resilience and vulnerability to stress   158
9.3 Hydrogen sulfide: A resilient signaling molecule in brain disorders   160
9.4 Plant polyphenols improve resilience and brain health via "Vitagenes" 166
9.5 Conclusions and future perspectives   169
References   170

10. Different components of frailty in the aging subjects—The role of
sarcopenia   173

Paolina Crocco, Serena Dato, Francesca Iannone, Giuseppe Passarino,
and Giuseppina Rose

10.1 Frailty definition and assessment   173
10.2 Physical frailty and sarcopenia: two sides of the same coin   176

10.3 Cellular and molecular mechanisms of Sarcopenia   177

10.4 Genetic components of sarcopenia   185

10.5 Lifestyle risk factors for sarcopenia   188

10.6 Management of sarcopenia   189

10.7 Conclusions and future perspectives   192

References   193

11. Hormones in aging   207

Andrea Sansone and Francesco Romanelli

11.1 Introduction   207

11.2 Endocrine physiology: The role of the pituitary gland and hypothalamus   207

11.3 Gonadal function in aging   208

11.4 Growth hormone and aging   210

11.5 Adrenal function in aging   211

11.6 Thyroid function in aging   212

11.7 Conclusions and future perspective   214

References   214

12. Chronobiology and chrononutrition: Relevance for aging   219

Damiano Galimberti and Giuseppe Mazzola

12.1 Introduction   219

12.2 Biorhythms   220

12.3 Central oscillator and peripheral oscillators   222

12.4 Clock-controlled genes   225

12.5 Biological clock modulation and chronodisruption   229

12.6 Diet, circadian rhythm, aging, and longevity   234

12.7 Meals composition for successful aging   239

12.8 Conclusion and future perspectives   241

References   242

13. Nutraceutical approach to age-related diseases—The clinical evidence on cognitive decline   255

Arrrgo F.G. Cicero and Alessandro Colletti

13.1 Introduction   255

13.2 Data selection   256

13.3 The state of the art   257

13.4 Conclusions and future perspectives   266

References   267

14. Ways to become old: Role of lifestyle in modulation of the hallmarks of aging   **273**

Giulia Accardi and Anna Aiello

14.1 Introduction   273
14.2 Primary hallmarks and lifestyle   274
14.3 Antagonistic hallmarks and lifestyle   281
14.4 Integrative hallmarks and lifestyle   284
14.5 Conclusion and future perspectives   286
References   287

15. Nutritional biomarkers in aging research   **295**

Sergio Davinelli and Giovanni Scapagnini

15.1 Introduction   295
15.2 Minerals (zinc and selenium)   297
15.3 Vitamins   299
15.4 Polyunsaturated fatty acids   301
15.5 Carotenoids   303
15.6 Polyphenols   305
15.7 Molecular biomarkers of aging and nutrition   307
15.8 Conclusions and future perspectives   310
References   311

16. The role of cytomegalovirus in organismal and immune aging   **319**

Christopher P. Coplen, Mladen Jergović, and Janko Nikolich-Žugich

16.1 Introduction   319
16.2 Host immune response to CMV   320
16.3 CMV, longevity, and chronic diseases   322
16.4 CMV and immune aging   324
16.5 Conclusion and future perspectives   324
References   325

17. Ethics of aging   **329**

Lucia Craxì

17.1 Population aging: The challenges   329
17.2 Moral and social attitudes to old age   330
17.3 Ethics of aging   332
17.4 Fair allocation of medical resources   333
17.5 Conclusion and future perspectives   337
References   338

18. Conclusions. Slowing aging and fighting age-related diseases, from bench to bedside?   **341**

Calogero Caruso and Giuseppina Candore

18.1 Introduction   341
18.2 Aging and gender medicine   343
18.3 The role of immune-inflammatory responses in aging and age-related diseases, and therapeutic interventions   346
18.4 Slowing aging and fighting age-related diseases   348
18.5 Conclusions and future perspectives   351
References   352

Index   *355*

Contributors

Giulia Accardi
Laboratory of Immunopathology and Immunosenescence, Department of Biomedicine, Neurosciences and Advanced Diagnostics, University of Palermo, Palermo, Italy

Anna Aiello
Laboratory of Immunopathology and Immunosenescence, Department of Biomedicine, Neurosciences and Advanced Diagnostics, University of Palermo, Palermo, Italy

Sara De Biasi
Department of Medical and Surgical Sciences of Children and Adults, University of Modena and Reggio Emilia, Modena, Italy

Filippa Bono
Department of Economics, Business and Statistics, University of Palermo, Palermo, Italy

Ewa Bryl
Department of Pathology and Experimental Rheumatology, Medical University of Gdansk, Gdansk, Poland

Annalisa Busetta
Department of Economics, Business and Statistics, University of Palermo, Palermo, Italy

Edward Calabrese
Department of Environmental Health Sciences, University of Massachusetts, Amherst, MA, United States

Vittorio Calabrese
Department of Biomedical and Biotechnological Sciences, School of Medicine, University of Catania, Catania, Italy

Giuseppina Candore
Laboratory of Immunopathology and Immunosenescence, Department of Biomedicine, Neurosciences and Advanced Diagnostics, University of Palermo, Palermo, Italy

Ciriaco Carru
Department of Biomedical Sciences, University of Sassari, Hospital University (AOUSS), Sassari, Italy

Calogero Caruso
Laboratory of Immunopathology and Immunosenescence, Department of Biomedicine, Neurosciences and Advanced Diagnostics, University of Palermo, Palermo, Italy

Arrrgo F.G. Cicero

Medical and Surgery Sciences Department, Alma Mater Studiorum University of Bologna; Italian Nutraceutical Society (SINut), Bologna, Italy

Alessandro Colletti

Italian Nutraceutical Society (SINut), Bologna; Pharmacology Department, University of Turin, Turin, Italy

Christopher P. Coplen

Department of Immunobiology, University of Arizona Center on Aging, University of Arizona College of Medicine-Tucson, Tucson, AZ, United States

Andrea Cossarizza

Department of Medical and Surgical Sciences of Children and Adults, University of Modena and Reggio Emilia, Modena, Italy

Lucia Craxì

Department of Biomedicine, Neuroscience and Advanced Diagnostics (Bi.N.D.), Section of Pathology, University of Palermo, Palermo, Italy

Paolina Crocco

Department of Biology, Ecology and Earth Sciences, University of Calabria, Rende, Italy

Serena Dato

Department of Biology, Ecology and Earth Sciences, University of Calabria, Rende, Italy

Sergio Davinelli

Department of Medicine and Health Sciences "V. Tiberio", University of Molise, Campobasso, Italy

Lucia Fidanza

Department of Medical and Surgical Sciences of Children and Adults, University of Modena and Reggio Emilia, Modena, Italy

Tamas Fulop

Department of medicine, Research Center on Aging, Graduate Program in Immunology, Faculty of Medicine and Health Sciences, University of Sherbrooke, Sherbrooke, QC, Canada

Anna De Gaetano

Department of Life Sciences, University of Modena and Reggio Emilia, Modena, Italy

Damiano Galimberti

Italian Association of Anti-Ageing Physicians, Milano, Italy

Lara Gibellini

Department of Medical and Surgical Sciences of Children and Adults, University of Modena and Reggio Emilia, Modena, Italy

Francesca Iannone
Department of Biology, Ecology and Earth Sciences, University of Calabria, Rende, Italy

Mladen Jergović
Department of Immunobiology, University of Arizona Center on Aging, University of Arizona College of Medicine-Tucson, Tucson, AZ, United States

Mattia Emanuela Ligotti
Laboratory of Immunopathology and Immunosenescence, Department of Biomedicine, Neurosciences and Advanced Diagnostics, University of Palermo, Palermo, Italy

Domenico Lo Tartaro
Department of Medical and Surgical Sciences of Children and Adults, University of Modena and Reggio Emilia, Modena, Italy

Marco Mattioli
Department of Medical and Surgical Sciences of Children and Adults, University of Modena and Reggio Emilia, Modena, Italy

Giuseppe Mazzola
Italian Association of Anti-Ageing Physicians, Milano, Italy

Milena Nasi
Department of Surgery, Medicine, Dentistry and Morphological Sciences, University of Modena and Reggio Emilia, Modena, Italy

Janko Nikolich-Žugich
Department of Immunobiology, University of Arizona Center on Aging, University of Arizona College of Medicine-Tucson, Tucson, AZ, United States

Fabiola Olivieri
Department of Clinical and Molecular Sciences, DISCLIMO, Università Politecnica delle Marche; Center of Clinical Pathology and Innovative Therapy, IRCCS INRCA, Ancona, Italy

Maria Laura Ontario
Department of Biomedical and Biotechnological Sciences, School of Medicine, University of Catania, Catania, Italy

Giuseppe Passarino
Department of Biology, Ecology and Earth Sciences, University of Calabria, Rende, Italy

Graham Pawelec
Department of Immunology, University of Tübingen, Tübingen, Germany; Health Sciences North Research Institute, Sudbury, ON, Canada

Rosario Perrotta
Department of Plastic and Reconstructive Surgery, University of Catania, Catania, Italy

Marcello Pinti
Department of Life Sciences, University of Modena and Reggio Emilia, Modena, Italy

Antonio Domenico Procopio
Department of Clinical and Molecular Sciences, DISCLIMO, Università Politecnica delle Marche; Center of Clinical Pathology and Innovative Therapy, IRCCS INRCA, Ancona, Italy

Maria Rita Rippo
Department of Clinical and Molecular Sciences, DISCLIMO, Università Politecnica delle Marche, Ancona, Italy

Francesco Romanelli
Department of Experimental Medicine, Section of Medical Pathophysiology, Food Science and Endocrinology, Sapienza University of Rome, Rome, Italy

Giuseppina Rose
Department of Biology, Ecology and Earth Sciences, University of Calabria, Rende, Italy

Andrea Sansone
Chair of Endocrinology and Medical Sexology (ENDOSEX), Department of Systems Medicine, University of Rome Tor Vergata, Roma, Italy

Giovanni Scapagnini
Department of Medicine and Health Sciences "V. Tiberio", University of Molise, Campobasso, Italy

Maria Scuto
Department of Biomedical and Biotechnological Sciences, School of Medicine, University of Catania, Catania, Italy

Mario Tomasello
Department of Biomedical and Biotechnological Sciences, School of Medicine, University of Catania, Catania, Italy

Angela Trovato
Department of Biomedical and Biotechnological Sciences, School of Medicine, University of Catania, Catania, Italy

Sonya Vasto
Department of Biological, Chemical and Pharmaceutical Sciences and Technologies, University of Palermo, Palermo, Italy

Jacek M. Witkowski
Department of Pathophysiology, Medical University of Gdansk, Gdansk, Poland

Lu Wu
Institute of Synthetic Biology, Institutes of Advanced Technology, Chinese Academy of Sciences, Shenzhen, China

About the editors

Calogero Caruso, formerly Full Professor of General Pathology, is Professor Emeritus at the University of Palermo, Italy. He graduated from the School of Medicine, University of Palermo, in 1971 (magna cum laude). He was founder and editor-in-chief of the journal Immunity & Ageing (2004–18) and has authored 386 publications (13,347 citations), mostly on aging, age-related diseases, and longevity, indexed on Scopus (H index of 60). He has recently edited the book "*Centenarians. An example of positive biology*" for Springer Nature. He was national coordinator of Italian Ministerial Project on Centenarians and Longevity (2017–20). He was Coordinator of the PhD program in Molecular Medicine and Biotechnology (formerly Pathobiology) (1998–18). The national and international recognition of his scientific achievements in the field of Biogerontology is documented by funded projects, by the number of citations of his papers, by invitations to national and international congresses and meetings, by his activity as a project reviewer, and by organization of international conferences.

Giuseppina Candore, PhD, Associate Professor of General Pathology has authored 256 publications (9,198 citations), mostly on aging, age-related diseases, and longevity, indexed on Scopus (H index of 54). She is Coordinator of postgraduate School in Clinical Biochemistry and Pathology. She is responsible for the local unit of the project "Improved Vaccination Strategies for Older Adults (ISOLDA)" funded by European Commission (2020–24). The national and international recognition of her scientific achievements in the field of Biogerontology is documented by the number of citations of her papers and by organization of national and international conferences.

Preface

The extraordinary increase in older people underlines the importance of studies on aging and the need for a prompt dissemination of knowledge on aging in order to satisfactorily diminish the medical, economic, and social problems associated with advancing years, problems caused by the continuous increase in the number of older people at risk of frailty and age-related diseases. Hence, improving the quality of life of older people is becoming a priority. This makes the studies of the processes involved in aging of great importance.

Human Aging: From Cellular Mechanisms to Therapeutic Strategies, written by expert researchers in biogerontology who are actively involved in various fields within aging research, offers an exhaustive picture of all the biological aspects of human aging by describing the key mechanisms associated with human aging. Each chapter includes a summary of the salient points covered and future perspectives. The book provides readers with the information they need to acquire or deepen the skills needed to assess the mechanisms of aging and age-related diseases and to monitor the effectiveness of therapies aimed at slowing aging. Reading this book will inspire PhD and postdoc students, researchers, health professionals, and all other figures interested in the biology of aging to explore the fascinating and challenging questions about why and how we age and what can and cannot do about it.

A long life in a healthy, vigorous, and young body has always been one of humanity's greatest dreams. Antiaging strategies aimed not at rejuvenating but at slowing aging and delaying or avoiding the onset of age-related diseases are discussed in the book. It is emphasized that the goal of aging research is not to increase human longevity regardless of the consequences, but to increase active life free from disability and functional dependence.

Calogero Caruso
Giuseppina Candore

CHAPTER 1

Aging and longevity: An evolutionary approach

Giuseppina Candore and Calogero Caruso
Laboratory of Immunopathology and Immunosenescence, Department of Biomedicine, Neurosciences and Advanced Diagnostics, University of Palermo, Palermo, Italy

1.1 Introduction

Aging is most likely a component of life, which first emerged in economically developed countries and results from the disruption of self-organizing system and reduced ability to adapt to the environment. It is an inescapable natural phenomenon, which affects all cells, tissues, organs, and organisms. As people age, detrimental changes accumulate at the level of molecules, cells, and tissues responsible for the decline in normal physiological functions. That leads inexorably to a reduced ability of the individual to maintain adequate homeostasis, determining a greater susceptibility toward different kinds of stressors (Avery et al., 2014).

Aging processes are defined as those that amplify the vulnerability of individuals, as they age, to the factors that ultimately lead to death (Fig. 1.1, Table 1.1). An emerging concept is the difference between chronological and biological aging; the cells, tissues, and organs of the same individual may have a different rate of aging in contrast to the chronological age. Conversely, individuals with the same chronological age may have a different aging rate and a different biological age (Avery et al., 2014).

As for the term longevity, based on the demographics, longevity can be defined in relative and absolute terms. The term "relative" suggests that longevity must take into account the life expectancy of different populations that show great variability due to historical, anthropological, and socioeconomic factors. Thus, long-lived individuals refer to people belonging to the 5 percentile of the survival curve, that is, in the Western world, to those over ninety. In "absolute" terms, longevity could be defined according to the maximum life span attained and scientifically validated by human beings (Avery et al., 2014; Villa et al., 2019).

Aging is considered a multifactorial process that does not recognize a single responsible cause, but which is, rather, the result of numerous mechanisms that interact simultaneously at different levels. Many variables contribute to aging and longevity such as cultural, anthropological, and socioeconomic status; as well as gender and sex, women live longer than men (Caruso, 2019); and ethnic differences, explained by discrepancies

Human Aging
https://doi.org/10.1016/B978-0-12-822569-1.00002-0

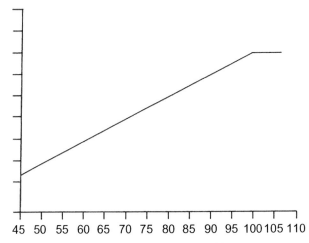

45 50 55 60 65 70 75 80 85 90 95 100 105 110

Fig. 1.1 Semi-log graph of mortality rates by age. As described by Gompertz, mortality rates increase exponentially with age with a doubling of mortality in adult life every eight years. At older age, the rate slows down, as indicated by the plateau of the curve. This phenomenon has been called demographic selection. The age is indicated on the abscissa, while the mortality rate log is shown in the ordinate.

Table 1.1 Characteristics of aging.

Changes at the level of biomolecules, cells, tissues, and organs
Reduced ability to respond to environmental stimuli, due to an ineffective homeostatic response
Increased risk of death

in health; environmental and economic condition; education, genetics, and kinds of job; as well as epigenetics and stochastic events (Accardi et al., 2019).

These variables also contribute to different kinds of aging, i.e., successful or unsuccessful aging. Although there is no universally accepted definition of "successful aging," one can refer to the World Health Organization definition, that is, "the process of developing and maintaining functional ability that enables well-being in older age." Some criteria have been developed that describe it. They include three main related components, i.e., low probability of developing disease-related diseases and disabilities, high cognitive and physical functional capacity, and active engagement with life. The combination of these components guarantees an active end of life that is the essential concept of successful aging. Reciprocally, the presence of age-related disease and disability, low cognitive and physical functional capacity, and reduced engagement with life should characterize unsuccessful aging (Bülow and Söderqvist, 2014; Aiello et al., 2019b).

Aging is generally thought to be a universal biological phenomenon that affects all living things, although prokaryotic kingdom seems to be not involved. The bacteria, in fact, divide in a symmetrical way, generating by division two identical individuals that do not seem to bear the signs of aging. The aging process therefore appears to be a characteristic of the eukaryotic kingdom. Brewer's yeast, *Saccharomyces cerevisiae*, one of the simplest single-cell organisms, splits by budding. At the end of the reproductive process, a parent cell can be identified, which has a maximum limit of cell divisions, and which progressively accumulates damages and loss of functions until death, i.e., it ages (Kaeberlein et al., 2007). In the gradual progress of the evolutionary scale, a greater complexity of the mechanisms regulating organism homeostasis is observed and, in parallel, also of those that cause a progressive deterioration.

Unlike what happens in single-cell organisms, in animals, in addition to a first level of cellular organization, there is a second organizational level, which is achieved through the regulation and modulation of the immune, endocrine, and nervous systems. Some studies on the biology of aging point out the prevalent role of systemic factors in the causes of aging, other studies promote the idea that the intrinsic changes that take place at the cell level are at the origin of aging (Zhang et al., 2013; Phillip et al., 2015). Probably, both conditions act simultaneously, and if the intrinsic cellular mechanisms are fundamental, their activation is closely regulated by extracellular factors, such as hormones, immune-inflammatory responses, and lifestyle.

1.2 Why does aging occur?

One question commonly arises: "Why does aging occur?" Aging from a biological point of view still remains a largely mysterious process. Over the years, various theories have been developed to attempt to explain the mechanisms that trigger and guide aging (more than 300 have been listed). As a consequence of the lack of adequate models for the study of human aging, and the frequent inability to distinguish between causes and effects, currently there is no general consensus on what causes aging, what determines the variability of longevity between species, what happens to a human being between 30 and 70 years of age who is responsible for the 30-fold increase in risk of death (Troen, 2003). In general, all theories tend toward two main lines of thought, the possibility of programmed aging, which triggers aging processes like a biological clock, and that of aging due to damage accumulated during life, which, over time, overwhelms repair capacities.

Old animals are practically not found in nature, but only in zoos and among domesticated species such as dogs and cats that are no longer prey or predators. There is the exception of elephants and turtles that can live very long, because for intuitive reasons they are not an easy prey. Since aging has a negligible impact on organisms in their natural environment due to predation, malnutrition, accidents, diseases, exposure to the elements, two corollaries can be immediately deduced; aging is not a genetically programmed event to control the size of the population (see later for the discussion of this point); the selection cannot exert a direct influence on the aging process.

In fact, nothing in biology makes sense except in light of evolution (Dobzhansky, 1973); therefore, we must try to frame the aging process in the context of the evolutionary limitations imposed by natural selection, in light of Darwinian medicine.

Aging theories can be grouped into two main groups, the first represented by the group of "genetically programmed," the second by the group of "error" theories or "stochastic." The "genetically programmed" theories consider aging as an event completely dependent on the biological clock, which regulates life expectancy through the fundamental stages of an individual life, such as growth, development, maturity, and old age. This regulation would depend on genes capable of inducing sequentially and regulated changes in the activation or inhibition signals of the various systems, responsible for maintaining homeostasis, triggering the various responses. Several researchers are attracted by the idea of the existence of a genetic program. It is hypothesized that aging (and consequent death) is necessary, either to prevent the overcrowding of the species environment or to promote evolutionary changes by accelerating the turnover of generations. This idea was proposed as early as 1891 by Weismann (References in Kowald and Kirkwood, 2016). Today, however, we know that the group selection on which this idea is based is much weaker than the selection at the individual level (Maynard Smith, 1976). A strong objection to the idea that aging is driven by a genetic program is the empirical finding that all model studies have never led to the discovery of gene mutations that abolish the aging process (Kirkwood and Melov, 2011). If such genes exist, as implicit in programmed aging, they would in fact be susceptible to inactivation by mutation. However, older age is not an adaptive phenomenon, it cannot be selected by evolution, because the selective pressure acts only on the characters that favor the reproduction of the species and cannot act toward the characters concerning the postreproductive life. In fact, human subjects and animals are selected in such a way as to guarantee survival until reproduction.

The theories of "error or damage" identify environmental insults within living organisms capable of inducing progressive damage at various levels. The best known is the free radical theory. It has greatly attracted scientific attention, presenting itself as a possible biological explanation of the entire aging process. This process, modifiable by genetic and environmental factors, is characterized by the accumulation of free oxygen radicals (ROS), which damage all biological molecules, causing, over time, a vicious circle. In fact, that causes, in turn, damage of mitochondrial DNA (mtDNA) as well as of the proteins of the electronic transport with a consequent increase in the production of ROS. ROS, exceeding the endogenous buffering capacities, are the cause of genotoxic damage, capable of eventually inducing cell senescence or apoptosis (Harman, 2003). Free radicals are implicated in the development of most age-related diseases, resulting in unsuccessful aging (Chapter 3). However, their significance in physiological aging remains unclear, mainly due to the limited effect, demonstrated in various animal models, of antioxidant systems in extending life span (Chapter 6).

Alternatively, the explanation for why aging occurs is thought to be found among three ideas all based on the principle that natural selection decreases during adult life (Medawar, 1952). There remains, therefore, only the explanation that the aging process is linked to the decreasing force of natural selection with chronological age, as recognized in 1952 by Medawar. This decline occurs because in progressively older age, the fraction of total future reproductive production, on which selection can act to discriminate between more suitable and less suitable genotypes, becomes progressively smaller.

The three evolutionary theories based on this insight, mutation accumulation (Medawar, 1952), antagonistic pleiotropy (Williams, 1957), and disposable soma (Kirkwood, 1977) are not mutually exclusive, but can be easily integrated. The theory of the accumulation of mutations (Medawar, 1952) assumes that over time there is a constant generation of deleterious mutations that, therefore, are expressed only after a certain age when the natural selection process will not be more efficient.

Williams (1957) suggests that a gene that has a benefit early in life, but is harmful later in life, may overall have a net positive effect and will be actively selected. Genes useful in the early stages of life such as, for example, those responsible for a powerful inflammatory response that protects against infections, become harmful in the late stages of postreproductive life and therefore not subjected to selective processes, as they are responsible for age-related inflammatory diseases. The immunoinflammatory responses represent typical adaptive responses resulting from evolutionary compromises. They represent in the modern world, cleaner from a microbial point of view, than in the past, the possible cause of allergic or autoimmune diseases or inflammatory diseases associated with old age. They are neither defensive nor pathological answers; they are teleonomic answers (Licastro et al., 2005).

The disposable soma theory (Kirkwood, 1977) is concerned with optimizing the allocation of resources between maintenance on the one hand and other processes such as growth and reproduction on the other. An organism that invests a larger fraction of its energy budget in preventing the accumulation of damage to its molecules, cells, and organs will have a slower aging rate, but will also have fewer resources available for growth and reproduction, and vice versa (Kirkwood, 2008). The theory of "disposable soma" adds a mechanistic specificity to the other two evolutionary theories of aging, which had previously been elaborated considering that the strength of natural selection tends to progressively reduce the effects of genes with age. The theory clearly emphasizes that aging is mainly driven by the impact, over the course of life, of molecular damages accumulated in cells. Aging appears as a process characterized, in its natural course, by the accumulation of damage from the early stages of life. The harmful events are completely random, as they affect a wide spectrum of targets within the cell. The rates at which they manifest and repair themselves are to some extent regulated through the evolutionary selection of the cellular machinery that serves macromolecular biosynthesis and repair (Kowald and Kirkwood, 2016). It is from the genetic specificity of certain mechanisms and from their level of functioning that the degree of heritability of longevity derives. The aging process clearly reveals how a variety of nongenetic factors can act to influence age-related morbidity and mortality. These factors include education, nutrition, lifestyle, socioeconomic status, sex and gender, and job.

1.3 Mechanisms of aging

Several studies have reported an age-related increase in somatic mutations and other forms of DNA damage, suggesting that DNA repair capacity plays a key role in cellular and molecular aging rates. When comparing species with different rates of longevity, it was found that there is a general relationship between longevity and DNA repair. The evidence for it is further strengthened by the accelerated aging phenotypes of DNA repair mice mutants and human progeroid syndromes (Hasty et al., 2003; Vijg and Suh, 2013; Oshima et al., 2018).

An important relationship between molecular stress and aging is given by the accumulation of mutations of mtDNA with increasing age. Cells in which mtDNA mutations are numerous are likely to produce reduced amounts of ATP with consequent decrease in tissue bioenergenesis, with therefore less energy available for maintenance and repair systems (Cui et al., 2012).

In addition to DNA, proteins are also prone to damage. The exchange of proteins is essential to preserve cellular functions, removing damaged proteins. Age-related impairment of protein turnover is evidenced by the accumulation of damaged proteins over time, which contributes to a wide range of age-related diseases, including cataracts, Alzheimer's disease, and Parkinson's disease. Protein exchange involves the chaperonins involved in the sequestration and wrapping (folding) of denatured proteins and the proteasomes involved in the recognition and degradation of damaged proteins. With aging, a functional decline in the activity of proteasomes and chaperonins as well as autophagy has been demonstrated. These functional reductions may be part of a more general malfunction, due to overloading, of the "waste disposal" cellular processes (Reeg and Grune, 2015; Vilchez et al., 2014; Barbosa et al., 2019).

Most researchers are convinced that age-related changes are characterized by an increase in entropy. This hypothesis is now supported by the reinterpretation of the Second Law of Thermodynamics, which would not apply only to closed systems. The entropy increase results in a random loss of molecular fidelity (Table 1.2) and builds up to slowly overwhelm maintenance systems. Entropy can be defined as the tendency of concentrated energy to disperse when it is not hindered. In biological systems, the hindrance is the relative strength of chemical bonds. In fact, preventing the breakdown of the chemical bond is absolutely essential for life. Natural selection has favored energy states that can maintain fidelity in most molecules until reproductive maturity, after which there is no species survival value for those energy states to be maintained indefinitely.

Table 1.2 Loss of molecular fidelity causes a wide spectrum of age-related changes.

Not dangerous: cataracts, gray hair, hearing loss, hyperpigmented skin spots, increased reaction times, presbyopia, and short-term memory loss

Potentially dangerous: molecular changes in cells, tissues, and organs, which increase vulnerability to Alzheimer's disease, cancer, cardiovascular disease, Parkinson's disease, and stroke

The dispersion of energy can result in a biologically inactive or malfunctioning molecules (Hayflick, 2007a,b).

Energy dispersal is never completely eliminated but can be buffered for some time by repair or replacement processes. After reproductive maturation, this balance slowly shifts into one where molecules that lose their biologically active energy states are less likely to be replaced or repaired. The aging process occurs because the change of energy states of the biomolecules makes them malfunctioning or inactive. Identical events obviously occur before the appearance of the aging phenotype, but the repair and replacement processes are able to maintain the balance in favor of the functioning of the molecules. The decrease in repair and replacement capacity is exacerbated by the fact that the enormously complex biomolecules that make up repair and replacement systems suffer the same fate as the biomolecules of their substrates. When the increasing loss of molecular fidelity eventually outweighs the ability to repair and replace, vulnerability to age-associated pathologies or diseases increases. In developed countries, the weakest links are the cells that make up the vascular system and those in which cancer is most likely to happen. The molecular instability that occurs in these cells is the weakest link that increases vulnerability to these two main causes of death (Hayflick, 2000, 2007b).

To understand how fundamental age changes occur might lead to a better knowledge of the etiology of all major causes of death (see Chapter 3).

1.4 Causality and chance in aging and longevity

The relationship between causality and chance is an open discussion in many disciplines. Often, the boundary among these events is thin to understand whether an occurrence is related to one or both. In particular, aging, the related diseases, and longevity are difficult to define as a consequence of causality, chance, or both (Accardi and Caruso, 2017).

Stochastic processes are accidental phenomena due to casual factors that play a key role in physiological and pathological events at the same level as genetics, epigenetics, and the environment. Indeed, stochastic processes contribute to the individuality of every living organism, including humans, influencing phenotypic variability, as suggested by the role of chance in the creation of the immunological repertoire and neuronal synapses (Accardi and Caruso, 2017). Living organisms are subject to nature laws and genetic programs where both Brownian random motion, i.e., the erratic random movement of particles in a fluid, as a result of continuous bombardment from molecules of the surrounding water molecules in the fluid, and crossing over contribute to leave space for the chance. Chance is just that, the random occurrence, i.e., an event happening not according to a plan (Accardi and Caruso, 2017).

There is evidence of the inherent stochastic nature of both gene expression and macromolecular biosynthesis. Several genes are in fact transcribed in minimal amounts of mRNA, which can cause large fluctuations in macromolecular biosynthesis. Genomic instability, which results in somatic mutations and chromosomal abnormalities, is another important source of intrinsic variability, as shown in aged mice, which have a mutation frequency up to 10^{-4} per gene per cell. Reciprocally, the relatively low level of chromosomal aberrations observed in older persons should be a consequence of their genomic stability, hence a contributing factor to their attainment of advanced age (Kirkwood et al., 2005; Kirkwood, 2008; Vijg and Suh, 2013).

Genetic control ensures and guarantees the functionality of metabolic and development systems with a considerable degree of reliability and reproducibility. Nonetheless, there are small variations that can add up and cause large effects, for example, the significant variations in the size of some organs observed between genetically identical organisms. Underlying the visible phenotypic variations, there is a variety of sources of intrinsic variability within organisms observable at the molecular, cellular, and organ, and system levels, due to stochastic and epigenetic processes. All this can contribute to the differences observed in life span even in identical animals. In fact, the role of stochastic processes in aging has been demonstrated by studies conducted on inbred mice, which, as is well known, have the same genome. Well, despite the housing conditions are the same, the animals show a different life span, even up to 50% more. This demonstrates that there is a stochastic component that results in very few fluctuations in the genetic, epigenetic, environmental, and interaction components. This involves continuous microvariations which, accumulating over time, amplify the differences between individuals, manifesting themselves clearly in older ages (Kirkwood et al., 2005).

Epigenetics is used to describe phenotypic variations that may occur in cells following a different expression of individual genes without altering the DNA sequence. These phenotypic changes are stable and inheritable from cell progeny, through DNA replication and cell division cycles. Thus, during the proliferation that occurs in the normal homeostatic replacement, a cell expresses the characteristic genes of the corresponding tissue and not that characteristic of another one. In a broader sense, epigenetic processes are called to explain the changes in the regulation of the transcription of individual genes. Considering this broader definition of epigenetics, it is not surprising that the profound changes that occur with age in cells and tissues are also due to epigenetic processes as well as accumulation of nuclear and mitochondrial DNA mutations (Peaston and Whitelaw, 2006).

To fully understand the mechanisms underlying epigenetic modifications is important to elucidate how environmental and genetic factors cooperate to determine the aging process and aging-related phenotypes and diseases. On the contrary, data are accumulating which show that epigenetic modifications may represent important tools to monitor the rate and the quality of aging, or to warn for the onset of age-related diseases (Bellizzi et al., 2019). The level of complexity is very high if we consider that not only the effect of each variation but also their combinations must be evaluated. For example, lysine methylation at positions 4 and 9 in H3 has opposite effects, i.e., the first increases the expression of genes and the second reduces this action. miRNAs are also involved in gene transcription. They also regulate histone modification processes. Recent studies have, then, revealed that human aging can be characterized by a profile of circulating microRNAs that is predictive of chronological age and that can be used as a biomarker of risk for age-related outcomes (Bannister and Kouzarides, 2011; Cammarata et al., 2019; Dellago et al., 2017).

A recent study has presented the first long-term, longitudinal characterization of expression and splicing changes as a function of age and genetics (Balliu et al., 2019). The findings indicate that although gene expression and alternative splicing and their genetic regulation are mostly stable late in life, a small subset of genes is dynamic and is characterized by changes in expression and splicing and a reduction in genetic regulation, most likely due to an increase of environmental variance.

1.5 Conclusions and future perspectives

The aging process is driven by a lifelong accumulation of molecular damage, resulting in a gradual increase in the fraction of cells carrying defects. After sufficient time has passed, the increasing levels of these defects interfere with the performance of tissues and organs, resulting in a breakdown of self-organizing system and a reduced ability to adapt to the environment.

In fact, species are not selected for aging, but to survive until the age of reproduction and any parental care; the only way to live long is paradoxically to grow old. The animal machine has been selected to ensure efficient reproduction and possibly the protection of offspring, at the expense of possible deterioration in the following years. Throughout a series of mechanisms, partly endogenous and partly exogenous, multiple cellular alterations occur throughout life. These alterations would lead very quickly to aging and death if our organism did not possess important repair mechanisms, the efficiency of which is largely under genetic control. From the balance between aggressive factors, mainly conditioned by the environment, and factors that try to neutralize them, mainly conditioned by genetics, a different life span derives. For this reason, the way we age is not unambiguous and is not predictable. Over time, these factors condition a progressive loss of molecular precision with an accumulation of damage in cells, tissues, and the whole organism, in different ways in different individuals, so determining various phenotypes. Accordingly, the analysis of the age and function curves of the different individuals demonstrate that people, getting older, show increasingly differentiated levels of performance determining successful or unsuccessful aging. Hence, aging and longevity are related to the ability to cope with a variety of stressors (Caruso, 2019).

Contributing factors are cultural, anthropological, socioeconomic, sexual, and gender, ethnic differences, health care, epigenetics, and life occupation. In the case of nutrition, for example, a proinflammatory diet, containing an excess of refined sugars, animal proteins and saturated fats, and poor in nutraceuticals, directly contributes to increasing cell damage, while a Mediterranean-type diet, poor in refined sugars, animal proteins and saturated fats, as well as rich in nutraceuticals, decreases cell damage (Accardi et al., 2019; Aiello et al., 2019a, 2020).

Therefore, the ability to survive is the result of maintaining a suitable response to stressors within a well-established range, compatible with physical well-being. This could be an expression of a Gaussian (normal) distribution, in which centenarians represent its extreme tail. The centenarians would represent; therefore, the best adapted individuals, because they are equipped on a genetic or even stochastic basis, with more efficient maintenance and repair systems, to the different environmental conditions present during their life, and consequently capable of maintaining a suitable response to stressors, including microorganisms (Caruso, 2019).

As discussed in the second chapter, life expectancy has increased significantly, but the maximum life span of humans has not changed (about 120 years). Eliminating cancer or Alzheimer's disease would improve the quality of life, but it would not make us immortal, nor would it allow us to live much longer. To prolong life, it is necessary to intervene directly in the aging process. Further studies are therefore needed on the specific molecular changes of aging in search of key molecular components whose destruction leads, in cascade, to other damages. If such key components exist, then we will have targets for targeted interventions. It must be clear, however, that such interventions can postpone but not evade the inevitable molecular deterioration linked to the laws of physics.

References

Accardi G, Caruso C. Causality and chance in ageing, age-related diseases and longevity. In: Accardi G, Caruso C, editors. Updates in pathobiology: causality and chance in ageing, age-related diseases and longevity. Palermo University Press; 2017. p. 13–23.

Accardi G, Aiello A, Vasto S, Caruso C. Chance and causality in ageing and longevity. In: Caruso C, editor. Centenarians. Cham: Springer; 2019. p. 1–21. https://doi.org/10.1007/978-3-030-20762-5_1.

Aiello A, Caruso C, Accardi G. Slow-ageing diets. In: Gu D, Dupre M, editors. Encyclopedia of gerontology and population aging. Cham: Springer; 2019a. https://doi.org/10.1007/978-3-319-69892-2_134-1.

Aiello A, Ligotti ME, Cossarizza A. Centenarian offspring as a model of successful ageing. In: Caruso C, editor. Centenarians. Cham: Springer; 2019b. p. 35–51. https://doi.org/10.1007/978-3-030-207 62-5_3.

Aiello A, Accardi G, Candore G, Caruso C. Effects of nutraceuticals of Mediterranean diet on aging and longevity. In: Preedy VR, Watson RR, editors. The Mediterranean Diet—second edition—An Evidence-Based Approach. Academic Press; 2020. p. 547–53.

Avery P, Barzilai N, Benetos A, Bilianou H, Capri M, Caruso C, Franceschi C, Katsiki N, Mikhailidis DP, Panotopoulos G, Sikora E, Tzanetakou IP, Kolovou G. Ageing, longevity, exceptional longevity and related genetic and non genetics markers: panel statement. Curr Vasc Pharmacol 2014;12:659–61. https://doi.org/10.2174/1570161111666131219101226.

Balliu B, Durrant M, Goede O, Abell N, Li X, Liu B, Gloudemans MJ, Cook NL, Smith KS, Knowles DA, Pala M, Cucca F, Schlessinger D, Jaiswal S, Sabatti C, Lind L, Ingelsson E, Montgomery SB. Genetic regulation of gene expression and splicing during a 10-year period of human aging. Genome Biol 2019;20:230. https://doi.org/10.1186/s13059-019-1840-y.

Bannister AJ, Kouzarides T. Regulation of chromatin by histone modifications. Cell Res 2011;21:381–95. https://doi.org/10.1038/cr.2011.22.

Barbosa MC, Grosso RA, Fader CM. Hallmarks of aging: an autophagic perspective. Front Endocrinol (Lausanne) 2019;9:790. https://doi.org/10.3389/fendo.2018.00790.

Bellizzi D, Guarasci F, Iannone F, Passarino G, Rose G. Epigenetics and ageing. In: Caruso C, editor. Centenarians. Cham: Springer; 2019. p. 99–133. https://doi.org/10.1007/978-3-030-20762-5_7.

Bülow MH, Söderqvist T. Successful ageing: a historical overview and critical analysis of a successful concept. J Aging Stud 2014;31:139–49. https://doi.org/10.1016/j.jaging.2014.08.009.

Cammarata G, Duro G, Chiara TD, Curto AL, Taverna S, Candore G. Circulating miRNAs in successful and unsuccessful aging. A Mini review. Curr Pharm Des 2019;25:4150–3. https://doi.org/10.2174/1381612825666191119091644.

Caruso C, editor. Centenarians. Cham: Springer; 2019. p. 1–179. https://doi.org/10.1007/978-3-030-20762-5.

Cui H, Kong Y, Zhang H. Oxidative stress, mitochondrial dysfunction, and aging. J Signal Transduct 2012;2012:646354. https://doi.org/10.1155/2012/646354.

Dellago H, Bobbili MR, Grillari J. MicroRNA-17-5p: at the crossroads of Cancer and aging—a mini-review. Gerontology 2017;63:20–8. https://doi.org/10.1159/000447773.

Dobzhansky T. Nothing in biology makes sense except in the light of evolution. Am Biology Teacher 1973;35:125–9. https://doi.org/10.2307/4444260.

Harman D. The free radical theory of aging. Antioxid Redox Signal 2003;5:557–61. https://doi.org/10.1089/152308603770310202.

Hasty P, Campisi J, Hoeijmakers J, van Steeg H, Vijg J. Aging and genome maintenance: lessons from the mouse? Science 2003;299:1355–9. https://doi.org/10.1126/science.1079161.

Hayflick L. The future of ageing. Nature 2000;408:267–9. https://doi.org/10.1038/35041709.

Hayflick L. Biological aging is no longer an unsolved problem. Ann N Y Acad Sci 2007a;1100:1–13. https://doi.org/10.1196/annals.1395.001.

Hayflick L. Entropy explains aging, genetic determinism explains longevity, and undefined terminology explains misunderstanding both. PLoS Genet 2007b;3:e220. https://doi.org/10.1371/journal.pgen.0030220.

Kaeberlein M, Burtner CR, Kennedy BK. Recent developments in yeast aging. PLoS Genet 2007;3: e84https://doi.org/10.1371/journal.pgen.0030084.

Kirkwood TB. Evolution of ageing. Nature 1977;270:301–4. https://doi.org/10.1038/270301a0.

Kirkwood TB. Understanding ageing from an evolutionary perspective. J Intern Med 2008;263:117–27. https://doi.org/10.1111/j.1365-2796.2007.01901.x.

Kirkwood TB, Melov. On the programmed/non-programmed nature of ageing within the life history. Curr Biol 2011;21:R701–7. https://doi.org/10.1016/j.cub.2011.07.020.

Kirkwood TB, Feder M, Finch CE, Franceschi C, Globerson A, Klingenberg CP, LaMarco K, Omholt S, Westendorp RG. What accounts for the wide variation in life span of genetically identical organisms reared in a constant environment? Mech Ageing Dev 2005;126:439–43. https://doi.org/10.1016/j.mad.2004.09.008.

Kowald A, Kirkwood TBL. Can aging be programmed? A critical literature review. Aging Cell 2016;15:986–98. https://doi.org/10.1111/acel.12510.

Licastro F, Candore G, Lio D, Porcellini E, Colonna-Romano G, Franceschi C, Caruso C. Innate immunity and inflammation in ageing: a key for understanding age-related diseases. Immun Ageing 2005;2:8. https://doi.org/10.1186/1742-4933-2-8.

Maynard Smith J. Group selection. Q Rev Biol 1976;51:277–83.

Medawar PB. An unsolved problem of biology. London: H.K. Lewis; 1952.

Oshima J, Kato H, Maezawa Y, Yokote K. RECQ helicase disease and related progeroid syndromes: RECQ2018 meeting. Mech Ageing Dev 2018;173:80–3. https://doi.org/10.1016/j.mad.2018.05.002.

Peaston AE, Whitelaw E. Epigenetics and phenotypic variation in mammals. Mamm Genome 2006;17:365–74. https://doi.org/10.1007/s00335-005-0180-2.

Phillip JM, Aifuwa I, Walston J, Wirtz D. The mechanobiology of aging. Annu Rev Biomed Eng 2015;17:113–41. https://doi.org/10.1146/annurev-bioeng-071114-040829.

Reeg S, Grune T. Protein oxidation in aging: does it play a role in aging progression? Antioxid Redox Signal 2015;23:239–55. https://doi.org/10.1089/ars.2014.6062.

Troen BR. The biology of aging. Mt Sinai J Med 2003;70:3–22.

Vijg J, Suh Y. Genome instability and aging. Annu Rev Physiol 2013;75:645–68. https://doi.org/10.1146/annurev-physiol-030212-183715.

Vilchez D, Saez I, Dillin A. The role of protein clearance mechanisms in organismal ageing and age-related diseases. Nat Commun 2014;5:5659. https://doi.org/10.1038/ncomms6659.

Villa F, Ferrario A, Puca AA. Genetic signatures of centenarians. In: Caruso C, editor. Centenarians. Cham: Springer; 2019. p. 87–97. https://doi.org/10.1007/978-3-030-20762-5_6.

Williams GC. Pleiotropy, natural selection and the evolution of senescence. Evolution 1957;11:398–411.

Zhang G, Li J, Purkayastha S, Tang Y, Zhang H, Yin Y, Li B, Liu G, Cai D. Hypothalamic programming of systemic ageing involving IKK-β, NF-κB and GnRH. Nature 2013;497:211–6. https://doi.org/10.1038/nature12143.

CHAPTER 2

Demographic aspects of aging

Annalisa Busetta and Filippa Bono
Department of Economics, Business and Statistics, University of Palermo, Palermo, Italy

2.1 Introduction

Population aging is, together with population growth, urbanization, and international migration, one of the four "megatrends" of modern societies. This phenomenon, which is unprecedented in human history, is occurring in almost all developed countries, albeit with differences in timing and intensity. Every country in the world is experiencing growth in both the number and the proportion of older people in the population. Traditionally, measures of population aging are based on people's chronological age. In particular, the United Nations' measures define "old" people as those aged 60 and over or those aged 65 and over. According to the UN 2019 Revision of the World Population Prospects (United Nations, 2020b), there are 728 million people aged 65 years and over in the global population in 2020 (Table 2.1), and this number will rise to more than 1810 billion by 2060 (compared to 262 million in 1980).

In high-income countries, the number of people aged 65 and older is projected to increase from 232 million in 2020 to 372 million in 2060, whereas in low-income countries, this number is projected to rise from 26 million to 118 million over the same period (compared to 108 million and eight million, respectively, in 1980).

The differences by income group of countries are also large if we look at the percentage of older people in the population. In 2020, the share of people aged 65 and over in the population is 18.4% in high-income countries and is projected to exceed 28% by 2060. In low-income countries, this share is 3.3% in 2020, and it is expected to more than double to 6.8% by 2060 (compared to 11.2% and 3.1%, respectively, in 1980).

From a demographic point of view, the aging process is the result of continued declined fertility, longer lives, and low immigration. Mainly due to the large differences in the speed and the intensity of fertility and mortality decline, the population aging trends in high-income countries vary considerably. In the ranking of high-income countries by the percentage of the population aged 65 and over in 2060, all four Southern European countries (Italy, Spain, Portugal, and Greece) and three East Asian countries (Japan, South Korea, and Singapore) occupy the highest positions, with more than one person out of three in these countries projected to be in the 65 and over age group in 2060. The lowest positions in this ranking are held by the United States and Australia,

Human Aging
https://doi.org/10.1016/B978-0-12-822569-1.00019-6

Table 2.1 Population aged 65 and over and aged 80 and over by country income level.

	(Pop 65 and over thousand and % on the total population)						(Pop 80 and over thousand and % on the total population)					
	1980	%	2020	%	2060	%	1980	%	2020	%	2060	%
Low-income countries	8,357	3.1	25,679	3.3	117,711	6.8	772	0.3	3,300	0.4	18,011	1.0
Medium-income countries	146,117	4.5	469,782	8.2	1,320,021	18.6	16,180	0.5	78,913	1.4	354,227	5.0
High-income countries	107,970	11.2	231,812	18.4	371,893	28.2	18,760	1.9	63,213	5.0	148,607	11.3
World	262,532	5.8	727,606	9.3	1,810,398	17.8	35,727	0.9	145,504	1.9	521,145	5.1

Source: Our calculation based on United Nations, DESA, Population Division. World population prospects: the 2019 revision; 2020b. https://population.un.org/wpp/Download/Standard/Population; data acquired via the Web site. Note: The country classification by income level is based on June 2018 gross national income per capita from the World Bank.

with fewer than one person in four in these countries projected to be aged 65 or older in 2060. It is not a surprise that all of the countries in the top positions are characterized by a lowest–low fertility, whereas those in the lowest positions have fertility around the replacement level.

As part of the same process, the older population will grow even older, with the number of people aged 80 and over undergoing the largest relative increase in the coming decades. This age group is worthy of special attention, since these oldest-old (men and particularly women) are at higher risk of having substantial disabilities, of having limited financial resources, and of needing nursing care. Globally, the number of people aged 80 and over is projected to increase from 146 million in 2020 to more than 520 million in 2060. In high-income countries, the number of people aged 80 and older is projected to increase from 63 million in 2020 to more than 149 million in 2060; whereas in low-income countries, this number is expected to increase from 3.3 million to 18 million over the same period (compared to 18.7 million and less than 0.8 million, respectively, in 1980). Again, the highest percentage values are projected in Japan and South Korea, followed by in the Southern European countries, at between 16% and 20%.

Even more impressive is the increase in the numbers of centenarians. Globally, the number of people aged 100 and over has risen from 45,320 in 1980 to 573,423 in 2020 and is projected to reach 5,455,712 in 2060 (Table 2.2). Almost all centenarians live in high-income and middle-income countries. Indeed, more than 50% of centenarians live in four countries: the United States, Japan, China, and India. After these four countries, the countries ranked as having the most centenarians in 2020 are Vietnam, Brazil, France, Germany, Italy, the United Kingdom, Spain, and Mexico.

The picture is different if we look at the share of centenarian per 10,000 people in the same countries. According to the UN population prospects, the country with the highest share of centenarians in 2020 is Japan (6.2 per 10,000), followed by France, the United States, Spain, Italy, Vietnam, the United Kingdom, and Germany (with values among 2.9 and 2.3), and Mexico and Brazil (both around 1.0). In China and India, despite the impressive current and estimated number of centenarians, the share of centenarian is limited (0.5 and 0.3, respectively).

Table 2.2 Population aged 100+ by country income level.

	1980	2020	2060	1980	2020	2060
	in thousands			per 10,000		
World	45	573	5,455	0.10	0.74	5.37
High-income countries	27	329	2,432	0.28	2.60	18.43
Middle-income countries	18	242	2,989	0.06	0.42	4.21
Low-income countries	0	2	26	0.01	0.03	0.15

Source: Our calculation based on United Nations, DESA, Population Division. World population prospects: the 2019 revision; 2020b. https://population.un.org/wpp/Download/Standard/Population; data acquired via the Web site.

Population aging implies that the population age distribution will change completely. As the share of older people in the global population has been increasing, the proportion of children under the age of 15 has been decreasing, from more than one-third in 1980 to just over one-quarter in 2020, and is projected to further decline to around one-fifth by 2060. However, the projected changes in the population age structures of countries are highly differentiated by income group. The population pyramids below visually compare the changes in the entire age distribution from 1980 to 2020 and to 2060 in high-income and low-income countries (Fig. 2.1). The low-income pyramids show a high and decreasing percentage of children and a low and increasing percentage of old people. The high-income pyramids also reflect the decreasing role of young people and the increasing role of older people, albeit with different intensities. The pyramid shape of the population in the low-income countries in 2020 is the result of decreases in infant mortality and in the risk of death throughout life that narrows toward the top, whereas the pyramid shape of the population in the high-income countries is the result of a long-lasting low fertility trend and continuous decreases in mortality at all ages.

Due to differences in mortality between men and women, aging is a gender-unequal phenomenon. As the average survival of men is projected to gradually move closer to that of women in high-income countries, the gender imbalance among older people in these countries will likely become smaller. According to UN World Population Prospects, in 2020 the femininity rate of the old population (i.e., the number of females aged 65 and over per 100 males of the same age group) is almost equal in high-income and low-income countries (127 and 131, respectively), but it is projected to decrease at different speeds (by 2060, the number of women for every 100 men is projected to be 114 in high-income countries and 124 in low-income countries). Moreover, the number of women for every 100 men aged 80 or older is projected to decline slightly from 165 in 2020 to 131 in 2060 in high-income countries, and from 164 to 157 in low-income countries.

Finally, it should be noted that the age structure of a population is also affected by international migration, especially for countries with low levels of fertility that receive significant numbers of immigrants. As Lee and Mason (2014) have emphasized, immigration can help to reduce population aging in the short and middle term, but it has controversial effects in the long term. While immigrants are, on average, relatively young when they arrive, their age distribution becomes similar to the age distribution of the receiving country over time as their fertility rates converge with those of the receiving country.

Up to this point, we have described aging by the changes in the population age structure over time. But aging also implies that life expectancy at birth or at certain ages is increasing. The aging process reflects important and very welcome advances in health and in the overall quality of life in societies across the world. The next map (Fig. 2.2) shows the large differences in life expectancy across the world in 2015–2020. In many

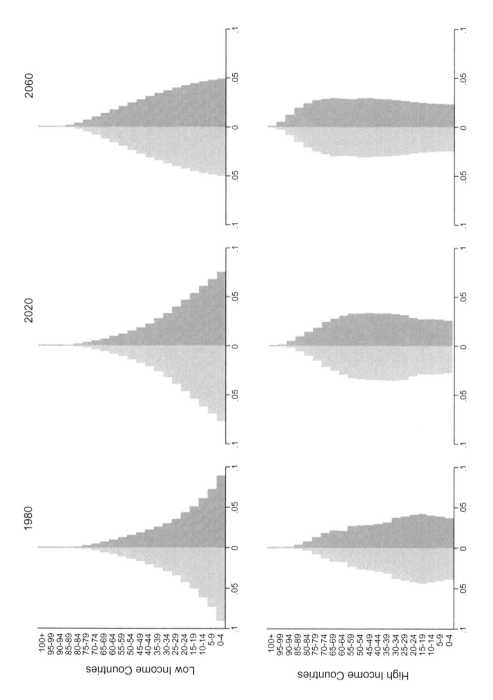

Fig. 2.1 Population pyramids in high-income and low-income countries (percentage). (*Source: Our calculations based on United Nations, DESA, Population Division. World population prospects: the 2019 revision; 2020b. https://population.un.org/wpp/Download/Standard/Population; data acquired via the Web site.*)

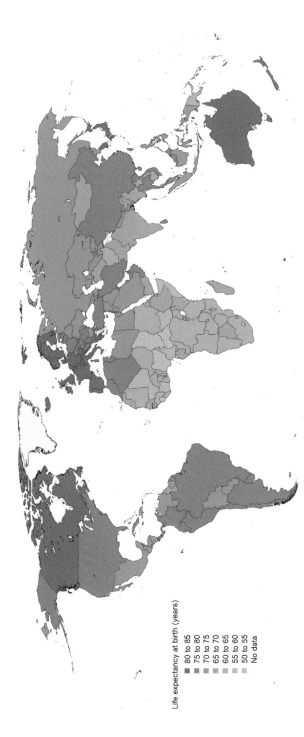

Fig. 2.2 Life expectancy at birth (both sexes, 2015–2020). (*Source: United Nations, DESA, Population Division. World population prospects: the 2019 revision; 2020b. https://population.un.org/wpp/Download/Standard/Population; data acquired via the Web site. Note: Life expectancy "describes what would happen to a hypothetical group if they moved through their lives experiencing the mortality rates observed for the country in any given year." It is defined as the measure of the average number of years to be lived by a group of people assuming that actual age-specific mortality levels in a single year remain constant.*)

Life expectancy at birth (years)

- 80 to 85
- 75 to 80
- 70 to 75
- 65 to 70
- 60 to 65
- 55 to 60
- 50 to 55
- No data

of the richest countries, life expectancy is over 80 years (in 2019, it was over 83 years in Spain, Switzerland, Italy, and Australia; and it was close to 85 years in Japan); whereas in the poorest countries, life expectancy is between 50 and 60 years (in 2019, it was only 53 years in the Central African Republic).

This situation is the result of the extraordinary progress in health that has occurred since the mid-19th century. From that point in time until today, life expectancy at birth has doubled across all regions of the world (from around 40 years to more than 81 years). This surge in life expectancy represents a completely new achievement in human history. In the United Kingdom (the country for which we have the longest series of data), life expectancy fluctuated between 30 and 40 years before the 19th century and is now higher than 80 years. In Italy and Japan (the countries that have the highest life expectancy currently), health started to improve later, but caught up quickly with the gains made in the United Kingdom, and surpassed them in the late 1960s (Fig. 2.3). The graph shows how low life expectancy was in some countries in the past. A century ago, life expectancy in India and South Korea was as low as 23 years. A century later, life expectancy has almost tripled in India and has almost quadrupled in South Korea. However, while there have

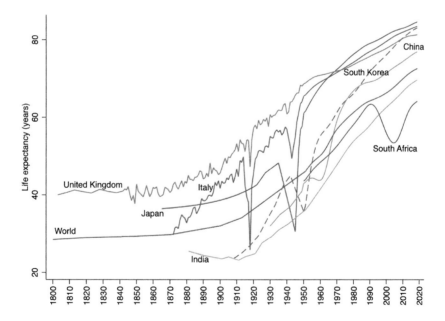

Fig. 2.3 Life expectancy from 1800 to 2015 in some selected countries. *(Source: data acquired via the Web site by Our World in Data based on estimates by Riley JC. Estimates of regional and global life expectancy, 1800–2001. Popul Dev Rev. 2005;31(3):537–43, Zijdeman R, Ribeira da Silva F. Life expectancy at birth (total). IISH Dataverse, V1; 2015. http://hdl.handle.net/10622/LKYT53, and the United Nations, DESA, Population Division. World population prospects: the 2019 revision; 2020b. https://population.un.org/wpp/Download/Standard/Population.)*

been tremendous increases in life expectancy in all countries of the world, there are still large differences between countries. For example, life expectancy in sub-Saharan countries is around 50 years, while it exceeds 80 years in Japan.

Globally, fewer and fewer people are dying at a young age. Of the 56 million deaths that occurred in 2017, nearly half were of people who were aged 70 or older (compared to one-third of all deaths in 1991), and only around one out of 10 were of children under the age of five (compared to nearly one-quarter of all deaths in 1990).

While it is often argued that life expectancy across the world has increased because child mortality has fallen, the data show that mortality rates have declined—and, consequently, life expectancy has increased—for all age groups (Roser et al., 2013). The average person can expect to live a longer life than in the past, irrespective of what age he or she currently is. For example, in 1870, a 10 year old in Italy could expect to live 46 years, whereas in 2019, he or she can expect to live 80 years (+34 years). Moreover, all of the countries of the world have also experienced extraordinary increases in life expectancy at older ages. For example, life expectancy at age 65 is projected to rise from 14 years in 1980 to more than 19 years in 2060. Once again, these values vary considerably depending on a country's income level. Live expectancy at age 65 is projected to increase from 16 to 24 years in high-income countries and from 12 years to more than 16 years in low-income countries. If we focus on the countries member of the Organization for Economic Cooperation and Development (OECD), we see that, on average, life expectancy at age 65 is 19.7 years, having increased 5.5 years from 1970 to 2017 (OECD, 2019).

2.2 Understanding the process: Browsing around the demographic transition theories

The broad change in the population age structure (described in paragraph 1) is a long-term process that completely reshapes a population's age structure from a young age structure to an old age structure. This shift—called the age structural transition (Pool and Wong, 2006; Vallin and Meslé, 2004)—seems to be an integral part of the well-known demographic transition theory (Notestein, 1945). In brief, the theory describes the change from a pretransition society with high levels of death and birth rates to a more modern post-transition society with low levels of both indicators, and the parallel changes in terms of population growth. In the post-transition societies, birth and death rates can be balanced, which leads to slow growth; or the birth rate can be lower than the death rate, which implies a negative natural change. While the original version of the theory linked these demographic trends to industrialization and modernization, more recently, the evolution of fertility and mortality in many developing countries has shown that modernization is not essential to cause birth and death rates to drop. Bangladesh, Sri Lanka, and some countries in sub-Saharan Africa are examples of improving mortality and falling fertility in the absence of any significant social and economic development.

Moreover, the demographic transition theory implicitly supposes that countries' life expectancy levels are converging toward the limits imposed by new epidemiological characteristics of modern societies. However, in some parts of the world, significant failures seem to have halted this convergence process in recent decades (e.g., the health crises in Eastern Europe and AIDS in Africa). In reality, some of these failures have not fundamentally challenged the theory. However, the unexpected and dramatic growth in cardiovascular diseases since the 1960s, which represents a new stage in a much more general process, has called this model into question.

The epidemiological transition theory (Omran, 1971) was the first to consider the extraordinary improvements that had occurred in industrialized countries since the 18th century. The theory describes the changes in the characteristics of the diseases that accompanied the overall improvement in health conditions from the end of the 19th to the first part of the 20th century. The theory focused on the evidence showing that, together with decreases in mortality and increases in life expectancy, populations experience shifts in the characteristics of the diseases from a period in which mortality was dominated by deaths from infectious diseases (e.g., diarrhea, pneumonia, and tuberculosis), to one in which chronic degenerative diseases were the leading causes of death (e.g., cancers, cardiovascular diseases, mental disorders, and dementia) as more people were surviving to older ages.

Omran (1971) argued that the epidemiological transition unfolds in three major stages (ages): (1) the age of pestilence and famine, i.e., the pretransition stage when mortality is high and fluctuating and is dominated by epidemics, famine, and wars; (2) the age of receding pandemics, i.e., the period in which mortality declines progressively, and epidemic crises become less common; and (3) the age of degenerative and man-made diseases, i.e., the period in which mortality continues to decline, and the shift from an infectious disease-dominated to a degenerative disease-dominated health profile is completed. The epidemiological transition represented not just the replacement of a series of problems with others, but an overall improvement in health (Meslè and Vallin, 2002; Robinson, 2003). According to Omran (1971), the countries of the world can be assigned to three different epidemiological transition models (McCracken and Phillips, 2017): the classical model, which is typical of Western countries (characterized by a gradual and progressive shift in mortality over a century or more); the accelerated model, which is typical of Japan (characterized by a shift that occurs in a shorter period of time); and the contemporary/delayed model, which is typical of Sri Lanka and Chile (characterized by a recent, and not yet completed transition). But the global HIV/AIDS pandemic and health reversals in several of the former Soviet republics clearly showed that a smooth, uninterrupted passage through the transition is not guaranteed. Since in most developed countries (and in many less developed countries), life expectancy rose to levels well beyond those anticipated by Omran, it appears that the third and final stage he posited is not the end point of epidemiological change.

In high-income nations, there has been a significant delay in deaths from these conditions to later ages. In these countries, a social pathology health profile has emerged in seriously disadvantaged subpopulation groups (due to violence, substance abuse, suicide, accidents, and HIV); and morbidity has increased, particularly in the more aged societies (epidemics of dementia).

Several decades after it was first proposed, the original epidemiological transition theory was deemed outdated and was therefore integrated into a broader health transition theory. The health transition model calls for a rethinking of the full transition process to take into account the ways in which different societies respond to the changes in health and vice versa (Caselli, 1995). Indeed, a health transition has accompanied the demographic and technological transitions in the developed countries of the world and is still underway in less developed societies.

This perspective is of particular interest because it explains how changes in the aging of individuals are influenced by the distribution of diseases in the population. An analysis of deaths by cause showed that in 2017, cardiovascular disease (CVD) was the leading cause of death, responsible for around one-third of deaths; and cancer was the second-largest cause of death, accounting for around 17% of deaths. In 2017, all the non-communicable diseases (NCDs) combined accounted for more than 73.4% (57.7% in 1990) of deaths, while the communicable, maternal, neonatal, and nutritional diseases were responsible for 18.6% (33.1% in 1990) of deaths; and injuries accounted for 8% (9.2% in 1990) of deaths. There are still some causes of death related to infectious diseases that can be greatly reduced through improved water quality, sanitation, hygiene, or oral rehydration salt packets. For example, around 3.2% of newborn deaths are due to complications at birth, 2.8% of deaths are from diarrheal diseases, and 1.1% of deaths are from malaria.

2.3 Aging inequalities

2.3.1 Differences by gender, education, and cause of death

Since the 20th century, women have been living longer than men in all countries of the world. While the literature on this gender gap has shown that it is caused by biological, behavioral, and environmental factors, there is still a debate about the relative contributions of each of these factors. Fig. 2.4 shows life expectancy at birth for men and women. In all of the countries above the diagonal parity line, life expectancy is higher for women than for men. The graph shows not only that the female advantage exists almost everywhere, but also that the crosscountry differences are large. In some countries (like Russia, South Africa, and Vietnam), women live many years longer than men; whereas in other countries (like in Bhutan, Guinea, Mali, and Sierra Leone), the gap is smaller.

The advantage of women in terms of life expectancy is due to women having lower mortality in childhood and throughout the life course (for a review of the literature, see

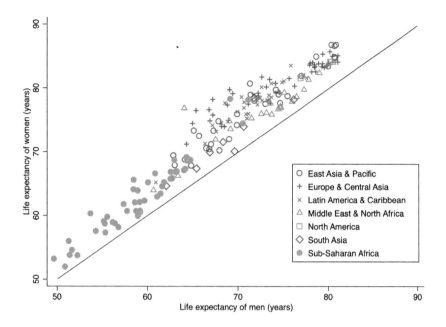

Fig. 2.4 Life expectancy at birth of women and men (2015). *(Source: United Nations, DESA, Population Division. World population prospects: the 2019 revision; 2020b. https://population.un.org/wpp/Download/ Standard/Population; data acquired via the Web site.)*

Roser et al., 2013). Currently, child mortality is higher among boys than girls in most countries of the world. The gender gap in infant and child mortality is obviously an important factor in life expectancy differences in poor countries (where child mortality is high), but it is less or not at all important in rich countries (where fewer children die and where sex differences in infant mortality are very small).

Globally, life expectancy at age 65 is more than 2.5 years higher for women than for men of the same age. This gender gap has not changed substantially since 1950, when life expectancy at age 65 was 2.6 years higher for women than for men. Moreover, there are gender differences in the increases in life expectancy at age 65 in recent decades (Table 2.3) that are larger in high-income countries (3.2 years) than in the low-income countries (1.7 years). There is no country where life expectancy at age 65 is higher than 25 years for men, but there are many developed countries where life expectancy at age 65 is higher than 25 years for women. Among the OECD countries in 2019, the countries with the highest life expectancy were Japan followed by France, Spain, Switzerland, South Korea, Italy, and Australia, while the countries with the lowest life expectancy were Eastern European countries (Russia, Hungary, Latvia, and the Slovak Republic) and Mexico.

Table 2.3 Life expectancy at age 65 and at age 80 (years), for men and women by country income level.

	1950–1955	2015–2020	2095–2100	1950–1955	2015–2020	2095–2100	1950–1955	2015–2020	2095–2100
	Men life expectancy at age 65 (years)			Women life expectancy at age 65 (years)			Difference among women and men (years)		
High-income countries	13.8	19.0	22.8	17.5	22.2	25.5	3.7	3.2	2.7
Middle-income countries	12.1	14.9	18.1	14.0	17.4	20.3	1.9	2.5	2.2
Low-income countries	11.1	12.9	15.0	12.1	14.6	17.2	1.0	1.7	2.2
World	12.7	15.9	18.4	15.3	18.7	20.7	2.6	2.8	2.3
	Men life expectancy at age 80 (years)			Women life expectancy at age 80 (years)			Difference among women and men (years)		
High-income countries	5.4	8.7	13.0	6.1	10.5	14.9	0.7	1.7	1.9
Middle-income countries	4.3	6.7	10.1	5.0	7.9	11.2	0.7	1.2	1.1
Low-income countries	4.1	5.5	7.8	4.4	6.2	9.0	0.3	0.7	1.2
World	4.8	7.4	10.4	5.5	8.8	11.5	0.7	1.4	1.2

Source: United Nations, DESA, Population Division. World population prospects: the 2019 revision; 2020b. https://population.un.org/wpp/Download/Standard/Population; data acquired via the Web site.

The analysis presented in the Health at a Glance 2019 by the OECD (OECD, 2019) showed that increased healthcare spending had a strong positive impact on life expectancy, with most OECD countries reporting increases in both health expenditures and life expectancy. In the last two decades, life expectancy increases in the OECD countries were driven mainly by health spending, followed by education and then by income. Lutz and Kebede (2018) tested whether educational attainment could be a better predictor of life expectancy than income at both micro- and macrolevels and concluded that the apparent positive association between health and income can largely be attributed to increasing educational attainment, which at the same time leads to rising incomes (Lutz et al., 2008) and better health outcomes.

Murtin et al. (2017) estimated that the differences in life expectancy by education account for about 10% of overall inequalities in ages at death. In all of the countries considered in the OECD analysis (OECD, 2019), both men and women at age 30 with the highest level of education can expect to live longer than their counterparts with the lowest level of education (+6.9 and +4 years, respectively). Among the explanations for this association are that individuals with higher educational levels also have better socioeconomic living conditions and that more educated people tend to adopt healthier lifestyles and have better access to appropriate health care. Smoking rates, excessive alcohol consumption among men, and obesity rates for both men and women are important contributors to gaps in life expectancy by education. The differences by education are particularly large among men and women in Central and Eastern European countries (the differences between the highest and the lowest educated individuals are more than 10 years among women and around six years among men), as in these countries, the prevalence of tobacco and alcohol use among men is high, and older people have lower levels of education. The differences by education are less pronounced in Canada, Turkey, and Sweden (around 3.5 among women and 2.5 among men).

The paper by Klenk et al. (2016) estimated the contributions of different age groups and disease-specific causes of death to the changes in life expectancy and identified some basic patterns. While life expectancy trends seem to be quite stable, the contributions of specific causes of death, and of specific age groups, have changed considerably by country and over time. Among women, the decline in cardiovascular disease, often in combination with changes in the 65 and over age group, was the dominant cause of reductions in the 1950s and 1960s in most of the developed countries (i.e., in North America, most Western European countries, Japan, Australia, and New Zealand). This large contribution of people over age 65 to the changes in life expectancy had already started in the 1950s in some Western European countries and Australia. For men, the developments in the age- and cause-specific contributions were similar but were delayed for up to two decades in most countries. In Japan, an accelerated epidemiologic transition in causes of death from infectious and respiratory diseases to CVD, and from younger age groups to older age groups, can be found. In Eastern Europe, Russia, and some other former Soviet republics, periods in which life expectancy declined, particularly among men, can be observed.

2.3.2 Does having a longer life also mean having a better life?

Despite the tremendous gains in life expectancy at age 65 of the last decades, a longer life is not always accompanied by a longer lifetime spent in good health. The number of additional years lived in good health at age 65 varies substantially across OECD countries. Using the question about disability included in the European Union Statistics on Income and Living Conditions (EU-SILC) survey, Eurostat calculates yearly the life expectancy free of disability, known as "healthy life years," or disability-free life expectancy. This indicator estimates that the additional number of years a person aged 65 is expected to live in the absence of limitations in functioning/disability. In the European Union, healthy life expectancy ranges from more than 15 years in Sweden and Norway to less than five years in Latvia and the Slovak Republic. On average across the OECD countries included in the survey, the number of healthy life years at age 65 is almost the same for men and women (9.6 for women and 9.4 for men). For both men and women, the number is highest in Norway, Sweden, and Iceland and is lowest in the Slovak Republic and Latvia. Gains in life expectancy at age 65 have slowed in recent years. In the OECD countries, life expectancy at age 65 increased, on average, by 11 months between 2002 and 2007 and by seven months between 2012 and 2017. This slowdown in the increase in life expectancy at age 65 can be partially explained by the severe influenza epidemic of 2014–2015, which affected frail and older populations in particular (OECD, 2019). The effect of the COVID-19 pandemic on life expectancy at age 65 can be assessed as soon as new data are available.

Even if life expectancy at age 65 has increased across the OECD countries, many adults spend a large proportion of their older years in poor or fair health. In 2017, more than half of the population aged 65 and over in the 35 OECD countries reported being in poor or fair health (Fig. 2.5). In particular, people aged 65 and older in Eastern European countries reported some of the highest rates of poor or fair health (more than three out of four reported that their health was fair, bad, or very bad). The patterns among older people in Portugal and South Korea were similar to that of the Eastern European countries. The lowest rates of poor or fair health were reported in Norway, Ireland, Switzerland, Sweden, and the Netherlands. There were small gender differences in self-assessed health, with women being slightly more likely than men to rate their health as poor or fair: on average across the OECD countries, 59% of women and 54% of men reported being in fair, bad, or very bad health.

Self-reported health is strongly linked to the socioeconomic status of individuals. Two out of three older people in the lowest income quintile, but fewer than one out of two older people in the highest income quintile, rated their health as poor or fair (OECD, 2019). In almost all of the OECD countries analyzed, the gap between self-reported poor or fair health among people in the lowest and highest income quintiles was larger than 14 percentage points.

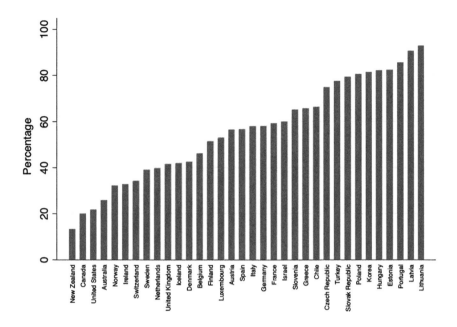

Fig. 2.5 Adults aged 65 and over rating their own health as fair, bad, or very bad in OECD countries (2017). *(Source: OECD data acquired via the Web site.)*

In the literature, different indicators had been proposed to measure the social protection and healthcare needs of older persons. In recent years, Lutz et al. (2018) proposed the new indicator "Years of Good Life" (YoGL) at age 50, which synthetizes the multiple dimensions of human well-being. It counts the "good" years of life spent above a threshold with respect to objectively observable conditions (being out of poverty, being without cognitive limitations, and having no serious physical disabilities), as well as subjective life satisfaction. Since mere survival does not capture well-being, the years of good life consider survival conditional on a minimum standard of life. The next graphs (Figs. 2.6 and 2.7) show that in European countries, the YoGL at age 50 for the EU–27 was 22.4 for women and 21.4 for men. Note that the gender differences in YoGL at age 50 are less pronounced than the gender differences in life expectancy at age 50 (in 2017, life expectancy at age 50 was 35 years for women and 30.2 years for men).

Another interesting indicator used to monitor the living conditions of older people is based on the category of limitations in daily activities and uses the Global Activity Limitation Indicator (GALI) question included in the EUSILC survey. It is noteworthy that in the OECD countries, almost half of people aged 65 and over reported having at least some (33%) or severe limitations (17%) in their daily activities (OECD, 2019). Again, the Slovak Republic, Latvia, and Estonia performed the worst on this indicator, while

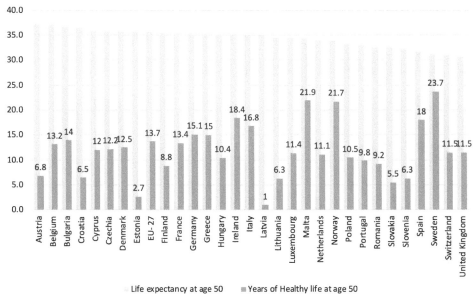

Fig. 2.6 Years of life expectancy at age 50 and years of good life at age 50 for men in EU-27 countries (2018). *(Source: Eurostat data acquired via the Web site. Note: Years of life are counted as "good" if they are spent above a threshold with respect to objectively observable conditions (being out of poverty, being without cognitive limitations, and having no serious physical disabilities), as well as subjective life satisfaction.)*

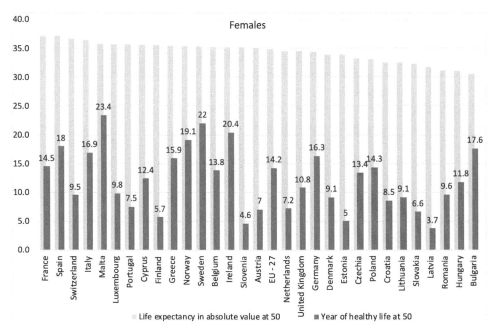

Fig. 2.7 Years of life expectancy at age 50 and years of good life at age 50 for women in EU-27 countries (2018). *(Source: Eurostat data acquired via the Web site. Note: Years of life are counted as "good" if they are spent above a threshold with respect to objectively observable conditions (being out of poverty, being without cognitive limitations, and having no serious physical disabilities), as well as subjective life satisfaction.)*

Sweden and Norway performed the best. Since people in institutions are not interviewed, the prevalence of disability in the population tends to be underestimated. Moreover, it should be noted that since health status is self-assessed, cultural differences among countries and/or among subpopulation groups can affect the indicator.

Mental health also declines with aging. According to WHO data, the number of older people with mental disorders is expected to double by 2030 due to population aging. Depending on the local context, certain individuals and groups in society may be at significantly higher risk of experiencing mental health problems (World Health Organization, 2011). Mental health policies, plans, and services need to take into account the health and social needs at all stages of the life course, including infancy, childhood, adolescence, adulthood, and older age.

2.3.3 Economics of population aging

Population aging is also monitored from an economic perspective. The old-age dependency ratio (OADR) is one of the most commonly used indicators for monitoring the economics of population aging. It measures the level of economic dependency associated with increases in the percentage of the population at older ages. As a result of the changes in mortality and fertility connected to the demographic transition, the OADR is projected to rise in most countries of the world. At the global level, the OADR is expected to increase from 16 in 2019 (i.e., 16 people aged 65 or older per 100 people aged 20–64 years) to 32 in 2060—but, as usual, there are large geographical differences in terms of both the intensity and the speed of these changes. The high-income countries have higher and increasing OADR values, while the low-income countries have lower and almost stable OADR values. The countries of Europe and North America have the highest values (from almost 30 in 2019 to more than 50 in 2060), while the countries of sub-Saharan Africa have the lowest values.

In the OADR, as in the other standard dependency ratio, the assumption is that a person is considered young or old based on his and her chronological age, net of whether he or she needs economic support. A large section of the United Nations World Population Aging 2019 report (2020a) was devoted to reviewing new measures of population aging that use different approaches. In particular, it compares the classical OADR with the prospective old-age dependency ratio (POADR) and the economic old-age dependency ratio (EOADR). The POADR is based on remaining life expectancy and suggests that the increase in old-age dependency will be slower for many countries than the traditional old-age dependency ratio estimates. The indicator provided by Sanderson and Scherbov (2007) found that at the global level, the POADR has declined slightly, from 13 per 100 in 1990 to 12 per 100 in 2019, but is projected to rise to 17 per 100 in 2050.

In addition to these measures based on chronological age and on prospective age, there are measures of population aging that use data on productivity and other components of life-cycle economic behavior (United Nations, 2013; Mason et al., 2017). Children and young people consume more and produce little or nothing because they are not yet of working age or are unemployed, whereas older people consume and do not produce because they are no longer of working age. The standard of living of children and older people depends on transfers from their families (population of working ages), but also from taxpayers via pensions or other economic supports. A life-cycle deficit occurs when consumption expenditures exceed labor income, and it is generally present in younger and older age groups. The UN (2020) reports the annual economic resources for children (aged 0–24) and the older (aged 65+) as a percentage of annual consumption. It is evident that in most European countries, public transfers play a marginal role among children aged 0–24, but are crucial for older people. In some countries, the older population are, thanks in part to the welfare system, quite wealthy and are able to provide economic resources to younger age groups. This is the case in France, Germany, Italy, and Sweden.

The economic old-age dependency ratio, provided by the United Nations (2020a) is based on the method outlined in Mason and others (Mason et al., 2017). Compared to the other measures, the advantage of this indicator is that it incorporates age-specific variations in the resource needs (namely, the consumption) of older people relative to the resources (namely, labor income) produced by all workers, irrespective of their age. An increasing EOADR reflects simultaneously the levels of consumption at older ages (compared to younger ages) and the percentage of older people in the population. The EOADR is projected to increase from 20 in 2019 to 34 in 2050. This means that the number of effective older consumers (ages 65+) is expected to increase from 20 to 34 per 100 effective workers (of all ages) (Table 2.4). In Europe and North America,

Table 2.4 Comparing three measures of old-age dependency by country income level.

	Old-age dependency ratio[a] (OADR)		Prospective old-age dependency ratio[b] (POADR)		Economic old-age dependency ratio[c] (EOADR)	
	2019	2050	2019	2050	2019	2050
High-income countries	30.2	50.5	16.0	23.6	43.8	72.3
Middle-income countries	13.5	28.1	11.6	18.5	15.1	30.9
Low-income countries	7.4	1.1	9.1	9.5	7.7	10.3
World	15.9	28.4	11.6	17.3	19.5	33.5

[a]Old-age dependency ratio: Number of persons aged 65 or over per 100 persons of working age 20–64.
[b]Prospective old-age dependency ratio: Number of persons above the age at which the remaining life expectancy is 15 years relative to the number of persons between age 20 years and the age at which the remaining life expectancy is 15 years.
[c]Economic old-age dependency ratio: Effective number of consumers aged 65 or over relative to the effective number of workers of all ages.
Source: United Nations, DESA, Population Division. World population prospects: the 2019 revision; 2020b. https://population.un.org/wpp/Download/Standard/Population.

which had the highest levels of the indicator in 2019, the EOADR is projected to increase from 42 in 2019 to 69 in 2050. The countries with values higher than 50 are Japan, followed by European countries (i.e., Finland, France, Germany, Denmark, Greece, Sweden, Italy, and the Netherlands). By 2050, the highest economic old-age dependency ratio values (more than 90) will be in Japan, the Southern European countries, Slovenia, Puerto Rico, Switzerland, Germany, and the Republic of Korea.

The EOADR includes information on economic consumption and production disaggregated by age. Another interesting measure is the "economic-demographic dependency ratio" (EDDR), proposed by Lau and Tsui (2020). It is based on a life-cycle model with schooling and retirement choices. The measure assumes that the cohort-invariant cutoff points in the conventional definition are misleading. People over age 65 who are considered resource dependents in a given year, such as in 1950, should not be considered dependent a century later, when people are generally living longer, are healthier, and may be able to work to higher ages. They argued that if individuals respond to rapid population aging by changing their schooling and retirement behavior in line with the life-cycle model, then the standard dependency ratio (which does not reflect these behavioral changes) may greatly overestimate the impact of population aging in the next few decades.

2.4 Conclusions and perspectives

An important and relevant consequence of the demographic, epidemiological, and health transitions is that the population age structure is changing from being primarily "youthful" to being primarily "aged" (McCracken and Phillips, 2017). Due to the combined effect of low fertility and increased life expectancy, the share of the population aged 65 and older has increased considerably in recent decades (from 5% in 1960 to more than 9% in 2020) and should reach 18% by 2060. These increases have been and will continue to be more rapid among the oldest-old. The share of the population aged 80 and older is projected to increase from 2% in 2020 to 5% in 2060 (compared to around 0.5% in 1980). Over the same period, the number of centenarians per 10,000 people is expected to grow from 0.7 to 5.4.

This shift in the population age structure poses profound challenges to public institutions that have to adapt their organizational capacities and services to a population with an older age structure, with a primary focus on ensuring the sustainability of pension funds and the ability of already overburdened healthcare systems to serve much higher numbers of people.

Due to the international relevance of this phenomenon, the United Nations have recommended policies to mitigate the consequences of population aging. But they have emphasized that "*there is no single best policy response to respond to population aging in all countries.*" Thus, public policies (such as the reorganization of social security systems;

and the changes in labor, immigration, and family policies) and interventions to promote an active and healthy lifestyle (to improve the health status of older people) should be selected based on the specific context of each country. "*How countries address population ageing depends on the fiscal space available to implement their tax and benefit programmes, the extent to which societies agree on the values of redistribution and intergenerational equity, and the role they assign to government, families and individuals in financing consumption, particularly during old-age*" (OECD, 2019, p. 27). The Programme of Action of the International Conference on Population and Development (ICPD), the Madrid International Plan of Action on Aging (MIPAA), and, most recently, the 2030 Agenda for Sustainable Development are only some of the international recommendations developed in recent years to mitigate the consequences of aging. Generally speaking, the aim of all of these recommendations is to ensure the sustainability of the welfare system (in terms of both the health and the pension system) and to suggest strategies and behaviors designed to help older people live vigorous and active lives (including, when possible, economically productive lives) until much higher ages than in the past.

When we consider the sustainability of health systems, it is noteworthy that although there have been gains in healthy life expectancy in recent years, population aging will likely lead to a greater demand for labor-intensive long-term care (LTC). Thus, to ensure access to and the quality of LTC services, adequate resources must be devoted to public long-term care systems. Population aging generally implies an increase in the prevalence of chronic diseases (Alzheimer's disease, cancer, cardiovascular, and cerebrovascular diseases, etc.) and a decrease in the ability to perform activities of daily living (ADL), with a consequent increase in disability in the population. It is not easy to get comparable data on how much informal care is provided to family members and friends, and how much time individuals devote to providing care and assistance. But the health surveys on people aged 50 and over, who report providing care and assistance to family members and friends, show that currently, family and friends are still the most important source of care for people with long-term care needs in OECD countries. In the OECD countries on average, almost 13% of people aged 50 and over have reported providing informal care at least weekly. This percentage ranges from around to 20% in Central European countries (the Czech Republic, Austria, France, and Germany), Belgium, and the United Kingdom to less than 10% in some Southern European countries (Portugal and Greece), Sweden, Poland, the United States, and Ireland (OECD, 2019). The observation that fewer people provide daily care in the countries where there is a stronger formal LTC system suggests that there is a trade-off between informal and formal care. Indeed, there are some OECD countries that have well-developed LTC services and public coverage and have implemented policies to support family carers with the aim of mitigating the negative effects associated with intensive caregiving. These policies provide for a range of benefits and services, including paid care leave, flexible work schedules, respite care, counseling/training services, cash benefits to family caregivers, cash-for-care allowances for recipients that can be used to pay informal caregivers, and periods of paid leave for informal carers. The implementation of such policies seems to be a good strategy for alleviating gender disparities (in some countries such as Greece and Portugal, over 70% of informal carers are women) and for dealing with the trends such as declining family size, increasing geographical mobility, and rising female labor market participation rates.

References

Caselli G. The key phases of the European health transition. Polish Population Review/Polish Demographic Society [and] Central Statistical office 1995;7:73–102.

Klenk J, Keil U, Jaensch A, Christiansen MC, Nagel G. Changes in life expectancy 1950–2010: contributions from age- and disease-specific mortality in selected countries. Popul Health Metr 2016;14(1) https://doi.org/10.1186/s12963-016-0089-x.

Lau SHP, Tsui AK. Economic-demographic dependency ratio in a life-cycle model. Macroecon Dyn 2020;24(7):1635–73.

Lee R, Mason A. Is low fertility really a problem? Population aging, dependency, and consumption. Science 2014;346(6206):229–34.

Lutz W, Kebede E. Education and health: redrawing the preston curve. Popul Dev Rev 2018;44:343–61. https://doi.org/10.1111/padr.12141.

Lutz W, Cuaresma JC, Sanserson WC. The demography of educational attainment and economic growth. Science 2008;319:1047–8. https://doi.org/10.1126/science.1151753.

Lutz W, Lijadi A, Strießnig E, Dimitrova A, Lima CBdS. Years of Good Life (YoGL): a new indicator for assessing sustainable progress. IIASA working paper.

Mason A, Lee R, Abrigo M, Lee SH. Support ratios and demographic dividends: estimates for the World, Technical Paper, https://iussp.confex.com/iussp/ipc2017/mediafile/Presentation/Paper2207/SR%20DD%20Draft.v8.pdf.

McCracken K, Phillips DR. Demographic and epidemiological transition. In: Douglas R, Castree N, Goodchild MF, Kobayashi A, Liu W, Marston RA, editors. International encyclopedia of geography: people, the earth, environment and technology. New York, NY John Wiley & Sons, Ltd; 2017.

Meslè F, Vallin J. La transition sanitaire: tendances e perspectives. In: Caselli G, Vallin J, Wunsch GJ, editors. Démographie: Les déterminantes de la mortalité, vol. 3. Paris: INED; 2002. p. 439–60.

Murtin F, Mackenbach J, Jasilionis D, d'Ercole MM. Inequalities in longevity by education in OECD countries: insights from new OECD estimates. OECD Statistics Working Papers, No. 2017/02, Paris: OECD Publishing; 2017https://doi.org/10.1787/6b64d9cf-en.

Notestein FW. Population—the long view. In: Food for the World Notestein. Chicago University Press; 1945. p. 37–57.

OECD. Health at a glance 2019: OECD indicators. Paris: OECD Publishing; 2019. https://doi.org/10.1787/4dd50c09-en.

Omran AR. The epidemiologic transition. A theory of the epidemiology of population change. Milbank Mem Fund Q 1971;49(4):509–38. https://doi.org/10.2307/3349375.

Pool I, Wong L. Age-structural transitions and policy: an emerging issue. Age-structural transitions, CICRED; 2006.3–19. http://www.cicred.org/Eng/Publications/pdf/AgeStructural-Book.pdf.

Riley JC. Estimates of regional and global life expectancy, 1800–2001. Popul Dev Rev 2005;31(3):537–43.

Robinson WC. Demographic history and theory as guides to the future of world population growth. Genus; 2003.

Roser M, Ortiz-Ospina E, Ritchie H. Life expectancy, Published Online at OurWorldInData.Org https://ourworldindata.org/life-expectancy; 2013.

Sanderson WC, Scherbov S. A new perspective on population aging. Demogr Res 2007;27–58. https://doi.org/10.4054/DemRes.2007.16.2.

United Nations. National transfer accounts manual: measuring and analysing the generational economy. UN: Sales No.: E.13.XIII.6; 2013.

United Nations, DESA, Population Division. World population ageing, https://www.un.org/en/development/desa/population/publications/pdf/ageing/WorldPopulationAgeing2019-Report.pdf; 2020.

United Nations, DESA, Population Division. World population prospects: the 2019 revision, https://population.un.org/wpp/Download/Standard/Population; 2020.

Vallin J, Meslé F. Convergences and divergences in mortality: a new approach of health transition. Demogr Res 2004;2:11–44.

World Health Organization. Global health and aging. National Institute on Aging, National Institutes of Health; 2011.

Zijdeman R, Ribeira da Silva F. Life expectancy at birth (total), In: IISH Dataverse, V1. 2015. http://hdl.handle.net/10622/LKYT53.

CHAPTER 3

Pathobiology of aging: An introduction to age-related diseases

Calogero Caruso[a], Giulia Accardi[a], Mattia Emanuela Ligotti[a], Sonya Vasto[b], and Giuseppina Candore[a]

[a]Laboratory of Immunopathology and Immunosenescence, Department of Biomedicine, Neurosciences and Advanced Diagnostics, University of Palermo, Palermo, Italy
[b]Department of Biological, Chemical and Pharmaceutical Sciences and Technologies, University of Palermo, Palermo, Italy

3.1 Introduction

As previously stated (Chapter 1), the aging process is considered a universal and inevitable process of physiological decline associated with a greater vulnerability to disease and death. Efforts to understand aging have suggested the need to distinguish aging from age-related diseases. Diseases affect a subset of older population and are linked to specific exogenous risk factors and pathophysiological mechanisms. However, chronic diseases increase with aging and, together, aging and diseases show mutual interactions in causing deterioration of health, physical and cognitive functions, and death. Therefore, it is unclear whether we can truly distinguish the effects of the diseases from those of aging itself. When we analyze the pathophysiology of age-related diseases, it appears that some fundamental biological processes are at the center of many diseases as well as of aging. Oxidative stress, inflammation, defective repair, and apoptosis are key features of both aging and many age-related pathological conditions. Molecular mechanisms are specific to individual tissues and various diseases, but tissues respond to injury with similar biological processes. The mechanisms are the same, but their different locoregional distribution likely depends on the genetic background, environmental influences, and chance. Potentially, these shared mechanisms could include aging processes (Newman and Ferrucci, 2009).

The study of centenarians, the best successful aging model, shows that these individuals also have age-related changes. The study of these individuals might help to understand what aspects of aging are distinct from the disease (Caruso, 2019). On the other hand, with the progress of medical knowledge, characteristics previously attributed to aging, such as arterial stiffness and insulin resistance, are now reclassified among risk factors of diseases. Therefore, the dichotomy between aging and disease is fluid and constantly evolving (Newman and Ferrucci, 2009).

Human Aging
https://doi.org/10.1016/B978-0-12-822569-1.00010-X

Anyway, people over the age of 70 have multiple chronic diseases. According to autopsy data, universal reservoir of latent prostate cancer that increases with age has been found (Bell et al., 2015). In addition, metabolic, immune, and cardiovascular dysfunctions affect the vast majority of older individuals. Although no disease is formally diagnosed, older people often develop preclinical signs linked to these dysfunctions. Aging is a key risk factor for human chronic diseases, many treatments and interventions aimed to delay the onset of age-related diseases can increase the lifespan of model organisms, and interventions that prolong the lifespan of models often delay disease of aging (Gladyshev and Gladyshev, 2016; Longo et al., 2015).

Many researchers do not consider aging a disease, but a normal, natural, inevitable process that, while predisposing individuals to the risk of disease, is different from a specific disease. On the contrary, researchers in the "aging as disease" field note many similarities between the two processes and support the designation of aging as a disease, thus suggesting treatment as a possibility. In fact, some believe that aging, similar to disease, can be cured and eventually stopped (Gladyshev and Gladyshev, 2016).

According to Rattan (2014), most biomedical research is dominated by disease-directed thinking. Some gerontologists have also succumbed to these disease-directed pressures by considering aging as another disease. However, as demonstrated by gerontological studies of the last years, the problems of aging, quality of life, and longevity cannot be successfully addressed with disease-oriented thinking. Understanding whether aging is a disease or a process that increases the possibility of disease outbreaks has implications for interventional strategies. Aging occurs progressively in each individual who survives beyond a certain duration of life within the evolutionary framework, so aging cannot be considered a disease. Therefore, a paradigm shift is needed to transform our approach to aging interventions from the so-called antiaging treatments to the development of strategies to maintain health, extend health lifespan, and improve public and social health, i.e., a health-oriented preventive approach (Rattan, 2014).

As stated by Hayflick (2004, 2007), we cannot deny a close relationship between the aging process and age-associated diseases, but the two phenomena are distinct in that the aging process simply increases the vulnerability to age-associated diseases. Unfortunately, this critical distinction is not universally appreciated, so it is believed that the resolution of age-associated diseases will improve our understanding of the aging process. It is not so. In fact, understanding different childhood diseases has not improved our understanding of childhood development. The inability to distinguish the biology of aging (gerontology) from age-associated diseases (geriatrics) is the most serious obstacle to our understanding of the aging process. No progress in geriatric medicine will increase our knowledge of the biology of aging (Hayflick, 2004, 2007).

To decide whether aging is a disease, the problem is complicated by the meaning given to the terms of normality (health) and disease that do not have universal definitions, varying with time and culture. It would be easy to say that disease is a deviation from the norm. However, the term "health" corresponds to a definition that is applied whenever the pathological deviations from the abstract model of normality do not exceed quantitatively or qualitatively certain limits established conventionally, both according to popular tradition and taking into account certain scientific parameters. From the point of view of an evolutionary logic, the ideal of optimal physiological adaptations, to which the statistic concepts of normality refer, does not actually exist. It is also necessary to take into account the medicalization process, a social process through which a previously normal condition becomes a medical problem. Here again, the main problem is what is considered normal vs. pathological. But often there is no right or wrong answer, since the answer is shaped by social attitudes, political forces, religious issues, and commercial interests and not only by medicine itself (Caruso and Candore, 2016).

However, if we accept the definition of British Medical Journal (Goodle, 2011) of health as the ability to adapt and self-management in the face of social, physical, and emotional challenges, then many old people are not healthy.

During our life, deleterious changes accumulate at all levels of the biological organization of the organism, changes influenced by genetic, environmental, and stochastic processes. As a reflection of the increasing amounts of different deleterious changes, dysfunctions can manifest themselves differently. For example, they can result from damage to macromolecules or metabolites, mitochondrial impairment, cell senescence, or homeostatic imbalance. Therefore, age-related disease such as heart disease, diabetes, and neurodegenerative diseases are an expression of macroscopic dysfunction. They are visible "markers" of an underlying process. In this paradigm, a chronic disease can simply be a feature of an age-related loss of function (Gladyshev and Gladyshev, 2016).

In the present chapter, we do not need to distinguish between aging and age-related diseases because, as discussed by Blagosklonny (2015), in protected environment, humans and animals die from age-related diseases, which are manifestation of aging.

In the nest paragraph, it is discussed another disputed topic concerning aging, i.e., whether the complexity of an older organism is greater or lesser than that of a young one.

3.2 Complexity

An organism is a highly complex structure. The concept of complexity was born in the field of physics, although today it is applied to various sciences, including biology. In this case, the complexity comes from the myriad of interactions and feedback controls that operate between the different structural units of an organism. They allow the body to cope with the numerous "stressful" factors with which cells, tissues, and organs are confronted daily. In biology, unlike other systems, an important component of complexity is that it is also the result of the evolutionary process, and therefore is a highly dynamic system. The degree of complexity depends on the number of variables and on the type of relationship existing between variables that have different degrees of hierarchy.

By changing the number of variables or the relationships existing between the various structural units, the complexity could be changed. The complexity of organisms is such that although many of the underlying physiological mechanisms are known as well as positive or negative feedback controls, not all phenomena that occur are predictable (Lipsitz and Goldberger, 1992) (Fig. 3.1).

Concerning aging, we must be aware that the "human" system has a component of complexity that requires researchers to be able to accurately interpret the observed phenomena in order to design any intervention that can interfere with what can potentially occur. As previously stated, this may not always be predictable both because a component of randomness is inherent in every system and because the mechanisms underlying a phenomenon may not be entirely known. The question that arises is whether the complexity

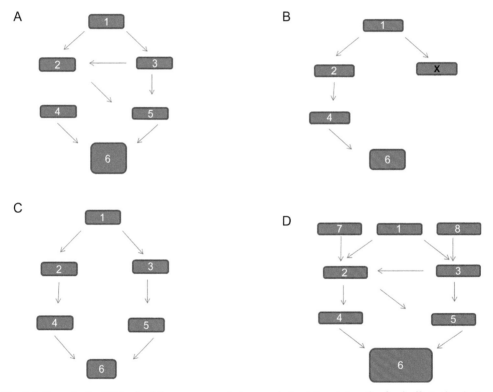

Fig. 3.1 Complexity. (A) Each of the components has a role in determining the final effect, thanks to direct or indirect interactions between them. (B) Component 3 of the system has been eliminated; the system is less complex both for the number and for the relationships that are established between the various components. The final result will be reduced. (C) The number of components is unchanged, but the relationships have changed; the system is less complex. The final result will be reduced. (D) Factors not controlled by the feedback systems intervene in the preexisting system; the system is more complex, but the result is pathological (e.g., hormonal overproduction).

of an older organism is greater or less than that of a young one. This is not simply a "philosophical" curiosity, since from the answer to this question one could conclude that, theoretically, by measuring the degree of complexity of an organism, we can obtain useful information for understanding the mechanisms underlying the aging process (and diseases) and its evolution over time. The idea is that aging leads to a decrease in complexity and the examples that support this hypothesis are numerous (Manor and Lipsitz, 2013).

Several studies, in fact, suggest that biological complexity diminishes with aging, as various tissues and organs, and their communication pathways, gradually break down, losing structural complexity. Both aging and age-related diseases are associated with a loss of complexity in the dynamics of many physiological systems. This loss of complexity may reduce the ability to adapt to hormetins (i.e., stressors able to give a positive hormetic effect, see Chapter 12) and stressors, hence leading to frailty and disease. For example, when the microscopic struts in bone tissue thin and disconnect, as occurs with osteoporosis, bones become brittle and are prone to fracturing. Likewise, the pruning of neural connections in the brain is associated with age-related neurodegenerative disorders, such as Alzheimer's disease (AD) (Lipsitz, 2004, 2016).

A further example is represented by presbycusis, that is, the typical hearing loss of older people. This is the loss of the ability to pick up high-frequency sounds (Lee, 2013). This pathology, which often progresses, and which modifies the ability to detect even the lowest frequencies, has a complex pathogenesis. Presbycusis primarily results from accumulated damage to the inner ear, particularly a loss of sensory hair cells in the cochlea (i.e., sensory presbycusis). In fact, the various components that participate in the perception of sound can be affected, i.e., hair cells, neurons connected to them, blood perfusion, the basilar membrane. When the complex relationships between all the protagonists of a biological phenomenon are interrupted, there can be manifestations ranging from those of physiological aging to those of the disease, and this depends on the extent of the loss of correlations and therefore of complexity. On the other hand, also in this case, we must remember the heterogeneity of the aging process. People start to suffer from hearing loss at age 60, while others at 90 have no signs of hearing loss. This happens because the chronological and biological age do not always coincide, being their ratio higher or equal to 1 even in successful aging less than 1 in case of unsuccessful aging.

As stated in Chapter 1, the chronological age is the mere age dictated by the calendar. Each subject results classified according to his date of birth. However, chronological age does not necessarily express the biological age of a person. Everyone has its biological age, which can be described as the age a person shows in relationship with the biological quality of its cells, tissues, and organs, and compared with standard values of reference (so-called normal). It is an expression of maturation processes, biological and external environmental influences (e.g., lifestyle and kind of job), with a genetically determined basis. The identification of a person's biological age would be an ideal tool for evaluating the effects of various antiaging therapies. Unfortunately, however, there are no common parameters and biomarkers, scientifically approved, to measure biological age (see Chapter 15 for the characteristics of an ideal biomarker).

Coming back to the concept of complexity, the Lorenz butterfly (Lorenz, 1972) is often mentioned, according to which a flapping of the wings of a butterfly in Brazil can trigger a hurricane in Texas. The metaphor suggests that even the slightest change can cause, over time, consequences that are not proportionate to the initial event. In other words, a complex system is very sensitive to the initial conditions. The smallest variation of these can give rise to a completely different dynamic evolution from that which would have occurred if the variation had not happened.

In this context, the metaphor helps to understand the strong instability of the "aging system" for which a small accident, a trivial infection can cause fatal effects.

Healthy physiological processes require the complex interaction of multiple control systems operating over multiple time scales. The output of these processes (for example, heart rate, blood pressure, hormonal rhythms, or postural sway) demonstrates a complex variability. Take, for instance, heart rate. Although average beats per minute may stay relatively constant over a person's lifespan, tiny variations in the timing between beats become more regular (less complex) with advancing age. Numerous studies have linked this change to cardiac disease and mortality. Similarly, neural activity produces electrical signals that appear less complex in older adults. As complexity declines, so do motor control and cognitive functions, including gait, attention, and memory (Lipsitz, 2004, 2016).

As discussed, aging is a process characterized by increasing structural disorder with an increase in entropy (see Chapter 1) and functional loss. The decrease in complexity, closely related to the increase in entropy, determines the reduction of the functional reserve of older people (Goldspink, 2005). If normally the human heart pumps 5 liters of blood with an average of 70 beats per minute, during an intense physical effort the same heart can pump 35 liters doubling the beats per minute. The aging process determines a progressive reduction of the functional reserve whereby the organism, if able to function well in ordinary conditions, can decompensate in the face of greater functional requirements. The heart of an old person may easily not be able to pump all the blood per minute required by intense physical exertion and an acute cardio-circulatory failure can result. The changes that accumulate with age, beyond a certain limit, make it impossible to adapt.

What determines healthy aging is the residual energy ability after the reproductive period. The aging organism shows a reduced ability to adapt to the environment as a result of a complex cascade of processes that lead to the progressive reduction of the functional reserve of the whole organism, of the individual organs and systems. The reduction rate of the functional reserve is, in turn, the result of the interaction of the genetic background with the environment. To maintain its integrity and functionality, the human body, in fact, must maintain its functional reserve constant, to cope with stressful situations capable of inducing a functional failure. In ordinary conditions, the body uses only a minimal part of its resources, and it keeps the remaining as a reserve and accesses it only in case of need.

3.3 Hallmarks of aging

A number of critical questions have arisen in the field of aging regarding the physiological sources of aging-causing damage, the compensatory responses that try to re-establish homeostasis, the interconnection between the different types of damage and compensatory responses, and the possibilities to intervene exogenously to delay aging. So, López-Otín et al. (2013) have attempted to identify and categorize cellular and molecular hallmarks of aging. They have characterized nine major hallmarks of aging: (1) genomic instability, (2) epigenetic alterations, (3) deregulated nutrient sensing pathways, (4) loss of proteostasis, (5) mitochondrial dysfunction, (6) telomere attrition, (7) cellular senescence, (8) stem cell exhaustion, and (9) altered intercellular communication. The hallmarks and their interconnectivity should serve as an evaluation tool to assess and prioritize interventions that can be deemed effective in slowing aging and preventing age-related diseases. As an example, inflammaging, the most important consequence of altered intercellular communication, plays a key role in the pathophysiology of inflammatory age-related diseases, and its knowledge provides insights into the measure that can target it. Although the contribution of each of these hallmarks towards the progression of biological aging is not yet fully elucidated, they need to be extensively studied to provide newer insights into age-related diseases and in the potential interventions. At this regard, in the paragraph on nutrient signaling (sensing) pathways, NSPs will be treated more extensively because, at the present state of our knowledge, it is the only one on which we can really intervene.

In the paragraph on mitochondrial dysfunctions, it will be treated the generation and the effects of radical oxygen species (ROS), since the age-related mitochondrial dysfunctions are responsible for their increased generation in aging.

At the end of the paragraphs on hallmarks of aging, cancer will be treated as a model of age-related disease, since it is a paradigmatic example of the convergent or divergent role of hallmarks in disease and aging.

3.4 Genomic instability

Genomic instability, an unavoidable phenomenon that can lead to nuclear DNA damage and that physiologically characterizes aging process, is the tendency of the genome to undergo alterations in DNA information content through a multitude of DNA alterations like mutations, insertions, deletions, and chromosomal rearrangements. Genomic instability, in fact, refers to the accumulation of genetic damage throughout lifespan, due to both endogenous (ROS, DNA replication errors, etc.) and exogenous (environmental and iatrogenic) agents. Such genetic damage involves oxidation of constituent DNA bases, crosslinking of DNA and protein as well as accumulation of DNA double-stranded breaks. Although the extent of lesions occurring daily in each somatic cell can be quite high, most of them are countervailed by the vast network of DNA repair mechanisms (Vijg and Suh, 2013).

Thus, lifespan and successful aging might be also a function of genomic integrity. According to this suggestion, the relatively low level of chromosomal aberrations observed in the "oldest old" individuals should be a consequence of their genomic stability, hence a contributing factor to their attainment of advanced age (Vijg and Suh, 2013). The evidence for it is based largely on accelerated aging phenotypes of DNA repair model mutants (Hasty et al., 2003) and human progeroid syndromes (Oshima et al., 2018). The progeroid syndromes, hereditary conditions, are characterized by defects in genes directly or indirectly involved in genomic stability by the restoration of DNA damages, causing the onset of premature aging phenotype, including typical age-related disorders, such as type 2 diabetes or cancer (Navarro et al., 2006). Furthermore, recent comparative analysis of 18 rodent species has identified a role for sirtuin-dependent DNA double-strand break repair as a major factor in organismal lifespan (Tian et al., 2019).

Sirtuins (Section 3.6.4), orthologues of Sir2 yeast protein (where Sir stands for silent information regulator since it silences certain genes), are a family of histone deacetylase enzymes, involved in cellular mechanisms regulation, identified as antiaging molecules in model organisms. In mammals, there are seven sirtuins (SIRT1-SIRT7) having different profiles of enzymatic activity and subcellular compartmentation. SIRT1 and SIRT6 play a role in slowing some aspects of aging at various levels, i.e., cell senescence, interaction with NSPs and with caloric restriction (CR), a diet model characterized by a strong decrease in the caloric intake without providing insufficient micronutrients and essential compounds (Chapters 12 and 14). All sirtuins contain 275-aminoacid catalytic subunits and use nicotinamide adenine dinucleotide (NAD^+) as a cosubstrate. NAD^+ is converted to nicotinamide, and its concentration is determined by the nutritional state of the cell (Lee et al., 2019).

The loss of sirtuin function is associated with genome instability and compromised organism viability. In particular, several data indicate that SIRT1 is able to perform multiple functions in different DNA repair pathways, highlighting its critical role in protecting against genomic instability (Choi and Mostoslavsky, 2014). Concerning a possible therapeutic approach, metformin, a drug used to control type II diabetes, has shown some encouraging results towards increasing longevity, targeting metabolic signaling pathway (Chapter 18). A recent study, performed on mononuclear cells from subjects with pre-diabetes, showed that, in addition, metformin increased SIRT1 gene/protein expression and SIRT1 promoter chromatin accessibility; hence, the drug should be able to promote genomic stability (de Kreutzenberg et al., 2015).

3.5 Epigenetic alteration

As stated in Chapter 1, the term "epigenetics" refers to the heritable alterations not due to changes in DNA sequence, which vary the individual phenotype by modulating the expression and the activity of genes. These alterations include DNA methylation, histone modification, and small noncoding RNAs (miRNA), all able to regulate patterns of gene expression. In fact, loss of histones, imbalance in histone modifications, changes in chromatin architecture, breakdown of the nuclear lamina, as well as DNA and histone methylation changes, are the characteristics of biological aging. Histone modifications such as acetylation and phosphorylation can lead to active transcription, loss of cellular homeostasis, and age-associated metabolic decline, whereas miRNA are active in the regulation of gene expression at the transcriptional and posttranscriptional level. (Sen et al., 2016).

Epigenetics is a bridge between genome and environment in the definition of phenotype, since environmental factors influence it. Throughout life, stochastic and environmental stimuli as well as the failure of the epigenetic machinery may induce random changes at certain loci, leading to a loss of phenotypic plasticity among individuals. Indeed, chance plays a role also in epigenetics as demonstrated by studies carried out in mono- and dizygotic twin pairs, because in monozygotic twins, there is a gradual age-related divergence in epigenetic marks. Epigenetic drift, i.e., the epigenetic dynamic changes during lifetime, deeply also affects the function of aged stem cells by limiting their plasticity and their differentiation potential. This contributes to the exhaustion of the stem cell pool (Bellizzi et al., 2019).

Literature data agree to consider epigenetics associated with biological mechanisms involved in aging and longevity. It has been assumed that most hyperage differentially methylated regions (DMRs) represent epigenetic perturbations inherent to the aging per se, while hypo-DMRs may be correlated with modifications associated with both aging per se and age-dependent modifications. Strong evidence shows that all epigenetic systems contribute to the lifespan control in various organisms. Similar to other cell systems, epigenome is prone to gradual degradation due to the genome damage, stressful agents, and other aging factors. However, unlike mutations and other kinds of the genome damage, age-related epigenetic changes could be fully or partially reversed to a "young" state (Ashapkin et al., 2017). Furthermore, DNA and histone methylation landscapes change during aging, generally leading to global hypomethylation and promoter-specific hypermethylation. These changes are determinants of conserved, tissue-specific, age-associated transcriptional changes that contribute to metabolic and inflammatory phenotypes of aging (Michalak et al., 2019).

Recently, several studies reported the presence of directional and nonstochastic changes occurring over time within clusters of consecutive CpG sites, regions of DNA where a cytosine nucleotide is followed by a guanine nucleotide in the linear sequence of bases along its $5' \rightarrow 3'$ direction, which occur with high frequency in genomic regions called CpG islands. Global genomic DNA hypomethylation is especially evident at these repetitive sequences. Heyn et al. (2012) reported the complete DNA methylomes of CD4+ T cells from newborns and centenarians. The samples of centenarians showed a lower correlation in terms of the methylation status of neighboring CpGs when compared with the samples of newborns. Indeed, these samples were more homogenously methylated in nearby located CpGs. The centenarian DNA hypomethylated CpGs covered all genomic compartments, such as promoters, exonic, intronic, and intergenic regions. Among the regulatory regions, the most hypomethylated sequences in the centenarian DNA were mainly observed at CpG-poor promoters and in tissue-specific genes, whereas a greater level of DNA methylation was observed in promoters of CpG island (Heyn et al., 2012).

With age, therefore, a significant decrease in overall DNA methylation levels is generally observed. Epigenome-wide association studies have identified the so-called clock CpGs, i.e., a large set of CpG markers whose methylation status measurement allows to build quantitative models for predicting the age of cells, tissues, or organs, called epigenetic age or DNAm age. DNAm age is thought to reflect both the chronological and biological age. Therefore, these biomarkers should identify individuals with the same chronological but with different biological age. This should allow to define a panel of measurements for successful aging and, maybe further, predict lifespan. So, a series of epigenetic clocks were developed and proposed by analyzing DNA methylation marks in single and multiple tissues (Bellizzi et al., 2019).

Horvath (2013) developed a multitissue predictor of age that allows to estimate the DNA methylation age of most tissues and cell types. The predictor was developed using 8000 samples from 82 Illumina DNA methylation array datasets, encompassing 51 healthy tissues and cell types. DNAm age was shown to have the following properties: (i) close to zero for embryonic and induced pluripotent stem cells; (ii) correlated with cell passage number; and (iii) rise to a highly heritable measure of age acceleration. Thus, DNA methylation age measures the cumulative effect of an epigenetic maintenance system.

On the other hand, Hannum et al. (2013) built a quantitative model of aging using measurements at more than 450,000 CpG markers from the whole blood of 656 human individuals, aged 19 to 101. The model measures the rate at which an individual's methylome ages, impacted by gender and genetic variants. Differences in aging rates help explain epigenetic drift and are reflected in the transcriptome. This model highlights specific components of the aging process and provides a quantitative readout for studying the role of methylation in age-related disease.

Currently, Hannum and Horvath clocks represent the most robust recognized models showing both a high age correlation ($r > 0.9$) and low mean error of the age prediction (4.9 and 3.6 years, respectively). Both models should be also able to predict all-cause mortality independently on several risk factors, including smoking, alcohol use, education, body mass index, and comorbidities (Jylhävä et al., 2017). More recently, while Hannum and Horvath clocks were primarily developed to predict chronological age, the epigenetic clock proposed by Levine et al. (2018), namely, "phenoAge," incorporating surrogate measures of biological age, is more directly aimed at predicting mortality and healthspan. In fact, it has been robustly associated with neurodegenerative diseases (Levine et al., 2018).

Thus, data are accumulating which show that epigenetic modifications may represent important tools to monitor the rate and the quality of aging, or to warn for the onset of age-related diseases (Bellizzi et al., 2019). Their main drawback is the limited accuracy when applied to the methylation profiles of older samples ($> \sim 60$ years) and in samples derived from certain tissues (El Khoury et al., 2019).

Epigenetic modifications play a fundamental role in gene expression. In fact, the increase and/or decrease of methylation of the cytosines belonging to the CpG islands determines a lesser or greater possibility of access to DNA for transcription factors. Both hypo- and hypermethylation have been associated with aging. This means that, in a tissue-dependent manner, the expression of certain genes changes with age. Aging is associated with a large variety of changes in gene expression, in different tissues. Studies conducted in models and in humans have shown that the expression of genes indicative of cell damage, such as those involved in the stress or inflammatory response, increases with increasing age. On the other hand, the expression of metabolic and biosynthetic genes decreases. Despite numerous studies, such as those conducted with the microarray technique that allows to compare gene expression in young and old tissues, it is still not possible today to outline a transcriptional pattern linked to aging and an exact correlation with physiological, biochemical, and pathological changes (Frenk and Houseley, 2018).

Puca et al. (2018) have recently revised how nutrition and dietary compounds perturb epigenetic modifications system towards a successful aging. An "epigenetic diet" based on consuming soy, grapes, cruciferous vegetables, and green tea can enhance the achievement of successful aging and the delay of the onset of age-related diseases. In particular, sulforaphane and epigallocatechin-3-gallate act by inhibiting DNA methyltransferase and histone acetyltransferase, or by modifying noncoding miRNA expression. Furthermore, evidence has suggested that the beneficial effects on successful aging and lifespan extension associated with CR are mediated by mechanisms involving sirtuins through (although not exclusively) epigenetic effect. In particular, it should be mediated by SIRT1-dependent, histone modifications in response to CR. On the other hand, a study on methylation profile of human liver demonstrated that high values of body mass index are associated with a profile of DNA similar to those of older individuals. Therefore, the phytochemicals, contained in the food, contribute to generate epigenetic modifications, and it is proven by many studies that underline, in vivo and in vitro, the possible effect of specific plant molecules in the attainment of successful aging and longevity (Daniel and Tollefsbol, 2015; Puca et al., 2018; Gensous et al., 2019).

3.6 Deregulated nutrient sensing pathways

Both in animal models and in humans, slow-aging diets, i.e., dietary interventions that can slow the aging process, delaying or preventing chronic age-related diseases (see Chapters 12 and 14), act through the modulation of NSPs. They are represented by the insulin/insulin-like growth factor-1 (Ins/IGF-1), the mechanistic (previously referred as mammalian) target of rapamycin (mTOR), the cellular energy sensor 5′ adenosine monophosphate-activated protein kinase (AMPK), and the sirtuin pathways. They, in turn, regulate autophagy (Johnson et al., 2013; Fontana et al., 2014; Aiello et al., 2017; Aiello et al., 2019a; Wong et al., 2020). (Fig. 3.2).

We know three types of autophagy with different modes of cargo delivery to the lysosome, i.e., macroautophagy, chaperone-mediated autophagy, and microautophagy. Chaperone-mediated autophagy involves chaperone-assisted translocation of substrate proteins across the lysosomal membrane, whereas microautophagy involves the direct engulfment of cytoplasmic contents by lysosomes. In macroautophagy, a portion of the cytoplasm is engulfed by a thin membrane cistern, which results in the formation of the autophagosome, a double-membrane organelle. Following the fusion of lysosomal membrane with autophagosomal one, lysosomal enzymes degrade the inner autophagosomal membrane as well as the enclosed material. Macroautophagy, hereafter referred to as autophagy, is now known to degrade selective cargoes, such as damaged mitochondria (mitophagy) (Mizushima and Levine, 2020).

NSPs are activated by carbohydrates or proteins that trigger signals resulting in a downstream activation of genes involved in aging. An excessive assumption of these

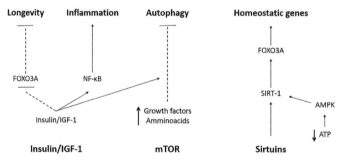

Fig. 3.2 Nutrient sensing (signaling) pathways. The figure shows some of the molecular events involved in insulin-IGF-1, mTOR, and sirtuin pathways, known as NSPs, and the possible effects on aging. The downregulation of IGF-1 and mTOR cascade or the upregulation of sirtuin one, through a low glycemic index and low animal protein intake, can extend lifespan in various model organisms, including mammals. The downregulation of IGF-1 and mTOR pathways activates the autophagy process, the homeostatic genes transcription, and the reduction of the inflammatory events as well as mitogenic ones (not reported in the figure).

nutrients can enhance these events, increasing the risk of unsuccessful aging and age-related diseases. In fact, the activation of these metabolic pathways is characterized by the trigger of inflammation and mitochondrial dysfunction with an increase of oxidative stress and a reduction of autophagy. Conversely, the downregulation of Ins/IGF-1 and mTOR cascades or the upregulation of sirtuins can extend successful aging and lifespan in various model organisms (Aiello et al., 2017). These effects are also obtained by the presence of specific single nucleotide polymorphisms in genes encoding proteins involved in NSPs, such as IGF-1 receptor (IGF-1R) and Forkhead box O3 (FOXO3). In particular, certain variants were found with higher frequency in centenarians, highlighting their role in successful aging probably due to an increased activity of FOXO3 (Suh et al., 2008; Di Bona et al., 2014).

3.6.1 FOXO3

The presence of a conserved forkhead DNA binding domain that allows the regulation of gene transcription characterizes the FOXO transcription family. These transcription factors are divided into subclasses and are present in all metazoans, from worms to humans. This underlines their role as key regulators of cell survival. In mammals, we find four FOXO proteins, FOXO1, 3, 4 and 6, which show a high sequence similarity. In human beings, the activity of the FOXO3 protein has been associated with age-related phenotypes and its role has been linked to a longer lifespan through the modulation of stress responses on oxidative stress, DNA damage, nutrient deficiency, and CR. It is noteworthy that functional changes in FOXO3 have been linked to degenerative diseases, premature aging, and poor prognosis in several types of cancer. Indeed, FOXO3 is a central regulator of cellular homeostasis, stress response, and longevity, as it can modulate a variety of cellular processes, integrating inputs from energy, growth factor, and stress signaling cascades (Fasano et al., 2019). It is one of the orthologues of daf-16 in *Caenorhabditis* (*C.*) *elegans* known to be involved in stress resistance and survival, through its action on homeostatic genes (Fontana et al., 2010).

FOXO3, as a key transcription factor, could integrate different signals from NSPs to modulate aging and longevity via shuttling from cytoplasm to nucleus. mTOR and Ins/IGF-1 have distinct effects on FOXO3 as well as its downstream target genes, although their activation inhibits FOXO3. Its phosphorylation by AMPK, instead, leads to the activation of FOXO3 transcriptional activity. In general, all these signals regulate their activity mainly through reversible posttranslational modifications, which include phosphorylation, acetylation, methylation, and ubiquitination (Sun et al., 2017).

FOXO3 induces a specific set of genes involved in the regulation of various cellular processes, including cell cycle progression, DNA repair, autophagy, ROS detoxification, and apoptosis. In fact, the cytoplasm FOXO3 is inactive and is transported either to the nucleus or to the mitochondria to exercise its transcriptional functions. FOXO3 modulates protein–protein interactions and orchestrates the spatial and temporal expression of genes, playing a key role in the stress response to maintain cellular homeostasis. Different types of stressors can differentially modulate nuclear and/or mitochondrial FOXO3 activity, allowing cells to implement the most appropriate stress response (Fasano et al., 2019).

3.6.2 Insulin/IGF-1 pathway

Ins/IGF-1 pathway affects lifespan in model organisms and, during evolution, has diverged from a single to multiple receptors, increasing the level of complexity in mammals. However, a series of genetic manipulations in mice have provided evidence that this pathway also affects aging and longevity in mammals. The Ins/IGF-1 signaling cascade starts from the binding of insulin or IGF-1 to the insulin or IGF-1 receptors. Consequently, inside the cell, the intracellular substrate proteins, known as insulin-responsive substrates, act as mediators, binding specific Src-homology-2 domain proteins. These include enzymes, such as phosphatidylinositol 3-kinase (PI3K) and other intracellular signaling molecules, as the adaptor protein growth factor receptor-bound protein 2, which is connected with the RAS pathway, a mitosis-stimulating protooncogene (Vasto et al., 2014; Aiello et al., 2017).

PI3K activates the second messenger phosphatidylinositol 3-phosphate (PIP3). PIP3 messenger leads to the activation of protein kinase B (PKB, also known as AKT) that inhibits FOXO3, preventing the transcription of homeostatic genes. Also, SIRT1 can act on FOXO3, through the deacetylation of FOXO3, modulating its response to oxidative stress. In addition, AKT activates the NF-kB pathway allowing the transcription of inflammatory genes. Moreover, it acts on glycogen synthase kinase 3 involved in cellular glucose uptake (Vasto et al., 2014; Aiello et al., 2017).

Moreover, this pathway, through the kinases PI3K and AKT, activates mTOR complex that has mitogenic and proinflammatory effects and inhibits autophagy and the transcription of homeostatic genes. Conversely, mTOR is inhibited by AMPK, a key sensor of energy status activated in response to low ATP levels (Vasto et al., 2014; Aiello et al., 2017).

In humans, aging is associated with lower IGF-1 circulating levels, and in long-living individuals, IGF-1R polymorphisms have been correlated with the modulation of human lifespan through the attenuation of IGF-1 signaling. However, IGF-1 levels decline both in successful and in unsuccessful aging, whereas in model organisms a relatively decreased signaling pathway extends longevity. It has been suggested that a constitutively reduced signaling implies a lower rate of cell growth and metabolism, and hence lower rates of cellular damage (Aiello et al., 2017).

So, continuous activation of this anabolic pathway by nutrients, i.e., proteins (IGF-1) and refined sugars with high glycemic index (GI) (insulin), is responsible for accelerating aging and age-related diseases, such as cancer, stimulating inflammation and cell proliferation and decreasing homeostatic responses. Slow-aging diets with low GI and with low protein intake may reduce the IGF-1 levels and may downregulate the Ins/IGF-1 pathway leading to the transcription of homeostatic genes through FOXO3 and stopping the mitogenic effect of RAS. Furthermore, decreased AKT activation determines the inhibition of mTOR complexes. That is also achieved by low-caloric diets which activate AMPK that inhibits mTOR. This favors successful aging and longevity (Vasto et al., 2014; Aiello et al., 2019b).

3.6.3 mTOR pathway

mTOR is a serine/threonine kinase belonging to the phosphatidylinositol kinase-related family, and it is structurally as well as functionally highly conserved throughout eukaryotes. In mammals, mTOR participates in the formation of mTORC1 and mTORC2. mTORC1 promotes anabolic processes and blocks catabolism and autophagy. As previously stated, slow-aging diets reduce growth factor signaling, which suppresses mTORC1 activity, leading to downregulation of metabolism and promotion of successful aging. On the other hand, mTORC2 phosphorylates and activates PKB, suggesting a regulation on mTORC1 (Hara et al., 1998; Sarbassov et al., 2005).

Stress conditions, such as hypoxia and osmotic stress, can also inhibit mTOR activity. Instead, nutrients, especially amino acids and growth factors, such as IGF-1, activate PKB through the PI3K and phosphoinositide-dependent kinase-1 cascade, acting, consequently, on mTOR pathway (Aiello et al., 2017).

The characterization of molecular interactions of mTOR permits to analyze its role in aging and age-related disease, as well as the possibility to use it as a target against unsuccessful aging. Amino acids as well as glucose can trigger the activation of mTORC1. Thus, an excess of nutrients intake can inhibit autophagic process, which regardless of them physiologically reduces their efficiency during aging, via mTORC1 hyperactivation. The consequence is an accumulation of damaged proteins, leading to proteotoxic stress and unbalanced proteostasis as well as to the accumulation of damaged organelles and other macromolecules from the cytoplasm, to a lack in the recycling of amino acids and cellular components, and to exacerbation of oxidative stress, thus increasing the rate of aging (Mizushima et al., 2008). Moreover, the hyperactivation of mTORC1 constitutes a potent anabolic signal with a key role in the onset of several age-related diseases, including neurodegenerative, metabolic, and cardiovascular ones. Conversely, CR and other dietary restriction (DR) (i.e., reduction of specific nutrients, such as specific amino acids) downregulate mTOR, other than Ins/IGF-1 pathway, extending healthy lifespan in different model organisms, including mammals, thanks to the counterbalance of the above-mentioned processes. CR and DR can produce an antiinflammatory effect, reducing the common inflammatory background of age-related pathologies. Hence, a reduction in chronic low-grade inflammation, typical of aging, is another feasible method by which mTORC1 downregulation could lead to successful aging (Omodei and Fontana, 2011).

Low cellular ATP levels or high AMP/ATP ratio, as a consequence of hypoxia or starvation, activates (AMPK), also triggered by CR. AMPK, in turn, inhibits mTORC1, as well as do genotoxic stress via p53, oxidizing agents, cigarette smoke, and glucocorticoids that act via the induction of the hypoxia–inducible factor (HIF)-1α.It has also been reported that mTORC1 stimulates the transcription of HIF-1α, which activates the expression of many genes essential for cellular processes and response to stress conditions. Moreover, mTORC1 inhibits autophagy in response to nutrients and growth factors, while upregulation of autophagy is observed during starvation or in response to oxidative stress (Gwinn et al., 2008; Majmundar et al., 2010; Yu et al., 2010; Kim and Guan, 2011).

mTORC1 also seems to be an important regulator of mitochondrial function. Inhibition of mTORC1 modifies mitochondrial phosphoproteome, and reduces mitochondrial membrane potential, oxygen consumption, and therefore cellular ATP levels and ROS production rate involved in oxidative stress (Schieke et al., 2006).

So, the mTOR kinase, which is part of two complexes, mTORC1 and mTORC2, has a central role in the regulation of metabolism, integrating numerous environmental signals and indicating whether conditions are favorable for anabolic processes. In fact, this pathway responds to stimuli, including insulin and IGF-1, the level of amino acids and glucose, the cellular energy status, and oxygen levels. These data again support the idea that intense anabolic activity is responsible for accelerating aging and age-related diseases (Kennedy and Lamming, 2016; López-Otín et al., 2013; Aiello et al., 2017).

The immunosuppressant and anticancer drug rapamycin is the inhibitor of mTOR, and studies in models clearly show the possibility of their use to slow aging and prevent age-related diseases (see Chapter 18). Unfortunately, it seems that the prolonged administration can inhibit mTORC2, leading to metabolic complications (i.e., abnormal glucose uptake by cells and impaired gluconeogenesis) (Lamming et al., 2012).

On the other hand, it is important to highlight that mTORC1 role is crucial in mitochondrial functions and biogenesis. So, if on the one side its inhibition could be beneficially and lifespan extending, on the other side, it could be important to preserve a basal function in aging of mTOR, at least in some tissues (Cunningham et al., 2007; Saxton and Sabatini, 2017).

3.6.4 Sirtuin pathway

Growing evidence suggests that SIRT1 and SIRT6 play a key role in delaying cellular senescence. The sirtuin effects on cellular senescence are mainly mediated through the prevention of telomere attrition and the promotion of DNA damage repair, both contributing to cellular senescence. Sirtuins, as previously stated, play vital roles in sustaining genome integrity, by contributing to maintain the normal chromatin condensation state, and responding to DNA damage and repair (Lee et al., 2019).

Sirtuins are known to interact with NSPs, such as AMPK, Ins/IGF-1, mTOR, and FOXO3, that are deacetylated by SIRT1. So, SIRT1 plays a crucial role in metabolic homeostasis, whereas SIRT6 transgenic mice express lower serum levels of IGF-1, higher levels of IGF-binding protein 1, and altered phosphorylation levels of major components of IGF-1 signaling. SIRT1 activates AMPK through the direct deacetylation of a regulator of AMPK, and AMPK is known to activate SIRT1 through the elevation of NAD+ levels (Price et al., 2012).

Although still controversial, it is believed that the beneficial effects of CR on lifespan extension and prevention of age-related diseases are mediated, at least in part, by sirtuins (Canto and Auwerx, 2009). In many animal models, the expression and activity of sirtuins was reported to be increased by CR and nutritional deprivation, and the activation of sirtuins by CR was mediated by the upregulation of AMPK and an increase in NAD+ levels (Bordone and Guarente, 2005; Sebastián et al., 2012).

So, sirtuins, which sense low energy states by detecting high NAD^+ levels, act as transcriptional effectors by controlling the acetylation state of histones, signaling nutrient scarcity, and catabolism. Thus, their action is in opposite to Ins/IGF-1 and mTOR, favoring successful aging and longevity (López-Otín et al., 2013).

Since sirtuins are commonly believed to mediate the beneficial effects of CR, the activators of SIRT1 are considered to mimic these beneficial effects and are hence attractive therapeutics for age-related diseases. High-throughput screening has identified over 14,000 sirtuin-activating compounds. The most potent activator of SIRT1 is resveratrol (3,5,4′-trihydroxystilbene), a polyphenol found in red wine, which is considered the prototype of CR mimetic (Lee et al., 2019) (see also Chapters 12, 13, and 14). An alternative approach to activate sirtuins is regulating NAD+ levels by activating enzymes involved in the biosynthesis of NAD, since, in model animals, their levels decrease with aging (Mouchiroud et al., 2013).

The low content of animal protein and the low GI of the Mediterranean diet (MedDiet) (Vasto et al., 2014) might directly modulate the insulin/IGF-1 and the mTOR pathways, with a downregulation of the signals that lead to the activation of FOXO3 and, consequently, to the transcription of homeostatic genes that favor longevity. The downregulation of both IGF-1 and mTORC1 also induces an anti-inflammatory effect. In addition, nutraceuticals from vegetables act as both hormetins and sirtuin activators (resveratrol). It is noteworthy that in Okinawa, where substantially more centenarians live with often good health, there exists a tradition of adhering to a vegetable-rich low GI, low caloric diet, rich in nutraceuticals, with high consumption of vegetables; with low assumption of meat, dairy products, and fat intake; and with the moderate consumption of fish products. Compared with Western diets, these diets do not overstimulate mTOR, insulin, and IGF-1 receptors (Aiello et al., 2019a).

3.6.5 Autophagy

Formation of the autophagosome is initiated by a protein kinase complex comprising several autophagy (ATG)-related proteins triggered by the activation of ATG1. This initiation complex may be activated by two kinds of signals. The first kind is mediated by environmental cues such as nutrient levels through a complex network of proteins and signaling pathways. AMPK and mTOR act in an opposing manner to modulate autophagy. Nutrient-depleted conditions driving a high AMP/ATP ratio induce autophagy by activating AMPK, which in turn phosphorylates and activates ATG1. Detection of elevated amounts of amino acids by mTOR as well as its activation by Ins/IGF-1 pathway actively promotes cell proliferation, biomass production, and represses autophagy by suppressing ATG1 activity. In addition, the increased presence of amino acids such as methionine, more represented in animal proteins compared with the vegetable ones, is also relevant, because it activates another kinase, Rag, which in turn stimulates mTOR. The mTOR protein can inhibit autophagy by transcriptional and epigenetic regulation as well as by direct posttranslational modification of ATG proteins. Under nutrient-deprived conditions or by pharmacological inhibition (rapamycin, see below) of mTOR, the initiation complex can translocate to the site of autophagosome formation, i.e., the endoplasmic reticulum. The second kind of signals involves potential autophagy cargoes, such as damaged mitochondria, which can activate the initiation complex by direct interaction with ATG proteins. ATG1/2 phosphorylates several other components, so generating autophagosomal precursor membranes and the process ends with the recognition of cargoes and fusion with lysosomes (Mizushima and Levine, 2020; Wong et al., 2020).

Thus, autophagy plays a key role in the adaptation to metabolic demands. One of the benefits of functional autophagy on metabolic homeostasis is the ability of this process to provide cells with a source of precursors for anabolic and energy-demanding pathways necessary for survival under nutrient-restricted conditions. In addition, autophagy plays homeostatic roles, particularly in postmitotic cells, in which obsolete and damaged material cannot be diluted by cell proliferation (Mizushima and Levine, 2020; Wong et al., 2020).

The expression of ATG genes, including SIRT1 (see below), decreases with age, and autophagy has a prominent role in determining the lifespan of many model organisms. Reduced autophagy has been associated with accelerated aging, whereas stimulation of autophagy might have potent slow-aging effects (Madeo et al., 2010). Autophagy is a crucial and versatile degradation mechanism for damaged macromolecules and dysfunctional organelles that accumulate during the process of aging. The age-related autophagy decline could easily explain the accumulation of altered membrane, mitochondria, and peroxisomes, and, therefore, the increase in oxidative stress and the alterations in signal transduction observed in aging. In particular, defective mitophagy is responsible for the release of toxic apoptotic mediators and ROS by dysfunctional mitochondria. So, autophagy may promote longevity by improving protein and organellar quality control, maintaining "stemness," promoting genomic stability, or a combination of these factors. Genetically engineered mice with increased autophagy show an improvement in age-related phenotypes, such as cardiac and renal fibrosis and spontaneous tumorigenesis, and live longer than normal mice. Studies in model animals prove that the autophagy pathway is essential for the most longevity states (Fernández et al., 2018; Hansen et al., 2018; Nakamura et al., 2019).

CR is the most physiological inducer of autophagy, and inhibition of autophagy prevents the antiaging effects of CR in all species investigated in this respect (Rubinsztein et al., 2011). CR studies in nonhuman primates have shown lifespan improvements and delayed onset of age-related disorders (Mattison et al., 2017). Treatment of mice or rats with the CR mimetics or metformin offers neuroprotection against aging-related oxidative stress with a concurrent increase in autophagy gene expression via the AMPK pathway (Garg et al., 2017). CR extends lifespan, in part, by increasing SIRT1 expression (or by activating its enzymatic activity). Resveratrol, which directly or indirectly activates SIRT1, can prolong lifespan, and this longevity-increasing effect is lost in models lacking its ortholog. Similarly, mice that are deficient for SIRT1 do not show some of the beneficial effects of CR related to longevity (Haigis and Sinclair, 2010), and the knockout or knockdown of ATG genes abolishes the lifespan-prolonging effect of CR, resveratrol, and SIRT1 overexpression (Morselli et al., 2010). Thus, autophagy is required for the lifespan-prolonging effect of SIRT1.

Autophagy regulation is central to metabolism and aging, and accumulating yet fragmentary evidence indicates that genetic or pharmacological manipulations, resulting in an increase in longevity, stimulate autophagy. Therefore, autophagy is a strong target for pharmacological strategies against aging and age-related diseases. However, physiological processes associated with the longevity of model organisms are likely to protect against diseases that afflict older people. Indeed, major causes of death in the Western world, such as heart disease, dementia, and cancer, may be facilitated by autophagy inhibition and thus perhaps mitigated by autophagy upregulation (Rubinsztein et al., 2011).

Age is a major risk factor for many neurodegenerative diseases such as AD and Parkinson's diseases (PD), frontotemporal lobar degeneration, and tauopathies. The age-dependent functional decline of autophagy (as well as of proteasome, see next paragraph) and the consequential attenuation of proteostasis and accrual of proteotoxicity over time are generally thought to have a significant contribution towards disease development and/or progression. Neurons are more susceptible to age-associated proteotoxicity, because postmitotic cells cannot segregate the proteotoxic damage from daughter cells upon mitosis. So, impaired autophagy plays a pathophysiological role. However, the exact mechanisms and roles of dysfunctional autophagy in each disease remain to be elucidated. In aging and neurodegeneration, defects in several steps of autophagy regulation and execution lead to the accumulation of damaged organelles and protein aggregates that adversely affect cell metabolism and homeostasis. This further exacerbates defective autophagy resulting in a vicious cycle that ends in cell death and neuronal loss (Wong et al., 2020).

Due to the ability of metformin to induce autophagy by the activation of AMPK, this drug is regarded as a potential hormesis-inducing agent with healthspan-promoting and prolongevity properties (Piskovatska et al., 2019). Some data suggest that interventions aimed at stimulating autophagy work even at a relatively mature age. Rapamycin treatment of middle-aged mice may still extend their lifespan, and 3 months of CR may improve verbal memory in healthy older humans, suggesting that proautophagic strategies may reduce the time-dependent deterioration (Rubinsztein et al., 2011).

3.7 Loss of proteostasis

The three key aspects of the proteostasis network, namely, protein synthesis and folding, maintenance of conformational stability, and protein degradation, are deteriorated with aging. Proteostasis indicates the protein homeostasis regulated by a network of >2000 proteins comprising molecular chaperones, proteolytic systems and regulators, and experimental perturbation of proteostasis can precipitate age-associated pathologies. Indeed, with cellular aging, there are complex changes both in the manufactured proteins, which may increase their propensity to become dysfunctional, unfolded, and aggregate, and in the decreased ability of the proteostatic machinery. Moreover, one of the most important causes of protein misfolding and aggregation is their oxidation by ROS (Hipp et al., 2019; Fernández del Río et al., 2016; López-Otín et al., 2013).

Protein degradation is performed both by autophagy and by proteasomes. Autophagy (previously described) degrades various macromolecules and organelles en bloc, whereas the proteasome degrades ubiquitinated proteins one by one. Proteasomes are protein complexes which degrade misfolded proteins by proteases. Proteins are tagged for degradation with a small protein called ubiquitin, reaction catalyzed by ubiquitin ligases. Once a protein is tagged with a single ubiquitin molecule, this is a signal to other ligases to attach additional ubiquitin molecules. The result is a polyubiquitin chain that is bound by the proteasome, allowing it to degrade the tagged protein. The degradation process yields peptides of about seven to eight amino acids long, which can then be further degraded into shorter amino acid sequences and used in synthesizing new proteins or presented on the membrane by class I histocompatibility antigens (HLA, see below immunoproteasome) (Fernández del Río et al., 2016; Kulkarni et al., 2020).

Proteostasis involves also mechanisms for stabilizing properly folded proteins, represented by heat shock proteins (HSPs), a family of proteins that are produced by cells in response to exposure to stressful condition. In particular, Hsp70 can be constitutively expressed or induced by stress. Constitutive Hsp70s has maintenance functions such as folding of nascent polypeptides, translocation of proteins between cellular compartments, and degradation of unstable and misfolded proteins. Stress-induced Hsp70s prevents the accumulation of proteins that have been denatured (unfolded) in response to various cellular stresses such as heat, radiation, ischemia, heavy metals. Stress-induced synthesis of HSP70 is significantly reduced in aging, and research with animal models supports the positive impact of chaperones on longevity (Fernández del Río et al., 2016; Kulkarni et al., 2020).

Refolding or degrading of altered proteins avoids the accumulation of aggregates and damaged fibrils with proteotoxic effect and associated with age-related diseases (e.g., neurodegenerative amyloidosis, such as AD and PD). Aging, age-associated pathologies, including AD and PD, as well as cataract, exhibit an impaired network of coordinated proteostasis, with the accumulation of unfolded, misfolded, or aggregated proteins resulting in intracellular damage (Kaushik and Cuervo, 2015; Powers et al., 2009; Hipp et al., 2019).

Immunoproteasomes contain replacements for the three catalytic subunits of standard proteasomes. In most cells, oxidative stress and proinflammatory cytokines are stimuli that lead to elevated production of immunoproteasomes. Immune system cells, especially antigen-presenting cells, express a higher basal level of immunoproteasomes. A well-described function of immunoproteasomes is to generate peptides with a hydrophobic C terminus that can be processed to fit in the groove of HLA class I molecules. This display of peptides on the cell surface allows surveillance by CD8 T cells (Ferrington and Gregerson, 2012).

It is noteworthy that a recent study has provided evidence that adherence to a Med-Diet may enhance proteasome and immunoproteasome activity in older people (Athanasopoulou et al., 2018).

3.8 Mitochondrial dysfunction

Decline in mitochondrial function with aging is known to contribute to age-associated dysregulation in energy homeostasis and increased predisposition to age-related diseases (Sun et al., 2016). Mitochondria perform key biochemical functions essential for metabolic homeostasis and are arbiters of cell death and survival. Mutations and deletions in aged mitochondrial (mt)DNA contribute to age-related mitochondrial dysfunctions, since mtDNA is a major target for somatic mutations due to the oxidative microenvironment of the mitochondria and the lack of protective histones in the mtDNA. In addition, telomere shortening, reduced mitochondrial biogenesis, defective mitophagy, and altered mitochondrial dynamics as well as protein oxidation and changes in the lipid composition of mitochondrial membranes, among others are causes of their dysfunction, occurring in aging. With an increased understanding of the role of mitochondria in addition to serving as the powerhouse of the cell, its dysfunction with aging can have implications on inflammation (because of the increased generation of ROS), senescence, autophagy, and retrograde nuclear signaling (López–Otín et al., 2013; Jang et al., 2018).

Free radicals are highly reactive atoms or molecules with one or more unpaired electrons in their outer shell and are formed when oxygen interacts with certain molecules. Since these radicals have an unpaired electron, they are chemically unstable and tend to initiate potentially harmful chain reactions. The terms "ROS" and "reactive nitrogen species (RNS)" refer, respectively, to reactive radicals and nonradical derivatives of oxygen and nitrogen. The production of free radicals has both endogenous and exogenous origins. Exogenous causes able to determine the formation of ROS and RNS are represented by pollutants, tobacco, alcohol, heavy or transition metals, drugs, foods (e.g., smoked meat, waste oil, and fat), and ionizing radiations. The most important endogenous source of ROS is represented by mitochondria. A part of the electrons transported in the mitochondrial respiratory chain inevitably deviate towards the continuous production of reactive intermediates, the ROS. In fact, mitochondria are considered the major source of ROS, the production of which is inevitable during normal aerobic metabolism. With advancing age, the production of ROS tends to increase significantly. ROS also originates from the enzymatic action of cytochromes P450 on substances destined to be transformed and eliminated, e.g., xenobiotics, drugs, catabolites, as well as from the action of myeloperoxidase and adenine dinucleotide phosphate (NADPH) oxidase in inflammatory cells. At low concentrations, ROS are involved in the fine regulation of important physiological cellular processes depicted in Fig. 3.3; at high concentration, they are among the main causes of cell damage. In fact, being highly oxidant, they can react with proteins, lipids, and DNA by donating them electrons. In particular, they can therefore be responsible for mutations in both nuclear and mtDNA. Mitochondrial mutations can make the mitochondrion less efficient, so, in turn, increasing the reduction of ROS (Table 3.1).

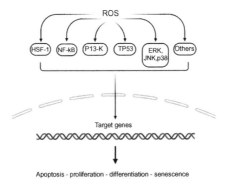

Fig. 3.3 Regulation by ROS of some physiological process. Some transcription factors modulated by reactive oxygen species (ROS).

Table 3.1 Production of ROS from iron, a redox metal.

Fenton reaction	$O_2\text{-}\bullet + Fe^{+++} \rightarrow Fe^{++} + O_2$
	$F^{++} + H_2O_2 \rightarrow Fe^{+++} + {}^{\bullet}OH + OH$
Haber–Weiss reaction	$O_2\text{-}\bullet + H_2O_2 \rightarrow OH^{-} + HO^{\bullet} + O_2$

The superoxide anion ($O_2^{\bullet}$) is formed by the one-electron reduction of molecular oxygen, with electrons supplied by NADPH, during cellular respiration. Superoxide dismutase (SOD) dismutes most of the $O_2^{\bullet}$ into the hydrogen peroxide (H_2O_2), which has no unpaired electrons, so it is not a free radical. However, it is able to form the highly reactive ROS hydroxyl ion ($OH^{\bullet}$) through the Fenton and Haber-Weiss reactions (Table 3.1). Nitric oxide (NO) is produced from l-arginine by three main isoforms of nitric oxide synthase (NOS): epithelial NOS, related to vasodilation and vascular regulation; neuronal NOS, linked to intracellular signaling; and inducible NOS, activated in response to various inflammatory signals. Finally, O_2 may react with NO to form another relatively reactive molecule, peroxynitrite ($ONOO^{-}$) (Table 3.2) (Kohen and Nyska, 2002; Valko et al., 2007; Liguori et al., 2018).

The effects of ROS and RNS are balanced by the action of antioxidants, i.e., by molecules having the ability to specifically neutralize free radicals, to chelate metals, to interact (regenerate) with other antioxidants within the antioxidant network, to modulate the

Table 3.2 Some reactive species.

Species	Name	Characteristics
O_2^{-}	Superoxide	Free radical produced in various types of redox reactions. It is estimated that mitochondria convert about 5% of oxygen into superoxide anion.
$HO^{\bullet}$	Hydroxyl	Very reactive free radical, product of the reaction of a transition metal (e.g., iron) with a peroxide or hydroperoxide.
$HO^{\bullet}$	Alkoxide	Radical containing an alky groups attached to the oxygen, produced in reactions with lipids.
$ROO^{\bullet}$	Peroxyl	Peroxyl radicals ($ROO\bullet$), formed by reactions of molecular oxygen with carbon-centered radicals.
$NO^{\bullet}$	Nitrogen monoxide (nitric oxide)	In human cells, such as endothelial cells, macrophages and neurons, it is formed from arginine by enzymatic reactions. Being very reactive, this radical reacts with superoxide to form peroxynitrite (ONOO-), 3 powerful oxidant.
$NO^{\bullet}$ $^{1}O_2$	Singlet oxygen	It is a very reactive non-radical species and a characteristic product of photosensitization reactions, but it can a so form in peroxide and hydroperoxide decay reactions. Elective target are both guanines and polyunsaturated fatty acids.

expression of some genes, to have relevant physiological concentrations in tissues and biofluids, and to be able to act in aqueous or membranous domains. Antioxidant systems efficiently neutralize free radicals, protecting biological structures from damage. They include enzymatic and nonenzymatic antioxidant mechanisms. In physiological conditions, there is a balance between the intracellular levels and the activities of enzymatic and nonenzymatic antioxidants. Their activation is in most cases induced by various signal transduction pathways, capable of controlling oxidative stress. This balance is essential for the survival of all organisms. The primary antioxidant enzymes are SOD, catalase (CAT), and glutathione peroxidase (GSH-Px). As previously stated, O_2 is converted by SOD to H_2O_2, which is decomposed to water and oxygen by CAT, hence preventing hydroxyl radical production. GSH-Px converts peroxides and hydroxyl radicals into nontoxic forms by the oxidation of reduced glutathione (GSH) into glutathione disulfide, which is then reduced to GSH by glutathione reductase. The nonenzymatic neutralization mechanisms (some of them act in a hydrophilic environment, others in the hydrophobic one, and still others in both) are represented, among others, by uric acid and bilirubin (which eliminate free radicals in the plasma), by α-tocopherol, the most powerful and active form of vitamin E (contained in cell membranes, reacts mainly with peroxyl radicals) and ascorbic acid, vitamin C, (inserted in redox reactions, it can regenerate vitamin E) (Kohen and Nyska, 2002; Valko et al., 2007; Liguori et al., 2018).

As previously stated, with this advancing age, the concentrations of ROS and RNS increase. The excessive amount of free radicals appears mainly to be determined by a functional decline in the electron transport chain, related to age and caused by changes in the gene expression profile. Another possible cause of the age-related increase in ROS and RNS could be the decline of antioxidant systems. However, the correlation between the decline of antioxidant enzymatic systems and aging remains debated. Oxidative stress is an oxidative condition, resulting from an imbalance between oxidant and antioxidant factors in favor of the former, capable of damaging biological molecules with consequent tissue damage. Oxidative stress damages DNA, lipids, and proteins. In DNA, there are transcription errors, base inversions, and strand breakage. The proteins, damaged in their secondary and tertiary structures, undergo a total or partial loss of their functionality. The peroxidation of the unsaturated lipids of the membranes causes a reduction in their fluidity with a consequent alteration in the mechanisms of transduction and transport (Kohen and Nyska, 2002; Valko et al., 2007; Liguori et al., 2018).

The oxidative stress is thought to be closely related to inflammaging (De la Fuente and Miquel, 2009). The increase of ROS is correlated with damage-associated molecular pattern (DAMPs) release, endogenous nuclear or cytosolic molecules released from injured and dying cells. DAMPs are able to activate inflammasome, through specific receptors, promoting the production of proinflammatory cytokines, such as interleukin (IL)-1ß, IL-6, and tumor necrosis factor (TNF)-α. In turn, these cytokines activate inflammatory cells, further potentiating ROS production, in a vicious cycle (Accardi et al., 2019).

Thus, oxidative stress is closely related to damage observed in age-related diseases. Concerning, instead, aging process, in models the reduced ROS production was not positively associated with an extended lifespan; meanwhile, an increased oxidative stress by the mutation of certain antioxidant enzymes did not imply shorter lifespan (Chapter 6).

Regarding possible therapeutic interventions, targeting mitochondria with organelle-specific agents or prodrugs might be an effective therapeutic strategy. More specifically, controlling the cellular ROS balance via selective delivery of an antioxidant "payload" into mitochondria is an elegant therapeutic concept (Frantz and Wipf, 2010). However, the most popular approach is represented by the intake of antioxidants foods or their derivatives (Davinelli and Scapagnini, 2019) (Chapters 13 and 15).

Finally, in Chapter 6 is discussed ROS interaction with gut barrier and dysbiosis of gut microbiota.

Protein glycation is considered one of the processes involved in the pathogenesis of many age-related diseases, such as the vascular complications of diabetes and AD. Glycation consists of the reaction between glucose or its metabolites (e.g., α-oxo-aldehydes) and the amino groups of proteins. The adducts formed are called Schiff bases that reorganize themselves to form fructosamine residues. The degradation of fructosamine and the direct reaction of proteins with the α-oxo-aldehydes lead to the formation of a large variety of advanced glycation end products (AGE) (Fig. 3.4). AGEs act at two levels: the first is due to structural changes at the protein level and consequently also in the DNA and lipids that alter their biochemical properties. Therefore, aging leads to an accumulation of these AGEs in the collagen, skin, cornea, nervous system, and arteries. AGEs are very resistant to the normal processes of destruction and renewal of normal tissues. The stiffening of collagen, which is a ubiquitous protein, obviously involves an alteration of the various tissues, in particular the cardiovascular system. The second level is through their binding to cellular receptors RAGE, generating ROS, and promoting inflammation. Oxidation and glycation reinforce each other in a vicious circle, while on the one hand,

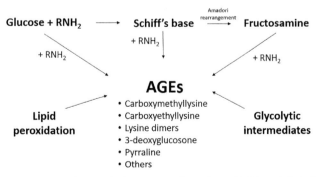

Fig. 3.4 AGE formation. Some advanced glycation end products (AGE) and their formation.

the glycated proteins act as free radical generators, and the glycation process fixes the damage caused by free radicals (Prasad and Mishra, 2018).

3.9 Telomere attrition

Telomeres are repetitive sequences of DNA at the ends of chromosomes that undergo shortening with each mitotic division. This is due to the lack of the telomerase enzyme, normally expressed only in germ cells and, to less extent, in adult stem cells. Although telomeric DNA loss is commonly attributed to the inability of normal DNA replication machinery to copy the ends of the filaments in the absence of telomerase, inflammation and oxidative damage as well as increased levels of different kinds of stressors have been found to have an even greater effect on the rate of telomere loss. The short telomeres therefore represent a marker of cumulative burden of inflammation and oxidative stress. Indeed, short telomeres have been shown to be associated with a higher risk of all-cause mortality, while individuals leading a healthy lifestyle have longer telomeres. In other words, telomere length should be considered a marker of biological age rather than of chronological one (Soares et al., 2014; Fernández del Río et al., 2016; Aubert and Lansdorp, 2008; Davinelli and De Vivo, 2019). Furthermore, telomere shortening is also shown to trigger cellular senescence, inflammation, and mitochondrial dysfunction via p53-mediated DNA damage response pathways (Zhu et al., 2019).

Healthy lifestyle is related to longer telomeres; therefore, it has been suggested that telomere attrition is modifiable, as substantial variability exists in the rate of telomere shortening that is independent on chronological age. Although telomere length is strongly influenced by genetic factors, variability of telomere length may be, at least in part, partially explained by lifestyle practices, including dietary patterns, and a healthy lifestyle is a remedy against telomere shortening (Chapter 14) (Davinelli and De Vivo, 2019).

3.10 Cellular senescence

Cellular senescence (Chapter 4) is a physiological phenotype aimed at permanent cell cycle arrest and is morphologically identified as flattening, increased size of nucleus and nucleoli, and the appearance of vacuoles in the cytoplasm (Cristofalo et al., 2004).

In that chapter are described the role of miRNA as well as the role of the senescence-associated secretory phenotype in aging, and possible therapeutic interventions. In the present chapter (Sections 3.9, 3.11, and 3.13) is discussed the role of telomere shortening as well as the INK4a/ARF locus. This locus encodes two proteins, p16INK4a and p14ARF, which, while acting on different targets, respectively, arrest of the cell cycle in G1 and stabilization of p53 via MDM2, contribute to the induction of cellular senescence.

3.11 Stem cell exhaustion

With aging, there is a systemic decrease in the regenerative capacity of tissues. Accordingly, the decrease in stem cell function associated with age has been observed in virtually all the different stem cell populations. Adult stem cells are characterized by the ability to self-renew and differentiate into multiple cell types within a tissue. Despite their heterogeneous aging phenotypes, adult stem cells have many characteristics in common. Their position at the base of cellular lineages makes their dysfunction potentially more impactful than that of other cell types. The exact mechanism responsible for the decreased replicative capacity of stem and progenitor cells (cells that, like stem cells, have a tendency to differentiate into specific types of cells, dividing only a limited number of times) is unclear, but the mechanisms may be common to the more general ones of cellular aging, i.e., the remaining seven hallmarks of aging (López-Otín et al., 2013; Ermolaeva et al., 2018).

To maintain their function, it is necessary that stem cells receive support from the cells that constitute the niche, since they reside in niches that regulate their behavior (Cheung and Rando, 2013). Therefore, aging of the cell niche stem cell negatively modulates stem cell function. Consistent with this, studies utilizing heterochronic transplantation and parabiosis experiments show that aging in stem cells is also linked to extrinsic mechanisms. Parabiosis experiments showed that the blood of younger mice seemed to have rejuvenating effects on older mice when they shared a circulatory system. Afterward, it was demonstrated that a protein in the blood of young mice, called growth differentiation factor 11 (GDF11), was able to quickly reverse symptoms of heart failure in older mice. GDF11 is a member of the transforming growth factor-β (TGF-β) superfamily that plays an essential role in mammalian development. GDF11 can affect the developmental process via exerting its effects at least partially by binding to TGF-β activin receptors. GDF11 has been shown to stimulate proliferation and angiogenesis of stem cells in ischemia/reperfusion models and to be implicated in the regulation of mature cell and adult organ in a variety of pathological conditions. Recent studies showing the rejuvenating effects of GDF11 on stem cells are one promising area of ongoing research, although this is not without controversy (Schultz and Sinclair, 2016; Zhao et al., 2020).

Concerning genomic instability and epigenetic alterations, an accumulation in DNA damage and mutations as well as epigenetic alterations has also been implicated in stem cell aging. In Section 3.4, it is reported that mice with defects in DNA damage repair display some aspects of premature aging, and enhancing DNA repair through increased expression of sirtuin increases lifespan. However, it is not yet known whether these effects are linked to stem cells, and, conversely, whether double-strand breaks are induced in stem cells, will aging be accelerated? It is clear that the consequences of damage are different in stem cells, having a high intrinsic division capacity, and in which

DNA damage can determine a high risk of malignant transformation. Presumably, this is the reason why some stem cell populations, for example, those of the intestinal epithelium, are very sensitive to low doses of DNA-damaging agents and easily go into apoptosis (Barker et al., 2008).

It will be interesting to determine whether long-lived animals and humans are relatively resistant to epigenetic changes induced by DNA damage and the resulting changes in stem cell lineage (Schultz and Sinclair, 2016). Regarding loss of proteostasis, it is well known that stem cells can divide either asymmetrically, producing a daughter stem cell and a differentiating cell, or symmetrically, producing two daughter cells with the same fate. They distribute their macromolecules asymmetrically during asymmetric cell divisions and mechanisms have evolved to enrich the daughter stem cell with undamaged components. In addition, they are able to enhance proteostasis. Stem cells use numerous mechanisms to prevent damage accumulation and epigenetic changes, including maintenance of relatively long telomeres, enhanced proteostasis, and avoidance of ROS production (Flores et al., 2008; Salemi et al., 2012).

Telomere attrition and hence cellular senescence also occur in stem cells, although telomeres shorten at a rate lower than that of the somatic cells because of telomerase expression, and they are the most important barrier to oncogenic transformation in stem cells. However, stem cells are defined by their ability to self-renew and differentiate; therefore, the induction of senescence in a stem cell clearly compromises its function (Schultz and Sinclair, 2016).

The role of NSPs in stem cells is suggested by results showing that CR increases the abundance of stem cells in muscle and improves the function of many stem cell populations in models. So, some of the benefits of CR may be exerted through altered stem cell phenotypes. Metabolic status and the formation of damaging ROS are also thought to play a role in stem cell aging. Mitochondrial dysfunctions in stem and progenitor cells might also underlie age-related changes in stem cell function. A potential road to stem cell aging reversal is through the manipulation of longevity pathways such as mTOR inhibition by rapamycin and sirtuin activation by CR mimetic (Tilly and Sinclair, 2013; Cerletti et al., 2012; Chen et al., 2003).

Finally, senescent niche cells might also affect neighboring stem cells by secreting tumor-promoting mitogens and proinflammatory cytokines that negatively affect stem cell function through paracrine signaling (Doles et al., 2012).

Stem cell exhaustion unfolds as the integrative consequence of multiple types of aging-associated damages and likely constitutes one of the ultimate culprits of tissue and organismal aging. Possible future strategies are represented by cellular and genetic therapies, including genetic reprogramming. In particular, genetically reprogramming cells into induced pluripotent stem cells might rejuvenate any tissue type (Stahl and Brown, 2015; Schultz and Sinclair, 2016).

3.12 Altered intercellular communication

Endocrine, neuronal, and neuroendocrine pathways provide cues to the cells to respond effectively to environmental changes, pathogens, tissue disruptors, and mechanical stressors (López-Otín et al., 2013). Thus, aging leads to a systemic dysregulation of effective cell–cell connectivity and its associated response, disrupting the maintenance of intercellular communication. One of the most important intercellular communication events is inflammation. Inflammation is not per se a negative phenomenon, since it is the response of the immune system to the invasion of viruses or bacteria and other pathogens. During evolution, the human organism was set to live 40 or 50 years; today, however, the immune system must remain active for much a longer time. This very long activity leads to a chronic inflammation, called inflammaging, that slowly but inexorably damages one or several organs. This is a typical phenomenon linked to aging, and it is considered the major risk for age-related chronic diseases (Licastro et al., 2005). Immunosenescence represents the most important contributor to inflammaging, in turn, contributing to impaired immune responses. In fact, inflammaging is responsible for a high expression of micro-RNAs that interfere with B-cell activation, driving TNF-α production and inhibiting B-cell activation as measured in vitro. Increased serum levels of TNF-α are also linked to a defective T-cell response, in part due to reduced expression of activation marker CD28 (Aiello et al., 2019b. Importantly, inflammation is shown to trigger or respond to multiple other hallmarks, including epigenetic alterations, stem cell dysfunction, and loss of proteostasis, thereby further contributing to the pathogenesis of age-related diseases and accelerated aging (Kulkarni et al., 2020). Pathophysiology of inflammaging and immunosenescence is exhaustively treated in Chapters 5 and 7, respectively. The strategies for achieving optimal functioning of the immune system and controlling inflammaging are, instead, discussed in Chapters 5, 14, and 16.

3.13 Cancer and aging

In all industrialized countries, cancer is the second leading cause of death, after cardiovascular disease. Epidemiological studies have shown that cancer mortality increases with age up to the ninetieth year, where it reaches the plateau and then declines. Cancer mortality is about 40% between 50 and 70 years and less than 4% in centenarians. Indeed, in autopsy studies carried out on a group of Japanese centenarians, it has been clearly demonstrated that in centenarians the mortality from cancer is lower than expected; the same data was obtained in autopsies conducted on Italian centenarians. The centenarians have, in fact, a greater "chance" of avoiding the main age-related diseases, including cancer. So, advanced age, but not extreme longevity, represents the most powerful of all "carcinogens"; however, the meaning of this association is not yet clear. The increased exposure during a longer life to carcinogens contributes to the phenomenon of the progressive increase in cancer mortality with advancing age, although as previously mentioned, in long-lived individuals (LLIs) the prevalence of cancer does not increase, but declines (Vasto et al., 2009). This is not strange because LLIs are known to delay or avoid almost all inflammatory age-related diseases (Caruso, 2019). As an example of possible mechanism, centenarian proinflammatory status is counterbalanced by a higher levels of anti-inflammatory molecules, such as TGF-β, IL-1 receptor antagonist, or cortisol (Chapter 7). This example is appropriate because if genetic damage is the "match that lights the fire" of cancer, inflammation may provide the "fuel that feeds the flames" (Balkwill and Mantovani, 2001). Finally, obviously, the possibility that the increased prevalence of cancer in older people is linked to immunosenescence must not be excluded (Aiello et al., 2019b).

Recent studies indicate that both convergent and divergent mechanisms can correlate cancer and aging with each other. In the first case, molecular pathways simultaneously provide susceptibility or protection versus cancer and aging by acting on specific agents involved in the generation of or protection from cellular, genetic, or epigenetic damage. The main source of damage to cellular macromolecules, including DNA, is metabolism with the production of ROS. An excess of nutrients determines a low energy request to the mitochondria with a consequent suboptimal performance—from this reduced efficiency derives a greater production of ROS (Section 3.8). Closely associated with ROS production is autophagy (Section 3.6.5); the reduced removal of damaged mitochondria (mitophagy) could link autophagy, aging, and cancer, since it involves an increased production of ROS. It should also be remembered that senescent cells produce inflammatory mediators and therefore increase the production of ROS with a vicious circle in which inflammation plays a role (see below). As discussed in Section 3.6.2, high levels of IGF-1 play a key role in unsuccessful aging and cancer, because of promitotic, proinflammatory, and antihomeostatic effects. P53 is normally inactive, as a result of its binding to MDM2. A huge series of stresses stabilize and modify p53, activating it. A key role is played by the tumor suppressor ARF, which inhibits MDM2, stabilizing p53. Once p53 is activated, it triggers a series of processes whose ultimate goal is to maintain genomic integrity, protecting from cancer and unsuccessful aging. Therefore, these mechanisms are at play in both aging and cancer (Campisi and Yaswen, 2009; Finkel et al., 2007) (Fig. 3.5).

Other mechanisms are divergent; that is, they can have opposite effects on cancer and aging, specifically, protecting against cancer but promoting aging, and vice versa. Cells can use two pathways to limit their replicative potential, telomere shortening, and hyperregulation of INK4a/ARF locus. Both pathways have an effect on the proliferative potential of stem cells, especially in older age, and on the proliferation induced by oncogenes. All dividing cells are subject to shortening of the telomeres. In embryonic stem cells, but not in adult stem cells, this process is completely avoided by the presence of telomerase. Protection from cancer comes at the expense of the possibility of tissue regeneration in old age, thus promoting aging. So, the shortening of telomeres, inducing

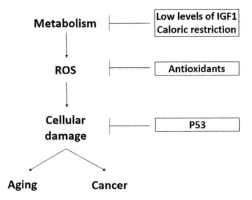

Fig. 3.5 Convergent mechanisms that correlate cancer and aging with each other. (For explanation, see text.)

cellular senescence, represents a powerful cancer suppression mechanism, but in a period of life not foreseen by evolution, such as the postreproductive one, this mechanism instead contributes to the aging process. Indeed, transgenic mice with increased telomerase activity simultaneously show an increased incidence of cancer and a slight increase in lifespan. In fact, the control of cell proliferation determines contrasting effects on cancer and aging; while it promotes protection from cancer, longevity is somewhat altered. Paradigmatic is the case of stem cells (Section 3.11). Both cancer and aging are stem cell diseases—cancer as an effect of further mutations and aging as a consequence of depletion of the stem cell pool. This is in line with the disposable soma theory (Chapter 1), according to which there are no genes that promote aging but only genes that cannot give protection to individuals who have passed the reproductive period. A similar evolutionary significance can be hypothesized for the INK4a/ARF locus, whose levels correlate with the chronological age of essentially all tissues analyzed both in mice and in humans. INK4a/ARF (involved in cellular senescence) can be regarded as a beneficial compensatory response aimed at avoiding the propagation of damaged cells and its consequences on aging and cancer. However, when damage is pervasive, the regenerative capacity of tissues can be exhausted or saturated, and under these extreme conditions, INK4a/ARF responses can become deleterious and accelerate aging. Counter-proof is the strong selective pressure to eliminate the locus during the carcinogenesis process. In conclusion, these divergent mechanisms do not normally limit the lifespan of animals in their natural environment, but in a "civilized" environment, they contribute to aging (Campisi and Yaswen, 2009; Finkel et al., 2007) (Fig. 3.6).

A subtle balance between converging and diverging mechanisms could therefore play a key role in achieving successful aging. An example is given by cellular senescence that we have discussed in both convergent and divergent mechanisms. The senescence response is widely recognized as a potent tumor-suppressive mechanism. However,

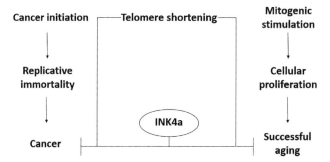

Fig. 3.6 Divergent mechanisms that correlate cancer and aging with each other. (For explanation, see text.)

recent evidence strengthens the idea that it also drives both degenerative and progressive pathologies, most likely by promoting chronic inflammation. Thus, the senescence response may be the result of antagonistically pleiotropic gene action (Campisi, 2013).

3.14 Conclusion and future perspectives

In this chapter, addressing the various mechanisms of aging and age-related diseases, we have discussed possible therapeutic interventions. We have always written about them as interventions to delay aging and age-related diseases; obviously, we have not discussed how to avoid death.

But what does one die of in old age, what does the colloquial expression "to die of old age" mean? This colloquial expression "dying of old age" reveals our ignorance of the biological basis of this event. Obviously, respiratory or circulatory arrest causes irreversible damage to vital organs. However, insisting that aging of the heart or lungs is the cause of death does not solve the problem but displaces it. Postmortem examination of people who have died of old age will reveal nonsignificant biological changes and possibly even pathological changes that were only mildly symptomatic or even asymptomatic in life. Why then do people die? Actually, we cannot define "dying of old age" in biological terms. In the absence of an acute event or a devastating disease, the aging process affects all tissues and cells, postmitotic, actively mitotic or quiescent. The result is an evident progressive change for which the risk of the increased probability of death is related to pathophysiological variations that can be observed empirically, but without any clue as to the mechanism. It can only be assumed that death occurs when the open system represented by a living being, human or animal, becomes unable to stop the dissipation of energy in the inorganic world in which it is immersed. It is clear, therefore, that death is related to the inevitable molecular deterioration linked to the laws of physics.

Thus, it is clear that therapeutic interventions can only be directed at enabling a healthy, vigorous, and youthful-looking body.

References

Accardi G, Ligotti ME, Candore G. Phenotypic aspects of longevity. In: Caruso C, editor. Centenarians. Cham: Springer; 2019. p. 23–34. https://doi.org/10.1007/978-3-030-20762-5_2.

Aiello A, Accardi G, Candore G, Gambino CM, Mirisola M, Taormina G, Virruso C, Caruso C. Nutrient sensing pathways as therapeutic targets for healthy ageing, Expert Opin Ther Targets 2017;21(4):371–80. https://doi.org/10.1080/14728222.2017.1294684. Epub 2017 Feb 17 28281903.

Aiello A, Caruso C, Accardi G. Slow-ageing diets. In: Gu D, Dupre M, editors. Encyclopedia of gerontology and population aging. Cham: Springer; 2019a. https://doi.org/10.1007/978-3-319-69892-2_134-1.

Aiello A, Farzaneh F, Candore G, Caruso C, Davinelli S, Gambino CM, Ligotti ME, Zareian N, Accardi G. Immunosenescence and its hallmarks: how to oppose aging strategically? a review of potential options for therapeutic intervention. Front Immunol 2019b;10:2247. https://doi.org/10.3389/fimmu.2019.02247 PMID: 31608061; PMCID: PMC6773825.

Ashapkin VV, Kutueva LI, Vanyushin BF. Aging as an epigenetic phenomenon. Curr Genomics. 2017;18 (5):385–407. https://doi.org/10.2174/1389202918666170412112130. PMID: 29081695; PMCID: PMC5635645.

Athanasopoulou S, Chondrogianni N, Santoro A, Asimaki K, Delitsikou V, Voutetakis K, Fabbri C, Pietruszka B, Kaluza J, Franceschi C, Gonos ES. Beneficial effects of elderly tailored Mediterranean diet on the proteasomal proteolysis. Front Physiol 2018;9:457. https://doi.org/10.3389/fphys.2018.00457 PMID: 29765333; PMCID: PMC5938393.

Aubert G, Lansdorp PM. Telomeres and aging, Physiol Rev 2008;88(2):557–79. https://doi.org/10.1152/physrev.00026.2007.18391173.

Balkwill F, Mantovani A. Inflammation and cancer: back to Virchow?, Lancet 2001;357(9255):539–45. https://doi.org/10.1016/S0140-6736(00)04046-0.11229684.

Barker N, van de Wetering M, Clevers H. The intestinal stem cell. Genes Dev 2008;22:1856–64. https://doi.org/10.1101/gad.1674008.

Bell KJ, Del Mar C, Wright G, Dickinson J, Glasziou P. Prevalence of incidental prostate cancer: A systematic review of autopsy studies. Int J Cancer 2015;137(7):1749–57. https://doi.org/10.1002/ijc.29538 Epub 2015 Apr 21. PMID: 25821151; PMCID: PMC4682465.

Bellizzi D, Guarasci F, Iannone F, Passarino G, Rose G. Epigenetics and ageing. In: Caruso C, editor. Centenarians. Cham: Springer; 2019. p. 99–133. https://doi.org/10.1007/978-3-030-20762-5_7.

Blagosklonny MV. Rejuvenating immunity: "anti-aging drug today" eight years later. Oncotarget 2015; 6(23):19405–12. https://doi.org/10.18632/oncotarget.3740 PMID: 25844603; PMCID: PMC4637294.

Bordone L, Guarente L. Calorie restriction, SIRT1 and metabolism: understanding longevity, Nat Rev Mol Cell Biol 2005;6(4):298–305. https://doi.org/10.1038/nrm1616.15768047.

Campisi J. Aging, cellular senescence, and cancer. Annu Rev Physiol 2013;75:685–705. https://doi.org/10.1146/annurev-physiol-030212-183653 Epub 2012 Nov 8. PMID: 23140366; PMCID: PMC4166529.

Campisi J, Yaswen P. Aging and cancer cell biology, 2009, Aging Cell 2009;8(3):221–5. https://doi.org/10.1111/j.1474-9726.2009.00475.x. Epub 2009 Mar 27 19627264.

Canto C, Auwerx J. Caloric restriction, SIRT1 and longevity. Trends Endocrinol Metab 2009;20:325–31. https://doi.org/10.1016/j.tem.2009.03.008.

Caruso C, editor. Centenarians. Cham: Springer; 2019. p. 1–179. https://doi.org/10.1007/978-3-030-20762-5.

Caruso C, Candore G. La Malattia: dagli Sciamani alla Medicina di precisione. Un'introduzione alla Patologia General. Palermo: Medical Books; 2016.

Cerletti M, Jang YC, Finley LW, Haigis MC, Wagers AJ. Short-term calorie restriction enhances skeletal muscle stem cell function. Cell Stem Cell 2012;10(5):515–9. https://doi.org/10.1016/j.stem.2012.04.002 PMID: 22560075; PMCID: PMC3561899.

Chen J, Astle CM, Harrison DE. Hematopoietic senescence is postponed and hematopoietic stem cell function is enhanced by dietary restriction, Exp Hematol 2003;31(11):1097–103. https://doi.org/10.1016/s0301-472x(03)00238-8.14585375.

Cheung TH, Rando TA. Molecular regulation of stem cell quiescence. Nat Rev Mol Cell Biol 2013; 14(6):329–40. https://doi.org/10.1038/nrm3591 PMID: 23698583; PMCID: PMC3808888.

Choi JE, Mostoslavsky R. Sirtuins, metabolism, and DNA repair. Curr Opin Genet Dev 2014;26:24–32. https://doi.org/10.1016/j.gde.2014.05.005 Epub 2014 Jul 5. PMID: 25005742; PMCID: PMC4254145.

Cristofalo VJ, Lorenzini A, Allen RG, Torres C, Tresini M. Replicative senescence: a critical review. Mech Ageing Dev 2004;125:827–48. https://doi.org/10.1016/j.mad.2004.07.010.

Cunningham JT, Rodgers JT, Arlow DH, Vazquez F, Mootha VK, Puigserver P. mTOR controls mitochondrial oxidative function through a YY1-PGC-1alpha transcriptional complex, Nature 2007; 450(7170):736–40. https://doi.org/10.1038/nature06322.18046414.

Daniel M, Tollefsbol TO. Epigenetic linkage of aging, cancer and nutrition. J Exp Biol 2015; 218(Pt 1):59–70. https://doi.org/10.1242/jeb.107110 PMID: 25568452; PMCID: PMC4286704.

Davinelli S, De Vivo I. Lifestyle choices, psychological stress and their impact on ageing: the role of telomeres. In: Caruso C, editor. Centenarians. An Example of Positive Biology. Switzerland: Springer; 2019. p. 135–48.

Davinelli S, Scapagnini G. Lifespan and Healthspan extension by nutraceuticals: An overview. In: Caruso C, editor. Centenarians. Cham: Springer; 2019. https://doi.org/10.1007/978-3-030-20762-5_11.

de Kreutzenberg SV, Ceolotto G, Cattelan A, Pagnin E, Mazzucato M, Garagnani P, Borelli V, Bacalini MG, Franceschi C, Fadini GP, Avogaro A. Metformin improves putative longevity effectors in peripheral mononuclear cells from subjects with prediabetes. A randomized controlled trial, Nutr Metab Cardiovasc Dis 2015;25(7):686–93. https://doi.org/10.1016/j.numecd.2015.03.007. Epub 2015 Mar 24 25921843.

De la Fuente M, Miquel J. An update of the oxidation-inflammation theory of aging: the involvement of the immune system in oxi-inflamm-aging, Curr Pharm Des 2009;15(26):3003–26. https://doi.org/10.2174/138161209789058110.19754376.

Di Bona D, Accardi G, Virruso C, Candore G, Caruso C. Association between genetic variations in the insulin/insulin-like growth factor (Igf-1) signaling pathway and longevity: a systematic review and meta-analysis, Curr Vasc Pharmacol 2014;12(5):674–81. https://doi.org/10.2174/1570161111666131218152807.24350933.

Doles J, Storer M, Cozzuto L, Roma G, Keyes WM. Age-associated inflammation inhibits epidermal stem cell function. Genes Dev 2012;26(19):2144–53. https://doi.org/10.1101/gad.192294.112 Epub 2012 Sep 12. PMID: 22972935; PMCID: PMC3465736.

El Khoury LY, Gorrie-Stone T, Smart M, Hughes A, Bao Y, Andrayas A, Burrage J, Hannon E, Kumari M, Mill J, Schalkwyk LC. Systematic underestimation of the epigenetic clock and age acceleration in older subjects. Genome Biol 2019;20(1):283. https://doi.org/10.1186/s13059-019-1810-4 PMID: 31847916; PMCID: PMC6915902.

Ermolaeva M, Neri F, Ori A, Rudolph KL. Cellular and epigenetic drivers of stem cell ageing, Nat Rev Mol Cell Biol 2018;19(9):594–610. https://doi.org/10.1038/s41580-018-0020-3.29858605.

Fasano C, Disciglio V, Bertora S, Lepore Signorile M, Simone C. FOXO3a from the nucleus to the mitochondria: a round trip in cellular stress response. Cells 2019;8(9):1110. https://doi.org/10.3390/cells8091110 PMID: 31546924; PMCID: PMC6769815.

Fernández del Río L, Gutiérrez-Casado E, Varela-López A, Villalba JM. Olive oil and the hallmarks of aging. Molecules 2016;21(2):163. https://doi.org/10.3390/molecules21020163 PMID: 26840281; PMCID: PMC6273542.

Fernández ÁF, Sebti S, Wei Y, Zou Z, Shi M, McMillan KL, He C, Ting T, Liu Y, Chiang WC, Marciano DK, Schiattarella GG, Bhagat G, Moe OW, Hu MC, Levine B. Disruption of the beclin 1-BCL2 autophagy regulatory complex promotes longevity in mice. Nature 2018;558(7708):136–40. https://doi.org/10.1038/s41586-018-0162-7 Epub 2018 May 30. Erratum in: Nature. 2018 Sep; 561(7723):E30. PMID: 29849149; PMCID: PMC5992097.

Ferrington DA, Gregerson DS. Immunoproteasomes: structure, function, and antigen presentation. Prog Mol Biol Transl Sci 2012;109:75–112. https://doi.org/10.1016/B978-0-12-397863-9.00003-1 PMID: 22727420; PMCID: PMC4405001.

Finkel T, Serrano M, Blasco MA. The common biology of cancer and ageing, Nature 2007;448(7155): 767–74. https://doi.org/10.1038/nature05985.17700693.

Flores I, Canela A, Vera E, Tejera A, Cotsarelis G, Blasco MA. The longest telomeres: a general signature of adult stem cell compartments. Genes Dev 2008;22(5):654–67. https://doi.org/10.1101/gad.451008 Epub 2008 Feb 18. PMID: 18283121; PMCID: PMC2259034.

Fontana L, Partridge L, Longo VD. Extending healthy life span—from yeast to humans. Science 2010; 328(5976):321–6. https://doi.org/10.1126/science.1172539 PMID: 20395504; PMCID: PMC3607354.

Fontana L, Kennedy BK, Longo VD, Seals D, Melov S. Medical research: treat ageing, Nature 2014;511 (7510):405–7. https://doi.org/10.1038/511405a.25056047.

Frantz MC, Wipf P. Mitochondria as a target in treatment. Environ Mol Mutagen 2010;51(5):462–75. https://doi.org/10.1002/em.20554 PMCID: PMC2920596.

Frenk S, Houseley J. Gene expression hallmarks of cellular ageing. Biogerontology 2018;19(6):547–66. https://doi.org/10.1007/s10522-018-9750-z Epub 2018 Feb 28. PMID: 29492790; PMCID: PMC6223719.

Garg G, Singh S, Singh AK, Rizvi SI. Antiaging effect of metformin on brain in naturally aged and accelerated senescence model of rat, Rejuvenation Res 2017;20(3):173–82. https://doi.org/10.1089/rej.2016.1883. Epub 2017 Jan 9 27897089.

Gensous N, Franceschi C, Santoro A, Milazzo M, Garagnani P, Bacalini MG. The impact of caloric restriction on the epigenetic signatures of aging. Int J Mol Sci 2019;20(8):2022. https://doi.org/10.3390/ijms20082022 PMID: 31022953; PMCID: PMC6515465.

Gladyshev TV, Gladyshev VN. A disease or not a disease? Aging as a pathology. Trends Mol Med 2016;22 (12):995–6. https://doi.org/10.1016/j.molmed.2016.09.009 Epub 2016 Oct 25. PMID: 27793599; PMCID: PMC5540438.

Goldspink DF. Ageing and activity: their effects on the functional reserve capacities of the heart and vascular smooth and skeletal muscles, Ergonomics 2005;48(11–14):1334–51. https://doi.org/10.1080/00140130500101247.16338704.

Goodle F. What is health? BMJ 2011;343:d4817.

Gwinn DM, Shackelford DB, Egan DF, Mihaylova MM, Mery A, Vasquez DS, Turk BE, Shaw RJ. AMPK phosphorylation of raptor mediates a metabolic checkpoint. Mol Cell 2008;30(2):214–26. https://doi.org/10.1016/j.molcel.2008.03.003 PMID: 18439900; PMCID: PMC2674027.

Haigis MC, Sinclair DA. Mammalian sirtuins: biological insights and disease relevance. Annu Rev Pathol 2010;5:253–95. https://doi.org/10.1146/annurev.pathol.4.110807.092250 PMID: 20078221; PMCID: PMC2866163.

Hannum G, Guinney J, Zhao L, Zhang L, Hughes G, Sadda S, Klotzle B, Bibikova M, Fan JB, Gao Y, Deconde R, Chen M, Rajapakse I, Friend S, Ideker T, Zhang K. Genome-wide methylation profiles reveal quantitative views of human aging rates. Mol Cell 2013;49(2):359–67. https://doi.org/10.1016/j.molcel.2012.10.016 Epub 2012 Nov 21. PMID: 23177740; PMCID: PMC3780611.

Hansen M, Rubinsztein DC, Walker DW. Autophagy as a promoter of longevity: insights from model organisms. Nat Rev Mol Cell Biol 2018;19(9):579–93. https://doi.org/10.1038/s41580-018-0033-y Erratum in: Nat Rev Mol Cell Biol. 2018 Jul 25;: PMID: 30006559; PMCID: PMC6424591.

Hara K, Yonezawa K, Weng QP, Kozlowski MT, Belham C, Avruch J. Amino acid sufficiency and mTOR regulate p70 S6 kinase and eIF-4E BP1 through a common effector mechanism, J Biol Chem 1998; 273(23):14484–94. https://doi.org/10.1074/jbc.273.23.14484. Erratum in: J Biol Chem 1998 Aug 21;273(34):221609603962.

Hasty P, Campisi J, Hoeijmakers J, van Steeg H, Vijg J. Aging and genome maintenance: lessons from the mouse?, Science 2003;299(5611):1355–9. https://doi.org/10.1126/science.1079161.12610296.

Hayflick L. The not-so-close relationship between biological aging and age-associated pathologies in humans, J Gerontol A Biol Sci Med Sci 2004;59(6):B547–50. discussion 551-3 https://doi.org/10.1093/gerona/59.6.b54715215261.

Hayflick L. Entropy explains aging, genetic determinism explains longevity, and undefined terminology explains misunderstanding both. PLoS Genet 2007;3(12)e220. https://doi.org/10.1371/journal.pgen.0030220 PMID: 18085826; PMCID: PMC2134939.

Heyn H, Li N, Ferreira HJ, Moran S, Pisano DG, Gomez A, Diez J, Sanchez-Mut JV, Setien F, Carmona FJ, Puca AA, Sayols S, Pujana MA, Serra-Musach J, Iglesias-Platas I, Formiga F, Fernandez AF, Fraga MF, Heath SC, Valencia A, Gut IG, Wang J, Esteller M. Distinct DNA methylomes of newborns and centenarians. Proc Natl Acad Sci U S A 2012;109(26):10522–7. https://doi.org/10.1073/pnas.1120658109 Epub 2012 Jun 11. PMID: 22689993; PMCID: PMC3387108.

Hipp MS, Kasturi P, Hartl FU. The proteostasis network and its decline in ageing, Nat Rev Mol Cell Biol 2019;20(7):421–35. https://doi.org/10.1038/s41580-019-0101-y.30733602.

Horvath S. DNA methylation age of human tissues and cell types. Genome Biol 2013;14(10)R115https://doi.org/10.1186/gb-2013-14-10-r115 Erratum in: Genome Biol. 2015;16:96. PMID: 24138928; PMCID: PMC4015143.

Jang JY, Blum A, Liu J, Finkel T. The role of mitochondria in aging. J Clin Invest 2018;128(9):3662–70. https://doi.org/10.1172/JCI120842 Epub 2018 Jul 30. PMID: 30059016; PMCID: PMC6118639.

Johnson SC, Rabinovitch PS, Kaeberlein M. mTOR is a key modulator of ageing and age-related disease. Nature 2013;493(7432):338–45. https://doi.org/10.1038/nature11861 PMID: 23325216; PMCID: PMC3687363.

Jylhävä J, Pedersen NL, Hägg S. Biological age predictors. EBioMedicine 2017;21:29–36. https://doi.org/10.1016/j.ebiom.2017.03.046 Epub 2017 Apr 1. PMID: 28396265; PMCID: PMC5514388.

Kaushik S, Cuervo AM. Proteostasis and aging, Nat Med 2015;21(12):1406–15. https://doi.org/10.1038/nm.4001.26646497.

Kennedy BK, Lamming DW. The mechanistic target of rapamycin: the grand conductor of metabolism and aging. Cell Metab 2016;23(6):990–1003. https://doi.org/10.1016/j.cmet.2016.05.009 PMID: 27304501; PMCID: PMC4910876.

Kim J, Guan KL. Amino acid signaling in TOR activation, Annu Rev Biochem 2011;80:1001–32. https://doi.org/10.1146/annurev-biochem-062209-094414.21548787.

Kohen R, Nyska A. Oxidation of biological systems: oxidative stress phenomena, antioxidants, redox reactions, and methods for their quantification, Toxicol Pathol 2002;30(6):620–50. https://doi.org/10.1080/01926230290166724.12512863.

Kulkarni AS, Gubbi S, Barzilai N. Benefits of metformin in attenuating the hallmarks of aging. Cell Metab 2020;32(1):15–30. https://doi.org/10.1016/j.cmet.2020.04.001 Epub 2020 Apr 24. PMID: 32333835; PMCID: PMC7347426.

Lamming DW, Ye L, Katajisto P, Goncalves MD, Saitoh M, Stevens DM, Davis JG, Salmon AB, Richardson A, Ahima RS, Guertin DA, Sabatini DM, Baur JA. Rapamycin-induced insulin resistance is mediated by mTORC2 loss and uncoupled from longevity, Science 2012;335(6076):1638–43. https://doi.org/10.1126/science.1215135. 22461615 PubMed Central PMCID: PMC3324089.

Lee KY. Pathophysiology of age-related hearing loss (peripheral and central). Korean J Audiol 2013;17(2):45–9. https://doi.org/10.7874/kja.2013.17.2.45 Epub 2013 Sep 24. PMID: 24653905; PMCID: PMC3936539.

Lee SH, Lee JH, Lee HY, Min KJ. Sirtuin signaling in cellular senescence and aging. BMB Rep 2019;52(1):24–34. https://doi.org/10.5483/BMBRep.2019.52.1.290 PMID: 30526767; PMCID: PMC6386230.

Levine ME, Lu AT, Quach A, Chen BH, Assimes TL, Bandinelli S, Hou L, Baccarelli AA, Stewart JD, Li Y, Whitsel EA, Wilson JG, Reiner AP, Aviv A, Lohman K, Liu Y, Ferrucci L, Horvath S. An epigenetic biomarker of aging for lifespan and healthspan. Aging (Albany NY) 2018;10(4):573–91. https://doi.org/10.18632/aging.101414 PMID: 29676998; PMCID: PMC5940111.

Licastro F, Candore G, Lio D, Porcellini E, Colonna-Romano G, Franceschi C, Caruso C. Innate immunity and inflammation in ageing: a key for understanding age-related diseases. Immun Ageing 2005;2:8. https://doi.org/10.1186/1742-4933-2-8 PMID: 15904534; PMCID: PMC1166571.

Liguori I, Russo G, Curcio F, Bulli G, Aran L, Della-Morte D, Gargiulo G, Testa G, Cacciatore F, Bonaduce D, Abete P. Oxidative stress, aging, and diseases. Clin Interv Aging 2018;13:757–72. https://doi.org/10.2147/CIA.S158513 PMID: 29731617; PMCID: PMC5927356.

Lipsitz LA. Physiological complexity, aging, and the path to frailty, Sci Aging Knowledge Environ 2004;2004(16)pe16. https://doi.org/10.1126/sageke.2004.16.pe1615103055.

Lipsitz LA. The real secret of youth is complexity, http://nautil.us/issue/36/aging/the-real-secret-of-youth-is-complexity; 2016.

Lipsitz LA, Goldberger AL. Loss of 'complexity' and aging. Potential applications of fractals and chaos theory to senescence, JAMA 1992;267(13):1806–9.1482430.

Longo VD, Antebi A, Bartke A, Barzilai N, Brown-Borg HM, Caruso C, Curiel TJ, de Cabo R, Franceschi C, Gems D, Ingram DK, Johnson TE, Kennedy BK, Kenyon C, Klein S, Kopchick JJ, Lepperdinger G, Madeo F, Mirisola MG, Mitchell JR, Passarino G, Rudolph KL, Sedivy JM, Shadel GS, Sinclair DA, Spindler SR, Suh Y, Vijg J, Vinciguerra M, Fontana L. Interventions to slow aging in humans: Are we ready? Aging Cell 2015;14(4):497–510. https://doi.org/10.1111/acel.12338 Epub 2015 Apr 22. PMID: 25902704; PMCID: PMC4531065.

López-Otín C, Blasco MA, Partridge L, Serrano M, Kroemer G. The hallmarks of aging. Cell 2013; 153(6):1194–217. https://doi.org/10.1016/j.cell.2013.05.039 PMID: 23746838; PMCID: PMC3836174.

Lorenz. Lorenz: "Predictability". AAAS 139th meeting, 1972 Archived 2013-06-12 at the Wayback Machine Retrieved May 22, 2015.

Madeo F, Tavernarakis N, Kroemer G. Can autophagy promote longevity? Nat Cell Biol 2010;12:842–6. https://doi.org/10.1038/ncb0910-842.

Majmundar AJ, Wong WJ, Simon MC. Hypoxia-inducible factors and the response to hypoxic stress. Mol Cell 2010;40(2):294–309. https://doi.org/10.1016/j.molcel.2010.09.022 PMID: 20965423; PMCID: PMC3143508.52.

Manor B, Lipsitz LA. Physiologic complexity and aging: implications for physical function and rehabilitation. Prog Neuropsychopharmacol Biol Psychiatry 2013;45:287–93. https://doi.org/10.1016/j.pnpbp.2012.08.020 Epub 2012 Sep 15. PMID: 22985940; PMCID: PMC3568237.

Mattison JA, Colman RJ, Beasley TM, Allison DB, Kemnitz JW, Roth GS, Ingram DK, Weindruch R, de Cabo R, Anderson RM. Caloric restriction improves health and survival of rhesus monkeys. Nat Commun 2017;8:14063. https://doi.org/10.1038/ncomms14063 PMID: 28094793; PMCID: PMC5247583.

Michalak EM, Burr ML, Bannister AJ, Dawson MA. The roles of DNA, RNA and histone methylation in ageing and cancer, Nat Rev Mol Cell Biol 2019;20(10):573–89. https://doi.org/10.1038/s41580-019-0143-1. Epub 2019 Jul 3 31270442.

Mizushima N, Levine B. Autophagy in human diseases, N Engl J Med 2020;383(16):1564–76. https://doi.org/10.1056/NEJMra2022774.33053285.

Mizushima N, Levine B, Cuervo AM, Klionsky DJ. Autophagy fights disease through cellular self-digestion, Nature 2008;451(7182):1069–75. https://doi.org/10.1038/nature06639. Review 18305538 PMC2670399.

Morselli E, Maiuri MC, Markaki M, Megalou E, Pasparaki A, Palikaras K, Criollo A, Galluzzi L, Malik SA, Vitale I, Michaud M, Madeo F, Tavernarakis N, Kroemer G. Caloric restriction and resveratrol promote longevity through the Sirtuin-1-dependent induction of autophagy. Cell Death Dis 2010;1(1)e10. https://doi.org/10.1038/cddis.2009.8 PMID: 21364612; PMCID: PMC3032517.

Mouchiroud L, Houtkooper RH, Moullan N, Katsyuba E, Ryu D, Cantó C, Mottis A, Jo YS, Viswanathan M, Schoonjans K, Guarente L, Auwerx J. The NAD(+)/sirtuin pathway modulates longevity through activation of mitochondrial UPR and FOXO signaling. Cell 2013;154(2):430–41. https://doi.org/10.1016/j.cell.2013.06.016 PMID: 23870130; PMCID: PMC3753670.

Nakamura S, Oba M, Suzuki M, Takahashi A, Yamamuro T, Fujiwara M, Ikenaka K, Minami S, Tabata N, Yamamoto K, Kubo S, Tokumura A, Akamatsu K, Miyazaki Y, Kawabata T, Hamasaki M, Fukui K, Sango K, Watanabe Y, Takabatake Y, Kitajima TS, Okada Y, Mochizuki H, Isaka Y, Antebi A, Yoshimori T. Suppression of autophagic activity by Rubicon is a signature of aging. Nat Commun 2019;10(1):847. https://doi.org/10.1038/s41467-019-08729-6 PMID: 30783089; PMCID: PMC6381146.

Navarro CL, Cau P, Lévy N. Molecular bases of progeroid syndromes, Hum Mol Genet 2006;15(2): R151–61. https://doi.org/10.1093/hmg/ddl214.16987878.

Newman AB, Ferrucci L. Call for papers: aging versus disease, J Gerontol A Biol Sci Med Sci 2009; 64(11):1163–4. https://doi.org/10.1093/gerona/glp039. Epub 2009 Apr 9 19359442.

Omodei D, Fontana L. Calorie restriction and prevention of age-associated chronic disease, FEBS Lett 2011;585(11):1537–42. https://doi.org/10.1016/j.febslet.2011.03.015. Epub 2011 Mar 12. Review 21402069 PMC3439843.

Oshima J, Kato H, Maezawa Y, Yokote K. RECQ helicase disease and related progeroid syndromes: RECQ2018 meeting. Mech Ageing Dev 2018;173:80–3. https://doi.org/10.1016/j.mad.2018.05.002 Epub 2018 May 9. PMID: 29752965; PMCID: PMC6217841.

Piskovatska V, Stefanyshyn N, Storey KB, Vaiserman AM, Lushchak O. Metformin as a geroprotector: experimental and clinical evidence, Biogerontology 2019;20(1):33–48. https://doi.org/10.1007/s10522-018-9773-5. Epub 2018 Sep 25 30255224.

Powers ET, Morimoto RI, Dillin A, Kelly JW, Balch WE. Biological and chemical approaches to diseases of proteostasis deficiency, Annu Rev Biochem 2009;78:959–91. https://doi.org/10.1146/annurev.biochem.052308.114844.19298183.

Prasad K, Mishra M. AGE-RAGE stress, stressors, and antistressors in health and disease. Int J Angiol 2018; 27(1):1–12. https://doi.org/10.1055/s-0037-1613678 Epub 2017 Dec 28. PMID: 29483760; PMCID: PMC5825221.

Price NL, Gomes AP, Ling AJ, et al. SIRT1 is required for AMPK activation and the beneficial effects of resveratrol on mitochondrial function. Cell Metab 2012;15:675–90. https://doi.org/10.1016/j.cmet.2012.04.003.

Puca AA, Spinelli C, Accardi G, Villa F, Caruso C. Centenarians as a model to discover genetic and epigenetic signatures of healthy ageing. Mech Ageing Dev 2018;174:95–102. https://doi.org/10.1016/j.mad.2017.10.004 Epub 2017 Oct 31. PMID: 29096878.0.

Rattan SI. Aging is not a disease: implications for intervention. Aging Dis 2014;5(3):196–202. https://doi.org/10.14336/AD.2014.0500196 PMID: 24900942; PMCID: PMC4037311.

Rubinsztein DC, Mariño G, Kroemer G. Autophagy and aging, Cell 2011;146(5):682–95. https://doi.org/10.1016/j.cell.2011.07.030.21884931.

Salemi S, Yousefi S, Constantinescu MA, Fey MF, Simon HU. Autophagy is required for self-renewal and differentiation of adult human stem cells. Cell Res 2012;22(2):432–5. https://doi.org/10.1038/cr.2011.200 Epub 2011 Dec 20. PMID: 22184008; PMCID: PMC3271583.

Sarbassov DD, Guertin DA, Ali SM, Sabatini DM. Phosphorylation and regulation of Akt/PKB by the rictor-mTOR complex, Science 2005;307(5712):1098–101. https://doi.org/10.1126/science.1106148.15718470.

Saxton RA, Sabatini DM. mTOR signaling in growth, metabolism, and disease. Cell 2017;168(6):960–76. https://doi.org/10.1016/j.cell.2017.02.004 Erratum in: Cell. 2017 Apr 6;169(2):361–371. PMID: 28283069; PMCID: PMC5394987.

Schieke SM, Phillips D, McCoy JP, et al. The mammalian target of rapamycin (mTOR) pathway regulates mitochondrial oxygen consumption and oxidative capacity. J Biol Chem 2006;281:27643–52.

Schultz MB, Sinclair DA. When stem cells grow old: phenotypes and mechanisms of stem cell aging. Development 2016;143(1):3–14. https://doi.org/10.1242/dev.130633 PMID: 26732838; PMCID: PMC4725211.

Sebastián C, Satterstrom FK, Haigis MC, Mostoslavsky R. From sirtuin biology to human diseases: an update. J Biol Chem 2012;287(51):42444–52. https://doi.org/10.1074/jbc.R112.402768 Epub 2012 Oct 18. PMID: 23086954; PMCID: PMC3522245.

Sen P, Shah PP, Nativio R, Berger SL. Epigenetic mechanisms of longevity and aging. Cell 2016;166 (4):822–39. https://doi.org/10.1016/j.cell.2016.07.050 PMID: 27518561; PMCID: PMC5821249.

Soares JP, Cortinhas A, Bento T, Leitão JC, Collins AR, Gaivão I, Mota MP. Aging and DNA damage in humans: a meta-analysis study. Aging (Albany NY) 2014;6(6):432–9. https://doi.org/10.18632/aging.100667 PMID: 25140379; PMCID: PMC4100806.

Stahl EC, Brown BN. Cell therapy strategies to combat immunosenescence. Organogenesis 2015; 11(4):159–72. https://doi.org/10.1080/15476278.2015.1120046 Epub 2015 Nov 20. PMID: 26588595; PMCID: PMC4879890.

Suh Y, Atzmon G, Cho MO, Hwang D, Liu B, Leahy DJ, Barzilai N, Cohen P. Functionally significant insulin-like growth factor I receptor mutations in centenarians. Proc Natl Acad Sci U S A 2008; 105(9):3438–42. https://doi.org/10.1073/pnas.0705467105 Epub 2008 Mar 3. PMID: 18316725; PMCID: PMC2265137.

Sun N, Youle RJ, Finkel T. The mitochondrial basis of aging. Mol Cell 2016;61(5):654–66. https://doi.org/10.1016/j.molcel.2016.01.028 PMID: 26942670; PMCID: PMC4779179.

Sun X, Chen WD, Wang YD. DAF-16/FOXO transcription factor in aging and longevity. Front Pharmacol 2017;8:548. https://doi.org/10.3389/fphar.2017.00548 PMID: 28878670; PMCID: PMC5572328.

Tian X, Firsanov D, Zhang Z, Cheng Y, Luo L, Tombline G, Tan R, Simon M, Henderson S, Steffan J, Goldfarb A, Tam J, Zheng K, Cornwell A, Johnson A, Yang JN, Mao Z, Manta B, Dang W, Zhang Z, Vijg J, Wolfe A, Moody K, Kennedy BK, Bohmann D, Gladyshev VN, Seluanov A, Gorbunova V. SIRT6 is responsible for more efficient DNA double-strand break repair in long-lived species. Cell 2019;177(3)https://doi.org/10.1016/j.cell.2019.03.043 622–638.e22. PMID: 31002797; PMCID: PMC6499390.

Tilly JL, Sinclair DA. Germline energetics, aging, and female infertility. Cell Metab 2013;17(6):838–50. https://doi.org/10.1016/j.cmet.2013.05.007 PMID: 23747243; PMCID: PMC3756096.

Valko M, Leibfritz D, Moncol J, Cronin MT, Mazur M, Telser J. Free radicals and antioxidants in normal physiological functions and human disease, Int J Biochem Cell Biol 2007;39(1):44–84. https://doi.org/10.1016/j.biocel.2006.07.001. Epub 2006 Aug 4 16978905.

Vasto S, Carruba G, Lio D, Colonna-Romano G, Di Bona D, Candore G, Caruso C. Inflammation, ageing and cancer, Mech Ageing Dev 2009;130(1–2):40–5. https://doi.org/10.1016/j.mad.2008.06.003. Epub 2008 Jul 10 18671998.

Vasto S, Buscemi S, Barera A, Di Carlo M, Accardi G, Caruso C. Mediterranean diet and healthy ageing: a Sicilian perspective, Gerontology 2014;60(6):508–18. https://doi.org/10.1159/000363060. Epub 2014 Aug 23 25170545.

Vijg J, Suh Y. Genome instability and aging, Annu Rev Physiol 2013;75:645–68. https://doi.org/10.1146/annurev-physiol-030212-183715.23398157.

Wong SQ, Kumar AV, Mills J, Lapierre LR. Autophagy in aging and longevity. Hum Genet 2020;139 (3):277–90. https://doi.org/10.1007/s00439-019-02031-7 Epub 2019 May 30. PMID: 31144030; PMCID: PMC6884674.

Yu L, McPhee CK, Zheng L, Mardones GA, Rong Y, Peng J, Mi N, Zhao Y, Liu Z, Wan F, Hailey DW, Oorschot V, Klumperman J, Baehrecke EH, Lenardo MJ. Termination of autophagy and reformation of lysosomes regulated by mTOR. Nature 2010;465(7300):942–6. https://doi.org/10.1038/nature09076 Epub 2010 Jun 6. PMID: 20526321; PMCID: PMC2920749.

Zhao Y, Zhu J, Zhang N, Liu Q, Wang Y, Hu X, Chen J, Zhu W, Yu H. GDF11 enhances therapeutic efficacy of mesenchymal stem cells for myocardial infarction via YME1L-mediated OPA1 processing. Stem Cells Transl Med 2020;9(10):1257–71. https://doi.org/10.1002/sctm.20-0005 Epub ahead of print. PMID: 32515551; PMCID: PMC7519765.

Zhu Y, Liu X, Ding X, Wang F, Geng X. Telomere and its role in the aging pathways: telomere shortening, cell senescence and mitochondria dysfunction, Biogerontology 2019;20(1):1–16. https://doi.org/10.1007/s10522-018-9769-1. Epub 2018 Sep 18 30229407.

CHAPTER 4

Cellular senescence and senescence-associated secretory phenotype (SASP) in aging process

Fabiola Olivieri[a,b], Antonio Domenico Procopio[a,b], and Maria Rita Rippo[a]
[a]Department of Clinical and Molecular Sciences, DISCLIMO, Università Politecnica delle Marche, Ancona, Italy
[b]Center of Clinical Pathology and Innovative Therapy, IRCCS INRCA, Ancona, Italy

4.1 Introduction

Cellular senescence first reported in 1961 by Hayflick and Moorehead is a persistent cell condition characterized mainly by the lack of replicative ability, resistance to apoptosis and the acquisition of the so-called senescence-associated secretory phenotype (SASP) (Hayflick and Moorhead, 1961). For the role of telomere shortening in senescence induction, see Chapter 3. The acquisition of SASP entails the secretion of myriads of bioactive molecules, mainly involved in the modulation of inflammatory process, including inflammatory cytokines, chemokines, growth factors, and proteases that can markedly affect the behavior of neighboring cells (Basisty et al., 2020). Therefore, numerous activities of senescent cells depend on the aptitude of these cells to secrete bioactive molecules, and SASP supports cell-autonomous functions like the senescence-associated growth arrest and mediates paracrine interactions between senescent cells and their surrounding microenvironment. The biological role played by senescent cells is complex; senescent cells can have prohealth functions during development, tissue remodeling, and wound. However, the accumulation of senescent cells also drives aging and age-related diseases. In this framework, the characterization of specific biomarkers of cellular senescence and the strategies to kill senescent cells are cutting-edge issues in aging research (Sikora, 2012). The senescent phenotype is multifaceted, senescent cells manifest various senescence markers and SASP components can be different depending on triggers and cell types. However, the best-characterized hallmarks of cellular senescence are cell flattening, increased senescence-associated β galactosidase (SA-β gal) activity, increased expression of negative cell-cycle regulators p53, p21(Waf1) and p16(Ink4a), specific chromatin reorganization including DNA segments with chromatin alterations, DNA-SCARS, and mitochondria dysfunctions (Hernandez-Segura et al., 2018). These features are related to a persistent DNA damage response (DDR) and SASP can be considered as an active transcriptional program activated in response to persistent DDR.

4.2 Signaling pathway stimulating the appearance of SASP

Epigenetic mechanisms play key roles in the implementation of senescent phenotypes. In senescent cells, DDR induces the proteasomal degradation of histone H3K9 mono- and dimethyltransferases, G9a and GLP, causing a global decrease in H3K9 dimethylation. Interestingly, the induction of Interleukin (IL)-6 and IL-8, major players of SASP, is correlated with a decline of H3K9 dimethylation around the respective gene promoters (Takahashi et al., 2012). Several histone deacetylase inhibitors robustly activated SASP in the absence of DNA breaks, suggesting that DDR-dependent SASP activation can occur also in response to chromatin remodeling as well as in response to physical breaks in DNA (Pazolli et al., 2012). In senescent cells, disorganization of the nuclear architecture and loss of perinuclear heterochromatin are common findings. Because heterochromatin governs gene silencing and genomic stability, loss of this epigenetic mechanism during senescence may lead to increased transcription of various genes (Tominaga and Pereira-Smith, 2012). Moreover, hypermethylation is another general mechanism for transcriptional suppression and during senescence, global DNA methylation is known to decrease in the genome. In senescent cells, therefore, loss of these suppressive mechanisms may lead to a global increase in transcriptional activity and thereby induce the secretion of a wide range of molecules. Some histone variants, such as H3.3 and macroH2A, increase as cells undergo senescence, suggesting histone variants and their associated chaperones could be important in chromatin structure maintenance in senescent cells (Nacarelli et al., 2017). Emerging data have revealed that nuclear factor kappa-light-chain-enhancer of activated B cells (NF-κB) signaling is the major signaling pathway that stimulates the appearance of SASP (Salminen et al., 2012). One of the earliest events in the acquisition of SASP is, in fact, an increase in the expression of IL-1α, a cytokine that is rarely secreted, but rather is membrane-associated where it binds its juxtaposed receptor (juxtacrine signaling). IL-1α receptor engagement triggers a signaling cascade that ultimately activates the NF-κB transcription factor that transcribes the genes for many of the proinflammatory components of the SASP. Moreover, environmental stress and chronic inflammation can stimulate mammalian p38 mitogen-activated protein kinase (p38MAPK) inducing cellular senescence with proinflammatory responses (Freund et al., 2011). P38MAPK pathway can be activated by a wide variety of cellular stresses, e.g., oxidative, metabolic, and endoplasmic stresses, DNA damage, heat shock, and mechanical damage. In addition, cytokines and other inflammatory mediators activate the p38MAPK signaling pathway in order to regulate chemotaxis by enhancing the expression of several chemokines, e.g., macrophage inhibitor protein (MIP)-1α/β, monocyte chemoattractant protein (MCP-1), and chemokine (C-C motif) ligand (CCL)5. The age-related increase in ceramides is also a potent inducer of NF-κB signaling. Ceramide is a bioactive sphingolipid that has been suggested to regulate toll-like receptor-4-induced NF-κB signaling. Ceramide led also to the induction of the inducible cyclooxygenase-2 and subsequent increase of prostaglandin E2 (Doyle et al., 2011).

It seems that epigenetic changes, in both cellular senescence and organismal aging, can enhance the transcription of NF-κB-dependent inflammatory genes, e.g., via changes in high mobility group box 1 (HMGB1) and sirtuin 6 function. Genomic instability evoked by cellular stress triggers epigenetic changes, e.g., release of HMGB1 proteins, which are also potent enhancers of inflammatory responses. Downregulation of cytoplasmic deoxyribonuclease (DNase)-2 and three prime repair exonuclease 1 (TRX1) that target double-stranded (ds)DNA and single-stranded (ss)DNA (cytoplasmic chromatin fragments, CCFs) for degradation, respectively, in senescent cells causes accumulation into the cytoplasm of nuclear-derived DNA fragments; this phenomenon in turn induces aberrant upregulation and activation of cGAS-STING cytoplasmic DNA sensor, a component of the innate immune system that functions to detect the presence of cytosolic DNA, provoking SASP through induction of interferon-β (Takahashi et al., 2018). Interestingly, it has been shown that the accumulation of these DNA fragments is caused by an extended reactive oxygen species (ROS)-c-Jun N-terminal kinase (JNK) mitochondria-to-nucleus retrograde signaling triggered by downregulation of nuclear-encoded mitochondrial oxidative phosphorylation genes (Vizioli et al., 2020). An interesting perspective suggests a role for mitochondria in integrating signals deriving from nuclear-encoded microRNAs (miRNAs) and the inflammatory response: In practice, some hyperexpressed miRNAs in the senescent cells are transported from the nucleus to the mitochondria where they regulate the expression of different proteins encoded by the mitochondrial genome and important in the cellular metabolism, thus affecting the energetic, oxidative, and inflammatory status of senescent cells (Giuliani et al., 2017a). The persistent accumulation of CCFs in senescent cells is partially associated with a defect in DNA-degrading activity in autolysosomes and reduced abundance of activated DNase 2α (Han et al., 2020). All endoplasmic reticulum (ER) stress sensors are able to induce a specific transcriptional program, aimed to restore ER proteostasis and cellular homeostasis. Some of the unfolded protein response downstream signaling events, yet unknown, control the establishment and/or maintenance of the main senescence hallmarks, including cell-cycle arrest, DNA repair capacity, morphological changes, metabolic changes, the secretory pathway, and changes in membrane lipid composition (Hetz et al., 2020) (Table 4.1).

Table 4.1 Signaling pathway stimulating the appearance of SASP.

Signaling pathway	References
NF-κB	Salminen et al. (2012)
P38MAPK	Salminen et al. (2012)
Ceramide	Salminen et al. (2012)
DNase-2 and TREX1/cGAS-STING	Takahashi et al. (2018)
ROS-JNK retrograde mitochondria-to-nucleus signaling	Vizioli et al. (2020)
miRNA/mitochondria proteins	Giuliani et al. (2017a)
Unfolded protein response	Hetz et al. (2020)

4.3 SASP components

SASP is promoted via the activation of general machinery responsible for transcription, translation, and/or protein secretion, and a number of protocols were proposed to identify the component of the SASP (Rodier, 2013). Inflammatory and immune-modulatory cytokines and chemokines, e.g., IL-6, IL-7, and IL-8, MCP-2, and MIP-3a, growth factors, e.g., growth-related oncogene (GRO), hepatocyte growth factor (HGF), and insulin growth factor binding proteins (IGFBPs), shed cell surface molecules, e.g., intercellular adhesion molecules, urokinase receptor, and tumor necrosis factor (TNF) receptors, and survival factors can be considered as components of SASP of different cell types (Coppé et al., 2008). It is thus conceivable that some secretory reporter protein could serve as an indicator for SASP. IL-6, IL-8, and GROα are among the most highly and consistently secreted human SASP factors and overlap with the similarly secreted mouse SASP factors (Coppé et al., 2010). Interestingly, most of the SASP components are coded by gene clusters that are coordinately induced upon senescence, suggesting that the increased expression of human SASP genes can involve large chromosomal segments. IL-1 family members (human chromosome 2), several members of the chemokine (C-X. C motif), ligand locus (human chromosome 4), and CCL locus (human chromosome 17) are important components of SASP, as well as matrix metalloproteinases (MMPs) genes, such as MMP1, MMP3, MMP10, and MMP12 (human chromosome 11). Genes codifying for other important SASP components, such as IL-6 and TNF-α, are located in chromosome 6 and 7, respectively. The composition of the SASP may vary as time progresses after the initiation of senescence and might partly depend on the mechanism through which senescence is induced. Recently, gene expression profiling of diploid fibroblasts and aging Sprague-Dawley rat skin tissues revealed four stages of senescence, such as early, middle, advanced, and very advanced, with specific gene expression modules governing each stage. Interestingly, SASP-related genes also displayed a stage-specific expression pattern with three unique features during senescence: differential expression of interleukin isoforms, differential expression of interleukins and their receptors, and differential expression of matrix metalloproteinases and their inhibitory proteins (Kim et al., 2013). Many factors that characterize SASP are not properly specific to aging: SASP-related cytokines, chemokines, and MMPs are also produced in response to inflammatory and noninflammatory stimuli. Recently, a number of techniques were proposed to facilitate the discovery of novel functions and regulators of the SASP. It was proposed that SASP-responsive alkaline phosphatase can serve as a selective, convenient, and general marker for detection and monitoring of SASP during cellular senescence. However, to date, selective, universal, and convenient assays for SASP have not been established yet.

4.4 MiRNA and extracellular vesicles as new regulators and components of SASP

During the last years, miRNAs have emerged as an important epigenetic mechanism of SASP regulation mainly modulating the NF-κB pathway; however, their role as essential components of secretoma and their paracrine effect on the surrounding microenvironment have not been extensively investigated. Here we discuss recent findings on the role of specific miRs in targeting NF-κB signaling and of those that, once secreted as components of the SASP, can or could have paracrine effects on tissue or even systemic dysfunction. In relation to miRs belonging to and regulating NF-κB pathway, miR–146 and miR–21 are the most studied in aging cell models. Upregulated miR–146a has been associated with cellular senescence in various cell types, including human fibroblasts (Bhaumik et al., 2009), monocytes (Ong et al., 2018), and endothelial cells (ECs) where it indicates a senescence-associated proinflammatory status (Olivieri et al., 2013a), In particular, miR–146a, its target protein IL-1 receptor-associated kinase (IRAK-1), and released IL-6, indicates SASP acquisition by primary human umbilical vein ECs (Olivieri et al., 2013b). Recent findings have expanded the biological contexts in which miR–146a/b modulates inflammatory responses. Cells undergoing senescence without induction of a robust SASP did not express miR–146a/b, suggesting that miR–146a/b is involved in the appearance of SASP. MiR–146 is able to suppress IL-6 and IL-8 secretion and to downregulate IRAK1, a crucial component of the IL-1 receptor signal transduction pathway, in primary human fibroblasts. Further, IL-1α neutralizing antibodies abolished both miR–146a/b expression and IL-6 secretion, suggesting that IL-1 receptor signaling initiates both miR–146a/b upregulation and cytokine secretion and that miR–146a/b is expressed in response to rising inflammatory cytokine levels as part of a negative feedback loop that restrains excessive SASP activity. Interestingly, the same miR–146a, which is nuclear-coded along with other miRs involved in the aging process, may localize in the mitochondria of senescent cells, regulate several proteins involved in the respiratory chain and apoptosis and thus affect their energetic, oxidative status and amplify the inflammatory process and the release of SASP-related compounds (Giuliani et al., 2017b; Rippo et al., 2014). MiR–21 is overexpressed in aging cells and biological fluids and tissues of old subjects (Olivieri et al., 2012). MiR–21 targets some molecules belonging to TLR signaling pathway, such as myeloid differentiation factor 88 and IRAK1, are key modulators of inflammation. Furthermore, it was reported that miR-21 overexpression reduces the replicative life span, whereas its stable knockdown extends the replicative life span of normal ECs (Dellago et al., 2013). Interestingly, circulating miR-21 is, among those miRNAs involved in the regulation of inflammation (inflamma-miRs), endowed with the strongest ability to provide information about the trajectories of healthy and unhealthy aging (Olivieri et al., 2017). MiRNAs, as already mentioned, in addition to acting within the cell can be released and transported to the external environment and into the bloodstream; this can be done through different mechanisms, all effective in promoting their paracrine and systemic action. They can be found inside vesicles, such as exosomes, microparticles, and apoptotic bodies (Valadi et al., 2007), bound to RNA binding proteins such as Argonaute 2 (Arroyo et al., 2011) or to HDL/LDL (Vickers and Remaley, 2012). A number of different types of membrane-bound extracellular vesicles (EVs) have been identified in tissues and biological fluids of all organisms, and they were supposed to be involved in cell-to-cell cross talk (Maas et al., 2017). Exosomes are EVs produced in the multivesicular body being part of endosomal compartment of most eukaryotic cells. Recent reports show that senescent cells produce higher amounts of EVs and that senescent cell-associated miRNAs can be selectively found within them and considered as an integral member of the SASP (Fafián-Labora and O'Loghlen, 2020; Wallis et al., 2020). Exosomes can be secreted from cells in the extracellular environment, where they can interact with other recipient cells, fusing with the plasma membrane of the target cells and leading to the release of their content into the target cell. In vitro experiments suggested that cargoes carried by exosomes released from senescent cells can spread prosenescence signals to younger cells, affecting DNA methylation and cell replication (Mensà et al., 2020). Parabiosis experiments in mice demonstrated that a young environment could partially rejuvenate multiple tissues of old organisms. EVs isolated from plasma of young mice increase life span in old mice, suggesting that circulating vesicles may mediate the beneficial effect of a young milieu on aging (Prattichizzo et al., 2019; Wang et al., 2018). The biological effects of EVs can be due, almost in part, to the specific pattern of miRNAs. For example, senescent dermal human fibroblasts produce fourfold increase of EVs compared to their younger counterparts with specific differences in miRNA composition mostly targeting antiapoptotic proteins, whereas other upregulated miRNAs are retained within the cells (Terlecki-Zaniewicz et al., 2018). Similarly, miRNAs shuttled by sEVs released from senescent ECs can propagate prosenescent signals; in particular, sEVs from senescent cells enriched in miR-21-5p and miR-217 can target DNA (cytosine-5)-methyltransferase 1 and sirtuin 1 expression in normal younger recipient cells and induce the reduction of proliferation markers, the acquisition of a senescent phenotype, and a partial demethylation of the locus encoding for miR-21. Overall, these data suggest an active role of miRNAs carried by senescent cell-produced EVs (Mensà et al., 2020).

4.5 SASP profile in different cell types

Important features of the SASP include the fact that it is conserved between human and mouse cells and occurs in vitro and in vivo in a variety of proliferative cell types, such as fibroblasts, epithelial cells, ECs, astrocytes, vascular smooth muscle cells, and mesenchymal stem/stromal cells (MSCs). However, SASP is not fixed but assumes distinct quantitative or qualitative characteristics depending on the cell type and the context in which the cell itself is located. Therefore, wide-ranging profiles have been observed over these years and increasing efforts are devoted to identify common components of SASP (Coppé et al., 2010; Özcan et al., 2016). Moreover, it was demonstrated that SASP is largely influenced by O_2 concentration in in vitro cultured cellular models. MSCs are self-renewing cells present throughout vascularized organs, as well as bone marrow, showing reparative and regenerative abilities being able to differentiate into different cytotypes of mesenchymal origin and because they are able to modulate the microenvironment in which they are found in addition to having remarkable immunomodulatory properties. For these reasons, MSCs have a wide range of potential clinical applications. However, these reservoirs could be subjected to cellular senescence, characterized by alteration of their differentiating potential and loss of proliferative capacities, both in vitro and in vivo, thus having a detrimental impact on their availability for therapies and tissue engineering and tissue homeostasis (Wang et al., 2013). MSC senescence is considered a critical step for the exhaustion of cellular repair and regeneration capacity during aging and for age-related disease progression, ultimately affecting the overall function of tissues and organs. However, experimental pieces of evidence suggest a double mechanism through which the senescent MSCs alter tissue homeostasis: the loss of regenerative capacity due to reduced self-renewal and differentiation, as just mentioned, and the acquisition of the SASP with the production of molecules that promote inflammation and degrade the matrix with also systemic effects. Senescent MSC secretome is also composed of a number of extracellular growth factors, including transforming growth factor-β, platelet-derived growth factor, HGF, epidermal growth factor, IGFBPs, decoy receptors, and receptor antagonists (Freund et al., 2010). These SASP components can induce senescent phenotype in other cells in an autocrine and paracrine manner (Kuilman et al., 2010). Senescent MSCs show many morphological and biochemical characteristics common to other cytotypes, such as the telomeric shortening and the DDR. However, one of the pathways involved in MSC senescence seems to be the Wnt/β-catenin signaling (Wang et al., 2013). Wnt/β-catenin signaling could induce MSC aging through promoting the intracellular production of ROS, and ROS may be the main mediators of MSC aging induced by excessive activation of Wnt/β-catenin signaling (Zhang et al., 2013). Interestingly, chronic inflammatory conditions like obesity can impair MSC function by inducing cellular senescence (Conley et al., 2020). A number of studies investigated the phenomenon of cellular senescence in ECs, as a main culprit of endothelial dysfunction. Alterations in EC metabolism have been identified in age-related diseases (ARDs) associated with a dysfunctional vasculature, including cardiovascular diseases (CVD) and type 2 diabetes mellitus (T2DM). In particular, higher production of reactive oxygen species deriving from a variety of enzymatic sources, including uncoupled endothelial nitric oxide synthase and the electron transport chain, causes DNA damage and activates the NAD^+-consuming enzymes poly(ADP-ribose) polymerase 1. These nonphysiological mechanisms drive the impairment of the glycolytic flux and the diversion of glycolytic intermediates into many pathological pathways (Sabbatinelli et al., 2019). Of note, the accumulation of senescent ECs has been reported in the context of ARDs. Through their prooxidant, proinflammatory, vasoconstrictor, and prothrombotic activities, they negatively impact vascular physiology, promoting both onset and development of ARDs. Recently, it has been shown that factors secreted by senescent ECs in vivo could have a relevant role in the platelet activation observed in older people or in patients undergoing therapeutic stress (Venturini et al., 2020). In addition, these factors, released as active components of SASP, have not only a pathogenic role but can be used as markers of biological age, frailty, and risk of disease onset. Recently, common components released from different senescent cell types (ECs, fibroblasts, preadipocytes, epithelial cells, and myoblasts) were identified and tested for their usefulness as circulating predictive biomarkers of adverse events in a community-based sample of people aged 20–90 years; results showed that a set of SASP proteins (growth differentiation factor 15, FAS, osteopontin, TNF receptor 1, ACTIVIN A, CCL3, and IL-15) predicted postsurgery outcome markedly better than a single SASP protein or age (Schafer et al., 2020).

4.6 Cellular senescence, SASP, and aging

Cellular senescence can be induced by two main types of triggers: stimuli inducing cellular replication and harmful environmental factors interaction (reviewed in Herranz and Gil, 2018). Both types of triggers increase during aging process, so that the burden of senescent cells increases with aging "per se." Since senescent cells are characterized by a proinflammatory phenotype, their accumulation in tissues of aged animal models and humans has been associated with aging per se (López-Otín et al., 2013) and with increased levels of "inflammaging" (Franceschi et al., 2000). Inflammaging is the chronic, systemic, low-grade proinflammatory status that characterizes human aging and it is considered the main risk factor for the development of the most common age-related diseases (Fulop et al., 2018a). At molecular level, cellular senescence fuel inflammaging since it is characterized by the persistent activation of the NF-κB signaling, one of the major pathways that stimulate the appearance of SASP. Even if SASP is not a fixed phenotype but rather a wide-ranging profile that varies with cell type and/or senescence inducers; however, IL-6, TNF-a, and IL-8 are considered hallmarks of senescent secretome of different types of cells. The rate of accumulation of senescent cells increases during aging and it appears to be accelerated in the presence of the most common age-related diseases (ARDs), including CVD, T2DM, neurodegenerative diseases, and cancers and their related risk factors (obesity, hypertension). A number of studies demonstrated that the persistent accumulation of senescent cells and the increased SASP activity during aging is due almost in part to the impairment of innate and adaptive immune responses, a phenomenon named "immunosenescence" (Fulop et al., 2018b). Attracted and activated by

SASP factors, immune cells, especially those of innate immunity, such as natural killer cells, macrophages, and granulocytes, mediate the clearance of senescent cells. However, the interaction with SASP factors can make immune cells senescent and dysfunctional, leading to a vicious circle characterized by a persistent accumulation of senescent cells (Kale et al., 2020). Interestingly, the acquisition of SASP can have beneficial or detrimental effects in vivo, depending on the local tissue environment. Transient activation of senescence can play a prohealthy role, promoting wound repair in skin (Demaria et al., 2014), limb patterning in developing embryos, and suppressing tumorigenesis. In this framework, SASP can allow damaged cells to communicate their compromised state to the surrounding tissue. On the contrary, persistent activation of SASP can have detrimental effects promoting the disruption of normal tissue structures and function, and the malignant phenotypes in nearby cells. This view on senescent cells as carriers and inducers of damage puts new light on senescence, considering it as a significant contributor to the rise in organismal damage (Ogrodnik et al., 2019). The role played by senescent cells in tumorigenesis is paradigmatic of the complex function of cellular senescence. Despite the general consensus that senescence plays a tumor-suppressive role, the long-term implications of senescent cells are potentially protumorigenic, promoting neoplastic conversion, SASP-related chronic inflammation, and cell-cycle reentry of stemness-reprogrammed senescent cancer cells (Chan and Narita, 2019; Milanovic et al., 2018). Notably, the secretome of senescent cells can induce dysfunction in neighboring cells, suggesting that the messages spread by the secretoma of senescent cells have the ability to modify the biologic behaviors of the receiving cells. Considering the richness of secretomes of different types of senescent cells, increasing efforts have been devoted to gain insights into their individual components. In this framework, recent findings suggest that the circulating SASP may serve as a clinically useful candidate biomarker of age-related health and a powerful tool for interventional human studies (Schafer et al., 2020). Although senescent cells can appear at any point during life in vertebrates, they accumulate in multiple tissues during aging. An increased burden of senescent cells was observed at etiological sites of multiple human diseases. Senescent cells accumulate in adipose tissue in diabetes and obesity, in the hippocampi and frontal cortex in Alzheimer's disease, the substantia nigra in Parkinson's disease, bone and marrow in age-related osteoporosis, lungs in idiopathic pulmonary fibrosis, liver in cirrhosis, retinae in macular degeneration, plaques in psoriasis, kidneys in diabetic kidney disease, endothelium in preeclampsia, and the heart and major arteries in cardiovascular disease, among many other conditions.

4.7 Conclusions and future perspectives

Senescent cells can accumulate at etiological sites of the most common age-related diseases throughout the life span, and probably, there is a lag between senescent cell accumulation and development of functional effects. Increasing efforts were devoted to identify drugs that specifically target senescent cells by inducing apoptosis, drugs that were named senolytics (Kirkland and Tchkonia, 2015). Natural or synthetic compounds with senolytic activity, i.e., the ability to selectively remove senescent cells by inducing apoptosis, and immunotherapies using immune cell-mediated clearance of senescent cells are currently the most promising strategies to fight aging and human age-related diseases (Gurău et al., 2018). Recent results support the premise that senolytic therapy can delay disease recurrence following conventional cancer therapy (Saleh et al., 2020). Because senescent cells do not divide, they are unlikely to develop drug resistance, a problem encountered with compounds that target dividing cancer cells. Also, finding compounds or antibodies that target 100% of senescent cells and 0% of normal cells might not be necessary to achieve clinical benefit (Tchkonia et al., 2013). Thus, pharmacological strategies aimed to reverse cellular senescence and SASP are hot topics in the framework of antiaging strategy. A number of drugs that target gene products that protect senescent cells from apoptosis were tested for their ability to eliminate senescent cells in vitro. Dasatinib (D) and quercetin (Q) showed particular promise in clearing senescent cells; D is a multiple tyrosine kinase inhibitor, used for treating cancers, whereas Q is a natural flavonol, inhibiting phosphatidylinositol 3-kinases, other kinases, and serpins (Kirkland and Tchkonia, 2020). Preliminary results of open-label phase 1 pilot studies suggest that oral administration of D and Q to subjects affected by diabetic kidney disease or idiopathic pulmonary fibrosis significantly decreases senescent cell burden in humans (Hickson et al., 2019; Justice et al., 2019). Recently, it was reported that chimeric antigen receptor (CAR) T cells can be redirected to target senescent cells in animal models. CAR T cells that specifically target senescent cells could be a novel senolytic treatment strategy for senescence-associated diseases. Urokinase-type plasminogen activator receptor was chosen as cell surface protein able to mark senescent cells in vitro and in vivo (Abbadie and Pluquet, 2020). In mice, senescence-targeted CAR T cells extended survival in lung cancer and improved liver fibrosis (Amor et al., 2020).

References

Abbadie C, Pluquet O. Unfolded protein response (UPR) controls major senescence hallmarks. Trends Biochem Sci 2020;45:371–4. https://doi.org/10.1016/j.tibs.2020.02.005.

Amor C, Feucht J, Leibold J, Ho YJ, Zhu C, Alonso-Curbelo D, Mansilla-Soto J, Boyer JA, Li X, Giavridis T, Kulick A, Houlihan S, Peerschke E, Friedman SL, Ponomarev V, Piersigilli A, Sadelain M, Lowe SW. Senolytic CAR T cells reverse senescence-associated pathologies. Nature 2020;583:127–32. https://doi.org/10.1038/s41586-020-2403-9.

Arroyo JD, Chevillet JR, Kroh EM, Ruf IK, Pritchard CC, Gibson DF, Mitchell PS, Bennett CF, PogosovaAgadjanyan EL, Stirewalt DL, Tait JF, Tewari M. Argonaute2 complexes carry a population of circulating microRNAs independent of vesicles in human plasma. Proc Natl Acad Sci U S A 2011;108:5003–8.

Basisty N, Kale A, Jeon OH, Kuehnemann C, Payne T, Rao C, Holtz A, Shah S, Sharma V, Ferrucci L, Campisi J, Schilling B. A proteomic atlas of senescence-associated secretomes for aging biomarker development. PLoS Biol 2020;18:e3000599. https://doi.org/10.1371/journal.pbio.3000599.

Bhaumik D, Scott GK, Schokrpur S, Patil CK, Orjalo AV, Rodier F, Lithgow GJ, Campisi J. MicroRNAs miR-146a/b negatively modulate the senescence-associated inflammatory mediators IL-6 and IL-8. Aging (Albany NY) 2009;1:402–11.

Chan ASL, Narita M. Short-term gain, long-term pain: the senescence life cycle and cancer. Genes Dev 2019;33:127–43. https://doi.org/10.1101/gad.320937.118.

Conley SM, Hickson LJ, Kellogg TA, McKenzie T, Heimbach JK, Taner T, Tang H, Jordan KL, Saadiq IM, Woollard JR, Isik B, Afarideh M, Tchkonia T, Kirkland JL, Lerman LO. Human obesity induces dysfunction and early senescence in adipose tissue-derived mesenchymal stromal/stem cells. Front Cell Dev Biol 2020;26(8):197. https://doi.org/10.3389/fcell.2020.00197. eCollection 2020.

Coppé JP, Patil CK, Rodier F, Sun Y, Muñoz DP, Goldstein J, Nelson PS, Desprez PY, Campisi J. Senescence-associated secretory phenotypes reveal cell–nonautonomous functions of oncogenic RAS and the p53 tumor suppressor. PLoS Biol 2008;6:2853–68. https://doi.org/10.1371/journal.pbio.0060301.

Coppé JP, Patil CK, Rodier F, Krtolica A, Beauséjour CM, Parrinello S, Hodgson JG, Chin K, Desprez PY, Campisi J. A human-like senescence-associated secretory phenotype is conserved in mouse cells dependent on physiological oxygen. PLoS One 2010;5:e9188. https://doi.org/10.1371/journal.pone.0009188.

Dellago H, Preschitz-Kammerhofer B, Terlecki-Zaniewicz L, et al. High levels of oncomiR-21 contribute to the senescence-induced growth arrest in normal human cells and its knock-down increases the replicative lifespan. Aging Cell 2013;12:446–58. https://doi.org/10.1111/acel.12069.

Demaria M, Ohtani N, Youssef SA, Rodier F, Toussaint W, Mitchell JR, Laberge R-M, Vijg J, Van Steeg H, Dollé MET, Hoeijmakers JHJ, de Bruin A, Hara E, Campisi J. An essential role for senescent cells in optimal wound healing through secretion of PDGF-AA. Dev Cell 2014;31:722–33.

Doyle T, Chen Z, Muscoli C, Obeid LM, Salvemini D. Intraplantar-injected ceramide in rats induces hyperalgesia through an NF-κB- and p38 kinase-dependent cyclooxygenase 2/prostaglandin E2 pathway. ASEB J 2011;25:2782–91. https://doi.org/10.1096/fj.10-178095.

Fafián-Labora JA, O'Loghlen A. Classical and nonclassical intercellular communication in senescence and ageing. Trends Cell Biol 2020;30:628–39. https://doi.org/10.1016/j.tcb.2020.05.003.

Franceschi C, Bonafè M, Valensin S, Olivieri F, De Luca M, Ottaviani E, De Benedictis G. Inflamm-aging. An evolutionary perspective on immunosenescence. Ann N Y Acad Sci 2000;908:244–54. https://doi.org/10.1111/j.1749-6632.2000.tb06651.x.

Freund A, Orjalo AV, Desprez PY, Campisi J. Inflammatory networks during cellular senescence: causes and consequences. Trends Mol Med 2010;16:238–46. https://doi.org/10.1016/j.molmed.2010.03.003.

Freund A, Patil CK, Campisi J. p38MAPK is a novel DNA damage response-independent regulator of the senescence-associated secretory phenotype. EMBO J 2011;30:1536–48. https://doi.org/10.1038/emboj.2011.69.

Fulop T, Witkowski JM, Olivieri F, Larbi A. The integration of inflammaging in age-related diseases. Semin Immunol 2018a;40:17–35. https://doi.org/10.1016/j.smim.2018.09.003.

Fulop T, Larbi A, Dupuis G, Le Page A, Frost EH, Cohen AA, Witkowski JM, Franceschi C. Immunosenescence and inflamm-aging as two sides of the same coin: friends or foes? Front Immunol 2018b;10(8):1960. https://doi.org/10.3389/fimmu.2017.01960. eCollection 2017.

Giuliani A, Prattichizzo F, Micolucci L, Ceriello A, Procopio AD, Rippo MR. Mitochondrial (Dys) function in inflammaging: do mitomirs influence the energetic, oxidative, and inflammatory status of senescent cells? [published correction appears in Mediators Inflamm. 2019 Aug 14;2019:8716351]. Mediators Inflamm 2017a;2309034. https://doi.org/10.1155/2017/2309034.

Giuliani A, Prattichizzo F, Micolucci L, Ceriello A, Procopio AD, Rippo MR. Mitochondrial (Dys) function in inflammaging: do mitomirs influence the energetic, oxidative, and inflammatory status of senescent cells? [published correction appears in Mediators Inflamm. 2019:8716351]. Mediators Inflamm 2017b;2309034. https://doi.org/10.1155/2017/2309034.

Gurău F, Baldoni S, Prattichizzo F, Espinosa E, Amenta F, Procopio AD, Albertini MC, Bonafè M, Olivieri F. Anti-senescence compounds: a potential nutraceutical approach to healthy aging. Ageing Res Rev 2018;46:14–31. https://doi.org/10.1016/j.arr.2018.05.001.

Han X, Chen H, Gong H, Tang X, Huang N, Xu W, Tai H, Zhang G, Zhao T, Gong C, Wang S, Yang Y, Xiao H. Autolysosomal degradation of cytosolic chromatin fragments antagonizes oxidative stress-induced senescence. J Biol Chem 2020;295:4451–63. https://doi.org/10.1074/jbc.RA119.010734).

Hayflick L, Moorhead PS. The serial cultivation of human diploid cell strains. Exp Cell Res 1961;25:585–621. https://doi.org/10.1016/0014-4827(61)90192-6.

Hernandez-Segura A, Nehme J, Demaria M. Hallmarks of cellular senescence, Trends Cell Biol 2018;28:436–53. https://doi.org/10.1016/j.tcb.2018.02.001.

Herranz N, Gil J. Mechanisms and functions of cellular senescence. J Clin Invest 2018;128:1238–46. https://doi.org/10.1172/JCI95148.

Hetz C, Zhang K, Kaufman RJ. Mechanisms, regulation and functions of the unfolded protein response. Nat Rev Mol Cell Biol 2020;21:421–38. https://doi.org/10.1038/s41580-020-0250-z.

Hickson LJ, Langhi Prata LGP, Bobart SA, Evans TK, Giorgadze N, Hashmi SK, Herrmann SM, Jensen MD, Jia Q, Jordan KL, Kellogg TA, Khosla S, Koerber DM, Lagnado AB, Lawson DK, LeBrasseur NK, Lerman LO, McDonald KM, McKenzie TJ, Passos JF, Pignolo RJ, Pirtskhalava T, Saadiq IM, Schaefer KK, Textor SC, Victorelli SG, Volkman TL, Xue A, Wentworth MA, Wissler Gerdes EO, Zhu Y, Tchkonia T, Kirkland JL. Senolytics decrease senescent cells in humans: preliminary report from a clinical trial of Dasatinib plus quercetin in individuals with diabetic kidney disease. EBioMedicine 2019;47:446–56. https://doi.org/10.1016/j.ebiom.2019.08.069.

Justice JN, Nambiar AM, Tchkonia T, LeBrasseur NK, Pascual R, Hashmi SK, Prata L, Masternak MM, Kritchevsky SB, Musi N, Kirkland JL. Senolytics in idiopathic pulmonary fibrosis: results from a first-in-human, open-label, pilot study. EBioMedicine 2019;40:554–63. https://doi.org/10.1016/j.ebiom.2018.12.052.5).

Kale A, Sharma A, Stolzing A, Desprez P-Y, Campisi J. Role of immune cells in the removal of deleterious senescent cells. Immun Ageing 2020;17:16. Published online 2020 Jun 3. https://doi.org/10.1186/s12979-020-00187-9.

Kim YM, Byun HO, Jee BA, Cho H, Seo YH, Kim YS, Park MH, Chung HY, Woo HG, Yoon G. Implications of time-series gene expression profiles of replicative senescence. Aging Cell 2013;17. https://doi.org/10.1111/acel.12087.

Kirkland JL, Tchkonia T. Clinical strategies and animal models for developing senolytic agents. Exp Gerontol 2015;68:19–25.

Kirkland JL, Tchkonia T. Senolytic drugs: from discovery to translation. J Intern Med 2020. https://doi.org/10.1111/joim.13141.

Kuilman T, Michaloglou C, Mooi WJ, Peeper DS. The essence of senescence. Genes Dev 2010;24:2463–79. https://doi.org/10.1101/gad.1971610.

López-Otín C, Blasco MA, Partridge L, Serrano M, Kroemer G. The hallmarks of aging. Cell 2013;153:1194–217. https://doi.org/10.1016/j.cell.2013.05.039.

Maas SLN, Breakefield XO, Weaver AM. Extracellular vesicles: unique intercellular delivery vehicles. Trends Cell Biol 2017;27:172–88. https://doi.org/10.1016/j.tcb.2016.11.003.

Mensà E, Guescini M, Giuliani A, Bacalini MG, Ramini D, Corleone G, Ferracin M, Fulgenzi G, Graciotti L, Prattichizzo F, Sorci L, Battistelli M, Monsurrò V, Bonfigli AR, Cardelli M, Recchioni R, Marcheselli F, Latini S, Maggio S, Fanelli M, Amatori S, Storci G, Ceriello A, Stocchi V, De Luca M, Magnani L, Rippo MR, Procopio AD, Sala C, Budimir I, Bassi C, Negrini M, Garagnani P, Franceschi C, Sabbatinelli J, Bonafè M, Olivieri F. Small extracellular vesicles deliver miR-21 and miR-217 as pro-senescence effectors to endothelial cells. J Extracell Vesicles 2020;9:1725285. https://doi.org/10.1080/20013078.2020.1725285. eCollection 2020.

Milanovic M, Fan DNY, Belenki D, Däbritz JHM, Zhao Z, Yu Y, Dörr JR, Dimitrova L, Lenze D, Monteiro Barbosa IA, Mendoza-Parra MA, Kanashova T, Metzner M, Pardon K, Reimann M, Trumpp A, Dörken B, Zuber J, Gronemeyer H, Hummel M, Dittmar G, Lee S, Schmitt CA. Senescence-associated reprogramming promotes cancer stemness. Nature 2018;553:96–100. https://doi.org/10.1038/nature25167.

Nacarelli T, Liu P, Zhang R. Epigenetic basis of cellular senescence and its implications in aging. Genes (Basel) 2017;8:343. https://doi.org/10.3390/genes8120343.

Ogrodnik M, Salmonowicz H, Gladyshev VN. Integrating cellular senescence with the concept of damage accumulation in aging: relevance for clearance of senescent cells. Aging Cell 2019;18(1):e12841. https://doi.org/10.1111/acel.12841. Published online 2018 Oct 22. Correction in: Aging Cell. 18(2019) e12942.

Olivieri F, Spazzafumo L, Santini G, Lazzarini R, Albertini MC, Rippo MR, Galeazzi R, Abbatecola AM, Marcheselli F, Monti D, Ostan R, Cevenini E, Antonicelli R, Franceschi C, Procopio AD. Age-related differences in the expression of circulating microRNAs: miR-21 as a new circulating marker of inflammaging. Mech Ageing Dev 2012;133:675–85.

Olivieri F, Lazzarini R, Recchioni R, et al. MiR-146a as marker of senescence-associated pro-inflammatory status in cells involved in vascular remodelling. Age (Dordr) 2013a;35:1157–72. https://doi.org/10.1007/s11357-012-9440-8.

Olivieri F, Lazzarini R, Babini L, et al. Anti-inflammatory effect of ubiquinol-10 on young and senescent endothelial cells via miR-146a modulation. Free Radic Biol Med 2013b;63:410–20. https://doi.org/10.1016/j.freeradbiomed.2013.05.033.

Olivieri F, Capri M, Bonafè M, et al. Circulating miRNAs and miRNA shuttles as biomarkers: perspective trajectories of healthy and unhealthy aging. Mech Ageing Dev 2017;165:162–70. https://doi.org/10.1016/j.mad.2016.12.004).

Ong SM, Hadadi E, Dang TM, et al. The pro-inflammatory phenotype of the human non-classical monocyte subset is attributed to senescence. Cell Death Dis 2018;9:266. https://doi.org/10.1038/s41419-018-0327-1.

Özcan S, Alessio N, Acar MB, Mert E, Omerli F, Peluso G, Galderisi U. Unbiased analysis of senescence associated secretory phenotype (SASP) to identify common components following different genotoxic stresses. Aging (Albany NY) 2016;8:1316–29. https://doi.org/10.18632/aging.100971.

Pazolli E, Alspach E, Milczarek A, Prior J, Piwnica-Worms D, Stewart SA. Chromatin remodeling underlies the senescence-associated secretory phenotype of tumor stromal fibroblasts that supports cancer progression. Cancer Res 2012;72:2251–61. https://doi.org/10.1158/0008-5472.CAN-11-3386.

Prattichizzo F, Giuliani A, Sabbatinelli J, Mensà E, De Nigris V, La Sala L, de Candia P, Olivieri F, Ceriello A. Extracellular vesicles circulating in young organisms promote healthy longevity. J Extracell Vesicles 2019;8:1656044. https://doi.org/10.1080/20013078.2019.1656044. eCollection 2019.

Rippo MR, Olivieri F, Monsurrò V, Prattichizzo F, Albertini MC, Procopio AD. MitomiRs in human inflamm-aging: a hypothesis involving miR-181a, miR-34a and miR-146a. Exp Gerontol 2014;56:154–63. https://doi.org/10.1016/j.exger.2014.03.002.

Rodier F. Detection of the senescence-associated secretory phenotype (SASP). Methods Mol Biol 2013;965:165–73. https://doi.org/10.1007/978-1-62703-239-1_10.

Sabbatinelli J, Prattichizzo F, Olivieri F, Procopio AD, Rippo MR, Giuliani A. Where metabolism meets senescence: focus on endothelial cells. Front Physiol 2019;10:1523. Published 2019 Dec 18. https://doi.org/10.3389/fphys.2019.01523.

Saleh T, Carpenter VJ, Tyutyunyk-Massey L, Murray G, Leverson JD, Souers AJ, Alotaibi MR, Faber A, Reed J, Harada H, Gewirtz DA. Clearance of therapy-induced senescent tumor cells by the senolytic ABT-263 via interference with BCL-XL -BAX interaction. Mol Oncol 2020. https://doi.org/10.1002/1878-0261.12761.

Salminen A, Kauppinen A, Kaarniranta K. Emerging role of NF-κB signaling in the induction of senescence-associated secretory phenotype (SASP). Cell Signal 2012;24:835–45. https://doi.org/10.1016/j.cellsig.2011.12.006.

Schafer MJ, Zhang X, Kumar A, et al. The senescence-associated secretome as an indicator of age and medical risk. JCI Insight 2020;5:133668. https://doi.org/10.1172/jci.insight.133668.

Sikora E. Rejuvenation of senescent cells-the road to postponing human aging and age-related disease? Exp Gerontol 2012. https://doi.org/10.1016/j.exger.2012.09.008. pii: S0531-5565(12)00275-6.

Takahashi A, Imai Y, Yamakoshi K, Kuninaka S, Ohtani N, Yoshimoto S, Hori S, Tachibana M, Anderton E, Takeuchi T, Shinkai Y, Peters G, Saya H, Hara E. DNA damage signaling triggers degradation of histone methyltransferases through APC/C(Cdh1) in senescent cells. Mol Cell 2012;45:123–31. https://doi.org/10.1016/j.molcel.2011.10.018.

Takahashi A, Loo TM, Okada R, et al. Downregulation of cytoplasmic DNases is implicated in cytoplasmic DNA accumulation and SASP in senescent cells. Nat Commun 2018;9:1249. Published 2018 Mar 28. https://doi.org/10.1038/s41467-018-03555-8.

Tchkonia T, Zhu Y, van Deursen J, Campisi J, Kirkland JL. Cellular senescence and the senescent secretory phenotype: therapeutic opportunities. J Clin Invest 2013;123:966–72. https://doi.org/10.1172/JCI64098.

Terlecki-Zaniewicz L, Lämmermann I, Latreille J, et al. Small extracellular vesicles and their miRNA cargo are anti-apoptotic members of the senescence-associated secretory phenotype. Aging (Albany NY) 2018;10:1103–32. https://doi.org/10.18632/aging.101452.

Tominaga K, Pereira-Smith OM. The role of chromatin reorganization in the process of cellular senescence. Curr Drug Targets 2012;13:1593–602.

Valadi H, Ekström K, Bossios A, Sjöstrand M, Lee JJ, Lötvall JO. Exosome-mediated transfer of mRNAs and microRNAs is a novel mechanism of genetic exchange between cells. Nat Cell Biol 2007;9:654–9.

Venturini W, Olate-Briones A, Valenzuela C, et al. Platelet activation is triggered by factors secreted by senescent endothelial HMEC-1 cells in vitro. Int J Mol Sci 2020;21:3287. https://doi.org/10.3390/ijms21093287.

Vickers KC, Remaley AT. Lipid-based carriers of microRNAs and intercellular communication. Curr Opin Lipidol 2012;23:91–7.

Vizioli MG, Liu T, Miller KN, et al. Mitochondria-to-nucleus retrograde signaling drives formation of cytoplasmic chromatin and inflammation in senescence. Genes Dev 2020;34:428–45. https://doi.org/10.1101/gad.331272.119.

Wallis R, Mizen H, Bishop CL. The bright and dark side of extracellular vesicles in the senescence-associated secretory phenotype. Mech Ageing Dev 2020;189:111263. https://doi.org/10.1016/j.mad.2020.111263.

Wang Y, Chen T, Yan H, Qi H, Deng C, Ye T, Zhou S, Li FR. Role of histone deacetylase inhibitors in the aging of human umbilical cord mesenchymal stem cells. J Cell Biochem 2013. https://doi.org/10.1002/jcb.24569.

Wang W, Wang L, Ruan L, Oh J, Dong X, Zhuge Q, Su DM. Extracellular vesicles extracted from young donor serum attenuate inflammaging via partially rejuvenating aged T-cell immunotolerance. FASEB J 2018;32:fj201800059R. https://doi.org/10.1096/fj.201800059R.

Zhang DY, Pan Y, Zhang C, Yan BX, Yu SS, Wu DL, Shi MM, Shi K, Cai XX, Zhou SS, Wang JB, Pan JP, Zhang LH. Wnt/β-catenin signaling induces the aging of mesenchymal stem cells through promoting the ROS production. Mol Cell Biochem 2013;374:13–20. https://doi.org/10.1007/s11010-012-1498-1.

CHAPTER 5

The role of inflammaging in the development of chronic diseases of older people

Jacek M. Witkowski[a], Ewa Bryl[b], and Tamas Fulop[c]

[a]Department of Pathophysiology, Medical University of Gdansk, Gdansk, Poland
[b]Department of Pathology and Experimental Rheumatology, Medical University of Gdansk, Gdansk, Poland
[c]Department of medicine, Research Center on Aging, Graduate Program in Immunology, Faculty of Medicine and Health Sciences, University of Sherbrooke, Sherbrooke, QC, Canada

5.1 Introduction

With passing time and accumulating years of age, human beings assist to the statistically inevitable increase in certain (mostly chronic inflammatory) diseases, including, metabolic, especially type 2 diabetes mellitus (T2DM), respiratory diseases, frailty syndrome, coronary vascular disease, and myocardial infarction, and stroke, as well as the neurodegenerative diseases (mainly Alzheimer's disease and Parkinson's disease). With the conditional exception of neurodegenerative diseases mentioned earlier (which will be elaborated upon later), these diseases can be observed also in the middle-aged, young adults and under certain conditions also in children. However, highest frequency of clinically new cases (incidence) of most of them can be observed among people older than 60 years of age. This incidence is rising in the even older age groups, with the characteristic exception of malignancies. On the other hand, prevalence of all of these diseases increases quasi-linearly with age. Thus, together, these diseases are called the aging-related diseases or ARDs.

5.2 Basic mechanisms: Cellular senescence, inflammaging, molecular inflammation, and senoinflammation

5.2.1 Cellular senescence

Cellular senescence is a state occurring to the cells with their (and organismal) advancing age, on condition that they do not die, but rather clog the organs impairing their functions. It is characterized by cessation of cellular divisions associated usually with extremely shortened telomeres, enlarged size compared to "young," functional cells of the same type, at least partial loss of typical functions of the cell type in question, and the acquisition of relatively high activity of the lysosomal enzyme called the senescence-associated

Human Aging
https://doi.org/10.1016/B978-0-12-822569-1.00014-7

beta–galactosidase (SA–β–gal). Apparently the activity of this enzyme, which is detectable at pH 6, while in the young cells the same enzyme works only around pH 4.5, is not a part of mechanism of cellular senescence but rather its marker (Lee et al., 2006). However, what is much more important for the message of this chapter is the senescence-associated secretory phenotype (SASP) which is displayed by senescent cells universally, i.e., independently from their tissue or organ of origin. This phenotype involves the ability to secrete significant amounts of proinflammatory cytokines, including Tumor necrosis factor-α, interleukin (IL)-1 and IL-6, as well as growth factors, chemokines, and matrix metalloproteinases (MMPs) among other mediators. Thus cellular senescence may be one of the important factors causing the chronic state underlying the chronic inflammatory diseases of older people called the inflammaging (see also Chapters 3 and 4).

5.2.2 Inflammaging

Inflammaging proposed here as the underlying mechanism of the ARDs, was first defined by Franceschi two decades ago and the concept was extensively developed since, as exemplified in a few related papers (Cevenini et al., 2013; Franceschi et al., 2000, 2007, 2018, 2017a,b; Franceschi and Campisi, 2014). Of course even earlier there were numerous observations of not only that aging is associated with chronic inflammatory diseases, but also that even in apparently healthy older people the levels of some proinflammatory cytokines (notably IL-6) may be elevated (Mysliwska et al., 1998). Inflammaging is thus defined as aging-associated (pro)inflammatory state, manifesting itself not as any overt, chronic inflammatory disease, but as paucisymptomatic state where the immune and other cells of the aged individuals secrete excessive amount of proinflammatory cytokines, including IL-1β, IL-2, IL-6, IL-8, IL-18, Chemokine (C—C motif) ligand 5 (CCL5), Monocyte chemoattractant protein 1 (MCP-1) and other, effectively increasing their extracellular concentrations throughout the aging body (Fulop et al., 2016b, 2019, 2018b; Mysliwska et al., 1997, 1998, 1999). These cytokines exert their activity not only on the immune cells of an older person, but also on many extra-immune tissues and organs, including the intestine, liver, heart, and blood vessels as well as peripheral and central nervous system, which they "prime" for chronic deterioration and disease in case of further challenges. After acknowledging that inflammaging is an increased level of proinflammatory cytokines in the body fluids of at least a proportion of older people, an obvious question is where they come from and why. It is clear that most of these cytokines are secreted by activated immune/inflammatory cells; however, activation of these cells requires a challenge in the form of pathogen- or danger-associated molecular patterns (PAMPs, DAMPs), as well as antigens, i.e., requires prior contact with external or internal pathogen. Reaction to external PAMPs would (at any age) be an "ordinary," highly desirable inflammatory/immune reaction to the pathogen, aimed at its neutralization and usually symptomatic, at least at first contacts. However, it is possible that reactivation of certain "stubborn" pathogens, like the cytomegalovirus or other herpes-viruses to mention just a couple, may still trigger the system but remain asymptomatic, becoming a source of the "cytokines of inflammaging." We can assume, that such challenges occur through life and some of these increase in frequency with aging (e.g., DAMPs and autoantigens released from rising numbers of damaged and dying cells) (Tang et al., 2011). However, this is only a partial explanation. The other one is of course the cellular senescence described before. Increased numbers of senescent cells mean increased effect of their cumulated SASP phenotype. Cell senescence does not leave the cell of the immune system behind, rendering some of them truly senescent and exhibiting the SASP, but in part also contributing to the state of immunosenescence. Inflammaging cannot be fully comprehended without the concept of immunosenescence. Immunosenescence itself can be defined as the state of the immune system of an older individual, where functionalities of cells of adaptive immunity T and B lymphocytes are decreased, resulting (in very simple words) in their decreased proliferation, cytokine (notably IL-2), and antibody secretion in response to antigenic or cytokine stimulation (Fulop et al., 2016b, 2017a, 2014b; Witkowski et al., 2019). These functional modifications are in part a consequence of known shifts in the proportions of different subpopulations of mainly T cells, including rising predominance of memory T cells over naïve ones, as well as accumulation of regulatory T cells (Bryl and Witkowski, 2004; Fulop et al., 2014b; Gruver et al., 2007; Hong et al., 2004; Larbi et al., 2008; Pawelec, 2006; Trzonkowski et al., 2006). At the subcellular level, many characteristic changes occur in multiple signal transduction pathways leading to or characterizing immunosenescent cells; their detailed description is beyond the scope of this chapter (Fulop et al., 2014a, 2017b; Le Page et al., 2018) (see Chapter 7). One cannot miss yet another factor, which significantly contributes to inflammaging, i.e., the changing composition and proportion of various component microorganisms of our (mainly gut) microbiome leading to its deleterious state called dysbiosis (Biragyn and Ferrucci, 2018; Jones et al., 2019; Kim and Jazwinski, 2018; Laing et al., 2018; Maffei et al., 2017; Riaz Rajoka et al., 2018; Ticinesi et al., 2019). Dysbiosis leads to the shifts in the proportions and functions of gut mucosa-resident immune cells, notably leading to decreased proportions and function of Tregs and to increased proportions of proinflammatory Th1 and Th17 cells. The role of gut microbiome modifications in the older people (stemming from changing diet, existing comorbidities, and associated polypragmasia) and its association with inflammaging was recently convincingly summarized in a systematic review of available literature on the topic (Shintouo et al., 2020) (see Chapter 6). One final part of the puzzle leading to inflammaging is the compensatory increase of innate immunity in the older subjects. It has its benefits; when adaptive immunity fails due to immunosenescence, an aging organism is left to cope with pathogens using the less specific, but otherwise quite effective ways of eliminating them, i.e., the innate immunity. Indeed, this predominance of innate immune cell functions in the geriatric population can be regarded as an adaptation of aging immune system, eventually, under certain circumstances, promoting longevity (Fulop et al., 2016a, 2017a, 2020a; Witkowski et al., 2019). However, benefits of this process may be overcome by its deleterious counterparts; namely, excessive activity of innate immune cells overlaid on the actual immunobiography of an individual (the totality of challenges to the immune system, its reactions, and associated modifications over life) and current comorbidities promote the development and progression of ARDs (Franceschi et al., 2017b; Fulop et al., 2016b, 2017a; Witkowski et al., 2019). Recently two concepts appeared, attempting at broadening our understanding of aging-associated increased inflammatory state beyond the concepts of inflammaging and immunosenescence elaborated before: molecular inflammation and senoinflammation.

5.2.3 Molecular inflammation

The proponents of the molecular inflammation concept (oxy-inflammaging) focus on the redox-sensitive transcription factors, NF-κB and FOXO, as essential players in the expression of proinflammatory mediators and antioxidant enzymes, respectively. Their tenet is that accumulation of Reactive Oxygen Species (ROS, i.e., oxidative stress) associated with advancing age activates NF-κB. In the quiescent state, NF-κB is sequestered in the cytoplasm by inhibitor of NF-κB (IκB). Upon activation, IκB is phosphorylated by the IκB kinase (IKK) complex. Phosphorylation of IκB leads to its degradation, and subsequently, nuclear transport of NF-κB proteins initiates the downstream transcription of target genes. However, the p38 MAPK is a family of serine/threonine kinases that form an integral component of proinflammatory signaling cascades in various cell types. Blocking p38 with specific inhibitors diminishes NF-κB-driven transcriptional activity and attenuates the expression of NF-κB target genes. Transcription controlled by NF-κB concerns proinflammatory cytokines, adhesion molecules, inducible nitric oxide synthase, lipoxygenase, and cyclooxygenase-2 and in consequence induces inflammation. NFκB-induced products are then responsible for amplifying the release of ROS which closes the vicious circle, eventually leading to cellular aging and ARDs (Chung et al., 2002, 2011, 2006). A bit surprisingly, increased apoptosis is also mentioned in this concept as a proinflammatory factor (Chung et al., 2019); the authors do not elaborate, but the only way it could be is that apoptotic bodies shed from apoptotically dying cells might be endocytosed and eventually stimulate the innate immune cells. Albeit interesting and originally proposed almost two decades ago, this concept seems too limited to explain everything about the inflammaging, chronic inflammatory ARDs and aging, not taking into consideration other facets of the problem discussed here.

5.2.4 Senoinflammation

On the other hand the concept of senoinflammation, although directly stemming from the molecular inflammation concept briefly characterized before, seems to go beyond its limitations and treating it as mechanistic explanation only. Senoinflammation encompasses three stages: the "core" which in fact is pure mechanism of molecular inflammation, i.e., stimulation of NF-kB by ROS and stimulation of relevant genes transcription by the NF-kB; second, cellular stage where "core stage" products—induced activation of innate and adaptive immune cells occurs, leading to release of inflammatory factors, and where senescent cells accumulate; and finally the third stage corresponding to systemic, disease-oriented level, where the authors of this concept include everything from the roles of DAMPs, Toll-like Receptor (TLRs), activation of inflammasomes and SASP, activities of proinflammatory miRNAs, to macrophage M1/M2 polarization, augmented autophagy and their overall impact on age-related pathologies, including namely inflammation of adipose tissue, endoplasmic reticulum (ER) stress, insulin resistance, and lipid accumulation (Chung et al., 2019; Kim et al., 2020). Thus the concept of senoinflammation attempts to be more inclusive than that of inflammaging. Consequently, senoinflammation is not only an aging-associated state of increased proinflammatory readiness (as is the case for strictly defined inflammaging), but also profound effects of this increased inflammatory reactivity in the older people on their metabolism, including insulin resistance, lipid accumulation (replacement of muscle by fat), stemming from the ER stress, dysregulation of autophagy and redox state and multiple altered intracellular signaling pathways, of which the NF-κB pathway seems to be the most essential. Recently, several studies have documented that (excessive) NF-κB signaling increases the expression of proinflammatory cytokine, chemokine, and adhesion molecule genes and consequently production of IL-1β, IL-2, IL-6, IL-8, CCL5, MCP-1, and other observed during aging, thus promoting the inflammaging (Chung et al., 2019, 2011; Franceschi, 2007; Franceschi et al., 2007). Still, in a sense, as an explanation of pathomechanisms of the ARDs senoinflammation is very similar if not identical with inflammaging.

5.3 Is chronic inflammatory state a common denominator of ARDs?

When a group of apparently unrelated diseases is put together under a common banner (in this case ARDs) it is only natural to seek some underlying conditions that decided about their increasing incidence with increasing age above 60 years of age, i.e., in the older part of human population. The most obvious one is the factor of time, the age itself. Long (and not so long) ago it was obvious for everybody, including medical professionals of that times, that when somebody reaches a ripe age of 50, 60, or 70 years that person - *must* show signs and symptoms of one or more diseases that we currently call ARDs, simply because this person has so many years. Yet, even at ancient times there were individuals credibly recorded as extremely healthy and long living, e.g., Terentia, wife of Marcus Tullius Cicero reported as his great (if not greatest) supporter and help, born in 98 BCE and deceased 6 CE (Pliny the Elder, Naturalis Historia, lib. Vii, 158). Even if the dates of her birth and death are off by a year or two, she still had to be a centenarian; so, her body had to have enough stamina to survive that long. Centenarians, although generally not free from chronic diseases, may still be considered as presenting a healthy aging and biologically fit for long life (Bonafe et al., 2002; Candore et al., 2006; Capri et al., 2008; Caruso et al., 2004; Chevanne et al., 2007; Franceschi, 2007; Franceschi and Bonafe, 2003; Franceschi et al., 2000, 1999, 1995; Garagnani et al., 2013; Lio et al., 2004; Marini et al., 2004; Monti et al., 2000; Nasi et al., 2006; Salvioli et al., 2009, 2013; Sgarbi et al., 2014). These reports that even so long ago, when average expected lifespan at birth was between 20 and 30 years, there were "occasional" cases of centenarians has obtained a statistical support from John R. Wilmoth who calculated that the emergence of centenarians occurred once world population rose to about 100 million, around 4500 years ago (Wilmoth, 2000). What all the earlier mentioned teaches us? Namely that age (number of years survived), although important, is not the only factor deciding about further survival and the occurrence of diseases, especially those dubbed the ARDs. So, we need other elements for explanation of the ARDs. One such component will be the extended chronic inflammatory characteristics and background of all of the ARDs defined before as inflammaging/senoinflammation (Chung et al., 2019). Is it so? Let us consider each of the ARDs mentioned before and ask for their inflammatory roots and mechanisms.

5.3.1 T2DM

The T2DM is notorious for slow, frequently undetected, rise in blood sugar concentration, due to decreased insulin sensitivity of muscle (skeletal and cardiac) and adipose tissue. This increased level of free glucose in plasma leads to the augmentation of the process of nonenzymatic glycosylation or glycation of proteins and other molecules. As a consequence, apart from malfunction of glycated molecules, a category of substances called the Advanced Glycation Endproducts (AGEs) is being made in amounts higher than physiologically. These substances, e.g., derivatives of glyceraldehyde called the toxic AGEs (TAGEs) and other, including carboxymethyl lysine, carboxyethyl lysine, pentosidine, glucosepane, methylglyoxal lysine dimer, glyoxal lysine dimer, and glycolic acid lysine amide (Henning and Glomb, 2016) may exert adverse metabolic effects on cells (and, at least in case of TAGEs, be cytotoxic) and so are the target for removal by macrophages, due to their binding with devoted pattern-recognition receptors for AGEs (RAGEs) (Takeuchi, 2016). RAGE-dependent intracellular signaling in monocytes/macrophages leads to their activation and proinflammatory events, including secretion of different proinflammatory cytokines and ROS and fueling the inflammatory reactions. These eventually affect the arteries of a patient, facilitating the development of atherosclerosis and its consequences (Senatus and Schmidt, 2017). Also other DAMPs, including the islet amyloid polypeptide, palmitate, ceramide, and endocannabinoids, are released in the diabetic environment, trigger the inflammasome, and released proinflammatory cytokines affect the liver, muscle, adipose tissue, and pancreas, being conductive to the symptoms and complications of T2DM (Shin et al., 2015). Increased amounts of

AGEs directly affect also other cell types, e.g., leading to vascular calcification typical for aged arteries (Kay et al., 2016). Deposits of calcium phosphates stimulate the immune/inflammatory cells which amplifies inflammation. Another facet of relations between T2DM and inflammation is the accumulation of senescent cells with strongly proinflammatory SASP phenotype. As described before, these features fit both under the heading of inflammaging as well as in the apparently broader (but very similar), recently proposed term of senoinflammation (Kim et al., 2020).

5.3.2 Chronic aging-related respiratory diseases

They include the chronic obstructive pulmonary disease (COPD) and idiopathic pulmonary fibrosis (IPF). COPD was meta-analytically proven to have a systemic inflammatory component already at the beginning of this century (Gan et al., 2004). The mechanism of the disease involves damage to the alveolar septa due to toxic and proinflammatory effects of cigarette smoking, overlaying chronic infection of lung parenchyma, imbalance between serum proteolytic enzymes and their inhibitors (notably α1-antitrypsin) and asthma. Both cellular damage, via the release of DAMPS, and infection (in turn bringing PAMPs into the picture) are obviously proinflammatory, as both PAMPS and DAMPs will be recognized by pattern recognition receptors of the TLR and NOD-like receptor families on monocyte/macrophages and other immune-inflammatory cells. IPF is a chronic, fibrosing lung disease, as the name calls still a bit mysterious, with poorly known etiopathogenesis and poor prognosis (Pardo and Selman, 2020). However, its fibrosing mechanism (which is undisputable as the definition of this disease) must involve lung fibroblast and myofibroblast hyperactivity. The latter is usually a consequence of an acute or chronic inflammation and postpneumonia fibrosis of lung parenchyma is frequent, but no etiological proinflammatory factor is so far known for IPF. Interestingly, a very recent (published online on September 29, 2020) paper by Yao et al. proves that the source of inflammatory cytokines and other factors stimulating lung fibroblasts is driven by the senescence of type 2 alveolar cells (Yao et al., 2020).

5.3.3 Atherosclerosis

Atherosclerosis, a necessary background condition for the development of coronary vascular disease and myocardial infarction, stroke, as well as the ischemic neurodegeneration, was long ago defined as a chronic proliferative/destructive inflammation of the internal (t. intima) and partially middle (t. media) layers of large arteries. Based on what was already written here, it may be a direct effect of inflammaging (whatever its cause as discussed before), as endothelia get activated by proinflammatory cytokines, and respond to them by presenting on their surface more adhesion molecules (integrins and selectins) which facilitate binding and trapping passing-by leukocytes. Increased amounts of proinflammatory cytokines may also be the effect of chronic infections, autoimmune diseases, and microbiome dysbiosis. Release of some bonds holding the endothelial layer together, with concomitant contraction of endothelial cells leaves open spaces between them, allowing for leukocyte diapedesis (migration to subendothelial environment). This process leads to infiltration by neutrophils, monocyte/macrophages, and later by lymphocytes and is quite beneficial if it happens in the capillaries, e.g., near infection or tissue damage, but deleterious in the limited space inside the internal layers of arteries. Both its proliferative part (executed mainly by myofibroblasts and fibroblasts and leading to the formation of stable atherosclerotic plaque), and destructive part where macrophages (both active M1 cells and in a form of cholesterol- and lipid-laden foam cells) together with cytotoxic lymphocytes release enzymes (MMPs, granzymes) which damage both the nearby cells and extracellular matrix of the plaque, eventually leading to its disruption and initiating blood coagulation, are the final causes of chronic (angina) and acute ischemia (MI, stroke). Formation of foam cells is facilitated also by increased endocytosis of droplets of cholesterol- and lipid-rich plasma by cytokine-activated endothelia; cholesterol crystals are a known stimulant activating the inflammasome (Baragetti et al., 2020). Thus atherosclerosis and its consequences are clearly the result of proinflammatory conditions in the organism, including the inflammaging (Liu et al., 2020).

5.3.4 Frailty

The frailty syndrome seems to be an opposition to successful aging (Rolfson, 2018). We and others have over the last decade demonstrated and discussed that frailty is in part a consequence of (sub) inflammatory processes occurring in an aging body (Fulop et al., 2010, 2015) (see Chapter 10).

5.3.5 Alzheimer's disease (AD)

Obvious for T2DM, chronic respiratory diseases, frailty, and atherosclerosis and its complications, as described earlier, it seems not so obvious for neurodegenerative diseases. For AD, the amyloid cascade hypothesis of its pathogenesis still holds strong, putting as the perpetrator of neuronal death the accumulating aggregates of beta-amyloid. Similarly, aggregating α-synuclein is believed to be the cause of degeneration of dopaminergic neurons in substantia nigra and other regions of the brain. However, more recent and already quite numerous studies indicate that both these diseases are in fact at least in part neuroinflammatory. Regarding the AD, a nagging question is what evolutionary pressure led to the creation of enzymes β- and γ-secretases which cleave the amyloid precursor protein in such a way that a dangerous, aggregating, microglia-stimulating β-amyloid is made, eventually leading to the formation of senile plaques and cognitive collapse? Now we know at least a partial response; β-amyloid is an antimicrobial peptide protecting the CNS from spread of infection by neurotropic viruses, bacteria and fungi (Bourgade et al., 2016a, 2015, 2016b; Fulop et al., 2018a; Itzhaki et al., 2016; Perrotte et al., 2020).

Unfortunately, the same pathogen-binding property of β-amyloid aggregates makes it a potent stimulant for microglia and possibly infiltrating immune/inflammatory cells from the periphery, initiating the state of neuroinflammation (Fulop et al., 2020b; Jozwik et al., 2012). Obviously, the more such β-amyloid-inducing responses to (e.g., reactivating) neurotropic pathogen, the more amyloid aggregates will appear in the brain. However, this process may also be amplified in the case of peripheral increase of proinflammatory cytokine levels. These cytokines, by acting first on the endothelia component, and later in concert with similar proinflammatory cytokines from activated microglia also on the astrocyte component of the blood–brain barrier (BBB), induce its breach and increase permeability. This in turn facilitates the movement of proinflammatory cells and factors into the brain, amplifying neurodegeneration and progress toward the AD (Fulop et al., 2020b). Thus clearly the AD can be considered to be fueled and in part made possible by peripheral state of inflammation/inflammaging.

5.3.6 Parkinson's disease (PD)

Considering PD, increasing amount of accumulated data convincingly shows that here also inflammaging plays an important role for the disease pathomechanism (Boyko et al., 2017; Calabrese et al., 2018a,b; Meszaros et al., 2020). Whatever cause of accumulation of cytokines in the periphery, they will have the same effect on the BBB as in the case of AD. However, the excessive oxidative stress in the substantia nigra associated with the production of neuromelanin will damage dopaminergic neurons (initially the local, but with progression of the disease also in more remote parts of the brain) leading to locally initiated neuroinflammation and characteristic symptoms of the disease. Again, as in case of AD, this neuroinflammation will lead to BBB breach from the cerebral side, closing the vicious circle (Boyko et al., 2017; Calabrese et al., 2018a,b; Meszaros et al., 2020).

5.4 The case of COVID-19

Summarizing, the earlier mentioned convincingly shows that all of the ARDs discussed before are at least in part a result of chronic underlying low-grade proinflammatory condition. We can speculate that the same may be true also for other ARDs. A special case is here the recent (less than a year old at the time of writing) pandemics of new infective disease COVID-19, following infection with a β-coronavirus SARS-CoV-2. Already after the outbreak of the disease in China it was recognized that the disease is more severe and deadly in the aged part of the population. This observation was eventually confirmed throughout other countries with the spread of the pandemics (Bonafe et al., 2020; Vellas et al., 2020). Studies from last year (2020) confirm that the background for this higher incidence of symptomatic COVID-19 and its increased severity may be in fact inflammaging and immunosenescence which together facilitate a cytokine storm which damages not only lungs, but also blood vessels, heart, liver, and kidney of the older patient, especially already having comorbidities (Blagosklonny, 2020; Cunha et al., 2020; Domingues et al., 2020; Kuo et al., 2020; Marcon et al., 2020; Meftahi et al., 2020; Pence, 2020; Robinson and Pierce, 2020).

5.5 Proposed interventions to prevent ARDs

Another similarity is that the pathomechanisms responsible for all of the ARDs likely start many years and sometimes decades prior to their overt, symptomatic, clinical manifestation. This suggests that proinflammatory environment, as described before conductive to all the ARDs discussed here, slowly develops over time, with multiple and/or recurrent pathogen and other challenges to the immune/inflammatory cells and slow accumulation of senescent cells. One could say that youth leads inevitably to aging, and, perhaps avoidably, to the ARDs; for both the underlying processes (inflammaging, cellular senescence, immunosenescence, etc.) are similar, but likely occur at different rates (Franceschi et al., 2018). This similarity should also be considered when contemplating new approaches to prevention of ARDs. For instance, recently proposed "allo-priming," which aims at boosting T Helper Cell Type 1/Cytotoxic T Lymphocyte responses in the older COVID-19 patients and likely retain this effect for future new viral infections, should be considered very carefully, as it may likely accelerate inflammaging in some patients (Har-Noy and Or, 2020). As discussed before, the ARDs are characterized by common background in which inflammaging/chronic inflammation plays the major causative role. Therefore, in our opinion, also the preventive measures should be early and complex, resulting in the lowering of proinflammatory (inflammaging) state in the older and so in complex protection from the ARDs. In our opinion, this goal can be achieved by: (I) Early poly-vaccination against pathogens known to recur or reinfect leading to inflammatory response; such pathogens should be identified; (II) if an (especially chronic or recurrent) infection is already present, it should be treated with the use of antiviral drugs, antibiotics, antimycotic drugs and when relevant, phage therapy to decrease the pathogen load and the amount of inflammatory response; (III) reasonable and personalized use of antiinflammatory treatment even in healthy older exhibiting elevated levels of proinflammatory cytokines; (IV) preventing and curing microbiome dysbiosis (if detected, which asks for cheap, easy tests which could be performed population-wide) with the use of pre- and probiotics and if needed, healthy microbiome transplantation or supplementation; and (V) preventing accumulation of senescent cells and disrupting these already present with the use of senolytics (see Chapter 4). We have recently proposed this strategy for prevention of AD (Fulop et al., 2020b). However, for obvious reason of abovementioned similarities in the inflammatory background and pathomechanisms of all ARDs, we believe it can be extended as a potential prevention also of other ARDs Fig. 5.1. However, the most efficient interventions would seem to be the multimodal lifelong lifestyle interventions, including nutrition (e.g., Mediterranean diet), healthy psychological

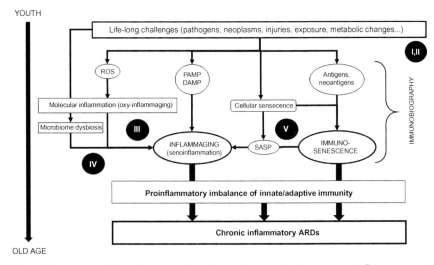

Fig. 5.1 Simplified chart of time- (aging-) dependent changes in the immune/inflammatory status of an individual leading to the development of ARDs. I, II, HI, IV, V—different preventive/therapeutic approaches as delineated in the text. SASP, senescence-associated secretory phenotype; ROS, reactive oxygen species; PAMP, pathogen-associated molecular patterns; DAMP, danger-associated molecular patterns.

approach (e.g., mitigate life stress), exercise (e.g., walking 30 min each day), and satisfactory social interactions (e.g., family life). These should start already in utero (by proxy) and last until the end of life.

5.6 Conclusion and future perspective

It is a common sense that the incidence and prevalence of ARDs increase in older adults but not all physiological and pathophysiological changes occurring with age could be rendered responsible for that increase. Most of them are the consequences of the lifelong and constant challenges that life imposes on an organism and against which they should fight. This fight of course may result in deleterious consequences leading with time to the emergence of the clinically diagnosed ARD. This conceptualization of ARD leads to a different approach to these ARDs, implying a very early prevention strategy in life. Blaming aging per se for all these ARD reduces the efficiency of our fight against these ARD. Recognizing the lifelong nature of their development as mentioned before will help to implement meaningful strategies for a longer health- and function-span for older adults.

References

Baragetti A, Catapano AL, Magni P. Multifactorial activation of NLRP3 inflammasome: relevance for a precision approach to atherosclerotic cardiovascular risk and disease. Int J Mol Sci 2020;21:4459.

Biragyn A, Ferrucci L. Gut dysbiosis: a potential link between increased cancer risk in ageing and inflammaging. Lancet Oncol 2018;19:e295–304.

Blagosklonny MV. From causes of aging to death from COVID-19. Aging (Albany NY) 2020;12:10004–21.

Bonafe M, Barbi C, Storci G, Salvioli S, Capri M, Olivieri F, Valensin S, Monti D, Gonos ES, De Benedictis G, Franceschi C. What studies on human longevity tell us about the risk for cancer in the oldest old: data and hypotheses on the genetics and immunology of centenarians. Exp Gerontol 2002;37:1263–71.

Bonafe M, Prattichizzo F, Giuliani A, Storci G, Sabbatinelli J, Olivieri F. Inflamm-aging: why older men are the most susceptible to SARS-CoV-2 complicated outcomes. Cytokine Growth Factor Rev 2020;53:33–7.

Bourgade K, Garneau H, Giroux G, Le Page AY, Bocti C, Dupuis G, Frost EH, Fulop Jr. T. Beta-amyloid peptides display protective activity against the human Alzheimer's disease-associated herpes simplex virus-1. Biogerontology 2015;16:85–98.

Bourgade K, Dupuis G, Frost EH, Fulop T. Anti-viral properties of amyloid-beta peptides. J Alzheimers Dis 2016a;54:859–78.

Bourgade K, Le Page A, Bocti C, Witkowski JM, Dupuis G, Frost EH, Fulop Jr. T. Protective effect of amyloid-beta peptides against herpes simplex virus-1 infection in a neuronal cell culture model. J Alzheimers Dis 2016b;50:1227–41.

Boyko AA, Troyanova NI, Kovalenko EI, Sapozhnikov AM. Similarity and differences in inflammation-related characteristics of the peripheral immune system of patients with Parkinson's and Alzheimer's diseases. Int J Mol Sci 2017;18:2633.

Bryl E, Witkowski JM. Decreased proliferative capability of CD4(+) cells of elderly people is associated with faster loss of activation-related antigens and accumulation of regulatory T cells. Exp Gerontol 2004;39:587–95.

Calabrese V, Santoro A, Monti D, Crupi R, Di Paola R, Latteri S, Cuzzocrea S, Zappia M, Giordano J, Calabrese EJ, Franceschi C. Aging and Parkinson's disease: inflammaging, neuroinflammation and biological remodeling as key factors in pathogenesis. Free Radic Biol Med 2018a;115:80–91.

Calabrese V, Santoro A, Trovato Salinaro A, Modafferi S, Scuto M, Albouchi F, Monti D, Giordano J, Zappia M, Franceschi C, Calabrese EJ. Hormetic approaches to the treatment of Parkinson's disease: perspectives and possibilities. J Neurosci Res 2018b;96:1641–62.

Candore G, Balistreri CR, Listi F, Grimaldi MP, Vasto S, Colonna-Romano G, Franceschi C, Lio D, Caselli G, Caruso C. Immunogenetics, gender, and longevity. Ann N Y Acad Sci 2006;1089 (516–37):516–37.

Capri M, Salvioli S, Monti D, Caruso C, Candore G, Vasto S, Olivieri F, Marchegiani F, Sansoni P, Baggio G, Mari D, Passarino G, De Benedictis G, Franceschi C. Human longevity within an evolutionary perspective: the peculiar paradigm of a post-reproductive genetics. Exp Gerontol 2008;43:53–60.

Caruso C, Lio D, Cavallone L, Franceschi C. Aging, longevity, inflammation, and cancer. Ann N Y Acad Sci 2004;1028(1–13):1–13.

Cevenini E, Monti D, Franceschi C. Inflamm-ageing. Curr Opin Clin Nutr Metab Care 2013;16:14–20.

Chevanne M, Calia C, Zampieri M, Cecchinelli B, Caldini R, Monti D, Bucci L, Franceschi C, Caiafa P. Oxidative DNA damage repair and parp 1 and parp 2 expression in Epstein-Barr virus-immortalized B lymphocyte cells from young subjects, old subjects, and centenarians. Rejuvenation Res 2007;10:191–204.

Chung HY, Kim HJ, Kim KW, Choi JS, Yu BP. Molecular inflammation hypothesis of aging based on the anti-aging mechanism of calorie restriction. Microsc Res Tech 2002;59:264–72.

Chung HY, Sung B, Jung KJ, Zou Y, Yu BP. The molecular inflammatory process in aging. Antioxid Redox Signal 2006;8:572–81.

Chung HY, Lee EK, Choi YJ, Kim JM, Kim DH, Zou Y, Kim CH, Lee J, Kim HS, Kim ND, Jung JH, Yu BP. Molecular inflammation as an underlying mechanism of the aging process and age-related diseases. J Dent Res 2011;90:830–40.

Chung HY, Kim DH, Lee EK, Chung KW, Chung S, Lee B, Seo AY, Chung JH, Jung YS, Im E, Lee J, Kim ND, Choi YJ, Im DS, Yu BP. Redefining chronic inflammation in aging and age-related diseases: proposal of the senoinflammation concept. Aging Dis 2019;10:367–82.

Cunha LL, Perazzio SF, Azzi J, Cravedi P, Riella LV. Remodeling of the immune response with aging: immunosenescence and its potential impact on COVID-19 immune response. Front Immunol 2020;11:1748.

Domingues R, Lippi A, Setz C, Outeiro TF, Krisko A. SARS-CoV-2, immunosenescence and inflammaging: partners in the COVID-19 crime. Aging (Albany NY) 2020;12:18778–89.

Franceschi C. Inflammaging as a major characteristic of old people: can it be prevented or cured? Nutr Rev 2007;65:S173–6.

Franceschi C, Bonafe M. Centenarians as a model for healthy aging. Biochem Soc Trans 2003;31:457–61.

Franceschi C, Campisi J. Chronic inflammation (inflammaging) and its potential contribution to age-associated diseases. J Gerontol A Biol Sci Med Sci 2014;69(Suppl 1):S4–9.

Franceschi C, Monti D, Sansoni P, Cossarizza A. The immunology of exceptional individuals: the lesson of centenarians. Immunol Today 1995;16:12–6.

Franceschi C, Mondello C, Bonafe M, Valensin S, Sansoni P, Sorbi S. Long telomeres and well preserved proliferative vigor in cells from centenarians: a contribution to longevity? Aging (Milano) 1999;11:69–72.

Franceschi C, Bonafe M, Valensin S, Olivieri F, De Luca M, Ottaviani E, De Benedictis G. Inflamm-aging. An evolutionary perspective on immunosenescence. Ann N Y Acad Sci 2000;908(244–54):244–54.

Franceschi C, Capri M, Monti D, Giunta S, Olivieri F, Sevini F, Panourgia MP, Invidia L, Celani L, Scurti M, Cevenini E, Castellani GC, Salvioli S. Inflammaging and anti-inflammaging: a systemic perspective on aging and longevity emerged from studies in humans. Mech Ageing Dev 2007;128:92–105.

Franceschi C, Garagnani P, Vitale G, Capri M, Salvioli S. Inflammaging and 'Garb-aging'. Trends Endocrinol Metab 2017a;28:199–212.

Franceschi C, Salvioli S, Garagnani P, de Eguileor M, Monti D, Capri M. Immunobiography and the heterogeneity of immune responses in the elderly: a focus on inflammaging and trained immunity. Front Immunol 2017b;8:982.

Franceschi C, Garagnani P, Morsiani C, Conte M, Santoro A, Grignolio A, Monti D, Capri M, Salvioli S. The continuum of aging and age-related diseases: common mechanisms but different rates. Front Med (Lausanne) 2018;5:61.

Fulop T, Larbi A, Witkowski JM, McElhaney J, Loeb M, Mitnitski A, Pawelec G. Aging, frailty and age-related diseases. Biogerontology 2010;11:547–63.

Fulop T, Le Page A, Fortin C, Witkowski JM, Dupuis G, Larbi A. Cellular signaling in the aging immune system. Curr Opin Immunol 2014a;29:105–11.

Fulop T, Witkowski JM, Pawelec G, Alan C, Larbi A. On the immunological theory of aging. Interdiscip Top Gerontol 2014b;39:163–76.

Fulop T, McElhaney J, Pawelec G, Cohen AA, Morais JA, Dupuis G, Baehl S, Camous X, Witkowski JM, Larbi A. Frailty, inflammation and immunosenescence. Interdiscip Top Gerontol Geriatr 2015;41:26–40.

Fulop T, Dupuis G, Baehl S, Le Page A, Bourgade K, Frost E, Witkowski JM, Pawelec G, Larbi A, Cunnane S. From inflamm-aging to immune-paralysis: a slippery slope during aging for immune-adaptation. Biogerontology 2016a;17:147–57.

Fulop T, Dupuis G, Witkowski JM, Larbi A. The role of immunosenescence in the development of age-related diseases. Rev Invest Clin 2016b;68:84–91.

Fulop T, Larbi A, Dupuis G, Le Page A, Frost EH, Cohen AA, Witkowski JM, Franceschi C. Immunosenescence and inflamm-aging as two sides of the same coin: friends or foes? Front Immunol 2017a;8:1960.

Fulop T, Witkowski JM, Le Page A, Fortin C, Pawelec G, Larbi A. Intracellular signalling pathways: targets to reverse immunosenescence. Clin Exp Immunol 2017b;187:35–43.

Fulop T, Itzhaki RF, Balin BJ, Miklossy J, Barron AE. Role of microbes in the development of Alzheimer's disease: state of the art—an international symposium presented at the 2017 IAGG congress in San Francisco. Front Genet 2018a;9:362.

Fulop T, Witkowski JM, Olivieri F, Larbi A. The integration of inflammaging in age-related diseases. Semin Immunol 2018b;40:17–35.

Fulop T, Larbi A, Witkowski JM. Human inflammaging. Gerontology 2019;65:495–504.

Fulop T, Larbi A, Hirokawa K, Cohen AA, Witkowski JM. Immunosenescence is both functional/adaptive and dysfunctional/maladaptive. Semin Immunopathol 2020a;42:521–36.

Fulop T, Munawara U, Larbi A, Desroches M, Rodrigues S, Catanzaro M, Guidolin A, Khalil A, Bernier F, Barron AE, Hirokawa K, Beauregard PB, Dumoulin D, Bellenger JP, Witkowski JM, Frost E. Targeting infectious agents as a therapeutic strategy in Alzheimer's disease. CNS Drugs 2020b;34:673–95.

Gan WQ, Man SF, Senthilselvan A, Sin DD. Association between chronic obstructive pulmonary disease and systemic inflammation: a systematic review and a meta-analysis. Thorax 2004;59:574–80.

Garagnani P, Giuliani C, Pirazzini C, Olivieri F, Bacalini MG, Ostan R, Mari D, Passarino G, Monti D, Bonfigli AR, Boemi M, Ceriello A, Genovese S, Sevini F, Luiselli D, Tieri P, Capri M, Salvioli S, Vijg J, Suh Y, Delledonne M, Testa R, Franceschi C. Centenarians as super-controls to assess the biological relevance of genetic risk factors for common age-related diseases: a proof of principle on type 2 diabetes. Aging (Albany NY) 2013;5:373–85.

Gruver AL, Hudson LL, Sempowski GD. Immunosenescence of ageing. J Pathol 2007;211:144–56.

Har-Noy M, Or R. Allo-priming as a universal anti-viral vaccine: protecting elderly from current COVID-19 and any future unknown viral outbreak. J Transl Med 2020;18:196.

Henning C, Glomb MA. Pathways of the Maillard reaction under physiological conditions. Glycoconj J 2016;33:499–512.

Hong MS, Dan JM, Choi JY, Kang I. Age-associated changes in the frequency of naive, memory and effector CD8+ T cells. Mech Ageing Dev 2004;125:615–8.

Itzhaki RF, Lathe R, Balin BJ, Ball MJ, Bearer EL, Braak H, Bullido MJ, Carter C, Clerici M, Cosby SL, Del Tredici K, Field H, Fulop T, Grassi C, Griffin WS, Haas J, Hudson AP, Kamer AR, Kell DB, Licastro F, Letenneur L, Lovheim H, Mancuso R, Miklossy J, Otth C, Palamara AT, Perry G, Preston C, Pretorius E, Strandberg T, Tabet N, Taylor-Robinson SD, Whittum-Hudson JA. Microbes and Alzheimer's disease. J Alzheimers Dis 2016;51:979–84.

Jones L, Kumar J, Mistry A, Sankar Chittoor Mana T, Perry G, Reddy VP, Obrenovich M. The transformative possibilities of the microbiota and mycobiota for health, disease, aging, and technological innovation. Biomedicine 2019;7:24.

Jozwik A, Landowski J, Bidzan L, Fulop T, Bryl E, Witkowski JM. Beta-amyloid peptides enhance the proliferative response of activated CD4CD28 lymphocytes from Alzheimer disease patients and from healthy elderly. PLoS One 2012;7:e33276.

Kay AM, Simpson CL, Stewart Jr. JA. The role of AGE/RAGE signaling in diabetes-mediated vascular calcification. J Diabetes Res 2016;2016:6809703.

Kim S, Jazwinski SM. The gut microbiota and healthy aging: a mini-review. Gerontology 2018;64:513–20.

Kim DH, Bang E, Arulkumar R, Ha S, Chung KW, Park MH, Choi YJ, Yu BP, Chung HY. Senoinflammation: a major mediator underlying age-related metabolic dysregulation. Exp Gerontol 2020;134:110891.

Kuo CL, Pilling LC, Atkins JC, Masoli J, Delgado J, Tignanelli C, Kuchel G, Melzer D, Beckman KB, Levine M. COVID-19 severity is predicted by earlier evidence of accelerated aging. medRxiv 2020; https://doi.org/10.1101/2020.07.10.20147777.

Laing B, Barnett MPG, Marlow G, Nasef NA, Ferguson LR. An update on the role of gut microbiota in chronic inflammatory diseases, and potential therapeutic targets. Expert Rev Gastroenterol Hepatol 2018;12:969–83.

Larbi A, Franceschi C, Mazzatti D, Solana R, Wikby A, Pawelec G. Aging of the immune system as a prognostic factor for human longevity. Physiology (Bethesda) 2008;23:64–74.

Le Page A, Dupuis G, Larbi A, Witkowski JM, Fulop T. Signal transduction changes in CD4(+) and CD8(+) T cell subpopulations with aging. Exp Gerontol 2018;105:128–39.

Lee BY, Han JA, Im JS, Morrone A, Johung K, Goodwin EC, Kleijer WJ, DiMaio D, Hwang ES. Senescence-associated beta-galactosidase is lysosomal beta-galactosidase. Aging Cell 2006;5:187–95.

Lio D, Candore G, Crivello A, Scola L, Colonna-Romano G, Cavallone L, Hoffmann E, Caruso M, Licastro F, Caldarera CM, Branzi A, Franceschi C, Caruso C. Opposite effects of interleukin 10 common gene polymorphisms in cardiovascular diseases and in successful ageing: genetic background of male centenarians is protective against coronary heart disease. J Med Genet 2004;41:790–4.

Liu D, Richardson G, Benli FM, Park C, de Souza JV, Bronowska AK, Spyridopoulos I. Inflammageing in the cardiovascular system: mechanisms, emerging targets, and novel therapeutic strategies. Clin Sci (Lond) 2020;134:2243–62.

Maffei VJ, Kim S, Blanchard Et, Luo M, Jazwinski SM, Taylor CM, Welsh DA. Biological aging and the human gut microbiota. J Gerontol A Biol Sci Med Sci 2017;72:1474–82.

Marcon G, Tettamanti M, Capacci G, Fontanel G, Spano M, Nobili A, Forloni G, Franceschi C. COVID-19 mortality in Lombardy: the vulnerability of the oldest old and the resilience of male centenarians. Aging (Albany NY) 2020;12:15186–95.

Marini M, Lapalombella R, Canaider S, Farina A, Monti D, De Vescovi V, Morellini M, Bellizzi D, Dato S, De Benedictis G, Passarino G, Moresi R, Tesei S, Franceschi C. Heat shock response by EBV-immortalized B-lymphocytes from centenarians and control subjects: a model to study the relevance of stress response in longevity. Exp Gerontol 2004;39:83–90.

Meftahi GH, Jangravi Z, Sahraei H, Bahari Z. The possible pathophysiology mechanism of cytokine storm in elderly adults with COVID-19 infection: the contribution of "inflame-aging" Inflamm Res 2020;69:825–39.

Meszaros A, Molnar K, Nogradi B, Hernadi Z, Nyul-Toth A, Wilhelm I, Krizbai IA. Neurovascular inflammaging in health and disease. Cell 2020;9:1614.

Monti D, Salvioli S, Capri M, Malorni W, Straface E, Cossarizza A, Botti B, Piacentini M, Baggio G, Barbi C, Valensin S, Bonafe M, Franceschi C. Decreased susceptibility to oxidative stress-induced apoptosis of peripheral blood mononuclear cells from healthy elderly and centenarians. Mech Ageing Dev 2000;121:239–50.

Mysliwska J, Bryl E, Zorena K, Balon J, Foerster J, Mysliwski A. Overactivity of tumor necrosis factor-alpha but not interleukin 6 is associated with low natural killer cytotoxic activity in the elderly. Gerontology 1997;43:158–67.

Mysliwska J, Bryl E, Foerster J, Mysliwski A. Increase of interleukin 6 and decrease of interleukin 2 production during the ageing process are influenced by the health status. Mech Ageing Dev 1998;100:313–28.

Mysliwska J, Bryl E, Foerster J, Mysliwski A. The upregulation of TNF alpha production is not a generalised phenomenon in the elderly between their sixth and seventh decades of life. Mech Ageing Dev 1999;107:1–14.

Nasi M, Troiano L, Lugli E, Pinti M, Ferraresi R, Monterastelli E, Mussi C, Salvioli G, Franceschi C, Cossarizza A. Thymic output and functionality of the IL-7/IL-7 receptor system in centenarians: implications for the neolymphogenesis at the limit of human life. Aging Cell 2006;5:167–75.

Pardo A, Selman M. The interplay of the genetic architecture, aging, and environmental factors in the pathogenesis of idiopathic pulmonary fibrosis. Am J Respir Cell Mol Biol 2021;64(2):163–72.

Pawelec G. Immunity and ageing in man. Exp Gerontol 2006;41:1239–42.

Pence BD. Severe COVID-19 and aging: are monocytes the key? GeroScience 2020;42:1051–61.

Perrotte M, Haddad M, Le Page A, Frost EH, Fulop T, Ramassamy C. Profile of pathogenic proteins in total circulating extracellular vesicles in mild cognitive impairment and during the progression of Alzheimer's disease. Neurobiol Aging 2020;86:102–11.

Riaz Rajoka MS, Zhao H, Li N, Lu Y, Lian Z, Shao D, Jin M, Li Q, Zhao L, Shi J. Origination, change, and modulation of geriatric disease-related gut microbiota during life. Appl Microbiol Biotechnol 2018;102:8275–89.

Robinson LA, Pierce CM. Is 'inflammaging' fuelling severe COVID-19 disease? J R Soc Med 2020;113:346–9.

Rolfson D. Successful aging and frailty: a systematic review. Geriatrics (Basel) 2018;3:79.

Salvioli S, Capri M, Bucci L, Lanni C, Racchi M, Uberti D, memo M, Mari D, Govoni S, Franceschi C. Why do centenarians escape or postpone cancer? The role of IGF-1, inflammation and p53. Cancer Immunol Immunother 2009;58:1909–17.

Salvioli S, Monti D, Lanzarini C, Conte M, Pirazzini C, Bacalini MG, Garagnani P, Giuliani C, Fontanesi E, Ostan R, Bucci L, Sevini F, Yani SL, Barbieri A, Lomartire L, Borelli V, Vianello D, Bellavista E, Martucci M, Cevenini E, Pini E, Scurti M, Biondi F, Santoro A, Capri M, Franceschi C. Immune system, cell senescence, aging and longevity—inflamm-aging reappraised. Curr Pharm Des 2013;19:1675–9.

Senatus LM, Schmidt AM. The AGE-RAGE axis: implications for age-associated arterial diseases. Front Genet 2017;8:187.

Sgarbi G, Matarrese P, Pinti M, Lanzarini C, Ascione B, Gibellini L, Dika E, Patrizi A, Tommasino C, Capri M, Cossarizza A, Baracca A, Lenaz G, Solaini G, Franceschi C, Malorni W, Salvioli S. Mitochondria hyperfusion and elevated autophagic activity are key mechanisms for cellular bioenergetic preservation in centenarians. Aging (Albany NY) 2014;6:296–310.

Shin JJ, Lee EK, Park TJ, Kim W. Damage-associated molecular patterns and their pathological relevance in diabetes mellitus. Ageing Res Rev 2015;24:66–76.

Shintouo CM, Mets T, Beckwee D, Bautmans I, Ghogomu SM, Souopgui J, Leemans L, Meriki HD, Njemini R. Is inflammageing influenced by the microbiota in the aged gut? A systematic review. Exp Gerontol 2020;141:111079.

Takeuchi M. Serum levels of toxic AGEs (TAGE) may be a promising novel biomarker for the onset/progression of lifestyle-related diseases. Diagnostics (Basel) 2016;6:23.

Tang D, Kang R, Zeh 3rd HJ, Lotze MT. High-mobility group box 1, oxidative stress, and disease. Antioxid Redox Signal 2011;14:1315–35.

Ticinesi A, Tana C, Nouvenne A. The intestinal microbiome and its relevance for functionality in older persons. Curr Opin Clin Nutr Metab Care 2019;22:4–12.

Trzonkowski P, Szmit E, Mysliwska J, Mysliwski A. CD4 + CD25 + T regulatory cells inhibit cytotoxic activity of CTL and NK cells in humans-impact of immunosenescence. Clin Immunol 2006;119:307–16.

Vellas C, Delobel P, de Souto Barreto P, Izopet J. COVID-19, virology and geroscience: a perspective. J Nutr Health Aging 2020;24:685–91.

Wilmoth JR. Demography of longevity: past, present, and future trends. Exp Gerontol 2000;35:1111–29.

Witkowski JM, Bryl E, Fulop T. Should we try to alleviate immunosenescence and inflammaging—why, how and to what extent? Curr Pharm Des 2019;25:4154–62.

Yao C, Guan X, Carraro G, Parimon T, Liu X, Huang G, Mulay A, Soukiasian HJ, David G, Weigt SS, Belperio JA, Chen P, Jiang D, Noble PW, Stripp BR. Senescence of alveolar type 2 cells drives progressive pulmonary fibrosis. Am J Respir Crit Care Med 2020; https://doi.org/10.1164/rccm.202004-1274OC Epub ahead of print.

CHAPTER 6

A new perspective on ROS in aging with an integrated view of the gut microbiota

Lu Wu[a] and Ciriaco Carru[b]
[a]Institute of Synthetic Biology, Institutes of Advanced Technology, Chinese Academy of Sciences, Shenzhen, China
[b]Department of Biomedical Sciences, University of Sassari, Hospital University (AOUSS), Sassari, Italy

6.1 Introduction

The theory of free radical damage on aging has long been the dominant theory of why we age and die. However, several studies conducted in recent years have raised doubts about its importance in determining the aging process and therefore the lifespan. However, it seems that the effect of lifespan is too strict a criterion according to which the theory should be evaluated. Rather, the role of oxidative stress on the healthspan could be a more appropriate method to assess its importance. This suggestion is perhaps simply a recognition of the fact that a complex biological phenomenon such as aging is unlikely to be explained by such a simplistic hypothesis. So in this chapter we discuss a new perspective on ROS in aging with an integrated view of their relationship with gut barrier, inflammation, and dysbiosis of gut microbiota. Chapter 3 shows the list of free radicals and the mechanisms of the damage they can cause.

6.2 The reactive oxygen species

Reactive oxygen species (ROS) are mainly generated endogenously from leakage of electrons in the electron transport chain in the mitochondrion, with the process of oxidative phosphorylation; or in response to different exogenous stimulators such as chemotherapeutics, inflammatory cytokines, ionizing radiation, UV, chemical oxidants. Moreover, it has recently been demonstrated that human commensal microbiota could induce the generation of physiological levels of ROS within human epithelial cells (Sommer and Bäckhed, 2015). ROS may be generated in other cellular organelles such as peroxisomes or in the cytosol via a variety of enzymes which, including NADPH oxidases (NOXs), lipoxygenases, and cytochrome P450 systems, can induce increased ROS in mitochondria as well as in cytosol in a variety of cell types. Take NOXs for example, NOXs are a group of transmembrane enzymes (NOX1, NOX2, NOX3, NOX4, NOX5, DUOX1, and DUOX2) with oxidase activity that have been observed to be deeply involved in ROS production and redox signaling during aging (Manea, 2010;

Human Aging
https://doi.org/10.1016/B978-0-12-822569-1.00004-4

Krause, 2007). Meanwhile, the body has particular defense mechanisms in place to protect against excess ROS, including endogenous enzymatic and nonenzymatic molecules and some exogenous nonenzymatic molecules from the diet, diet-derived metabolites, and medication. Enzymes such as superoxide dismutase (SOD), catalase (CAT), and glutathione peroxidase; small nonenzymatic molecule antioxidants (Vitamin E, pyruvate, bilirubin, flavonoids, carotenoids and perhaps most importantly, glutathione) played a crucial role in regulating oxidative stress and redox signaling. Some of those small molecules are important second metabolites from gut microbiota. Evidence showed that the antioxidant enzymes, in particular SOD-2 and CAT, were transcriptionally regulated by the forkhead box class O (FOXO) family of winged-helix transcription factors (Klotz et al., 2015). Besides, ROS also can cross regulate the expression of FOXO proteins via multiple mechanisms, including posttranslational modifications (Klotz et al., 2015; Van Der Horst and Burgering, 2007). Interestingly, FOXO3 is consistently annotated as a human longevity gene due to the homeostatic activity of its product (Flachsbart et al., 2009; Greer and Brunet, 2005). In response to oxidative stress, the nuclear transcription factor Nrf2 (nuclear factor erythroid-derived 2-like 2) binds to AREs, mediating transcriptional activation of its responsive genes, such as Peroxiredoxin 6 (Prdx6) driven antioxidant defense mechanisms play a critical role in the induction of endogenous antioxidant enzymes acting against oxidative damage (Jian et al., 2011; Suzuki and Yamamoto, 2017). Previous studies have found some gut microbes and microbe-derived metabolites can modulate the gut barrier function by targeting the Nrf2 pathways (Jones et al., 2015; Singh et al., 2019). Within physiological levels, ROS are in a biological redox steady state and facilitate the maintenance of cellular homeostasis and function. "Oxidative stress" is defined as the situation when an imbalance between the production of ROS and the biological systems' ability to counteract or detoxify their harmful effects through neutralization by antioxidants or repair the resulting systemic damage. Although high levels of ROS are generated by the immune cells in defense against certain pathogens, ROS overproduction could be harmful. Oxidative stress could directly target and modify the cellular macromolecules that cause DNA damage, lipid peroxidation, and protein oxidation, and which ultimately lead to cellular injury and death and even impact the aging process. Moreover, the oxidative stress could also affect the biological systems indirectly through activation of the redox signaling pathways. A previous study showed that about 5- to 10-fold higher superoxide anion level is present in the mitochondrial matrix than in the cytosolic and nuclear space (Cadenas and Davies, 2000) making the mitochondria primary target for ROS-induced damages. DNA, especially the mitochondrial DNA is the most sensitive to ROS damage; numerous studies have demonstrated the age-related mitochondrial DNA damages in both human and animal models (Liang and Godley, 2003; Lu et al., 1999; Barja, 1998; Berneburg et al., 1997). Moreover, mitochondria were also found play a critical role in maintaining the intestinal homeostasis, suggesting the connection between the gut microbiome and some common age-related disease phenotypes might be due to underlying decline in mitochondrial function (Yardeni et al., 2019). the cytosol and nuclear area (Cadenas and Davies, 2000) making the mitochondria primary target for ROS-induced damages. DNA, especially the mitochondrial DNA is the most sensitive to ROS damage; dozens of researches have demonstrated the age-related mitochondrial DNA damages in both human and animal models (Liang and Godley, 2003; Lu et al., 1999; Barja, 1998; Berneburg et al., 1997). Moreover, mitochondria were also found to have played a critical role in maintaining the intestinal homeostasis, suggesting the connection between the gut microbiome and some common age-related disease phenotypes might be due to underlying decline in mitochondrial function (Yardeni et al., 2019).

6.3 The biological function of ROS

Although the conventional image of the biological effects of ROS, perhaps, is those that occur when cellular antioxidant defenses are overwhelmed and ROS reacts directly with cellular components, causing age-related cell damage and death, more evidence shows that ROS is also a critical signaling molecule. ROS is not only a by-product from aerobic respiration which may cause damages to the cell components but also a key signaling pathway molecule with both physiological and pathological functions. ROS levels within the physiological range are critical signaling molecules for many redox-dependent signaling processes, including gene expression, metabolic regulation, the hypoxic response, inflammatory and immune response, stem cell proliferation and differentiation, cancer pathogenesis, as well as aging. As for the pathological function, take the ROS in the gastrointestinal tract for example, regarding defense against infections, extracellular secretion of ROS by mucosal-resident immune cells such as leukocytes and macrophages, as well epithelial cells can act directly against the pathogens and meantime elicits further immune responses against the pathogens.

6.4 The oxidative stress theory of aging

The Mitochondrial Free Radical Theory of Aging (MFRTA) was a well-known theory in aging research since the 1950s (Harman, 1956). It proposed that aging results from the accumulation of oxidative damage, which is greatly contributed by the release of reactive oxygen species from mitochondria (Harman, 2009). Evidence sustains this theory includes: (a) A positive correlation between chronological age and the level of ROS production and accumulation of oxidative damage to lipid, DNA, and protein in several species (Barja, 1998; Bokov et al., 2004). It has been shown the critical role of oxidative damage in causing the mitochondrial dysfunction of aging and age-related diseases. Oxidants generated by mitochondria appear to be the major source of their oxidative lesions that accumulate with age. Accordingly, several mitochondrial functions decline during aging process and in several age-related diseases, such as Type 2 diabetes, atherosclerosis, Parkinson (Shigenaga et al., 1994; Lowell and Shulman, 2005; Madamanchi and Runge, 2007; Hattingen et al., 2009). (b) Some studies with animal models with extended lifespan and human with improved health status showed that they were more resistant to oxidative stress and reduced oxidative damage (Yu, 1996; Murakami et al., 2003; Honda and Honda, 1999; Buchowski et al., 2012). However, despite the close association observed between ROS and aging, it should be pointed out that simple correlation does not imply causation. A complicated process such as aging and longevity is clearly inappropriate to be explained by such a simplistic hypothesis, evaluating only ROS as a solo-indicator or causal factor. Conflicting evidence including (a) Animal model experiments have greatly challenged the MFRTA theory in recent years, questioning the causal effect of ROS in aging and lifespan. In *Mus musculus*, *Drosophila melanogaster*, *Caenorhabditis elegans*, and *Heterocephalus glaber* the reduced ROS production was not always positively associated with an extended lifespan, meanwhile, an increased oxidative stress by the mutation of certain antioxidant enzymes did not imply shorter lifespan (Miwa et al., 2004; Van Raamsdonk and Hekimi, 2009; Andziak et al., 2006). (b) Antioxidant supplements in animals and humans failed to extend lifespan and/or improve health status (Herrera et al., 2009; Lin et al., 2009; Lee et al., 2004; Bjelakovic et al., 2007). However, a recent study performed in flies showed that sleep deprivation leads to accumulation of ROS and consequent oxidative stress, specifically in the gut. ROS are not just correlates of sleep deprivation but drivers of death: their neutralization prevents oxidative stress and allows flies to have a normal lifespan with little to no sleep (Vaccaro, 2020).

6.5 ROS signaling in gut barrier, inflammation, and dysbiosis of gut microbiota in aging

The ROS signaling is critical in the maintenance of gut homeostasis (Aviello and Knaus, 2017). Both the gastrointestinal mucosa and the associated immune cells, as well as gut microbiota (both the commensal and the pathogens), are the sources of ROS. Meanwhile, they also can sense and respond to the ROS signaling. In the gastrointestinal tract, the commensals protect against pathogens through oxygen competition (Litvak et al., 2019). ROS are generated by epithelial cells, and, more importantly, from the innate immune cells to implement mucosal defense against the pathogens ROS can inhibit of prolyl hydroxylase enzymes and can stabilize hypoxia-inducible factor-1α, up-regulating barrier protective genes (Litvak et al., 2018; Van Welden et al., 2017). Moreover, the loss of homeostasis in redox state is associated with gut barrier dysfunction and the progression of inflammation (Aviello and Knaus, 2017). It is well known that chronic inflammation is a prognostic marker of increased disease risk and mortality during aging process (Franceschi and Campisi, 2014). Some aging-related inflammatory diseases are found to be related to the accommodation of ROS, such as cardiovascular diseases

(Ochoa et al., 2018). Meanwhile, aging-related gut microbiota changes have been fully described in previous studies (O'Toole and Jeffery, 2015; Wu et al., 2019) showing that age-associated microbial dysbiosis is also correlated with the increase of system inflammation (Thevarajan, 2018). In turn, inflammation is responsible for high amounts of ROS in gut lumen that cause pathogen booms in gut (Tian et al., 2017). What's more, derived from intestinal microbial fermentation of the indigestible fibers or mucus protein, short-chain fatty acids, particularly butyrate, are the main energy source of colonocytes for β-oxidation which, subsequently, consumes oxygen in the colonocytes, maintaining the epithelial hypoxia (Byndloss et al., 2017). The hypoxia condition, in gut, enables the dominance of the anaerobic microbes in gut which is critical for maintaining gut health. In addition, ROS signaling in the intestinal mucosa consumes a large amount of oxygen both in the gut mucosa and gut lumen, which subsequently cause physiological hypoxia in gut which plays an important role in shaping the gut microbiota.

6.6 Conclusion and future perspective

We propose an in-depth perspective on ROS and redox signaling in the gastrointestinal tract in homeostasis, infectious diseases, and intestinal inflammation. Finally, we argue that ROS production is a key event in the progression of aging and maintenance of health, rather than a single causative and determinant factor of aging.

Acknowledgments

Dr. WU Lu contributed to this study during a Visiting Professorship at the University of Sassari. Funding support was provided by Ministero dell'Istruzione, dell'Università e della Ricerca (MIUR, Italy) PRIN 20157ATSLF_002, and Consiglio Nazionale delle Ricerche Flagship InterOmics (code PB05).

References

Andziak B, O'Connor TP, Qi W, Dewaal EM, Pierce A, Chaudhuri AR, Van Remmen H, Buffenstein R. High oxidative damage levels in the longest-living rodent, the naked mole-rat. Aging Cell 2006;5(6): 463–71. https://doi.org/10.1111/j.1474-9726.2006.00237.x.

Aviello G, Knaus UG. ROS in gastrointestinal inflammation: rescue or sabotage? Br J Pharmacol 2017; 174(12):1704–18. https://doi.org/10.1111/bph.13428.

Barja G. Mitochondrial free radical production and aging in mammals and birds, Ann N Y Acad Sci 1998;854:224–38. New York Academy of Sciences https://doi.org/10.1111/j.1749-6632.1998. tb09905.x.

Berneburg M, Gattermann N, Stege H, Grewe M, Vogelsang K, Ruzicka T, Krutmann J. Chronically ultraviolet-exposed human skin shows a higher mutation frequency of mitochondrial DNA as compared to unexposed skin and the hematopoietic system. Photochem Photobiol 1997;66(2):271–5. https://doi.org/10.1111/j.1751-1097.1997.tb08654.x.

Bjelakovic G, Nikolova D, Gluud LL, Simonetti RG, Gluud C. Mortality in randomized trials of antioxidant supplements for primary and secondary prevention: systematic review and meta-analysis. JAMA 2007;297(8):842–57. https://doi.org/10.1001/jama.297.8.842.

Bokov A, Chaudhuri A, Richardson A. The role of oxidative damage and stress in aging. Mech Ageing Dev 2004;125(10−11):811–26. https://doi.org/10.1016/j.mad.2004.07.009.

Buchowski MS, Hongu N, Acra S, Wang L, Warolin J, Roberts LJ. Effect of modest caloric restriction on oxidative stress in women, a randomized trial. PLoS One 2012;7(10). https://doi.org/10.1371/journal.pone.0047079.

Byndloss MX, Olsan EE, Rivera-Chávez F, Tiffany CR, Cevallos SA, Lokken KL, Torres TP, Byndloss AJ, Faber F, Gao Y, Litvak Y, Lopez CA, Xu G, Napoli E, Giulivi C, Tsolis RM, Revzin A, Lebrilla CB, Bäumler AJ. Microbiota-activated PPAR-γ signaling inhibits dysbiotic Enterobacteriaceae expansion. Science 2017;357(6351):570–5. https://doi.org/10.1126/science.aam9949.

Cadenas E, Davies KJA. Mitochondrial free radical generation, oxidative stress, and aging. Free Radic Biol Med 2000;29(3–4):222–30. https://doi.org/10.1016/S0891-5849(00)00317-8.

Flachsbart F, Caliebe A, Kleindorp R, Blanché H, Von Eller-Eberstein H, Nikolaus S, Schreiber S, Nebel A. Association of FOXO3A variation with human longevity confirmed in German centenarians. Proc Natl Acad Sci U S A 2009;106(8):2700–5. https://doi.org/10.1073/pnas.0809594106.

Franceschi C, Campisi J. Chronic inflammation (Inflammaging) and its potential contribution to age-associated diseases. J Gerontol A Biol Sci Med Sci 2014;69:S4–9. https://doi.org/10.1093/gerona/glu057.

Greer EL, Brunet A. FOXO transcription factors at the interface between longevity and tumor suppression. Oncogene 2005;24(50):7410–25. https://doi.org/10.1038/sj.onc.1209086.

Harman D. Aging: a theory based on free radical and radiation chemistry. J Gerontol 1956;11(3):298–300. https://doi.org/10.1093/geronj/11.3.298.

Harman D. Origin and evolution of the free radical theory of aging: a brief personal history, 1954-2009. Biogerontology 2009;10(6):773–81. https://doi.org/10.1007/s10522-009-9234-2.

Hattingen E, Magerkurth J, Pilatus U, Mozer A, Seifried C, Steinmetz H, Zanella F, Hilker R. Phosphorus and proton magnetic resonance spectroscopy demonstrates mitochondrial dysfunction in early and advanced Parkinson's disease. Brain 2009;132(12):3285–97. https://doi.org/10.1093/brain/awp293.

Herrera E, Jiménez R, Aruoma OI, Hercberg S, Sánchez-García I, Fraga C. Aspects of antioxidant foods and supplements in health and disease. Nutr Rev 2009;67(1):S140–4. https://doi.org/10.1111/j.1753-4887.2009.00177.x.

Honda Y, Honda S. The daf-2 gene network for longevity regulates oxidative stress resistance and Mn-superoxide dismutase gene expression in *Caenorhabditis elegans*. FASEB J 1999;13(11):1385–93. https://doi.org/10.1096/fasebj.13.11.1385.

Jian Z, Li K, Liu L, Zhang Y, Zhou Z, Li C, Gao T. Heme oxygenase-1 protects human melanocytes from H_2O_2-induced oxidative stress via the Nrf2-ARE pathway. J Invest Dermatol 2011;131(7):1420–7. https://doi.org/10.1038/jid.2011.56.

Jones RM, Desai C, Darby TM, Luo L, Wolfarth AA, Scharer CD, Ardita CS, Reedy AR, Keebaugh ES, Neish AS. Lactobacilli modulate epithelial cytoprotection through the Nrf2 pathway. Cell Rep 2015;12(8):1217–25. https://doi.org/10.1016/j.celrep.2015.07.042.

Klotz LO, Sánchez-Ramos C, Prieto-Arroyo I, Urbánek P, Steinbrenner H, Monsalve M. Redox regulation of FoxO transcription factors. Redox Biol 2015;6:51–72. https://doi.org/10.1016/j.redox.2015.06.019.

Krause KH. Aging: a revisited theory based on free radicals generated by NOX family NADPH oxidases. Exp Gerontol 2007;42(4):256–62. https://doi.org/10.1016/j.exger.2006.10.011.

Lee CK, Pugh TD, Klopp RG, Edwards J, Allison DB, Weindruch R, Prolla TA. The impact of α-lipoic acid, coenzyme Q10, and caloric restriction on life span and gene expression patterns in mice. Free Radic Biol Med 2004;36(8):1043–57. https://doi.org/10.1016/j.freeradbiomed.2004.01.015.

Liang FQ, Godley BF. Oxidative stress-induced mitochondrial DNA damage in human retinal pigment epithelial cells: a possible mechanism for RPE aging and age-related macular degeneration. Exp Eye Res 2003;76(4):397–403. https://doi.org/10.1016/S0014-4835(03)00023-X.

Lin J, Cook NR, Albert C, Zaharris E, Gaziano JM, Van Denburgh M, Buring JE, Manson JE. Vitamins C and E and beta carotene supplementation and cancer risk: a randomized controlled trial. J Natl Cancer Inst 2009;101(1):14–23. https://doi.org/10.1093/jnci/djn438.

Litvak Y, Byndloss MX, Bäumler AJ. Colonocyte metabolism shapes the gut microbiota. Science 2018;362(6418). https://doi.org/10.1126/science.aat9076.

Litvak Y, Mon KKZ, Nguyen H, Chanthavixay G, Liou M, Velazquez EM, Kutter L, Alcantara MA, Byndloss MX, Tiffany CR, Walker GT, Faber F, Zhu Y, Bronner DN, Byndloss AJ, Tsolis RM, Zhou H, Bäumler AJ. Commensal enterobacteriaceae protect against salmonella colonization through oxygen competition. e5 Cell Host Microbe 2019;25(1):128–39. https://doi.org/10.1016/j.chom.2018.12.003.

Lowell BB, Shulman GI. Mitochondrial dysfunction and type 2 diabetes. Science 2005;307(5708):384–7. https://doi.org/10.1126/science.1104343.

Lu CY, Lee HC, Fahn HJ, Wei YH. Oxidative damage elicited by imbalance of free radical scavenging enzymes is associated with large-scale mtDNA deletions in aging human skin. Mutat Res Fundam Mol Mech Mutagen 1999;423(1–2):11–21. https://doi.org/10.1016/S0027-5107(98)00220-6.

Madamanchi NR, Runge MS. Mitochondrial dysfunction in atherosclerosis. Circ Res 2007;100(4):460–73. https://doi.org/10.1161/01.RES.0000258450.44413.96.

Manea A. NADPH oxidase-derived reactive oxygen species: involvement in vascular physiology and pathology. Cell Tissue Res 2010;342(3):325–39. https://doi.org/10.1007/s00441-010-1060-y.

Miwa S, Riyahi K, Partridge L, Brand MD. Lack of correlation between mitochondrial reactive oxygen species production and life span in Drosophila, Ann N Y Acad Sci 2004;1019:388–91. New York Academy of Sciences https://doi.org/10.1196/annals.1297.069.

Murakami S, Salmon A, Miller RA. Multiplex stress resistance in cells from long-lived dwarf mice. FASEB J 2003;17(11):1565–6. https://doi.org/10.1096/fj.02-1092fje.

O'Toole PW, Jeffery IB. Gut microbiota and aging. Science 2015;350(6265):1214–5. https://doi.org/10.1126/science.aac8469.

Ochoa CD, Wu RF, Terada LS. ROS signaling and ER stress in cardiovascular disease. Mol Aspects Med 2018;63:18–29. https://doi.org/10.1016/j.mam.2018.03.002.

Shigenaga MK, Hagen TM, Ames BN. Oxidative damage and mitochondrial decay in aging. Proc Natl Acad Sci U S A 1994;91(23):10771–8. https://doi.org/10.1073/pnas.91.23.10771.

Singh D, Reeta KH, Sharma U, Jagannathan NR, Dinda AK, Gupta YK. Neuro-protective effect of monomethyl fumarate on ischemia reperfusion injury in rats: role of Nrf2/HO1 pathway in peri-infarct region. Neurochem Int 2019;126:96–108. https://doi.org/10.1016/j.neuint.2019.03.010.

Sommer F, Bäckhed F. The gut microbiota engages different signaling pathways to induce Duox2 expression in the ileum and colon epithelium. Mucosal Immunol 2015;8(2):372–9. https://doi.org/10.1038/mi.2014.74.

Suzuki T, Yamamoto M. Stress-sensing mechanisms and the physiological roles of the Keap1–Nrf2 system during cellular stress. J Biol Chem 2017;292(41):16817–24. https://doi.org/10.1074/jbc.R117.800169.

Thevarajan N. Age-associated microbial dysbiosis promotes intestinal permeability, systemic inflammation, and macrophage dysfunction. Cell Host Microbe 2018;23(4):570.

Tian T, Wang Z, Zhang J. Pathomechanisms of oxidative stress in inflammatory bowel disease and potential antioxidant therapies. Oxid Med Cell Longev 2017;2017. https://doi.org/10.1155/2017/4535194.

Vaccaro A. Sleep loss can cause death through accumulation of reactive oxygen species in the gut. Cell 2020;181: 1307–1328.

Van Der Horst A, Burgering BMT. Stressing the role of FoxO proteins in lifespan and disease. Nat Rev Mol Cell Biol 2007;8(6):440–50. https://doi.org/10.1038/nrm2190.

Van Raamsdonk JM, Hekimi S. Deletion of the mitochondrial superoxide dismutase sod-2 extends lifespan in Caenorhabditis elegans. PLoS Genet 2009;5(2). https://doi.org/10.1371/journal.pgen.1000361.

Van Welden S, Selfridge AC, Hindryckx P. Intestinal hypoxia and hypoxia-induced signalling as therapeutic targets for IBD. Nat Rev Gastroenterol Hepatol 2017;14(10):596–611. https://doi.org/10.1038/nrgastro.2017.101.

Wu L, Zeng T, Zinellu A, Rubino S, Kelvin DJ, Carru C. A cross-sectional study of compositional and functional profiles of gut microbiota in Sardinian centenarians. mSystems 2019;4(4)e00325-19.

Yardeni T, Tanes CE, Bittinger K, Mattei LM, Schaefer PM, Singh LN, Wu GD, Murdock DG, Wallace DC. Host mitochondria influence gut microbiome diversity: a role for ROS. Sci Signal 2019;12(588). https://doi.org/10.1126/scisignal.aaw3159.

Yu BP. Aging and oxidative stress: modulation by dietary restriction. Free Radic Biol Med 1996; 21(5):651–68. https://doi.org/10.1016/0891-5849(96)00162-1.

CHAPTER 7

Aging of immune system

Marcello Pinti[a], Sara De Biasi[b], Lara Gibellini[b], Domenico Lo Tartaro[b], Anna De Gaetano[a], Marco Mattioli[b], Lucia Fidanza[b], Milena Nasi[c], and Andrea Cossarizza[b]

[a]Department of Life Sciences, University of Modena and Reggio Emilia, Modena, Italy
[b]Department of Medical and Surgical Sciences of Children and Adults, University of Modena and Reggio Emilia, Modena, Italy
[c]Department of Surgery, Medicine, Dentistry and Morphological Sciences, University of Modena and Reggio Emilia, Modena, Italy

7.1 Introduction

Life expectancy has increased dramatically over the last century (for instance, in Italy it increased from 43 years in 1900 to the current 82.8 years (United Nations, Department of Economic and Social Affairs, Population Division, 2017). In the first stage, this increase was mainly due to the fall of infant mortality. At present, gain in life expectancy is achieved mainly through improvement in presenile and senile ages. This increase in life expectancy is highlighted by the rapid increase of centenarians (that is, subjects that are 100 years old or more) worldwide; nowadays, their number is higher than 450,000, a threefold increase from the first monitoring in 1990 (Robine and Cubaynes, 2017). As an overwhelming majority of individuals survives until advanced ages, we can expect a growing number of individuals with chronic noncommunicable diseases and disability (such as diabetes, neurodegenerative diseases, atherosclerosis, and autoimmunity among others), which in turn determines the allocation of progressively increasing economic and social resources to the health organizations. This extraordinary societal challenge is reflected by the increasing interest of the scientific community for the topic of aging, in order to understand why we age, which mechanisms are at the basis of the process of aging, and how we can counteract them.

Like any other organs and systems of the body, the immune system undergoes a process of aging. Indeed, the continuous adaptation of the body to the challenge from bacterial/viral infections and other stressors is considered the basis of the process of immunosenescence, a profound age-related remodeling of the immune system, characterized by increased morbidity and mortality due to infections and age-related pathologies (Cossarizza and Frasca, 2014; Pinti et al., 2016) It is important to stress that immunosenescence does not imply a simple deterioration of immune functions, but it is rather a multifaceted process, in which some functions undergo a steep decline with aging, others are preserved or even increased(Franceschi et al., 1995; Nasi et al., 2006; Pinti et al., 2004, 2010). In this chapter, we describe the most relevant changes that the immune system undergoes with aging in humans.

Human Aging
https://doi.org/10.1016/B978-0-12-822569-1.00008-1

7.2 Changes in the adaptive immunity

Most of the attention of the scientific community has been devoted to changes of the T and B cell compartment, which are characterized by a more pronounced decline, if compared to innate immunity, and that are responsible of most of the immune defects that we can observe in older people: the increased incidence of infections, the reduced response to vaccination, and the higher frequency of autoimmune diseases (Pinti et al., 2016).

7.2.1 T cells

Concerning T cells, an overwhelming amount of data clearly shows a progressive decline of their number and function with aging, as well as a striking change in their phenotype and functions. The number of T cells has been shown to diminish both in absolute number and percentage among lymphocytes (Valiathan et al., 2016). This strong reduction has been shown in different populations, including healthy subjects or patients affected by chronic diseases, and the trend has been observed even in subjects who reached the extreme limits of human life, such as centenarians or ultra-centenarians. With regard to phenotype, T cells undergo a progressive shift toward a more differentiated phenotype, that is, a striking reduction in the percentage of naïve T cells and a progressive increase in the number of effector memory T cells reexpressing CD45RA (TEMRA), a phenomenon that is more evident in the CD8 + T cell subset (Thome et al., 2014). Furthermore, an age-related restriction in the T cell repertoire diversity can be observed with aging (Qi et al., 2014; Britanova et al., 2014), indicating a reduced plasticity and capability to recognize novel antigens.

Two main phenomena drive this shift: the decline of T cell output by the thymus, and the continuous stimulation of peripheral T cell pool by chronic infections, particularly by cytomegalovirus (CMV). Concerning the former, it is well known that thymus undergoes a progressive involution with age that leads to an almost complete disappearance of thymic stroma at the age of 70–80, although some residual thymic activity can be detected even in centenarians (Nasi et al., 2006; Pinti et al., 2010). The evolutionary reasons of this involution are far for being clear (Aronson, 1991; Dowling and Hodgkin, 2009), but several observations indicate that thymic involution is not a mere loss of function due to unpredicted events, but it is, at least in part, a programmed process, and involves both the thymic stroma and the developing thymocytes (Shanley et al., 2009; Billard, et al., 2011; Sutherland et al., 2005). From the molecular point of view, thymic involution is driven at least by five different classes of mediators, namely type-1 interferons (IFNs), estrogens and androgens, progesterone, glucocorticoids, and adipocyte-derived factors (Sutherland et al., 2005; Anz et al., 2009; Olsen et al., 2001; Staples et al., 1999; Tibbetts et al., 1999; Ashwell et al., 2000). IFNs, sex hormones, and progesterone act directly on thymic epithelial cells, while glucocorticoids act by suppressing thymocyte development (Ashwell et al., 2000).

Adipocytes act in three different ways: (i) by growing at the expense of fibroblasts, which are one of the main sources of several growth factors needed for thymocytes and thymic epithelial cells (Yang et al., 2009); (ii) by producing molecules with thymo-suppressive activity, such as leukemia inhibitory factor and interleukin(IL)-6 (Sempowski, et al., 2000); and (iii) by forming a barrier that impedes the entry of thy-mocyte progenitors into the thymus stroma via post capillary venules (Mori et al., 2007; Cavallotti et al., 2008). It is interesting to note that IFN-α production, likely by plasma-cytoid dendritic cells present in the thymus, is induced by the recognition of virus-derived pathogen–associated molecular patterns, suggesting that viral infections can be a major cause of thymic involution (Schmidlin et al., 2006).

The second phenomenon that drives T cell compartment toward a highly differenti-ated, senescent phenotype is the chronic infection by viruses, including CMV. In older persons, the prevalence of CMV infection is about 70% in high-income countries, and almost 100% in developing nations. Once established, CMV infection is maintained silent by a competent immune system, with periodic, asymptomatic virus reactivation (Fang et al., 2016). The continuous stimulation of the immune system by viral antigens deter-mines an expansion of CMV-specific T cell clones that tend to occupy most if not the entire immunological space (that is, the amount of space in the body available for T cells to reside and proliferate), thus shrinking the space that less differentiated or immature cells can occupy (Ouyang et al., 2004). This phenomenon, present at some extent also in other chronic infections such as that caused by Epstein-Barr virus (Ouyang et al., 2003), has an extraordinary dimension in the case of CMV, as up to 30% of CD4+ and 50% of CD8+ peripheral T cells can be CMV-specific (Pourgheysari et al., 2007). The impact of CMV on T cell compartment is dramatic not only in quantitative, but also in qualitative terms. Indeed, it determines a huge expansion of TEMRA cells—particularly evident in the case of CD8+ T cells—not observed for other chronic viral infections, and leads to important functional alterations, including a reduced capacity to proliferate, and an impair-ment in producing cytokines such as IFN-γ after antigen challenge (Mekker et al., 2012). It is interesting to note that since those cells are phenotypically terminally differentiated and functionally senescent, the single cell response to viral antigen is decreased but the overall response is increased because of the high number of CMV-specific cells (Ouyang et al., 2004). This determines a status of chronic activation of the immune system, likely contrib-uting to the phenomenon of inflammaging (see later).

Thus the final result of the chronic CMV infection in the older people is the presence of few T cell clones, comprising mostly differentiated, senescent cells, which represent the large part of T cell repertoire (Vescovini et al., 2007; Griffiths et al., 2013; Hosie et al., 2017). As naïve T cells are few and most of expanded T cells are senescent, the immune response in older people is characterized by an impaired ability to respond to new antigens, by an unsustained memory response and by a greater propensity to auto-immunity (Pinchera et al., 1995; Prelog, 2006; Goronzy and Weyand, 2012).

7.2.2 B cells

Decline in T cell-mediated response is paralleled by a similar decline in the B cell compartment (Ademokun et al., 2010). Indeed, B cells undergo a progressive reduction in absolute number and percentage within lymphocytes, and show a progressive shift toward a late/exhausted phenotype, at the expense of naïve and switched memory cells (Ademokun et al., 2010; Frasca et al., 2012). As the latter population is the most important for obtaining a protective, optimal humoral response, the functional consequence of this shift is the humoral, antibody-mediated response is less effective in the older adults, as evidenced by the reduction in the efficacy of vaccinations (Frasca et al., 2015a; Frasca et al., 2015b; Kaml et al., 2006; Wolters et al., 2003). From the functional point of view, aging is also characterized by an impaired capability of B cells to undergo class switch and somatic hypermutation, which in turn determines the production of low-affinity antibodies that are less effective in binding the antigens (Frasca and Blomberg, 2014). As in the case of T cells, several studies have investigated the functional mechanisms at the basis of these quantitative and qualitative changes. The reduction of naïve B cells has been attributed, at least in part, to the impaired capability of hemopoietic stem cells (HSC) to differentiate toward the B cell lineage, a phenomenon that is both intrinsic to HSC, and due to changes in the bone marrow stromal cells that drive differentiation of B cells (Guerrettaz et al., 2008).

The reduction of memory switched B cells and the increase of late/exhausted cells are due to the impaired expression of activation-induced cytidine deaminase (AID), the enzyme responsible for introducing DNA mutations in the Ig loci. Such an impaired expression is caused by the lower expression of E2A, the transcription factor needed to up-regulate AID, in response to CD40/CD154 binding and T cell-secreted cytokines, although an intrinsic, age-dependent impairment of CD40 signaling independent of E2A expression cannot be excluded (Frasca et al., 2008).

Less attention has been paid to B cell repertoire if compared to T cell one. As in the case of T cells, change in the B cell repertoire with aging has been analyzed at single clone level, showing some contrasting results, probably because of the limited number of genes analyzed and sequences in B cell clones, as well as the relatively low number of subjects enrolled in these studies (Banerjee et al., 2002; Kolar et al., 2006; Martin et al., 2015; Rosner et al., 2001).

7.3 Changes in the innate immunity

Studies on immunosenescence have been focused on adaptive immunity for decades, and innate immunity had been considered unaffected by aging for a long time. However, several studies performed in the last 30 years have highlighted important changes also in innate immunity, which undergoes a profound remodeling.

7.3.1 Neutrophils

Neutrophils represent a crucial component of the very first phases of the innate immune response and play a crucial role in controlling bacterial infections. Normally neutrophils are 50%–70% of circulating leukocytes; their number is maintained during aging but functional alterations have been observed (Brubaker et al., 2013; Butcher et al., 2001). From the functional point of view, neutrophils from older subjects display a reduced response to chemotactic signals, because of constitutive activation of the lipid kinase phosphoinositide 3-kinase, which is crucial in sensing the gradient of chemotactic molecules (Sapey et al., 2014). Phagocytosis is well preserved when nonopsonized particles are ingested, but not in the case of opsonized pathogens (Butcher et al., 2001). This could be due to the reduced expression of CD16 in neutrophils from older subjects (Butcher et al., 2001) It must be noted, however, that neutrophil functions are well preserved in centenarians (Alonso-Fernández et al., 2008).

7.3.2 Monocytes and macrophages

Monocytes and macrophages are mononuclear cells characterized by inflammatory, immunomodulatory, and tissue-repairing properties, which are crucial players of the innate immune response. Upon stimulation, they can secrete a wide range of cytokines and chemokines, which recruit and modulate the function of other immune cells (Navegantes et al., 2017). According to their expression of CD14 and CD16 on cell surface, monocytes are divided in classical ($CD14^{++}CD16^-$), nonclassical ($CD14^+CD16^{++}$), and intermediate ($CD14^{++}CD16^+$) (Ziegler-Heitbrock, 2015). The absolute number and frequency of monocytes does not change with aging but a marked reduction of classical, counterbalanced by an increase of intermediate and nonclassical monocytes occurs (Metcalf et al., 2015). A significant reduction of reactive oxygen species (ROS) production and phagocytosis capability has been observed (Hearps et al., 2012; McLachlan et al., 1995). A crucial function of monocytes and macrophages is the release of cytokines and chemokines upon stimulation. In particular, the synthesis of tumor necrosis factor(TNF)-α, interleukin(IL)-1**β** and IL-6, after toll-like receptor (TLR) engagement is crucial in mononuclear cell functionality. TLR1/2 engagement is severely impaired in monocytes from older subject and leads to lower production of TNF-α and IL-6. Conversely, TNF-α synthesis upon TLR4 stimulation is enhanced (Van Duin et al., 2007). Monocytes from older subjects show a higher release of IL-8 after stimulation of TLR1/2, TLR2/6, TLR4, or TLR5 (Qian et al., 2012). Such dysregulation is caused by either an altered expression of TLRs or by an impairment of signal transduction. As far as TLR expression is concerned, TLR2 has been shown to be constant, while TLR1 expression declines with age. TLR1/2 signaling, mediated by MAPK and ERK1/1, is severely reduced in cells from old subjects, whereas signaling downstream of TLR5 is enhanced (Qian et al., 2012; Van Duin et al., 2007).

7.3.3 Dendritic cells

Dendritic cells (DCs) are antigen-presenting cells, whose main function is to capture, process, and present antigens to T cells, and to provide costimulatory signals needed to a proper activation of the adaptive response. Different DC populations have been described: conventional DCs, also known as myeloid DCs (mDC), are characterized by the expression of TLR2 and TLR4, and by the secretion of IL-12, while plasmacytoid DCs (pDC) express TLR7 and 9 and release interferon-α in response to viral infections. Langerhans cells (LC) are a specialized macrophage subset resident in the skin that share properties with DCs and have been considered DCs for years (Doebel et al., 2017). For this reason, they are discussed in this section.

Data concerning the change of pDC and mDCs with aging are discordant and not conclusive (Della Bella et al., 2007; Jing et al., 2009; Metcalf et al., 2015; Pérez-Cabezas et al., 2007) while it has been widely reported that a marked reduction of LC occurs. Such reduction is likely associated with a higher risk of skin infection in older subjects (Laube, 2004). Discordant data has been also reported about the functionality of mDC with aging; some authors reported a higher capability to secrete of proinflammatory cytokines upon stimulation in old subjects, while others did not show any change (Janssen et al., 2016) or observed a decreased production (Agrawal et al., 2007). Phagocytosis appears preserved in pDC (do Nascimento et al., 2015) but not in mDC (Agrawal et al., 2007). Although mDC cells express constant levels of TLRs with aging (do Nascimento et al., 2015), a clear decrease of TNF-α, IL-6, and IL-12, as well as IFN-α, IFN-β, and IFN-γ can be observed when cells are stimulated via TLR, or using a viral stimulus suggesting that the signal transduction pathways activated by TLR engagement are impaired in pDCs from old subjects (Jing et al., 2009; Panda et al., 2010). Basal production of proinflammatory cytokine in the absence of TLR engagement is higher in pDCs from old subjects, suggesting dysregulation of cytokine production. Such dysregulation has been also shown in human monocyte-derived DCs, where an age-associated increase in proinflammatory cytokine production has been observed after TLR4 and TLR8 engagement (Agrawal et al., 2007; Bhushan et al., 2002). Animal models provide some hints concerning DC recruitment to secondary lymphoid organs, a topic difficult to be studied in humans. DC recruitment to lymphoid organs is impaired with aging but is not still clear if this phenomenon is due to a reduced capability to sense chemokines, or to a reduced capability of lymphoid organs to attract DCs, or to a combination of both.

7.3.4 Mast cells, eosinophils, and basophils

The effect of age on mast cells, eosinophils, and basophils has not been studied in detail. Mast cells have been only studied in mice, with contrasting results concerning their number (Hart et al., 1999; Nguyen et al., 2005). From the functional point of view, mast cells are more prone to degranulation in old mice (Nguyen et al., 2005).

In humans, data concerning mast cells have been reported only in young and old patients with asthma, and no change has been observed (Atsuta et al., 1999). Similarly, data concerning eosinophils has been obtained in asthmatic patients of different ages, showing an age-related decrease in degranulation and ROS production, upon stimulation with IL-5 (Mathur et al., 2008). Finally, no recent studies are available on basophils.

Taken together, these observations indicate that some functions of immune response are impaired with aging, while others are maintained or even enhanced. Nevertheless, a common trait of these complex changes is the increase in the production of proinflammatory cytokines, which have been often observed in old subjects (Cohen et al., 2003; Ferrucci et al., 1999).

7.4 Inflammaging

The complex remodeling of the immune system described before is associated with a low-grade, chronic state of inflammation accompanied by a reduced capacity to respond to different stressors, which has been dubbed "inflammaging" more than 20 years ago (Franceschi et al., 2000). Thus, inflammation, an evolutionary conserved, complex biological response aimed to eliminate a source of damage and to protect the organism from harmful stimuli, becomes detrimental in older people. Indeed, inflammaging represents a risk factor for most of the diseases commonly associated with aging, such as diabetes (Donath and Shoelson, 2011), Alzheimer's or Parkinson's disease (Liang et al., 2017), atherosclerosis, and cardiovascular diseases (Brüünsgaard and Pedersen, 2003; Skoog et al., 2002), even if it is still unknown whether inflammaging is the cause or rather the effect of the aging process.

Inflammaging is a very complex phenomenon, whose causes and mechanisms are largely more complex than expected when it has been defined, and several organs (such as brain, gut, and liver), tissues (such as adipose or muscular tissues), and cells (including innate immune cells, the main producers of proinflammatory cytokines) are involved in its onset (Calder et al., 2011; Cevenini et al., 2010; Franceschi et al., 2007). So far, several causes of inflammaging have been defined, both exogenous and endogenous; in the framework of this review, we will discuss those more strictly related to the impairment of the immune response, i.e. chronic viral infections, the production of damage-associated molecular patterns (DAMPS), the impairment of complement system, and the microbial components released by gut microbiota.

As stated before, chronic viral infections determine a continuous activation of the adaptive immune response, leading to progressive exhaustion of the T and B cell repertoire. This continuous stimulation, however, can also contribute to fuel inflammaging. Again, CMV is thought to play a prominent role in this phenomenon; CMV contributes to the expansion of late-differentiated CD28-T cells during aging, and these cells can contribute to inflammaging by releasing huge amounts of proinflammatory cytokines upon stimulation. Moreover, T cell activation toward a proinflammatory phenotype caused by a viral infection can in turn lead to the release of inflammatory mediators directly by immune cells or by other cell types (Bauer and De la Fuente, 2016; Lee et al., 2014; Yurochko et al., 1997). Since other persistent viral infections, such as human immunodeficiency virus (HIV) or hepatitis C virus, are characterized by accelerated immunosenescence and an increase of age-related, inflammatory diseases, we can speculate that also these viral infections can fuel inflammaging. It must be noted, however, that CMV is almost always present in HIV + patients, and so it is difficult to distinguish the relative contribution of each infection. In sharp contrast with these observations, a study excluded CMV infection as a primary causative factor of inflammaging, casting some doubts on the role of chronic infections in inflammaging (Bartlett et al., 2012) (see also Chapter 16).

The complement system induces local inflammatory reactions. Its age-related impairment has been observed in several neurodegenerative diseases, such as age-related macular degeneration (van Lookeren Campagne et al., 2016). Furthermore, the activation of alternative complement pathway has been proved to be involved in the pathogenesis of atherosclerosis, as increased deposition of C3 has been observed in the intima of atherosclerotic lesions, and the C5b-9 complex is present in atherosclerotic plaques from the earliest to advanced lesions (Torzewski et al., 1998).

In this context, mitochondria have emerged as a main source of molecules containing damage–associated molecular patterns, as many of their components are recognized as foreign molecules of bacterial origin by cells of the innate immunity, and trigger an inflammatory response (Grazioli and Pugin, 2018). Among these molecules, a main role is played by N–formyl peptides, which are chemoattractant for neutrophils (Carp, 1982), cardiolipin which can activate NLP3 inflammasome (Iyer et al., 2013) and mitochondrial (mt)DNA, which is rich in unmethylated CpG regions and is recognized by TLR9, a receptor specialized in binding viral or bacterial DNA (Fang et al., 2016). Concerning the latter, we have shown that its plasma levels increase with age and are correlated with levels of proinflammatory cytokines such as IL-6 or IL1-**β** (Pinti et al., 2014). As mtDNA is bound by TLR9, it can cause an inflammatory response (Zhang et al., 2010), and we demonstrated that mtDNA is able to elicit the release of proinflammatory cytokines by monocytes in vitro (Pinti et al., 2014).

Finally, gut microbiota is emerging as a crucial player in the process of inflammaging. Indeed, the antigenic load of foods and bacteria present in the gut deeply influences the homeostasis of the immune system, as shown in diseases with an important immune component (including diabetes, multiple sclerosis, Alzheimer's disease (Cantarel et al., 2015; Fujimura and Lynch, 2015; Rothhammer et al., 2018; Spielman et al., 2018) and represents an important source of inflammatory stimuli during aging (Biagi et al., 2010) (see also Chapter 6).

Inflammaging is a process observed in any older subject. However, it is interesting to note that centenarians seem to be equipped with gene variants that optimize the balance between pro- and anti-inflammatory molecules, so minimizing the detrimental effect of inflammaging (Ostan et al., 2008). In these exceptional subjects, the increased plasma levels of proinflammatory molecules, such as IL-6, IL-15, or IL-18, C-Reactive Protein, resistin, von Willebrand Factor, typically associated with inflammaging, are counterbalanced by higher levels of antiinflammatory molecules, such as transforming growth factor-β, 1, IL-1 receptor antagonist, or cortisol (Collino et al., 2013; Genedani et al., 2008; Gerli et al., 2000; Meazza et al., 2011; Morrisette-Thomas et al., 2014). Thus, if we take centenarians as a model to successful aging, we can argue that it is not a matter of reducing inflammation, but rather of counterbalancing it.

7.5 Conclusions and future perspectives

Studies in the last 40 years have shed light on important changes which the immune system undergoes with aging. Nevertheless, the scientific literature in the field of immunology of aging often includes contradictory results, which make sometimes difficult to draw a general and consistent picture. Most of these issues are due to the selection criteria of the samples used for the analysis, to the limited sample size, and to inconsistencies in the methods used for the analysis.

Nowadays, the -omics approach in the study of immunosenescence is providing new, precious information in its comprehension and has opened new challenges. For instance, glycomics—the study of the repertoire of glycan sugars produced by the organism—allowed getting new insights into the comprehension of the alterations of sugar side chains of antibodies observed during aging (Dall'Olio et al., 2013), a phenomenon discovered more than 30 years ago (Parekh et al., 1988). Similarly, the explosion of cytomics technologies in the last 20 years has provided immunologists with powerful tools that allow the analysis of a huge number of parameters at the same time, in a single cell, such as fluorescence-based flow cytometry, mass cytometry based on nonfluorescent markers (Bjornson et al., 2013), laser scanning cytometry (Megason and Fraser, 2003), among others. In this way, it is possible to integrate information that in the past were analyzed separately or not analyzed at all, such as surface marker phenotype, functional capacities (in terms of cytokine production), and antigen specificities (Newell et al., 2012). This approach is particularly useful in the field of immunology of aging, where samples are precious and often limited in size, and could help to clarify some crucial aspects of the evolution of the immune system with age. For example, several attempts have been done to identify an "immunological score" aimed at predicting frailty, morbidity, and mortality. In particular, the OCTO and NONA studies, conducted in Sweden, led to the definition of an "Immune Risk Profile" that can be used to predict mortality in older individuals, with a certain degree of success (Strindhall et al., 2007; Wikby et al.,

1998, 2002, 2006). However, such attempts suffered from the technological limits of the methods available at the time of the study: the number of parameters that could be analyzed was limited, and the role and the significance of some cell surface markers (for example, the checkpoint regulators) were far from being defined. The possibility to analyze at the same time many phenotypic and functional parameters will allow defining a more precise picture of immunosenescence, and likely to define a highly reliable "immunoscore," that could be applied not only for the study of physiological aging but also for studying the aging of the immune system in physiopathologic conditions.

Finally, a rigorous, highly standardized protocol for processing and analyzing samples, which must be shared among the scientific community, will help to obtain a consistent result. In this perspective, a further effort should be done in terms of selection and recruitment of older subjects and should be necessarily accompanied by the accurate collection of demographics, clinical and behavioral data, which are precious for correlating immunological profile with successful aging.

References

Ademokun A, Wu YC, Dunn-Walters D. The ageing B cell population: composition and function. Biogerontology 2010;11(2):125–37. https://doi.org/10.1007/s10522-009-9256-9.

Agrawal A, Agrawal S, Cao JN, Su H, Osann K, Gupta S. Altered innate immune functioning of dendritic cells in elderly humans: a role of phosphoinositide 3-kinase-signaling pathway. J Immunol 2007;178 (11):6912–22. https://doi.org/10.4049/jimmunol.178.11.6912.

Alonso-Fernández P, Puerto M, Maté I, Ribera JM, De La Fuente M. Neutrophils of centenarians show function levels similar to those of young adults. J Am Geriatr Soc 2008;56(12):2244–51. https://doi.org/10.1111/j.1532-5415.2008.02018.x.

Anz D, Thaler R, Stephan N, Waibler Z, Trauscheid MJ, Scholz C, Kalinke U, Barchet W, Endres S, Bourquin C. Activation of melanoma differentiation-associated gene 5 causes rapid involution of the thymus. J Immunol 2009;182(10):6044–50.

Aronson M. Hypothesis: involution of the thymus with aging—programmed and beneficial. Thymus 1991;18(1):7–13.

Ashwell JD, Lu FW, Vacchio MS. Glucocorticoids in T cell development and function*. Annu Rev Immunol 2000;18:309–45.

Atsuta R, Akiyama K, Shirasawa T, Okumura K, Fukuchi Y, Ra C. Atopic asthma is dominant in elderly onset asthmatics: possibility for an alteration of mast cell function by aging through Fc receptor expression. Int Arch Allergy Immunol 1999;120(1):76–81. https://doi.org/10.1159/000053600.

Banerjee M, Mehr R, Belelovsky A, Spencer J, Dunn-Walters DK. Age- and tissue-specific differences in human germinal center B cell selection revealed by analysis of IgVH gene hypermutation and lineage trees. Eur J Immunol 2002;32(7):1947–57. https://doi.org/10.1002/1521-4141(200207)32:7<1947:: AID-IMMU1947>3.0.CO;2-1.

Bartlett DB, Firth CM, Phillips AC, Moss P, Baylis D, Syddall H, Sayer AA, Cooper C, Lord JM. The age-related increase in low-grade systemic inflammation (Inflammaging) is not driven by cytomegalovirus infection. Aging Cell 2012;11(5):912–5. https://doi.org/10.1111/j.1474-9726.2012.00849.x.

Bauer ME, De la Fuente M. The role of oxidative and inflammatory stress and persistent viral infections in immunosenescence. Mech Ageing Dev 2016;158:27–37. https://doi.org/10.1016/j.mad.2016.01.001.

Bhushan M, Cumberbatch M, Dearman RJ, Andrew SM, Kimber I, Griffiths CEM. Tumour necrosis factor-α-induced migration of human Langerhans cells: the influence of ageing. Br J Dermatol 2002;146(1):32–40. https://doi.org/10.1046/j.1365-2133.2002.04549.x.

Biagi E, Nylund L, Candela M, Ostan R, Bucci L, Pini E, Nikkila J, Monti D, Satokari R, Franceschi C, Brigidi P, De Vos W. Through ageing and beyond: gut microbiota and inflammatory status in seniors and centenarians. PLoS One 2010;5:e10667.

Billard MJ, Gruver AL, Sempowski GD. Acute endotoxin-induced thymic atrophy is characterized by intrathymic inflammatory and wound healing responses. PLoS One 2011;6(3)e17940.

Bjornson ZB, Nolan GP, Fantl WJ. Single-cell mass cytometry for analysis of immune system functional states. Curr Opin Immunol 2013;25(4):484–94. https://doi.org/10.1016/j.coi.2013.07.004.

Brubaker AL, Rendon JL, Ramirez L, Choudhry MA, Kovacs EJ. Reduced neutrophil chemotaxis and infiltration contributes to delayed resolution of cutaneous wound infection with advanced age. J Immunol 2013;190(4):1746–57. https://doi.org/10.4049/jimmunol.1201213.

Britanova OV, Putintseva EV, Shugay M, Merzlyak EM, Turchaninova MA, Staroverov DB, Bolotin DA, Lukyanov S, Bogdanova EA, Mamedov IZ, Lebedev YB, Chudakov DM. Age-related decrease in TCR repertoire diversity measured with deep and normalized sequence profiling. J Immunol 2014;192 (6):2689–98.

Brüünsgaard H, Pedersen BK. Age-related inflammatory cytokines and disease. Immunol Allergy Clin North Am 2003;23(1):15–39. https://doi.org/10.1016/S0889-8561(02)00056-5.

Butcher SK, Chahal H, Nayak L, Sinclair A, Henriquez NV, Sapey E, O'Mahony D, Lord JM. Senescence in innate immune responses: reduced neutrophil phagocytic capacity and CD16 expression in elderly humans. J Leukoc Biol 2001;70(6):881–6.

Calder PC, Ahluwalia N, Brouns F, Buetler T, Clement K, Cunningham K, Esposito K, Jönsson LS, Kolb H, Lansink M, Marcos A, Margioris A, Matusheski N, Nordmann H, O'Brien J, Pugliese G, Rizkalla S, Schalkwijk C, Tuomilehto J, … Winklhofer-Roob BM. Dietary factors and low-grade inflammation in relation to overweight and obesity. Br J Nutr 2011;106(3):S5–S78. https://doi.org/10.1017/s0007114511005460.

Cantarel BL, Waubant E, Chehoud C, Kuczynski J, Desantis TZ, Warrington J, Venkatesan A, Fraser CM, Mowry EM. Gut microbiota in multiple sclerosis: possible influence of immunomodulators. J Invest Med 2015;63(5):729–34. https://doi.org/10.1097/JIM.0000000000000192.

Carp H. Mitochondrial n-formylmethionyl proteins as chemoattractants for neutrophils. J Exp Med 1982;155(1):264–75. https://doi.org/10.1084/jem.155.1.264.

Cavallotti C, D'Andrea V, Tonnarini G, Cavallotti C, Bruzzone P. Age-related changes in the human thymus studied with scanning electron microscopy. Microsc Res Tech 2008;71(8):573–8.

Cevenini E, Caruso C, Candore G, Capri M, Nuzzo D, Duro G, Rizzo C, Colonna-Romano G, Lio D, Di Carlo D, Palmas MG, Scurti M, Pini E, Franceschi C, Vasto S. Age-related inflammation: the contribution of different organs, tissues and systems. How to face it for therapeutic approaches. Curr Pharm Des 2010;16(6):609–18. https://doi.org/10.2174/138161210790883840.

Cohen HJ, Harris T, Pieper CF. Coagulation and activation of inflammatory pathways in the development of functional decline and mortality in the elderly. Am J Med 2003;114(3):180–7. https://doi.org/10.1016/S0002-9343(02)01484-5.

Collino S, Montoliu I, Martin FPJ, Scherer M, Mari D, Salvioli S, Bucci L, Ostan R, Monti D, Biagi E, Brigidi P, Franceschi C, Rezzi S. Metabolic signatures of extreme longevity in northern Italian centenarians reveal a complex remodeling of lipids, amino acids, and gut microbiota metabolism. PLoS One 2013;8(3)https://doi.org/10.1371/journal.pone.0056564.

Cossarizza A, Frasca D. Aging and longevity: an immunological perspective. Immunol Lett 2014;162 (1):279–80. https://doi.org/10.1016/j.imlet.2014.09.003.

Dall'Olio F, Vanhooren V, Chen CC, Slagboom PE, Wuhrer M, Franceschi C. N-glycomic biomarkers of biological aging and longevity: a link with inflammaging. Ageing Res Rev 2013;12(2):685–98. https://doi.org/10.1016/j.arr.2012.02.002.

Della Bella S, Bierti L, Presicce P, Arienti R, Valenti M, Saresella M, Vergani C, Villa ML. Peripheral blood dendritic cells and monocytes are differently regulated in the elderly. Clin Immunol 2007;122(2):220–8. https://doi.org/10.1016/j.clim.2006.09.012.

do Nascimento MPP, Pinke KH, Penitenti M, Ikoma MRV, Lara VS. Aging does not affect the ability of human monocyte-derived dendritic cells to phagocytose Candida albicans. Aging Clin Exp Res 2015;27 (6):785–9. https://doi.org/10.1007/s40520-015-0344-1.

Doebel T, Voisin B, Nagao K. Langerhans cells—the macrophage in dendritic cell clothing. Trends Immunol 2017;38(11):817–28. https://doi.org/10.1016/j.it.2017.06.008.

Donath MY, Shoelson SE. Type 2 diabetes as an inflammatory disease. Nat Rev Immunol 2011;11 (2):98–107. https://doi.org/10.1038/nri2925.

Dowling MR, Hodgkin PD. Why does the thymus involute? A selection-based hypothesis, Trends Immunol 2009;30(7):295–300.

Fang C, Wei X, Wei Y. Mitochondrial DNA in the regulation of innate immune responses. Protein Cell 2016;7(1):11–6. https://doi.org/10.1007/s13238-015-0222-9.

Ferrucci L, Harris TB, Guralnik JM, Tracy RP, Corti MC, Cohen HJ, Penninx B, Pahor M, Wallace R, Havlik RJ. Serum IL-6 level and the development of disability in older persons. J Am Geriatr Soc 1999;47 (6):639–46. https://doi.org/10.1111/j.1532-5415.1999.tb01583.x.

Franceschi C, Monti D, Sansoni P, Cossarizza A. The immunology of exceptional individuals: the lesson of centenarians. Immunol Today 1995;16(1):12–6. https://doi.org/10.1016/0167-5699(95)80064-6.

Franceschi C, Bonafè M, Valensin S, Olivieri F, De Luca M, Ottaviani E, De Benedictis G. Inflamm-aging. An evolutionary perspective on immunosenescence, Ann N Y Acad Sci 2000;908:244–54. New York Academy of Sciences. https://doi.org/10.1111/j.1749-6632.2000.tb06651.x.

Franceschi C, Capri M, Monti D, Giunta S, Olivieri F, Sevini F, Panourgia MP, Invidia L, Celani L, Scurti M, Cevenini E, Castellani GC, Salvioli S. Inflammaging and anti-inflammaging: a systemic perspective on aging and longevity emerged from studies in humans. Mech Ageing Dev 2007;128 (1):92–105. https://doi.org/10.1016/j.mad.2006.11.016.

Frasca D, Blomberg BB. B cell function and influenza vaccine responses in healthy aging and disease. Curr Opin Immunol 2014;29(1):112–8. https://doi.org/10.1016/j.coi.2014.05.008.

Frasca D, Landin AM, Lechner SC, Ryan JG, Schwartz R, Riley RL, Blomberg BB. Aging down-regulates the transcription factor e2a, activation-induced cytidine deaminase, and Ig class switch in human b cells. J Immunol 2008;180(8):5283–90. https://doi.org/10.4049/jimmunol.180.8.5283.

Frasca D, Diaz A, Romero M, Phillips M, Mendez NV, Landin AM, Blomberg BB. Unique biomarkers for B-cell function predict the serum response to pandemic H1N1 influenza vaccine. Int Immunol 2012;24 (3):175–82. https://doi.org/10.1093/intimm/dxr123.

Frasca D, Diaz A, Blomberg BB. Activation-induced cytidine deaminase and switched memory B cells as predictors of effective in vivo responses to the influenza vaccine, Methods Mol Biol 2015a;1343:107–14. Humana Press Inc. https://doi.org/10.1007/978-1-4939-2963-4_9.

Frasca D, Diaz A, Romero M, Landin AM, Blomberg BB. Cytomegalovirus (CMV) seropositivity decreases B cell responses to the influenza vaccine. Vaccine 2015b;33(12):1433–9. https://doi.org/10.1016/j. vaccine.2015.01.071.

Fujimura KE, Lynch SV. Microbiota in allergy and asthma and the emerging relationship with the gut microbiome. Cell Host Microbe 2015;17(5):592–602. https://doi.org/10.1016/j.chom.2015.04.007.

Genedani S, Filaferro M, Carone C, Ostan R, Bucci L, Cevenini E, Franceschi C, Monti D. Influence of f-MLP, ACTH(1-24) and CRH on in vitro chemotaxis of monocytes from centenarians. Neuroimmunomodulation 2008;15(4–6):285–9. https://doi.org/10.1159/000156472.

Gerli R, Monti D, Bistoni O, Mazzone A, Peri G, Cossarizza A, Gioacchino D, Cesarotti M, Doni M, Mantovani A, Franceschi A, Paganelli C. Chemokines sTNF-Rs and sCD30 serum levels in healthy aged people and centenarians. Mech Ageing Dev 2000;121(1–3):37–46.

Goronzy JJ, Weyand CM. Immune aging and autoimmunity. Cell Mol Life Sci 2012;69(10):1615–23.

Grazioli S, Pugin J. Mitochondrial damage-associated molecular patterns: from inflammatory signaling to human diseases. Front Immunol 2018;9. https://doi.org/10.3389/fimmu.2018.00832 MAY.

Griffiths SJ, Riddell NE, Masters J, Libri V, Henson SM, Wertheimer A, Wallace D, Sims S, Rivino L, Larbi A, Kemeny DM, Nikolich-Zugich J, Kern F, Klenerman P, Emery VC, Akbar AN. Age-associated increase of low-avidity cytomegalovirus-specific CD8+ T cells that re-express CD45RA. J Immunol 2013;190(11):5363–72.

Guerrettaz LM, Johnson SA, Cambier JC. Acquired hematopoietic stem cell defects determine B-cell repertoire changes associated with aging. Proc Natl Acad Sci U S A 2008;105(33):11898–902. https://doi. org/10.1073/pnas.0805498105.

Hart PH, Grimbaldeston MA, Hosszu EK, Swift GJ, Noonan FP, Finlay-Jones JJ. Age-related changes in dermal mast cell prevalence in BALB/c mice: functional importance and correlation with dermal mast cell expression of kit. Immunology 1999;98(3):352–6. https://doi.org/10.1046/j.1365-2567.1999.00897.x.

Hearps AC, Martin GE, Angelovich TA, Cheng WJ, Maisa A, Landay AL, Jaworowski A, Crowe SM. Aging is associated with chronic innate immune activation and dysregulation of monocyte phenotype and function. Aging Cell 2012;11(5):867–75. https://doi.org/10.1111/j.1474-9726.2012.00851.x.

Hosie L, Pachnio A, Zuo J, Pearce H, Riddell S, Moss P. Cytomegalovirus-specific T cells restricted by HLA-Cw*0702 increase markedly with age and dominate the CD8(+) T-cell repertoire in older people. Front Immunol 2017;8:1776.

Iyer SS, He Q, Janczy JR, Elliott EI, Zhong Z, Olivier AK, Sadler JJ, Knepper-Adrian V, Han R, Qiao L, Eisenbarth SC, Nauseef WM, Cassel SL, Sutterwala FS. Mitochondrial cardiolipin is required for Nlrp3 inflammasome activation. Immunity 2013;39(2):311–23. https://doi.org/10.1016/j.immuni.2013.08.001.

Janssen N, Derhovanessian E, Demuth I, Arnaout F, Steinhagen-Thiessen E, Pawelec G. Responses of dendritic cells to TLR-4 stimulation are maintained in the elderly and resist the effects of CMV infection seen in the young. J Gerontol A Biol Sci Med Sci 2016;71(9):1117–23. https://doi.org/10.1093/gerona/glv119.

Jing Y, Shaheen E, Drake RR, Chen N, Gravenstein S, Deng Y. Aging is associated with a numerical and functional decline in plasmacytoid dendritic cells, whereas myeloid dendritic cells are relatively unaltered in human peripheral blood. Hum Immunol 2009;70(10):777–84. https://doi.org/10.1016/j.humimm.2009.07.005.

Kaml M, Weiskirchner I, Keller M, Luft T, Hoster E, Hasford J, Young L, Bartlett B, Neuner C, Fischer KH, Neuman B, Würzner R, Grubeck-Loebenstein B. Booster vaccination in the elderly: their success depends on the vaccine type applied earlier in life as well as on pre-vaccination antibody titers. Vaccine 2006;24(47–48):6808–11. https://doi.org/10.1016/j.vaccine.2006.06.037.

Kolar GR, Mehta D, Wilson PC, Capra JD. Diversity of the Ig repertoire is maintained with age in spite of reduced germinal centre cells in human tonsil lymphoid tissue. Scand J Immunol 2006;64(3):314–24. https://doi.org/10.1111/j.1365-3083.2006.01817.x.

Laube S. Skin infections and ageing. Ageing Res Rev 2004;3(1):69–89. https://doi.org/10.1016/j.arr.2003.08.003.

Lee YL, Liu CE, Cho WL, Kuo CL, Cheng WL, Huang CS, Liu CS. Presence of cytomegalovirus DNA in leucocytes is associated with increased oxidative stress and subclinical atherosclerosis in healthy adults. Biomarkers 2014;19(2):109–13. https://doi.org/10.3109/1354750X.2013.877967.

Liang Z, Zhao Y, Ruan L, Zhu L, Jin K, Zhuge Q, Su DM, Zhao Y. Impact of aging immune system on neurodegeneration and potential immunotherapies. Prog Neurobiol 2017;157:2–28. https://doi.org/10.1016/j.pneurobio.2017.07.006.

Martin V, Wu B, Kipling Y, Dunn-Walters D. Ageing of the B-cell repertoire. Philos Trans R Soc Lond B Biol Sci 2005;370:20140237. https://doi.org/10.1098/rstb.2014.0237.

Mathur SK, Schwantes EA, Jarjour NN, Busse WW. Age-related changes in eosinophil function in human subjects. Chest 2008;133(2):412–9. https://doi.org/10.1378/chest.07-2114.

McLachlan JA, Serkin CD, Morrey KM, Bakouche O. Antitumoral properties of aged human monocytes. J Immunol 1995;154(2):832–43.

Meazza C, Vitale G, Pagani S, Castaldi D, Ogliari G, Mari D, Laarej K, Tinelli C, Bozzola M. Common adipokine features of neonates and centenarians. J Pediatr Endocrinol Metab 2011;24(11 – 12):953–7. https://doi.org/10.1515/JPEM.2011.373.

Megason SG, Fraser SE. Digitizing life at the level of the cell: high-performance laser-scanning microscopy and image analysis for in toto imaging of development. Mech Dev 2003;120(11):1407–20. https://doi.org/10.1016/j.mod.2003.07.005.

Mekker A, Tchang VS, Haeberli L, Oxenius A, Trkola A, Karrer U. Immune senescence: relative contributions of age and cytomegalovirus infection. PLoS Pathog 2012;8(8)e1002850.

Metcalf TU, Cubas RA, Ghneim K, Cartwright MJ, Grevenynghe JV, Richner JM, Olagnier DP, Wilkinson PA, Cameron MJ, Park BS, Hiscott JB, Diamond MS, Wertheimer AM, Nikolich-Zugich J, Haddad EK. Global analyses revealed age-related alterations in innate immune responses after stimulation of pathogen recognition receptors. Aging Cell 2015;14(3):421–32. https://doi.org/10.1111/acel.12320.

Mori K, Itoi M, Tsukamoto N, Kubo H, Amagai T. The perivascular space as a path of hematopoietic progenitor cells and mature T cells between the blood circulation and the thymic parenchyma. Int Immunol 2007;19(6):745–53.

Morrisette-Thomas V, Cohen AA, Fülöp T, Riesco E, Legault V, Li Q, Milot E, Dusseault-Bélanger F, Ferrucci L. Inflamm-aging does not simply reflect increases in pro-inflammatory markers. Mech Ageing Dev 2014;139(1):49–57. https://doi.org/10.1016/j.mad.2014.06.005.

Nasi M, Troiano L, Lugli E, Pinti M, Ferraresi R, Monterastelli E, Mussi C, Salvioli G, Franceschi C, Cossarizza A. Thymic output and functionality of the IL-7/IL-7 receptor system in centenarians: implications for the neolymphogenesis at the limit of human life. Aging Cell 2006;5(2):167–75. https://doi.org/10.1111/j.1474-9726.2006.00204.x.

Navegantes KC, Souza Gomes R, Pereira PAT, Czaikoski PG, Azevedo CHM, Monteiro MC. Immune modulation of some autoimmune diseases: the critical role of macrophages and neutrophils in the innate and adaptive immunity. J Transl Med 2017;15(1). https://doi.org/10.1186/s12967-017-1141-8.

Newell EW, Sigal N, Bendall SC, Nolan GP, Davis MM. Cytometry by time-of-flight shows combinatorial cytokine expression and virus-specific cell niches within a continuum of CD8 + T cell phenotypes. Immunity 2012;36(1):142–52. https://doi.org/10.1016/j.immuni.2012.01.002.

Nguyen MT, Pace AJ, Koller BH. Age-induced reprogramming of mast cell degranulation. J Immunol 2005;175(9):5701–7. https://doi.org/10.4049/jimmunol.175.9.5701.

Olsen NJ, Olson G, Viselli SM, Gu X, Kovacs WJ. Androgen receptors in thymic epithelium modulate thymus size and thymocyte development. Endocrinology 2001;142(3):1278–83.

Ostan R, Bucci L, Capri M, Salvioli S, Scurti M, Pini E, Monti D, Franceschi C. Immunosenescence and immunogenetics of human longevity. Neuroimmunomodulation 2008;15(4–6):224–40. https://doi.org/10.1159/000156466.

Ouyang Q, Wagner WM, Walter S, Muller CA, Wikby A, Aubert G, Klatt T, Stevanovic S, Dodi T, Pawelec G. An age-related increase in the number of CD8+ T cells carrying receptors for an immunodominant Epstein-Barr virus (EBV) epitope is counteracted by a decreased frequency of their antigen-specific responsiveness. Mech Ageing Dev 2003;124(4):477–85.

Ouyang Q, Wagner WM, Zheng W, Wikby A, Remarque EJ, Pawelec G. Dysfunctional CMV-specific CD8(+) T cells accumulate in the elderly. Exp Gerontol 2004;39(4):607–13.

Panda A, Qian F, Mohanty S, Van Duin D, Newman FK, Zhang L, Chen S, Towle V, Belshe RB, Fikrig E, Allore HG, Montgomery RR, Shaw AC. Age-associated decrease in TLR function in primary human dendritic cells predicts influenza vaccine response. J Immunol 2010;184(5):2518–27. https://doi.org/10.4049/jimmunol.0901022.

Parekh R, Roitt I, Isenberg D, Dwek R, Rademacher T. Age-related galactosylation of the N-linked oligosaccharides of human serum IgG. J Exp Med 1988;167(5):1731–6. https://doi.org/10.1084/jem.167.5.1731.

Pérez-Cabezas B, Naranjo-Gómez M, Fernández MA, Grífols JR, Pujol-Borrell R, Borràs FE. Reduced numbers of plasmacytoid dendritic cells in aged blood donors. Exp Gerontol 2007;42(10):1033–8. https://doi.org/10.1016/j.exger.2007.05.010.

Pinchera A, Mariotti S, Barbesino G, Bechi R, Sansoni P, Fagiolo U, Cossarizza A, Franceschi C. Thyroid autoimmunity and ageing. Horm Res 1995;43(1–3):64–8.

Pinti M, Troiano L, Nasi M, Bellodi C, Ferraresi R, Mussi C, Salvioli G, Cossarizza A. Balanced regulation of mRNA production for Fas and Fas ligand in lymphocytes from centenarians: how the immune system starts its second century. Circulation 2004;110(19):3108–14. https://doi.org/10.1161/01.CIR.0000146903.43026.82.

Pinti M, Nasi M, Lugli E, Gibellini L, Bertoncelli L, Roat E, De Biasi S, Mussini C, Cossarizza A. T cell homeostasis in centenarians: from the thymus to the periphery. Curr Pharm Des 2010;16(6):597–603. https://doi.org/10.2174/138161210790883705.

Pinti M, Cevenini E, Nasi M, De Biasi S, Salvioli S, Monti D, Benatti S, Gibellini L, Cotichini R, Stazi MA, Trenti T, Franceschi C, Cossarizza A. Circulating mitochondrial DNA increases with age and is a familiar trait: implications for \inflamm-aging\. Eur J Immunol 2014;44(5):1552–62. https://doi.org/10.1002/eji.201343921.

Pinti M, Appay V, Campisi J, Frasca D, Fülöp T, Sauce D, Larbi A, Weinberger B, Cossarizza A. Aging of the immune system: focus on inflammation and vaccination. Eur J Immunol 2016;46(10):2286–301. https://doi.org/10.1002/eji.201546178.

Pourgheysari B, Khan N, Best D, Bruton R, Nayak L, Moss PA. The cytomegalovirus-specific CD4+ T-cell response expands with age and markedly alters the CD4+ T-cell repertoire. J Virol 2007;81(14):7759–65.

Prelog M. Aging of the immune system: a risk factor for autoimmunity? Autoimmun Rev 2006;5(2):136–9.

Qi Q, Liu Y, Cheng Y, Glanville J, Zhang D, Lee JY, Olshen RA, Weyand CM, Boyd SD, Goronzy JJ. Diversity and clonal selection in the human T-cell repertoire. Proc Natl Acad Sci U S A 2014;111 (36):13139–44.

Qian F, Wang X, Zhang L, Chen S, Piecychna M, Allore H, Bockenstedt L, Malawista S, Bucala R, Shaw AC, Fikrig E, Montgomery RR. Age-associated elevation in TLR5 leads to increased inflammatory responses in the elderly. Aging Cell 2012;11(1):104–10. https://doi.org/10.1111/j.1474-9726.2011.00759.x.

Robine JM, Cubaynes S. Worldwide demography of centenarians. Mech Ageing Dev 2017;165:59–67. https://doi.org/10.1016/j.mad.2017.03.004.

Rosner K, Winter DB, Kasmer C, Skovgaard GL, Tarone RE, Bohr VA, Gearhart PJ. Impact of age on hypermutation of immunoglobulin variable genes in humans. J Clin Immunol 2001;21(2):102–15. https://doi.org/10.1023/A:1011003821798.

Rothhammer V, Borucki DM, Tjon EC, Takenaka MC, Chao CC, Ardura-Fabregat A, De Lima KA, Gutiérrez-Vázquez C, Hewson P, Staszewski O, Blain M, Healy L, Neziraj T, Borio M, Wheeler M, Dragin LL, Laplaud DA, Antel J, Alvarez JI, … Quintana FJ. Microglial control of astrocytes in response to microbial metabolites. Nature 2018;557(7707):724–8. https://doi.org/10.1038/s41586-018-0119-x.

Sapey E, Greenwood H, Walton G, Mann E, Love A, Aaronson N, Insall RH, Stockley RA, Lord JM. Phosphoinositide 3-kinase inhibition restores neutrophil accuracy in the elderly: toward targeted treatments for immunosenescence. Blood 2014;123(2):239–48. https://doi.org/10.1182/blood-2013-08-519520.

Schmidlin H, Dontje W, Groot F, Ligthart SJ, Colantonio AD, Oud ME, Schilder-Tol EJ, Spaargaren M, Spits H, Uittenbogaart CH, Blom B. Stimulated plasmacytoid dendritic cells impair human T-cell development. Blood 2006;108(12):3792–800.

Sempowski GD, Hale LP, Sundy JS, Massey JM, Koup RA, Douek DC, Patel DD, Haynes BF. Leukemia inhibitory factor, oncostatin M, IL-6, and stem cell factor mRNA expression in human thymus increases with age and is associated with thymic atrophy. J Immunol 2000;164(4):2180–7.

Shanley DP, Aw D, Manley NR, Palmer DB. An evolutionary perspective on the mechanisms of immunosenescence. Trends Immunol 2009;30(7):374–81.

Skoog T, Dichtl W, Boquist S, Skoglund-Andersson C, Karpe F, Tang R, Bond MG, De Faire U, Nilsson J, Eriksson P, Hamsten A. Plasma tumour necrosis factor-α and early carotid atherosclerosis in healthy middle-aged men. Eur Heart J 2002;23(5):376–83. https://doi.org/10.1053/euhj.2001.2805.

Spielman LJ, Gibson DL, Klegeris A. Unhealthy gut, unhealthy brain: the role of the intestinal microbiota in neurodegenerative diseases. Neurochem Int 2018;120:149–63. https://doi.org/10.1016/j.neuint.2018.08.005.

Staples JE, Gasiewicz TA, Fiore NC, Lubahn DB, Korach KS, Silverstone AE. Estrogen receptor alpha is necessary in thymic development and estradiol-induced thymic alterations. J Immunol 1999;163 (8):4168–74.

Strindhall J, Nilsson BO, Löfgren S, Ernerudh J, Pawelec G, Johansson B, Wikby A. No immune risk profile among individuals who reach 100 years of age: findings from the Swedish NONA immune longitudinal study. Exp Gerontol 2007;42(8):753–61. https://doi.org/10.1016/j.exger.2007.05.001.

Sutherland JS, Goldberg GL, Hammett MV, Uldrich AP, Berzins SP, Heng TS, Blazar BR, Millar JL, Malin MA, Chidgey AP, Boyd RL. Activation of thymic regeneration in mice and humans following androgen blockade. J Immunol 2005;175(4):2741–53.

Thome JJ, Yudanin N, Ohmura Y, Kubota M, Grinshpun B, Sathaliyawala T, Kato T, Lerner H, Shen Y, Farber DL. Spatial map of human T cell compartmentalization and maintenance over decades of life. Cell 2014;159(4):814–28.

Tibbetts TA, DeMayo F, Rich S, Conneely OM, O'Malley BW. Progesterone receptors in the thymus are required for thymic involution during pregnancy and for normal fertility. Proc Natl Acad Sci U S A 1999;96(21):12021–6.

Torzewski M, Klouche M, Hock J, Meßner M, Dorweiler B, Torzewski J, Gabbert HE, Bhakdi S. Immunohistochemical demonstration of enzymatically modified human LDL and its colocalization with the terminal complement complex in the early atherosclerotic lesion. Arterioscler Thromb Vasc Biol 1998;18 (3):369–78. https://doi.org/10.1161/01.ATV.18.3.369.

United Nations, Department of Economic and Social Affairs, Population Division. World Population Prospects: The 2017 Revision, Volume I: Comprehensive Tables (ST/ESA/SER.A/399). New York, NY: United Nations; 2017.

Valiathan R, Ashman M, Asthana D. Effects of ageing on the immune system: infants to elderly. Scand J Immunol 2016;83(4):255–66.

Van Duin D, Mohanty S, Thomas V, Ginter S, Montgomery RR, Fikrig E, Allore HG, Medzhitov R, Shaw AC. Age-associated defect in human TLR-1/2 function. J Immunol 2007;178(2):970–5. https://doi.org/10.4049/jimmunol.178.2.970.

van Lookeren Campagne M, Strauss EC, Yaspan BL. Age-related macular degeneration: complement in action. Immunobiology 2016;221(6):733–9. https://doi.org/10.1016/j.imbio.2015.11.007.

Wikby A, Maxson P, Olsson J, Johansson B, Ferguson FG. Changes in CD8 and CD4 lymphocyte subsets, T cell proliferation responses and non-survival in the very old: the Swedish longitudinal OCTO-immune study. Mech Ageing Dev 1998;102(2–3):187–98. https://doi.org/10.1016/S0047-6374(97)00151-6.

Wikby A, Johansson B, Olsson J, Löfgren S, Nilsson BO, Ferguson F. Expansions of peripheral blood CD8 T-lymphocyte subpopulations and an association with cytomegalovirus seropositivity in the elderly: the Swedish NONA immune study. Exp Gerontol 2002;37(2–3):445–53. https://doi.org/10.1016/S0531-5565(01)00212-1.

Wikby A, Nilsson BO, Forsey R, Thompson J, Strindhall J, Löfgren S, Ernerudh J, Pawelec G, Ferguson F, Johansson B. The immune risk phenotype is associated with IL-6 in the terminal decline stage: findings from the Swedish NONA immune longitudinal study of very late life functioning. Mech Ageing Dev 2006;127(8):695–704. https://doi.org/10.1016/j.mad.2006.04.003.

Wolters B, Junge U, Dziuba S, Roggendorf M. Immunogenicity of combined hepatitis A and B vaccine in elderly persons. Vaccine 2003;21(25–26):3623–8. https://doi.org/10.1016/S0264-410X(03)00399-2.

Yang H, Youm YH, Dixit VD. Inhibition of thymic adipogenesis by caloric restriction is coupled with reduction in age-related thymic involution. J Immunol 2009;183(5):3040–52.

Yurochko AD, Hwang ES, Rasmussen L, Keay S, Pereira L, Huang ES. The human cytomegalovirus UL55 (gB) and UL75 (gH) glycoprotein ligands initiate the rapid activation of Sp1 and NF-κb during infection. J Virol 1997;71(7):5051–9. https://doi.org/10.1128/jvi.71.7.5051-5059.1997.

Zhang Q, Raoof M, Chen Y, Sumi Y, Sursal T, Junger W, Brohi K, Itagaki K, Hauser CJ. Circulating mitochondrial DAMPs cause inflammatory responses to injury. Nature 2010;464(7285):104–7. https://doi.org/10.1038/nature08780.

Ziegler-Heitbrock L. Blood monocytes and their subsets: established features and open questions. Front Immunol 2015;6. https://doi.org/10.3389/fimmu.2015.00423.

CHAPTER 8

Vaccination in old age: Challenges and promises

Calogero Caruso[a], Anna Aiello[a], Graham Pawelec[b,c], and Mattia Emanuela Ligotti[a]
[a]Laboratory of Immunopathology and Immunosenescence, Department of Biomedicine, Neurosciences and Advanced Diagnostics, University of Palermo, Palermo, Italy
[b]Department of Immunology, University of Tübingen, Tübingen, Germany
[c]Health Sciences North Research Institute, Sudbury, ON, Canada

8.1 Introduction

Aging is one of the main health challenges worldwide, and promoting healthy aging is a key global priority. The health of older people is threatened by their increased susceptibility to infectious disease and associated complications, which is related to many factors, especially the dysregulation of immunity generally termed "immunosenescence." This is believed to adversely affect the efficacy of vaccines, thus reducing the protection provided by most current vaccines in older people. Identifying the key factors responsible for reduced vaccination efficiency in older adults, and devising countermeasures to solve this problem, is essential for improving the outcomes of vaccination. This will allow for better protection against infections in this growing segment of the population (Caruso and Vasto, 2016).

In general, most available vaccines are not tailored for specific age or risk groups, such as that represented by old people. To address this problem, different approaches such as employing higher doses of vaccine, boosting vaccinations, and using adjuvants (for example, based on squalene) are being tested. However, due to immunosenescence, none of these approaches has so far been fully satisfactory, with the exception of an adjuvanted recombinant zoster vaccine (McElhaney et al., 2019).

Immunosenescence is likely to be a major cause contributing to reduced potency of current vaccines as well as vaccine failure in the senior population. However, the example quoted earlier (McElhaney et al., 2019), consisting of properly selected recombinant antigens, effective adjuvants, and suitable doses, shows that there may be no intrinsic reason why vaccine formulations for aged people cannot function highly effectively and mediate strong and long-lasting protection. Nonetheless, shingles is of course caused by reactivation of a latent herpesvirus, and requirements for protection are likely to be different from immune requirements for protection against de novo infections with novel acute viruses. As discussed by Aiello et al. (Aiello et al., 2019), the hallmarks of immunosenescence include a reduced ability to respond to new antigens, the accumulation of memory

Human Aging
https://doi.org/10.1016/B978-0-12-822569-1.00020-2

B and T cells for previously encounter antigens, and chronic low-grade inflammation (known as "inflamm-aging"). Although our understanding of the role of immunosenescence in the vaccination response in the senior population remains incomplete, and will be different for different infectious agents and different human populations, strategies for improving the efficacy of vaccines for older people must be assessed in the context of the hallmarks of immunosenescence.

This chapter is divided into two parts. In the first part, we will discuss the state of the art of vaccination as a preventive measure against infections in older adults. We will discuss the more relevant infections of old people that can be controlled by vaccines, i.e., influenza, pneumonia caused by *Streptococcus* (*S.*) *pneumoniae*, and herpes zoster. In the second part of this chapter, we will review the potential options for therapeutic intervention to improve vaccine efficiency in older adults.

8.2 The state of the art

8.2.1 Adjuvants

A vaccine adjuvant (from the Latin verb *adjuvare*, meaning to help) is a component designed to enhance and/or shape the specific immune response to a vaccine antigen, and has been incorporated in human vaccine formulations for more than 90 years. The first evidence of the use of adjuvants in vaccine formulations dates back to the early 1900s. At that time, Gaston Ramon observed that coadministration of diphtheria toxin with other compounds, such as starch and breadcrumbs, resulted in a significant increase of the antitoxin response to diphtheria (Ramon, 1925). A few years later, Jules Freund developed a powerful adjuvant, the complete Freund's adjuvant (CFA), consisting of a water-in-mineral oil emulsion with heat-killed *Mycobacteria* (Opie and Freund, 1937). Immunization with CFA results in strong antibody responses and cellular immunity. However, the use of CFA has been associated with injection site granuloma and necrotic abscesses, which limits its use to animal research. Incomplete Freund's adjuvant (IFA) is composed of just the mineral oil component of CFA, thus without the mycobacterial components. This adjuvant is highly effective in enhancing cytokine production and antibody responses, and is less inflammatory than the complete form. However, in this case, severe local reactions, including sterile abscesses and persistent painful granulomas, have been observed, hampering its use in humans (Miller et al., 2005).

A variety of compounds with adjuvant properties have been discovered since then, and those employed in human vaccines licensed for use include aluminum salts (aluminum oxyhydroxide, aluminum phosphate), oil-in-water (O/W) emulsions (for example, MF59), virosomes (Moser et al., 2011), Toll-like receptor (TLR) agonists, or a combination of immunostimulants (Table 8.1). Those currently employed are discussed later, and the others considered in the section Challenges and Promises.

Table 8.1 Old and new vaccine adjuvants.

Adjuvant name	Composition	Immune effects
Aluminum salts	Aluminum oxyhydroxide, aluminum phosphate	Promotes immune cell recruitment to the injection site; Increases local inflammation; Increases antigen uptake and local cytokine secretion by APCs.
MF59	Squalene O/W emulsion stabilized with Tween 80 and Span 85	Promotes: innate cells recruitment; antigen uptake by APCs; migration of cells to lymph nodes; local production of cytokines.
AS01	MPL and QS-21 with liposomes	Promotes: immune cell recruitment to the injection site; local production of cytokines; APC activation.
AS03	Squalene O/W emulsion with polysorbate 80 and α-tocopherol	Promotes local production of cytokines; Promotes recruitment of immune cell; Increases antigen uptake and presentation in draining lymph nodes.
AS04	Alum plus MPL	Promotes: immune cell recruitment to the injection site; APC activation via TLR-4.
Virosomes	Liposomes plus influenza antigens	Increases antigen uptake, processing and presentation by APCs.
MPL	A monophosphoryl lipid A preparation from LPS from *Salmonella minnesota*	Promotes APCs activation via TLR4.
CpG ODN	Synthetic single stranded DNA containing CpG dinucleotides	Promotes APCs activation via TLR9.

Continued

Table 8.1 Old and new vaccine adjuvants—cont'd

Adjuvant name	Composition	Immune effects
polyI:C	Synthetic double-stranded RNA polymer analog	Promotes APCs activation via TLR3.
Imidazoquinoline compounds	Imiquimod (R837), resiquimod (848)	Promotes APCs activation via TLR7/8.

APCs, antigen-presenting cells; *O/W*, oil-in-water; *LPS*, lipopolysaccharide; *MPL*, monophosphoryl lipid A; *TLR*, toll-like receptor; *ODN*, oligodeoxynucleotide; *polyI:C*, polyinosinic–polycytidylic acid.
References in the text.

Although each of these numerous types of adjuvant mediates their effects through a different mechanism of action with different molecular targets, all adjuvants mime some aspects of the natural response to pathogens. In particular, adjuvants should act by enhancing directly or indirectly antigen-presenting cell (APC) responses, primarily dendritic cells (DCs), and regulate differences within the activated subset. These events are reflected in the quality and magnitude of the adaptive immune response specific to vaccine antigens. In the context of vaccination of older people, achieving these immune outcomes is particularly important, because DC's function is also altered with aging (Aiello et al., 2019).

DCs, the most potent APCs, are classified into three major subsets with different biomarkers and distinct functional activities, namely, two subsets of myeloid DCs (mDCs) and one subset of plasmacytoid DCs (pDCs). mDCs are the primary source of the interleukin (IL)-12, needed to drive a T helper type1 (Th1) response, and produce other cytokines, including IL-6 and tumor necrosis factor (TNF). In contrast, pDCs produce large amounts of type I interferons (IFN-α/β), critical for antiviral responses (Collin et al., 2013). Several studies of the effects of aging on the immune system in mice and humans support the idea that DCs from aged individuals are less effective in activating T-cell activations (Grolleau-Julius et al., 2008; You et al., 2013), showing a numerical and functional decline, especially the pDCs (Jing et al., 2009; Oh et al., 2019). Therefore, adjuvants would need to counteract the immunosenescence hallmark "reduced ability to respond to new antigens" at the DCs level (Aiello et al., 2019).

Despite the crucial immune-modulating role of adjuvants, most currently used vaccines are usually less effective at inducing an adequate cellular response in older than in younger adults (Weinberger, 2018). As an example, aluminum salts, commonly referred to as alum, represent one of the oldest and most widely used adjuvants in human vaccines (McKee and Marrack, 2017; Clements and Griffiths, 2002), although their precise mechanism of action is only now beginning to be elucidated. Initially, it was proposed that adsorption to alum increases antigen persistence, creating a depot effect at the site of injection (Kool et al., 2012; Marrack et al., 2009). However, several further studies have disproved this theory, showing that resection of the injection site 2 h after administration does not affect the magnitude and kinetics of antigen-specific adaptive immune responses (Hutchison et al., 2012). It transpired that, in mice, alum increases immune cell recruitment to the injection site (Goto et al., 1997), DC's antigen uptake and their migration to the draining lymph node (Morefield et al., 2005). In addition, although unable to directly activate DCs, at the site of injection, alum induces the release of some endogenous signal of cellular damage, activating the NOD-like receptor protein 3 (NLRP3) inflammasome with consequent caspase-1-dependent local, but not systemic, secretion of cytokines IL-1β and IL-18 by DCs (Eisenbarth et al., 2008; Del Giudice et al., 2018). Despite the long-term success as a vaccine adjuvant, also due to its safety profile and ability to enhance immune response to a range of antigens, aluminum salts do not seem to perform optimally in old subjects (Yam et al., 2016).

Two powerful adjuvants able to induce stronger immune responses are adjuvant system (AS)03 and MF59, O/W emulsions containing squalene, a precursor to cholesterol, both used in influenza vaccines as adjuvants (Wilkins et al., 2017). MF59 was the first novel adjuvant approved for use in human vaccines after alum. When injected into the muscle, MF59 promotes a local activation of neutrophils, eosinophils, monocytes, macrophages, and DCs that respond by producing several chemokines, including C—C motif chemokine ligands (CCL)2, CCL3, and CCL4. AS03 seems to have similar effects. This local chemokine gradient promotes the recruitment of monocytes and immature DCs from blood to the site of injection, resulting in a more efficient uptake, processing, and transportation of the antigen to the lymph nodes (O'Hagan et al., 2013; Dupuis et al., 2001). Studies in mouse models have shown that recruited cells also internalize MF59 and then transport it to the lymph node, where the adaptive immune response to vaccine antigen is triggered through the activation of T and B cells (Khurana et al., 2011).

Immune response activation by MF59 is more powerful with higher levels of antigen-specific antibody production, increased innate immune cell recruitment, and cytokine production than accomplished by alum. In addition, compared with nonadjuvanted influenza vaccines, MF59-adjuvanted trivalent influenza vaccine elicited a significantly greater antibody response both in the young and in adults ≥65 years of age (Frey et al., 2014), although several local and systemic adverse reactions, albeit mild and transient, were more common after MF59-adjuvanted vaccination (46%) compared with the nonadjuvanted vaccine group (33%) (Kostova et al., 2013).

The available evidence suggests that MF59 is more effective than nonadjuvanted vaccines or vaccines with alum, and it could be an excellent ally in enhancing the efficacy of vaccines. However, future studies need to investigate the safety and the risk of adverse reactions related to MF59-adjuvanted vaccines in populations particularly vulnerable and at risk of developing severe complications, such as old people.

8.2.2 Influenza

Older adults are particularly vulnerable to respiratory and emerging virus infections. Acute respiratory viral infections are a major cause of death and disability in old people. Among these, seasonal influenza is by far the most important also for complications such as pneumonia and cardiovascular disease, which require hospitalization and increase mortality during the winter season. Even in the current SARS-CoV-2 pandemic, ongoing at the time of writing, one should not lose sight of the fact that influenza is a highly dangerous pathogen. Thus, an estimated 50%–70% of influenza-related hospitalizations are of adults over the age of 65 each year, with 70%–90% of deaths related to influenza occurring among this same age group. Many old people hospitalized during ´flu outbreaks led active lives before but do not fully recover thereafter. In addition to severe acute disease and death, flu in older adults also causes functional disabilities as long-term sequelae. Up to 12% of hospitalized old patients will require, once discharged, a higher level of care in daily life than they required before the disease (Kostova et al., 2013).

Vaccines against influenza are usually either split-virus or subunit formulations that contain distinct antigens each year determined by the World Health Organization on the basis of surveillance data on predicted circulating strains. They are standardized according to the content of virus hemagglutinin (HA) for each of three different strains (A/H1N1, A/H3N2, B), or sometimes four. There are egg-, cell-based, and recombinant flu vaccines in common use. Recombinant baculovirus-expressed HA protein avoids the possible generations of mutations associated with virus egg adaptation. Current seasonal anti-flu vaccines confer beneficial effects on older people, reducing disease with typical flu symptoms by approximately 50%–60% and influenza-related complications by approximately 30%–50%, depending on the major circulating strains of the season and accuracy of prediction. However, in this age group, vaccine efficacy and efficiency is commonly lower than in younger adults, although there is some debate regarding how these statistics are reported (Osterholm et al., 2012; Russell et al., 2018; Treanor, 2016; Andrew et al., 2019). In fact, the clinical efficacy of influenza vaccines is difficult to evaluate, because many variables must be taken into account, such as population types, different epidemiological criteria, and virus-related factors such as the different virulence of the different strains and in particular the possible discrepancy between vaccine, and circulating viral strains. The issue of "original antigenic sin" and the state of health of the individual are also of crucial importance. It must be borne in mind that in most studies, laboratory flu diagnosis is not made, but flu-like disease is referred to such that certainty of influenza infection is not guaranteed (Weinberger, 2018). In addition, immune changes associated with human cytomegalovirus (HCMV) seropositivity might have a significant impact on vaccination, although this point is controversial because this relationship has not been observed in all studies (Aiello et al., 2019; Merani et al., 2017). However, on the basis of meta-analyses, it can be concluded that, in old people, protection is reduced when compared with younger ones (Beyer et al., 2013).

Humoral immunity is known to play an important role in preventing influenza virus transmission and infection, and immunogenicity of influenza vaccines is usually measured by HA inhibition (HAI) assay, which quantifies antibodies specific for the virus HA. A greater number of older adults fail to seroconvert, i.e., to have the fourfold increase in postvaccination antibody titer, relative to their younger counterparts that is one of WHO criteria for assigning responsiveness, with seroconversion rates ranging from 10% to 30% in older adults compared with 50%–75% in younger subjects (although it can sometimes be the case that older people already have a high antibody titer due to previous exposures and thus cannot increase titers 4-fold, resulting in erroneous classification as nonresponders). In particular, older adults may fail to generate protective HAI antibody titers compared with younger adults (Weinberger, 2018; Crooke et al., 2019). Cellular immunity is also strongly associated with protection against influenza, as some older adults have been shown to remain protected against infection even in the absence of robust antibody responses (McElhaney et al., 2016). However, further studies are needed in order to fully understand the effects of immunosenescence on cellular immunity to influenza (Crooke et al., 2019). In this context, it is likely that HCMV has a larger impact on T-cell immunity than on humoral immunity (Haq et al., 2017).

In an effort to improve clinical outcome following influenza vaccination, vaccine formulations have been licensed specifically for use in old people. That led to the licensing of trivalent vaccines containing the O/W emulsion adjuvant MF59 (adjuvanted vaccines) (Ansaldi et al., 2008), or higher doses of 60 µg instead of 15 µg of HA per dose (high-dose) (Robertson et al., 2016), and of a vaccine administered via the intradermal instead of the intramuscular route (Holland et al., 2008). Trivalent influenza vaccines contain an A(H1N1)-like influenza virus, an A(H3N2)-like influenza virus, and a B-like influenza virus. In addition, high-dose quadrivalent vaccines became available, as two different B strains had circulated in parallel for several years (Smetana et al., 2018). Annual vaccination against influenza is recommended, as the composition of the vaccine changes in order to reflect currently circulating virus strains and many virulent new strains that infect humans jump species barriers (Trucchi et al., 2017; Weinberger, 2018).

A large trial demonstrated that vaccination with adjuvanted vaccine reduced the risk of hospitalization for influenza or pneumonia in old people during the peak of the influenza season by 25% relative to vaccination with nonadjuvanted vaccine. The routine use of adjuvanted vaccine in older people would provide an important clinical benefit over the traditional nonadjuvanted vaccines (Mannino et al., 2012). In a study including residents of long-term care facilities, the risk of influenza-like disease was calculated for nonadjuvant *vs.* adjuvant vaccine recipients, also stratifying for chronic cardiovascular, respiratory, and renal diseases. The risk was higher for the nonadjuvant vaccine recipients and highest for those with respiratory disease and cardiovascular disease. Therefore, it was concluded that the MF59-adjuvanted vaccine provides superior clinical protection among old people, especially those with chronic diseases (Iob et al., 2005). While the mechanisms of action are not completely understood, MF59 is believed to enhance innate immune responses and stimulate germinal center reactions (Cioncada et al., 2017).

A recent meta-analysis of studies reporting influenza vaccine efficacy against laboratory-confirmed hospitalized influenza among adults showed that influenza vaccines provided a moderate protection against influenza-associated hospitalizations among adults. They seemed to provide low protection among old people in seasons where vaccine and circulating A(H3N2) strains were antigenically variant (Rondy et al., 2017).

Influenza viruses have a huge impact on public health. Current influenza vaccines need to be updated annually and protect poorly against antigenic drift or novel emerging subtypes. Vaccination against influenza can be improved by inducing more broadly protective immune responses. Numerous candidate vaccines are being investigated, which utilize different antigens, such as conserved regions of the surface proteins HA and neuraminidase or internal viral proteins. The most promising universal influenza vaccine candidates are likely those that induce both broad humoral and cell-mediated responses (Wiersma et al., 2015).

8.2.3 Streptococcus pneumoniae

Bacterial pneumonia is another respiratory disease often more likely to be fatal in older people. The disease results from infection with *S. pneumoniae*, a common bacterial commensal, which frequently colonizes the upper respiratory tract. There are more than 90 different strains, many of which cause disease, but only a few of which are responsible for invasive pneumococcal infections. Pneumococcal infections are contracted by transmission by the air, via direct contact with the respiratory secretions, or saliva, or contact with healthy carriers of this bacterium, which can nest in the back of the nose or in the throat. While colonization is generally benign, migration of *S. pneumoniae* into the lower respiratory tract may lead to a pronounced disease progression. Pneumonia, bacteremia and sepsis, and meningitis are the most serious forms of infections. The mortality rates associated with pneumococcal disease vary between 15% and 30% among old people (Crooke et al., 2019). It has been calculated that between 2004 and 2040, as the American population will increase by 38%, pneumococcal pneumonia hospitalizations will increase by 96%, because population growth is fastest in older age groups experiencing the highest rates of pneumococcal disease (Wroe et al., 2012).

Human studies have found that people >65 years old have significantly lower antibody titers against many of the common pneumococcal serotypes and diminished opsonization activity compared with younger adults. Therefore, antibody titers wane over time, and there may also be functional deficiencies in antibody responses against pneumococcal antigens. While humoral immunity is primarily thought to mediate protection from disease, there are conflicting reports regarding age-related changes of T-cell responses against pneumococcal infection (Crooke et al., 2019).

Both young children and older adults are affected by invasive pneumococcal disease. However, it represents only a fraction of the adult pneumococcal disease burden. By comparison, nonbacteremic community-acquired pneumonia (CAP) makes up the vast majority of pneumococcal disease in adults and older people. However, vaccination recommendations for *S. pneumoniae* are different in different countries. In fact, some countries still recommend the polysaccharide vaccine, while others recommend the conjugate vaccine alone or followed by the polysaccharide vaccine one year later (Weinberger, 2018).

For many years, the 23-valent polysaccharide vaccine has been used for older people. However, polysaccharides are T cell-independent antigens, so they induce an IgM response without adequate immunological memory. Conjugation can overcome this limitation via binding of the pneumococcal glycans to diphtheria toxoid (Musher et al., 2011). Therefore, conjugated vaccines have been developed for the vaccination of children. Around the year 2000, 7-valent and 10-valent conjugate vaccines were introduced for childhood immunization. Reduction in children of disease incidence and carriage of the serotypes included in the vaccines followed, which has determined decreased transmission of these serotypes and, therefore, also disease incidence in the older age group. However, serotype replacement also resulted in increases in the incidence of pneumococcal disease caused by other serotypes, not included in the conjugated vaccines, both in children and in older adults (Hanquet et al., 2010; Weinberger, 2018). Thus, a 13-valent conjugate vaccine has been introduced also for older adults. The polysaccharide is bound to carrier protein CRM197 and adsorbed on aluminum phosphate. CRM197 is a genetically detoxified form of diphtheria toxin. A single mutation at position 52, substituting glutamic acid for glycine, causes the ADP-ribosyltransferase activity of the native toxin to be lost (Malito et al., 2012). The vaccine induces both B- and T-cell responses as well as mucosal immunity, i.e., suppression of carriage of nasal serotypes by the vaccine. In a large randomized trial in older adults (> 65 years), 75.0% fewer first episodes of vaccine-type strain invasive pneumococcal disease and 45.6% fewer first episodes of vaccine-type strain CAP requiring hospitalization occurred in the vaccine group compared with placebo (Bonten et al., 2015). However, similar effects, i.e., serotype replacement, have been observed for the 13-valent vaccine (Esposito and Principi, 2015; Weinberger, 2018).

The phenomenon of serotype replacement has provided the impetus for the development of a new-generation recombinant protein and whole-cell pneumococcal vaccines with the potential to provide serotype-independent protection. To this end, universal pneumococcal vaccines would also be very useful, as there are approximately 90–100 serotypes of *S. pneumoniae*. Currently, vaccine manufacturers try to increase the number of serotypes included in conjugated vaccines, but antibody responses to polysaccharides will probably always be serotype-specific. Several pneumococcal proteins have been identified as potential universal vaccine candidates. They are highly conserved in all clinically relevant serotypes and elicit potent immune responses in animal models. Additionally, whole-cell inactivated vaccines, live-attenuated vaccines, and combinations of protein and polysaccharide components are being investigated (Feldman and Anderson, 2014).

8.2.4 Varicella zoster virus

Almost all adults are latently infected with varicella zoster virus (VZV). The primary infection, which usually occurs in childhood, manifests as chickenpox and live-long latency is established afterward. Before the introduction of routine childhood vaccination against VZV, nearly 100% of the adult population was exposed to VZV during the lifetime, establishing latent infection within dorsal root ganglia (Gershon and Gershon, 2013). Reactivation of the virus is usually controlled by virus-specific T-cell responses, whereas several observations suggest that antibody responses do not play a significant role in protective immunity against VZV (Crooke et al., 2019). Following immunosenescence, the possible reactivation of VZV is responsible for herpes zoster disease. Accordingly, herpes zoster incidence shows an age-related increase, and 50% of cases are in individuals over 85 years of age (Crooke et al., 2019; Haq et al., 2017; Weinberger, 2018). In a fraction of patients, acute episodes of herpes zoster are followed by postherpetic neuralgia (PHN), characterized by long-lasting severe pain after the resolution of the zoster rash. The incidence of this complication is higher in older zoster patients, where it occurs in approximately one-third of the cases. Particularly in older patients, PHN is associated with significant deficits in the ability to carry out daily activities (Mallick-Searle et al., 2016).

Vaccine development efforts have proven very effective in combating herpes zoster in older adults. Two vaccine formulations have been licensed for clinical use, but clinical responses differ significantly between the two vaccines (Crooke et al., 2019). A live-attenuated vaccine that induces both T-cell and antibody responses has been licensed since 2016 (Weinberger, 2018). In people older than 60 years, it reduces the incidence of herpes zoster by 51% and the incidence of PHN by 66% (Oxman et al., 2005). Unfortunately, efficacy was found to significantly decline as age at the time of vaccination increased, decreasing to 41% in adults >70 years of age and 18% in individuals ≥80 years of age. Moreover, established protection waned over time, dropping to 21.1% for the prevention of herpes zoster and 35.4% for PHN 7–10 years after vaccination (Schmader et al., 2012; Morrison et al., 2015). A second dose of the vaccine more than 10 years after the first dose resulted in a cellular immune response higher after boosting, suggesting that a repeated vaccination of older individuals at appropriate intervals could be beneficial (Weinberger, 2018).

Both the immunogenicity and safety of a vaccine containing the viral glycoprotein E (gE) in combination with the liposome-based AS01B (MPL and QS21, see Section 8.3.1) have been documented in older adults (Chlibek et al., 2013). The assessment of safety and efficacy of the recombinant zoster vaccine in older adults was evaluated using pooled data from almost 17,000 subjects aged at least 70 years old. The estimate of vaccine efficacy against herpes zoster was 90% in the 70- to 79-year-old subjects and 89% in those aged 80 years and older, and 88.8% for the prevention of PHN over a 3.7-year follow-up period (Cunningham et al., 2016). The excellent performance of this vaccine is probably due to the development of a specific memory Th1-type response against the viral gE stronger than those present in subjects receiving the live-attenuated vaccine, since this VZV subunit promotes cell-to-cell interactions that lead to cell fusion in the pathogenesis of virus infection and reactivation. (Levin et al., 2018; Cunningham et al., 2018). According to studies performed in mice, MPL, in both the AS04 and AS01B adjuvanted combinations, stimulates TLR4 on antigen-presenting cells. QS21 stimulates inflammasomes in macrophages of lymph nodes draining the muscle injection site, with a subsequent increase in the number of activated resident lymph node dendritic cells and those derived from immigrating blood monocytes. These dendritic cells present the gE antigen to T cells (Didierlaurent et al., 2017).

The high efficacy of this recombinant vaccine is exceptional among vaccines given to older adults. Compared with live vaccine, it is distinguished by boosting robust and persistent memory responses. The AS01B adjuvant is critical for the magnitude of the Th response and probably plays a role in its persistence (Takeda and Akira, 2015). This effect is likely due to its ability to improve antigen presentation; to reverse memory T-cell exhaustion, an immunosenescence hallmark; and, likely, to reduce regulatory T-cell activity (Gershon and Gershon, 2013).

8.3 Challenges and promises

8.3.1 TLR agonists

As previously mentioned, a promising strategy to increase vaccine efficiency in older adults seems to be the incorporation of TLR agonists in vaccine formulations. The human TLR family consists of 10 members expressed on the cell surface (TLR1, 2, 4, 5, 6, 10) and in intracellular compartments (TLR3, 7, 8, 9) on sentinel cells, especially DCs, which recognize specific highly conserved microbial components, called pathogen-associated molecular patterns (PAMPs) (Chen et al., 2016). Each TLR recognizes specific PAMPs, and, in particular, TLRs located on the cell surface bind lipids and proteins, while TLRs located on the endosomal compartments are responsible for the recognition of bacterial and viral nucleic acids (Chen et al., 2016). Examples of PAMPs include lipopolysaccharide (LPS) for TLR4, imidazoquinoline compounds, such as R848, for TLR7/8 and CpG motif of bacterial and viral DNA for TLR9 (Guy, 2007; Kornbluth and Stone, 2006).

TLRs agonist efficiencies as vaccine adjuvants rely mostly on the promotion of antigen uptake, presentation, and maturation of DCs, their cytokine secretion and activation of T cells. DC subsets show different patterns of expression of TLRs with different effects on activation but, generally, binding of PAMPs to TLRs triggers an intracellular MyD88 or TRIF-dependent signaling cascade that induces maturation of immature DCs to professional APCs, increasing expression of major histocompatibility complex (MHC) class II and costimulatory molecules, and secretion of several cytokines (Schreibelt et al., 2010). Due to their powerful immunostimulatory properties, TLR agonists are considered important vaccine adjuvant candidates and some of them are indeed currently in use or being tested as adjuvants (Table 8.1). In the context of aged individuals, however, an important consideration regarding their use in vaccines is the influence of aging on TLR responsiveness and expression levels. For example, TLR3 and TLR8 in mDCs and TLR7 in pDCs seem to be expressed at lower levels in healthy older people than in young individuals, whereas TLR2 and TLR4 surface expression in mDCs appears to be unchanged with increasing age (Jing et al., 2009; Panda et al., 2010). The different expression of some TLRs in old people is associated with an altered TLR-mediated immune response. Several studies show an age-associated reduction of TLR-induced type 1 IFN(—I) production in pDCs in response to influenza virus and CpG ODN (Weinberger, 2018) and other stimuli (Shodell and Siegal, 2002; Qian et al., 2011), as well as defects in TLR-induced IL-12p40, IL-6, and TNF-α production in mDCs (Panda et al., 2010), impairing host defense to viral infections. Age-related increased oxidative stress can also contribute to impaired TLR-induced IFN-I response by aged pDCs (Stout-Delgado et al., 2008). For these reasons, most TLR agonists are administered in a combined treatment, as in the case of the AS01, a liposome-based vaccine adjuvant containing the TLR4 agonist 3-O-desacyl-4′-monophosphoryl lipid A (MPL), a detoxified derivative of LPS extracted from *Salmonella minnesota*, and the saponin QS-21, extracted from *Quillaja saponaria molina* (Didierlaurent et al., 2017). The two components of AS01, MPL acting via TLR4 and QS-21 in an NLRP3 inflammasome-dependent manner, act synergistically to strongly stimulate innate responses and to induce the highest antigen-specific T-cell response. Following intramuscular injection in mice, AS01 coinjected with antigen induces a rapid and transient local production of cytokines, mostly chemokines attracting circulating monocytes and granulocytes, ≥ 8.7-fold higher than antigen injection alone. In addition, AS01 coadministered with antigen does not persist at the injection site but is rapidly transported to the lymph nodes where AS01 enhances the recruitment of innate immune cells from the injection site and increases expression of T-cell stimulatory molecules in both recruited and lymph node-resident DCs. No increase in antigen uptake by DCs has been observed following vaccination with AS01-adjuvanted vaccine. Thus, AS01 acts by recruiting a large number of activated and cytokine-secreting APCs at the injection site and lymph node level, ensuring a favorable environment particularly efficient in T-cell priming (Didierlaurent et al., 2017).

Another adjuvant system, AS04, is a combination of aluminum salts with MPL. In mouse models, AS04 induces a rapid transient cytokine production and innate immune cell recruitment. When injected with a specific antigen at the same injection site within 24 h, AS04 is able to promote an increased number of activated antigen-loaded DCs and monocytes, which then leads to the activation of antigen-specific immune adaptive responses. Although the induction of cytokine secretion by DCs is mainly due to the action of the TLR agonist, the presence of alum in the AS04 formulation seems to prolong the MPL-mediated cytokine response (Didierlaurent et al., 2009). When used in combination, therefore, MPL and alum synergize, ensuring a more effective response than either alone.

With this in mind, a new type of adjuvant formulation that could enhance vaccine efficacy in older people consisting of combined adjuvants for synergistic activation of cellular immunity (CASAC) was formulated, incorporating two TLR agonists, CpG-oligodeoxynucleotides (TLR9 agonist) and polyI:C (TL3 agonist), IFN-γ, and MHC-class I and II peptides in an O/W emulsion (Tye et al., 2015). Promising results were obtained in murine models where the simultaneous and synergistic activation of TLRs has enhanced DC activation, resulting in increased cellular responses to the antigen of interest. Immunosenescent old and young mice were vaccinated with CASAC or CFA/IFA adjuvant, with a class I epitope, and the specific CD8+ T-cell response was subsequently evaluated, assessing frequency, by MHC pentamer staining, cytotoxicity, and intracellular production of IFN-γ. These analyses have shown that, in both young and aged mice, vaccination with CASAC generated higher frequencies and stronger responses of antigen-specific CD8+ T cells compared with mice vaccinated with CFA/IFA (Tye et al., 2015). To assess if these promising observations in animal models can be translated to humans, the ability of two combined TLR ligands, R848 (TLR7 agonist) and MPL (TLR4 agonist), to enhance the activation of DCs isolated from human healthy aged and young donors has been investigated. In both groups, the stimulation of DCs with the combination of TLR7/8 and TLR4 agonists induced a higher production of IL-12/p40 and TNF-α by mDCs, confirming their role in boosting the innate response of TLR agonists in humans. However, when the production of cytokines was compared between the young and the old donors, an increase of 5- to 10-fold in the production of IL-12/p40 by mDCs, as well as an increased amount of TNF-α by mDCs and pDCs isolated from old people, was observed (Gambino et al., 2017). Based on these results, it is evident that, despite the differences in their composition, adjuvants can contribute significantly to improving immune responses to vaccines in immunosenescent individuals but, since most of studies in this field have been done in mouse models, further studies are needed to test which combinations can offer the same protective effects in humans.

8.3.2 Virosomes, viral vectors, reverse vaccinology

Virosomes are particles structurally and functionally similar to viruses, although they are assembled in vitro. Immunopotentiating reconstituted influenza virosomes (IRIVs), the most commonly used form of virosome, are spherical vesicles consisting of unilamellar phospholipid membrane incorporating virus (HA, by which the membrane fusion activity of the native virus is preserved, and neuraminidase) (Zurbriggen, 2003). Mimicking both viral morphology and antigenic presentation, but unable to replicate, virosomes can deliver these antigens directly to their targets, enhancing antigen uptake, processing, and presentation by APCs. Virosomal HA promotes binding at the APC surface followed by receptor-mediated endocytosis and fusion of viral and endosomal membranes. A wide range of antigen types have been combined successfully with influenza virosomes and their location in the vesicles determines the processing and presentation pathway. Vaccine antigens cross-linked on the surface remain in the endosomal compartment after the fusion event and are thereby presented to the CD4+ T cells by MHC class II molecules, generating primarily antibody responses. Antigens encapsulated within the lumen of the virosome are delivered to the cytosol upon membrane fusion and are presented to the CD8+ T cells via the MHC class I pathway (Schumacher et al., 2004). Virosomal technology is approved for use in humans, and there are several clinical trials that demonstrate the immunogenicity and the high safety profile of IRIV for all age groups, including older people and immunocompromised individuals (Glück et al., 1994; Zanetti et al., 2002).

Other promising tools for vaccines are viral vectors, designed to take advantage of the natural efficiency of viruses at entering and transducing their own genome into host cells for their replication. By replacing nonessential viral genes with exogenous genes of therapeutic interest, viral vectors are able to transduce their cellular target at the immunization site, resulting in high levels of synthesis of the immunogen, characteristics that give rise to robust humoral and cellular immune responses (Bouard et al., 2009). Compared with virosomes that deliver vaccine antigens directly into a host cell, viral vectors enable intracellular antigen expression, ensuring highly efficient induction of both humoral and T-cell responses. To ensure safety for human use, most viral vectors are rendered unable to replicate through targeted gene deletion. For example, in adenovirus-based vectors, the genomic regions required for viral replication are deleted and replaced with the target gene (Ura et al., 2014).

Adenovirus is one of the most commonly exploited viral vectors for vaccine development, but the main disadvantage of the clinical use of human adenovirus is the presence of preexisting specific immunity against the vector, due to a previous exposure to the virus and leading to the production of neutralizing antibodies that reduce vaccine efficacy. This obstacle has been surmounted by replacing the human virus with replication-deficient chimpanzee adenoviruses (Ewer et al., 2016). In addition to adenoviruses, several other viral vectors have been shown to be successful in clinical trials. Modified vaccinia Ankara (MVA) virus is a highly attenuated strain derived from the vaccinia virus, belonging to the poxvirus family. MVA was rendered replication-deficient by the deletion of roughly 15% of its original genome (Choi and Chang, 2013). When used as a vaccine vector, MVA shows an excellent safety profile and induces potent humoral and cellular antigen-specific responses, especially in the cytotoxic T subset (Ewer et al., 2016).

The particular characteristics of viral vectors have been exploited for the development of alternative vaccination strategies to elicit potent T-cell responses specific for highly conserved viral antigens. A similar approach could provide protection against antigenically distinct viruses. This strategy is particularly relevant for vaccinations against viruses with pandemic potential, as in the case of influenza virus where its rapid evolution, due to an accumulation of mutations within antigenic sites, causes an escape from serological host immunity conferred by a previous infection or vaccination (Coughlan et al., 2018). Current seasonal influenza vaccines induce subtype-specific responses and offer no heterosubtypic immunity against novel subtypes or other emerging viruses. An approach that aims to enhance preexisting memory T-cell responses against highly conserved influenza antigens, such as nucleoprotein (NP) and matrix protein 1 (M1), rather than to stimulate naïve lymphocytes de novo may be particularly beneficial in the older people, characterized by reduction in naïve T-cell output and expansion of selected memory T-cell clones (Olsson et al., 2000; Antrobus et al., 2012). Based on these premises, the safety and immunogenicity of the vaccine MVA expressing the NP and M1 antigens (MVA-NP + M1) in people aged 50–85 years was assessed. MVA-NP + M1 was shown to be immunogenic, boosting influenza-specific CD4 + and CD8 + memory T-cell responses. Surprisingly, no significant differences were observed in the induction of the immune responses between young and old subjects, unlike what was observed for seasonal influenza vaccination (Antrobus et al., 2012; Boraschi and Italiani, 2014). These observations provided the basis for a subsequent phase I study in which MVA-NP + M1 was tested in combination with a ChAdOx1 vector carrying the same influenza antigens (ChAdOx1 NP + M1). The authors showed that a two-dose MVA/ChAdOx1 regimen was highly immunogenic, increasing preexisting T-cell responses to influenza antigens in both young and older adults. Also, in this case, both vaccines were well tolerated, demonstrating again the safety of the use of viral vectors in older people (Antrobus et al., 2012; Pawelec and McElhaney, 2018).

A significant step forward that revolutionized the classic vaccine concept is reverse vaccinology, a genome-based approach to vaccine design. Sequencing the whole genome of the pathogen allows the identification of new candidate vaccine antigens, without the need for growing microorganisms. Once identified, antigens are expressed and screened in animal models (Rappuoli, 2000). The first recently licensed vaccine using reverse vaccinology is a vaccine against meningococcus B (MenB), which has shown to be highly effective in preventing MenB disease in infants (Parikh et al., 2016). This innovative approach breaks new ground for vaccine development and, in the future, could be exploited to enhance the immune responses of immunosenescent subjects.

8.3.3 Interleukin-7

One of the best-studied and accepted age-related immune alterations is the reduction in the number and repertoire diversity of peripheral naïve T cells, as a result of progressive thymic involution, characterized by thymic epithelial space atrophy and replacement with adipose tissue (Lynch et al., 2009). All these phenomena lead to a reduction in the T-cell reservoir necessary for protection against newly encountered microbial antigens, including vaccine antigens. Strategies reported to prevent or reverse thymic atrophy include the use of IL-7, a pleiotropic cytokine produced by thymic epithelial cells with a wide range of functions. In particular, IL-7 binding to its specific heterodimeric receptor activates an interconnected intracellular signaling network critical for the development of T-cell lineages in the thymus and for the development and survival of naive T cells and CD4 and CD8 memory cells in the periphery (Barata et al., 2019). Therapeutic benefit of the use of IL-7 has been widely demonstrated in animal experiments, where IL-7 administered pharmacologically induced a transient expansion of naïve and memory T cells with low toxicity (Phillips et al., 2004; Okoye et al., 2005). In a trial in HIV-infected individuals treated with antiretrovirals, but with persistently low CD4+ T-cell counts, IL-7 increased both naïve and memory CD4+ and mostly naïve CD8+ T-cell counts (Lévy et al., 2012).

Considering the critical role of IL-7 in maintaining T-cell compartment homeostasis, the possibility of using IL-7 therapy to improve vaccine responses has been studied. Administration of IL-7 during immunization in mice led to increased expansion of antigen-specific effector and memory cytotoxic T cells, accompanied by increased death of effector cells during the contraction phase (Melchionda et al., 2005; Colombetti et al., 2009). These studies in the mouse suggest that IL-7 could be used for improving the adaptive response to vaccine antigens. However, the progressive thymic involution observed with age places an obstacle to the immune-restorative action of this pluripotent cytokine. It is reasonable to think that, in the case of older people, early restoration of the thymic architecture is required so that the IL-7 can act to its full potential (Aiello et al., 2019).

8.3.4 Inhibitors of mitogen-activated protein and adenosine monophosphate-activated protein kinases as therapeutic interventions for vaccination improvement

The role of mitogen-activated protein kinase (MAPK) and adenosine monophosphate-activated protein kinase (AMPK) pathways has been recently demonstrated in the functional competence of the immune system (Chi and Flavell, 2010; Silwal et al., 2018). The first is mainly involved in the production of cytokines, as well as in the intracellular signaling cascades initiated when a cytokine binds to its corresponding receptor. The second is a pathway activated by energy depletion, i.e., by low levels of intracellular adenosine triphosphate, leading to the extension of healthy lifespan in model organisms. Both are implicated in the regulation of T-cell immunosenescence hallmarks that are the reduced ability to respond to new antigens and the accumulation of memory T cells. In particular, sestrins, the mammalian products of the Sesn1, Sesn2, and Sesn3 genes, are a family of stress-sensing proteins that can bind these kinases, inhibiting them (Lanna et al., 2017). It was proposed a possible role for sestrins in the control of the immune response, although this has not yet been fully determined. Indeed, sestrins show pro-aging activities in senescent T lymphocytes through the sestrin-dependent MAPK activation complex.

This simultaneously coordinates the activation of each MAPK that controls a functional response, and its knockout restored T-cell activity (antigen-specific proliferation and cytokine production) from older humans, and enhanced responsiveness to influenza vaccination in aged mice (Lanna et al., 2017). Ex vivo, the inhibition of AMPK and MAPK via these small-molecules overcomes human T-cell senescence in human HCMV-specific T cells, and restores T-cell proliferation, IL-2 production, and cytotoxicity (Lanna et al., 2014).

Other approaches include the role of small-molecule kinase inhibitors (SMKIs), consisting of four classes of kinase inhibitors targeting tyrosine, serine/threonine, dual (that can phosphorylate either tyrosine or serine/threonine residues), and lipid kinases. These are compounds that include marketed drugs and drugs in development that have the potential to exercise a dual synergistic effect as immune-system modifiers of influenza infection and vaccination in older adults (Wu et al., 2016). Along with those approved by Food and Drug Administration, a large number of other SMKIs have been enrolled in clinical trials at different phases for the treatment of human malignancies and neoplastic disorders with immune-stimulatory properties. In particular, five of these have already proven efficacious in restoring T-cell functions in preclinical models and human senescent T cells. These include MAPK inhibitors with known anti-T-cell aging capacities in vitro and in vivo in murine models of influenza vaccination and with potential anti-inflamm-aging properties (Lanna et al., 2017; Lanna et al., 2014). Moreover, some of these SMKIs have already demonstrated their potential for direct incorporation into (injectable) vaccine formulations. As an example, doramapimod, a MAPK-p38 inhibitor, binds p38α, has low inhibitory action on molecules that regulate T-cell activation and effector functions, such as Raf, Fyn, and Lck, and does not show a significant inhibition of ERK and Syk. Also, an ERK inhibitor, FR18024, possesses strong anti-T-cell senescence activities and was shown to improve clinical parameters in murine models of viral infections (Sreekanth et al., 2014).

The possibility of using bioactive compounds derived from olive oil as inhibitors of inflamm-aging, the third hallmark of immunosenescence (Aiello et al., 2019), might also be tested in vitro for the improvement of influenza vaccination response together with the inhibitors previously described (Gambino et al., 2018). It was demonstrated that inflamm-aging plays an important role in compromising the immune responses by way of inducing high expression of some microRNAs that interfere with B-cell activation. In vitro, this drives TNF-α production and inhibits B-cell activation (Frasca et al., 2015). Increased serum levels of TNF-α are also linked to a defective T-cell response, in part due to reduced expression of CD28 on T cells (Ponnappan and Ponnappan, 2011). Moreover, in monocytes, the prevaccination expression of genes related to inflammation and innate immune response is negatively correlated with vaccination-induced activation of influenza-specific antibody responses (Nakaya et al., 2015). Phenolic compounds of olive oil include around 30 molecules, some with strong antioxidant and anti-inflammatory properties. Their nutraceutical properties counteract the pathophysiology of age-related diseases, such as cardiovascular diseases, arthritis, neurodegenerative diseases, with a relevant role in many antiaging strategies. The mechanisms of action involve the scavenging of radical oxygen species, inhibition of mast cell degranulation, and inhibition of cyclooxygenases 1 and 2, positively counteracting chronic low-grade inflammation.

Hydroxytyrosol exhibits a promising antioxidant potential in protecting mononuclear cells against 2,3,7,8-tetrachlorodibenzo-p-dioxin, an external stressor. The acetyl derivatives of phenolic compounds are more efficient compared with native molecules, with a higher rate of cellular internalization. Peracetylation of oleuropein and its derivatives may improve their capacity to permeate the molecular membrane. In vivo, the application of olive oil derivatives by parenteral administration has been shown to mitigate inflammation and oxidative stress and the parenteral administration of oleuropein aglycone after challenge of the pleural cavity with carrageenin, a strong inflammatory agent, fully abrogated induced inflammation in a mouse model (Impellizzeri et al., 2011). It was also seen that parenteral treatment with oleic derivatives reduced malondialdehyde, a marker of lipid peroxidation, in rat brains (Rizzo et al., 2017).

8.4 Conclusion and future perspectives

This chapter summarizes data regarding the immunogenicity and efficacy of influenza, pneumococcal, and herpes zoster vaccines currently in use for older people, and provides a perspective on possible new methodological approach for future vaccine development specifically for this age group. Considering the differences between young and older peoples' immune systems, successful development of new vaccines specifically tailored for the aged population will require a deeper knowledge of the mechanisms of immunosenescence. This is not a simple challenge for vaccinologists, also in light of the fact that immune responses in aged people are significantly influenced by the individual past history of exposures and immunity. In fact, the immunological experience and other stresses that individuals encounter over their lifetimes shape their ability to respond to external stimuli, such as vaccinations. In addition, many other factors influence outcomes, including nutrition, physical exercise, drug treatments, the microbiota, and chronic diseases. On the other hand, methodological and technological advances allow modern vaccination approaches that are opening up the possibility of developing new vaccines against practically any pathogen. This is relevant to address major and upcoming global threats like antimicrobial resistance, responsible for at least 700,000 deaths each year. Vaccines would be a promising solution against antimicrobial resistance for several reasons.

Unfortunately, antibiotics become quickly obsolete because resistance emerges immediately after their introduction, while vaccines allow lasting protection against infections. In addition, while antibiotics target only certain metabolic targets, vaccines can elicit a broad multitarget immune response by reducing the likelihood of evolving resistant mutations (Andreano et al., 2019; Tagliabue and Rappuoli, 2018).

It is to be hoped that these new approaches will be soon available also for the preparation of SARS-CoV-2 vaccines to protect against COVID-19, an acute requirement at the time of writing. It is amazing the rapidity with which SARS-CoV-2 was sequenced and vaccine candidates produced for testing in a matter of weeks. Broadly speaking, these vaccines group into several different "platforms," among them old standbys such as inactivated or weakened whole viruses, genetically engineered proteins, and the newer mRNA technology. One such vaccine designated mRNA-1273 developed by NIAID scientists and their collaborators at the biotechnology company Moderna, Inc., based in Cambridge, Massachusetts, was the first to enter clinical trials. This investigational vaccine directs cells to express a viral protein that it is hoped will elicit a robust immune response. The mRNA-1273 vaccine has shown promise in animal models, and there is the first trial to examine it in humans (Cohen, 2020). Furthermore, a chimpanzee adenovirus-vectored vaccine, ChAdOx1, developed by Oxford University's Jenner Institute has been shown to induce a strong immune antibody response up to the 56th day of the ongoing testing. These are the preliminary results of phase 1–2 of testing which involved 1077 healthy adults. The vaccine showed an acceptable safety profile, and homologous boosting increased antibody responses. These results, together with the induction of both humoral and cellular immune responses, support large-scale evaluation of this candidate vaccine in an ongoing phase 3 program (Folegatti et al., 2020). However, vaccine design for the older adult will need to overcome the challenges of immunosenescence and aim to stimulate a broad T- and B-cell response tailored to restoring reduced immune function in the older population (Pawelec and Weng, 2020).

Note added in proof

By the beginning of 2021 in the United States, United Kingdom and European Union three anti-SARS-CoV-2 vaccines were in use, two based on mRNA technology and the third an adenovirus-vectored vaccine. The antigen is the protein Spike that plays a key role for viral entry into cells. Concerning the Pfizer/BioNtech BNT162b2 mRNA COVID-19 vaccine (Comirnaty), 37,706 subjects (16–91 years, median 52 years) completed the trial; a two-dose regimen of BNT162b2 conferred 95% protection against COVID-19. Safety over a median of 2 months was similar to that of other viral vaccines (Polack et al., 2020). As regards the Moderna mRNA-1273 SARS-CoV-2 Vaccine, 29,148 subjects (>18 years) completed the two-dose regimen trial. The vaccine showed 94.1% efficacy at preventing COVID-19, including severe disease. Aside from transient local and systemic reactions, no safety concerns were identified (Baden et al., 2021). Regarding the third vaccine, the Astra-Zeneca/Oxford University ChAdOx1 nCoV-19, overall vaccine efficacy across groups was 70.4% in 11,636 subjects (18–55 years) which completed the two-dose regimen trial. ChAdOx1 nCoV-19 had an acceptable safety profile and has been found to be efficacious against symptomatic COVID-19 in this interim analysis of ongoing clinical trials (Voysey et al., 2021).

Funding

This study was supported by grants from "Piano di incentivi per la Ricerca, Linea Intervento 2 and Linea Intervento 3 PIACERI, 2020–2022", University of Catania, Italy.

References

Aiello A, Farzaneh F, Candore G, Caruso C, Davinelli S, Gambino CM, et al. Immunosenescence and its hallmarks: how to oppose aging strategically? A review of potential options for therapeutic intervention. Front Immunol 2019;10:2247. https://doi.org/10.3389/fimmu.2019.02247.

Andreano E, D'Oro U, Rappuoli R, Finco O. Vaccine evolution and its application to fight modern threats. Front Immunol 2019;10:1722. https://doi.org/10.3389/fimmu.2019.01722.

Andrew MK, Bowles SK, Pawelec G, Haynes L, Kuchel GA, McNeil SA, et al. Influenza Vaccination in older adults: recent innovations and practical applications. Drugs Aging 2019;36:29–37. https://doi.org/10.1007/s40266-018-0597-4.

Ansaldi F, Bacilieri S, Durando P, Sticchi L, Valle L, Montomoli E, et al. Cross-protection by MF59-adjuvanted influenza vaccine: neutralizing and haemagglutination-inhibiting antibody activity against A(H3N2) drifted influenza viruses. Vaccine 2008;26:1525–9. https://doi.org/10.1016/j.vaccine.2008.01.019.

Antrobus RD, Lillie PJ, Berthoud TK, Spencer AJ, McLaren JE, Ladell K, et al. A T cell-inducing influenza vaccine for the elderly: safety and immunogenicity of MVA-NP+M1 in adults aged over 50 years. PLoS One 2012;7:e48322https://doi.org/10.1371/journal.pone.0048322.

Baden LR, et al. Efficacy and safety of the mRNA-1273 SARS-CoV-2 vaccine. N Engl J Med 2021; 384(5):403–16. https://doi.org/10.1056/NEJMoa2035389.

Barata JT, Durum SK, Seddon B. Flip the coin: IL-7 and IL-7R in health and disease. Nat Immunol 2019;20:1584–93. https://doi.org/10.1038/s41590-019-0479-x.

Beyer WE, McElhaney J, Smith DJ, Monto AS, Nguyen-Van-Tam JS, Osterhaus AD. Cochrane re-arranged: support for policies to vaccinate elderly people against influenza. Vaccine 2013;31: 6030–3. https://doi.org/10.1016/j.vaccine.2013.09.063.

Bonten MJ, Huijts SM, Bolkenbaas M, Webber C, Patterson S, Gault S, et al. Polysaccharide conjugate vaccine against pneumococcal pneumonia in adults. N Engl J Med 2015;372:1114–25. https://doi.org/10.1056/NEJMoa1408544.

Boraschi D, Italiani P. Immunosenescence and vaccine failure in the elderly: strategies for improving response. Immunol Lett 2014;162:346–53. https://doi.org/10.1016/j.imlet.2014.06.006.

Bouard D, Alazard-Dany D, Cosset FL. Viral vectors: from virology to transgene expression. Br J Pharmacol 2009;157:153–65. https://doi.org/10.1038/bjp.2008.349.

Caruso C, Vasto S. Immunity and aging. In: Ratcliffe MJH, editor. Encyclopedia of immunobiology, vol. 5. Oxford: Academic Press (2016). p. 127–32.

Chen JQ, Szodoray P, Zeher M. Toll-like receptor pathways in autoimmune diseases. Clin Rev Allergy Immunol 2016;50:1–17. https://doi.org/10.1007/s12016-015-8473-z.

Chi H, Flavell RA. Studies on MAP kinase signaling in the immune system. Methods Mol Biol 2010;661:471–80.

Chlibek R, Bayas JM, Collins H, de la Pinta ML, Ledent E, Mols JF, et al. Safety and immunogenicity of an AS01-adjuvanted varicella-zoster virus subunit candidate vaccine against herpes zoster in adults >=50 years of age. J Infect Dis 2013;208:1953–61. https://doi.org/10.1093/infdis/jit365.

Choi Y, Chang J. Viral vectors for vaccine applications. Clin Exp Vaccine Res 2013;2:97–105. https://doi.org/10.7774/cevr.2013.2.2.97.

Cioncada R, Maddaluno M, Vo HTM, Woodruff M, Tavarini S, Sammicheli C, et al. Vaccine adjuvant MF59 promotes the intranodal differentiation of antigen-loaded and activated monocyte-derived dendritic cells. PLoS One 2017;12:e0185843https://doi.org/10.1371/journal.pone.0185843.

Clements CJ, Griffiths E. The global impact of vaccines containing aluminium adjuvants. Vaccine 2002; 20(Suppl 3):S24–33.

Cohen J. Vaccine designers take first shots at COVID-19. Science 2020;368:14–6. https://doi.org/10.1126/science.368.6486.14.

Collin M, Mc Govern N, Haniffa M. Human dendritic cell subsets. Immunology 2013;140:22–30.

Colombetti S, Lévy F, Chapatte L. IL-7 adjuvant treatment enhances long-term tumor-antigen-specific CD8+ T-cell responses after immunization with recombinant lentivector. Blood 2009;113:6629–37. https://doi.org/10.1182/blood-2008-05-155309.

Coughlan L, Sridhar S, Payne R, Edmans M, Milicic A, Venkatraman N, et al. Heterologous two-dose Vaccination with simian adenovirus and poxvirus vectors elicits long-lasting cellular immunity to Influenza virus a in healthy adults. EBioMedicine 2018;29:146–54. https://doi.org/10.1016/j.ebiom.2018.02.011.

Crooke SN, Ovsyannikova IG, Poland GA, Kennedy RB. Immunosenescence and human vaccine immune responses. Immun Ageing 2019;16:25. https://doi.org/10.1186/s12979-019-0164-9.

Cunningham AL, Lal H, Kovac M, Chlibek R, Hwang SJ, Díez-Domingo J, et al. ZOE-70 study group. Efficacy of the herpes zoster subunit vaccine in adults 70 years of age or older. N Engl J Med 2016;375:1019–32. https://doi.org/10.1056/NEJMoa1603800.

Cunningham AL, Heineman TC, Lal H, Godeaux O, Chlibek R, Hwang SJ, et al. ZOE-50/70 study group. Immune responses to a recombinant glycoprotein E herpes zoster vaccine in adults aged 50 years or older. J Infect Dis 2018;217:1750–60. https://doi.org/10.1093/infdis/jiy095.

Del Giudice G, Rappuoli R, Didierlaurent AM. Correlates of adjuvanticity: a review on adjuvants in licensed vaccines. Semin Immunol 2018;39:14–21. https://doi.org/10.1016/j.smim.2018.05.001.

Didierlaurent AM, Morel S, Lockman L, Giannini SL, Bisteau M, Carlsen H, et al. AS04, an aluminum salt- and TLR4 agonist-based adjuvant system, induces a transient localized innate immune response leading to enhanced adaptive immunity. J Immunol 2009;183:6186–97. https://doi.org/10.4049/jimmunol.0901474.

Didierlaurent AM, Laupèze B, Di Pasquale A, Hergli N, Collignon C, Garçon N. Adjuvant system AS01: helping to overcome the challenges of modern vaccines. Expert Rev Vaccines 2017;16:55–63.

Dupuis M, Denis-Mize K, LaBarbara A, Peters W, Charo IF, McDonald DM, et al. Immunization with the adjuvant MF59 induces macrophage trafficking and apoptosis. Eur J Immunol 2001;31:2910–8.

Eisenbarth SC, Colegio OR, O'Connor W, Sutterwala FS, Flavell RA. Crucial role for the Nalp3 inflammasome in the immunostimulatory properties of aluminium adjuvants. Nature 2008;453:1122–6. https://doi.org/10.1038/nature06939.

Esposito S, Principi N. ESCMID vaccine study group. Direct and indirect effects of the 13-valent pneumococcal conjugate vaccine administered to infants and young children. Future Microbiol 2015; 10:1599–607. https://doi.org/10.2217/fmb.15.81.

Ewer KJ, Lambe T, Rollier CS, Spencer AJ, Hill AV, Dorrell L. Viral vectors as vaccine platforms: from immunogenicity to impact. Curr Opin Immunol 2016;41:47–54. https://doi.org/10.1016/j.coi.2016.05.014.

Feldman C, Anderson R. Review: current and new generation pneumococcal vaccines. J Infect 2014; 69:309–25. https://doi.org/10.1016/j.jinf.2014.06.006.

Folegatti PM, Ewer KJ, Aley PK, Angus B, Becker S, Belij-Rammerstorfer S, et al. Duncan Bellamy. Safety and immunogenicity of the ChAdOx1 nCoV-19 vaccine against SARS-CoV-2: a preliminary report of a phase 1/2, single-blind, randomised controlled trial. Lancet 2020;https://doi.org/10.1016/S0140-6736(20)31604-4 Online ahead of print.

Frasca D, Diaz A, Romero M, Ferracci F, Blomberg BB. MicroRNAs miR-155 and miR-16 decrease AID and E47 in B cells from elderly individuals. J Immunol 2015;195:2134–240. https://doi.org/10.4049/jimmunol.1500520.

Frey SE, Reyes MR, Reynales H, Bermal NN, Nicolay U, Narasimhan V, et al. Comparison of the safety and immunogenicity of an MF59®-adjuvanted with a non-adjuvanted seasonal influenza vaccine in elderly subjects. Vaccine 2014;32:5027–34. https://doi.org/10.1016/j.vaccine.2014.07.013.

Gambino CM, Vasto S, Ioannou K, Candore G, Caruso C, Farzaneh F. Accardi G, Caruso C, editors. Triggering of toll-like receptors in the elderly. A pilot study relevant for vaccination. Updates in pathobiology: Causality and chance in ageing, age-related diseases and longevity. Palermo University Press; 2017.

Gambino CM, Accardi G, Aiello A, Candore G, Dara-Guccione G, Mirisola M, et al. Procopio effect of extra virgin olive oil and table olives on the ImmuneInflammatory responses: potential clinical applications. Endocr Metab Immune Disord Drug Targets 2018;18:14–22. https://doi.org/10.2174/1871 530317666171114113822.

Gershon AA, Gershon MD. Pathogenesis and current approaches to control of varicella-zoster virus infections. Clin Microbiol Rev 2013;26:728–43. https://doi.org/10.1128/CMR.00052-13.

Glück R, Mischler R, Finkel B, Que JU, Scarpa B, Cryz Jr. SJ. Immunogenicity of new virosome influenza vaccine in elderly people. Lancet 1994;344:160–3.

Goto N, Kato H, Maeyama J, Shibano M, Saito T, Yamaguchi J, Yoshihara S. Local tissue irritating effects and adjuvant activities of calcium phosphate and aluminium hydroxide with different physical properties. Vaccine 1997;15:1364–71.

Grolleau-Julius A, Harning EK, Abernathy LM, Yung RL. Impaired dendritic cell function in aging leads to defective antitumor immunity. Cancer Res 2008;68:6341–9. https://doi.org/10.1158/0008-5472. CAN-07-5769.

Guy B. The perfect mix: recent progress in adjuvant research. Nat Rev Microbiol 2007;5:505–17.

Hanquet G, Kissling E, Fenoll A, George R, Lepoutre A, Lernout T, et al. Pneumococcal serotypes in children in 4 European countries. Emerg Infect Dis 2010;16:1428–39. https://doi.org/10.3201/eid1609.100102.

Haq K, Fulop T, Tedder G, Gentleman B, Garneau H, Meneilly GS, et al. Cytomegalovirus Seropositivity predicts a decline in the T cell but not the antibody response to Influenza in vaccinated older adults independent of type 2 diabetes status. J Gerontol A Biol Sci Med Sci 2017;72:1163–70. https://doi.org/10.1093/gerona/glw216.

Holland D, Booy R, De Looze F, Eizenberg P, McDonald J, Karrasch J, et al. Intradermal influenza vaccine administered using a new microinjection system produces superior immunogenicity in elderly adults: a randomized controlled trial. J Infect Dis 2008;198:650–8. https://doi.org/10.1086/590434.

Hutchison S, Benson RA, Gibson VB, Pollock AH, Garside P, Brewer JM. Antigen depot is not required for alum adjuvanticity. FASEB J 2012;26(3):1272–9. https://doi.org/10.1096/fj.11-184556.

Impellizzeri D, Esposito E, Mazzon E, Paterniti I, Di Paola R, Bramanti P, et al. The effects of oleuropein aglycone, an olive oil compound, in a mouse model of carrageenan-induced pleurisy. Clin Nutr 2011;30:533–40.

Iob A, Brianti G, Zamparo E, Gallo T. Evidence of increased clinical protection of an MF59-adjuvant influenza vaccine compared to a non-adjuvant vaccine among elderly residents of long-term care facilities in Italy. Epidemiol Infect 2005;133:687–93.

Jing Y, Shaheen E, Drake RR, Chen N, Gravenstein S, Deng Y. Aging is associated with a numerical and functional decline in plasmacytoid dendritic cells, whereas myeloid dendritic cells are relatively unaltered in human peripheral blood. Hum Immunol 2009;70:777–84. https://doi.org/10.1016/j.humimm.2009.07.005.

Khurana S, Verma N, Yewdell JW, Hilbert AK, Castellino F, Lattanzi M, et al. MF59 adjuvant enhances diversity and affinity of antibody-mediated immune response to pandemic influenza vaccines. Sci Transl Med 2011;3:85ra48. https://doi.org/10.1126/scitranslmed.3002336.

Kool M, Fierens K, Lambrecht BN. Alum adjuvant: some of the tricks of the oldest adjuvant. J Med Microbiol 2012;61(Pt 7):927–34.

Kornbluth RS, Stone GW. Immunostimulatory combinations: designing the next generation of vaccine adjuvants. J Leukoc Biol 2006;80:1084–102.

Kostova D, Reed C, Finelli L, Cheng PY, Gargiullo PM, Shay DK, et al. Influenza illness and hospitalizations averted by Influenza Vaccination in the United States, 2005–2011. PLoS One 2013;8:e66312. https://doi.org/10.1371/journal.pone.0066312.

Lanna A, Henson SM, Escors D, Akbar AN. The kinase p38 activated by the metabolic regulator AMPK and scaffold TAB1 drives the senescence of human T cells. Nat Immunol 2014;15:965–72.

Lanna A, Gomes DC, Muller-Durovic B, McDonnell T, Escors D, Gilroy DW, Lee JH, Karin M, Akbar AN. A sestrin-dependent Erk-Jnk-p38 MAPK activation complex inhibits immunity during aging. Nat Immunol 2017;18:354–63.

Levin MJ, Kroehl ME, Johnson MJ, Hammes A, Reinhold D, Lang N, et al. Th1 memory differentiates recombinant from live herpes zoster vaccines. J Clin Invest 2018;128:4429–40. https://doi.org/10.1172/JCI121484.

Lévy Y, Sereti I, Tambussi G, Routy JP, Lelièvre JD, Delfraissy JF, et al. Effects of recombinant human interleukin 7 on T-cell recovery and thymic output in HIV-infected patients receiving antiretroviral therapy: results of a phase I/IIa randomized, placebo-controlled, multicenter study. Clin Infect Dis 2012;55:291–300. https://doi.org/10.1093/cid/cis383.

Lynch HE, Goldberg GL, Chidgey A, Van den Brink MR, Boyd R, Sempowski GD. Thymic involution and immune reconstitution. Trends Immunol 2009;30:366–73. https://doi.org/10.1016/j.it.2009.04.003.

Malito E, Bursulaya B, Chen C, Lo Surdo P, Picchianti M, Balducci E, et al. Structural basis for lack of toxicity of the diphtheria toxin mutant CRM197. Proc Natl Acad Sci U S A 2012;109:5229–34. https://doi.org/10.1073/pnas.1201964109.

Mallick-Searle T, Snodgrass B, Brant JM. Postherpetic neuralgia: epidemiology, pathophysiology, and pain management pharmacology. J Multidiscip Healthc 2016;9:447–54.

Mannino S, Villa M, Apolone G, Weiss NS, Groth N, Aquino I, et al. Effectiveness of adjuvanted influenza vaccination in elderly subjects in northern Italy. Am J Epidemiol 2012;176:527–33 [Epub 2012 Aug 31].

Marrack P, McKee AS, Munks MW. Towards an understanding of the adjuvant action of aluminium. Nat Rev Immunol 2009;9:287–93.

McElhaney JE, Kuchel GA, Zhou X, Swain SL, Haynes L. T-cell immunity to Influenza in older adults: a pathophysiological framework for development of more effective vaccines. Front Immunol 2016;7:41. https://doi.org/10.3389/fimmu.2016.00041.

McElhaney JE, Verschoor C, Pawelec G. Zoster vaccination in older adults: efficacy and public health implications. J Gerontol A Biol Sci Med Sci 2019;74(8):1239–43. https://doi.org/10.1093/gerona/glz085.

McKee AS, Marrack P. Old and new adjuvants. Curr Opin Immunol 2017;47:44–51. https://doi.org/10.1016/j.coi.2017.06.005.

Melchionda F, Fry TJ, Milliron MJ, McKirdy MA, Tagaya Y, Mackall CL. Adjuvant IL-7 or IL-15 overcomes immunodominance and improves survival of the CD8+ memory cell pool. J Clin Invest 2005;115:1177–87. https://doi.org/10.1172/JCI23134.

Merani S, Pawelec G, Kuchel GA, McElhaney JE. Impact of aging and cytomegalovirus on immunological response to Influenza Vaccination and infection. Front Immunol 2017;8:784. https://doi.org/10.3389/fimmu.2017.00784.

Miller LH, Saul A, Mahanty S. Revisiting Freund's incomplete adjuvant for vaccines in the developing world. Trends Parasitol 2005;21:412–4.

Morefield GL, Sokolovska A, Jiang D, HogenEsch H, Robinson JP, Hem SL. Role of aluminum-containing adjuvants in antigen internalization by dendritic cells in vitro. Vaccine 2005;18(23):1588–95.

Morrison VA, Johnson GR, Schmader KE, Levin MJ, Zhang JH, Looney DJ, et al. Shingles prevention study group. Long-term persistence of zoster vaccine efficacy. Clin Infect Dis 2015;60:900–9. https://doi.org/10.1093/cid/ciu918.

Moser C, Amacker M, Zurbriggen R. Influenza virosomes as a vaccine adjuvant and carrier system. Expert Rev Vaccines 2011;10:437–46. https://doi.org/10.1586/erv.11.15.

Musher DM, Sampath R, Rodriguez-Barradas MC. The potential role for protein-conjugate pneumococcal vaccine in adults: what is the supporting evidence? Clin Infect Dis 2011;52:633–40. https://doi.org/10.1093/cid/ciq207.

Nakaya HI, Hagan T, Duraisingham SS, Lee EK, Kwissa M, Rouphael N, et al. Systems analysis of immunity to Influenza Vaccination across multiple years and in diverse populations reveals shared molecular signatures. Immunity 2015;1186–98. https://doi.org/10.1016/j.immuni.2015.11.012.

Oh SJ, Lee JK, Shin OS. Aging and the immune system: the impact of immunosenescence on viral infection, immunity and vaccine immunogenicity. Immune Netw 2019;19:e37. https://doi.org/10.4110/in.2019.19.e37.

O'Hagan DT, Ott GS, Nest GV, Rappuoli R, Giudice GD. The history of MF59(®) adjuvant: a phoenix that arose from the ashes. Expert Rev Vaccines 2013;12:13–30. https://doi.org/10.1586/erv.12.140.

Okoye AA, Rohankhedkar M, Konfe AL, Abana CO, Reyes MD, Clock JA, et al. Effect of IL-7 therapy on naive and memory T cell homeostasis in aged rhesus macaques. J Immunol 2005;195:4292–305.

Olsson J, Wikby A, Johansson B, Löfgren S, Nilsson BO, Ferguson FG. Age-related change in peripheral blood T-lymphocyte subpopulations and cytomegalovirus infection in the very old: the Swedish longitudinal OCTO immune study. Mech Ageing Dev 2000;121:187–201.

Opie EL, Freund J. An experimental study of protective inoculation with heat killed tubercle bacilli. J Exp Med 1937;66:761–88.

Osterholm MT, Kelley NS, Sommer A, Belongia EA. Efficacy and effectiveness of influenza vaccines: a systematic review and meta-analysis. Lancet Infect Dis 2012;12:36–44. https://doi.org/10.1016/S1473-3099(11)70295-X.

Oxman MN, Levin MJ, Johnson GR, Schmader KE, Straus SE, Gelb LD, et al. A vaccine to prevent herpes zoster and postherpetic neuralgia in older adults. N Engl J Med 2005;352:2271–84.

Panda A, Qian F, Mohanty S, van Duin D, Newman FK, Zhang L, et al. Age-associated decrease in TLR function in primary human dendritic cells predicts influenza vaccine response. J Immunol 2010; 184:2518–27. https://doi.org/10.4049/jimmunol.0901022.

Parikh SR, Andrews NJ, Beebeejaun K, Campbell H, Ribeiro S, Ward C, White JM, et al. Effectiveness and impact of a reduced infant schedule of 4CMenB vaccine against group B meningococcal disease in England: a national observational cohort study. Lancet 2016;388:2775–82. https://doi.org/10.1016/S0140-6736(16)31921-3.

Pawelec G, McElhaney J. Vaccines for improved cellular immunity to Influenza. EBioMedicine 2018; 30:12–3. https://doi.org/10.1016/j.ebiom.2018.03.001.

Pawelec G, Weng NP. Can an effective SARS-CoV-2 vaccine be developed for the older population? Immun Ageing 2020;17:8. https://doi.org/10.1186/s12979-020-00180-2.

Phillips JA, Brondstetter TI, English CA, Lee HE, Virts EL, Thoman ML. IL-7 gene therapy in aging restores early thymopoiesis without reversing involution. J Immunol 2004;173:4867–74.

Polack FP, et al. Safety and efficacy of the BNT162b2 mRNA COVID-19 vaccine. N Engl J Med 2020; 383(27):2603–15. https://doi.org/10.1056/NEJMoa2034577.

Ponnappan S, Ponnappan U. Aging and immune function: molecular mechanisms to interventions. Antioxid Redox Signal 2011;14:1551–85. https://doi.org/10.1089/ars.2010.3228.

Qian F, Wang X, Zhang L, Lin A, Zhao H, Fikrig E, et al. Impaired interferon signaling in dendritic cells from older donors infected in vitro with West Nile virus. J Infect Dis 2011;203:1415–24. https://doi.org/10.1093/infdis/jir048.

Ramon G. Sur l'augmentation anormale de l'antitoxine chez les chevaux producteurs de serum antidiphterique. Bull Soc Cetr Med Vet 1925;101:227–34.

Rappuoli R. Reverse vaccinology. Curr Opin Microbiol 2000;3:445–50.

Rizzo M, Ventrice D, Giannetto F, Cirinnà S, Santagati NA, Procopio A, et al. Antioxidant activity of oleuropein and semisynthetic acetyl-derivatives determined by measuring malondialdehyde in rat brain. J Pharm Pharmacol 2017;69:1502–12.

Robertson CA, DiazGranados CA, Decker MD, Chit A, Mercer M, Greenberg DP. Fluzone® High-Dose Influenza Vaccine. Expert Rev Vaccines 2016;15:1495–505 [Epub 2016 Nov 14].

Rondy M, El Omeiri N, Thompson MG, Levêque A, Moren A, Sullivan SG. Effectiveness of influenza vaccines in preventing severe influenza illness among adults: a systematic review and meta-analysis of test-negative design case-control studies. J Infect 2017;75:381–94. https://doi.org/10.1016/j.jinf.2017.09.010.

Russell K, Chung JR, Monto AS, Martin ET, Belongia EA, McLean HQ, et al. Influenza vaccine effectiveness in older adults compared with younger adults over five seasons. Vaccine 2018;36:1272–8. https://doi.org/10.1016/j.vaccine.2018.01.045.

Schmader KE, Oxman MN, Levin MJ, Johnson G, Zhang JH, Betts R, et al. Shingles prevention study group. Persistence of the efficacy of zoster vaccine in the shingles prevention study and the short-term persistence substudy. Clin Infect Dis 2012;55:1320–8. https://doi.org/10.1093/cid/cis638.

Schreibelt G, Tel J, Sliepen KH, Benitez-Ribas D, Figdor CG, Adema GJ, et al. Toll-like receptor expression and function in human dendritic cell subsets: implications for dendritic cell-based anti-cancer immunotherapy. Cancer Immunol Immunother 2010;59:1573–82. https://doi.org/10.1007/s00262-010-0833-1.

Schumacher R, Adamina M, Zurbriggen R, Bolli M, Padovan E, Zajac P, et al. Influenza virosomes enhance class I restricted CTL induction through CD4+ T cell activation. Vaccine 2004;22:714–23.

Shodell M, Siegal FP. Circulating, interferon-producing plasmacytoid dendritic cells decline during human ageing. Scand J Immunol 2002;56:518–21.

Silwal P, Kim JK, Yuk JM, Jo EK. AMP-activated protein kinase and host Defense against infection. Int J Mol Sci 2018;6:19.

Smetana J, Chlibek R, Shaw J, Splino M, Prymula R. Influenza vaccination in the elderly. Hum Vaccin Immunother 2018;14(3):540–9. https://doi.org/10.1080/21645515.2017.1343226.

Sreekanth GP, Chuncharunee A, Sirimontaporn A, Panaampon J, Srisawat C, Morchang A, et al. Role of ERK1/2 signaling in dengue virus-induced liver injury. Virus Res 2014;188:15–26. https://doi.org/10.1016/j.virusres.2014.03.025.

Stout-Delgado HW, Yang X, Walker WE, Tesar BM, Goldstein DR. Aging impairs IFN regulatory factor 7 up-regulation in plasmacytoid dendritic cells during TLR9 activation. J Immunol 2008;181: 6747–56.

Tagliabue A, Rappuoli R. Changing priorities in vaccinology: antibiotic resistance moving to the top. Front Immunol 2018;9:1068. https://doi.org/10.3389/fimmu.2018.01068.

Takeda K, Akira S. Toll-like receptors. Curr Protoc Immunol 2015;109:14.12.1–14.12.10. https://doi.org/10.1002/0471142735.im1412s109.

Treanor JJ. Clinical practice. Influenza vaccination. N Engl J Med 2016;375:1261–8. https://doi.org/10.1056/NEJMcp1512870.

Trucchi C, Alicino C, Orsi A, Paganino C, Barberis I, Grammatico F, et al. Fifteen years of epidemiologic, virologic and syndromic influenza surveillance: a focus on type B virus and the effects of vaccine mismatch in Liguria region. Italy Hum Vaccin Immunother 2017;13:456–63. https://doi.org/10.1080/21645515.2017.1264779.

Tye GJ, Ioannou K, Amofah E, Quartey-Papafio R, Westrop SJ, Krishnamurthy P, et al. The combined molecular adjuvant CASAC enhances the CD8+ T cell response to a tumor-associated self-antigen in aged, immunosenescent mice. Immun Ageing 2015;12(6)https://doi.org/10.1186/s12979-015-0033-0.

Ura T, Okuda K, Shimada M. Developments in viral vector-based vaccines. Vaccines (Basel) 2014; 2:624–41. https://doi.org/10.3390/vaccines2030624.

Voysey M, et al. Safety and efficacy of the ChAdOx1 nCoV-19 vaccine (AZD1222) against SARS-CoV-2: an interim analysis of four randomised controlled trials in Brazil, South Africa, and the UK. Lancet 2021;397(10269):99–111. https://doi.org/10.1016/S0140-6736(20)32661-1.

Weinberger B. Vaccines for the elderly: current use and future challenges. Immun Ageing 2018;15(3) https://doi.org/10.1186/s12979-017-0107-2.

Wiersma LC, Rimmelzwaan GF, de Vries RD. Developing universal Influenza vaccines: hitting the nail, not just on the head. Vaccines (Basel) 2015;3:239–62. https://doi.org/10.3390/vaccines3020239.

Wilkins AL, Kazmin D, Napolitani G, Clutterbuck EA, Pulendran B, Siegrist CA, et al. AS03- and MF59-Adjuvanted Influenza vaccines in children. Front Immunol 2017;8:1760. https://doi.org/10.3389/fimmu.2017.01760.

Wroe PC, Finkelstein JA, Ray GT, Linder JA, Johnson KM, Rifas-Shiman S, et al. Aging population and future burden of pneumococcal pneumonia in the United States. J Infect Dis 2012;205:1589–92. https://doi.org/10.1093/infdis/jis240.

Wu P, Nielsen TE, Clausen MH. Small-molecule kinase inhibitors: an analysis ofFDA-approved drugs. Drug Discov Today 2016;21:5–10.

Yam KK, Gupta J, Allen EK, Burt KR, Beaulieu É, Mallett CP, et al. Comparison of AS03 and alum on immune responses elicited by A/H3N2 split influenza vaccine in young, mature and aged BALB/c mice. Vaccine 2016;34:1444–51. https://doi.org/10.1016/j.vaccine.2016.02.012.

You J, Dong H, Mann ER, Knight SC, Yaqoob P. Ageing impairs the T cell response to dendritic cells. Immunobiology 2013;218(8):1077–84. https://doi.org/10.1016/j.imbio.2013.02.002.

Zanetti AR, Amendola A, Besana S, Boschini A, Tanzi E. Safety and immunogenicity of influenza vaccination in individuals infected with HIV. Vaccine 2002;20(Suppl 5):B29–32.

Zurbriggen R. Immunostimulating reconstituted influenza virosomes. Vaccine 2003;21:921–4.

CHAPTER 9

Resilience signaling and hormesis in brain health and disease

Vittorio Calabrese[a], Angela Trovato[a], Maria Scuto[a], Maria Laura Ontario[a], Mario Tomasello[a], Rosario Perrotta[b], and Edward Calabrese[c]

[a]Department of Biomedical and Biotechnological Sciences, School of Medicine, University of Catania, Catania, Italy
[b]Department of Plastic and Reconstructive Surgery, University of Catania, Catania, Italy
[c]Department of Environmental Health Sciences, University of Massachusetts, Amherst, MA, United States

9.1 Introduction

It is generally recognized that multiple factors, such as severity and number of traumatic events during prenatal neurodevelopment and adolescence, timing of exposure to adversity, developmental history, cognitive flexibility, and environmental changes such as toxicant exposure, compromise health and well-being over the lifespan, including aging. Operationalizing at cellular levels the neurobiology of stress resistance, cellular resilience, describes the ability of a cell to cope with detrimental metabolic and signaling conditions where cellular metabolism does not collapse immediately after the hit or enter into the cell death. It, rather, puts in motion a programmed cell life program through stress-responsive signaling which promotes a new homeostasis under stress. However, in spite of increasing research efforts, the process of reverting "back to normal" is still an elusive matter, being studied not so extensively at the cellular level. The defense and resilience programs include a number of cellular stress responses such as rearrangements in energy metabolism, oxidative stress responses, including hypoxia signaling via Hypoxia-inducible factor-1, the heat shock response via heat shock transcription factor 1 and the antioxidant response via Nuclear factor erythroid 2-related factor 2 (NRF-2) (vitagene network), stress kinase signaling via c-Jun N-terminal kinase (JNK) and activator protein-1 DNA, damage responses via p21 or BCL-2, and the unfolded protein response/amino acid starvation response via activating transcription factors-4/-6 activation of antiapoptotic pathways and DNA repair mechanisms. Efficient functioning of resilience processes, enhancing endogenous cellular redox homeostatic mechanisms, integrates adaptive stress responses via the heat shock and Nrf-2 related pathways, as well as sirtuins (Sirt) via mammalian target of rapamycin AKT/mTOR pathway signaling, thus representing a complex operational network under control of genes, termed vitagenes (Calabrese et al., 2010) (Fig. 9.1), as well as epigenetic changes that leave a molecular "memory/scar"—mirroring alterations that are consequence of the stress experienced by the cell.

Human Aging
https://doi.org/10.1016/B978-0-12-822569-1.00012-3

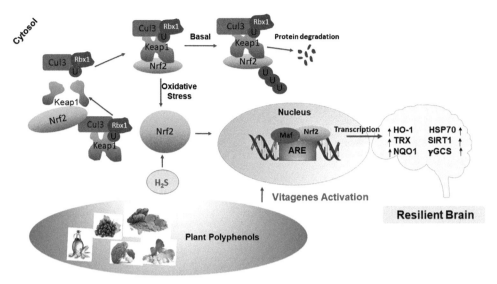

Fig. 9.1 Resilience signaling and stress response. Nrf2-vitagene pathway is modulated by plant polyphenols in the brain. In basal conditions Nrf2, bound to its inhibitor Kelch-like ECH-associated protein (Keap1), is restricted to the cytosol, where upon association with the Cul3-Rbx1-based E3/ ubiquitin ligase complex undergoes ubiquitination and proteasomal degradation. Under stress conditions, Nrf2 is released from Keap1, translocates into the nucleus where by binding to ARE sequences localized in the DNA promoter region in heterodimeric combination with Maf proteins initiates transcriptional activity. Plant polyphenols are small molecules that reverse oxidant insult by activating Nrf2 nuclear translocation and transcription of neuroprotective vitagenes. The upregulation of vitagene pathway such as heme-oxygenase 1(HO-1), heat shock protein (Hsp)-70, Trx, sirtuin Sirt1, NQO1, γ-GCS improves brain health resilience and confers protection against neurodegenerative damage. Nuclear factor-erythroid 2 p45-related factor 2 (Nrf2), (Keap1), antioxidant response element (ARE),HO-1, Hsp-70, thioredoxin (Trx), sirtuin 1 (Sirt1), NAD(P)H: quinone oxidoreductase 1 (NQO1), γ-glutamylcysteine synthetase (γ-GCS).

Brain is continuously challenged by perturbations in systemic homeostasis, and, in a resilient brain, integration of adaptive responses occurs covering areas of behavior associated with coping, fear, attention, cognitive flexibility, and emotional regulation. Thus high resilient individuals display more effective modulation of brain circuits involved in emotion and fear. A recent paradigm shift in operationalizing resilience has moved away from the focus on the nonemergence of pathology or symptoms after exposure to adversity, to include "resilient–conductive" factors such as personality traits, confidence, flexibility, optimism, or emotional liability, which can help promote positive subjective appraisal, negotiation, adaptation, or management of adverse situations with increased coping (Kalisch et al., 2015). Although precise information on the neurobiological correlates of complex psychosocial and spiritual factors is lacking, individual traits emerge, such as subjective well-being (both hedonic or eudaimonic) which show to be protective factors against adversity. Hedonic well-being refers to cognitive evaluation of life satisfaction and positive affect, whereas eudaimonic well-being is related to the determination of life-meaning and self-actualization resilience is related to both types of well-being (Di Fabio and Palazzeschi, 2015). Positive affect is thought to facilitate resilience by broadening one's attention and coping abilities, and by decreasing susceptibility to disease through increased vagal control. The overlap between measures of positive affect and resilience has also been observed in various conditions such as chronic pain or brain trauma. Endophenotypes are measurable markers of genetic vulnerability to current or future disorder, thus, identification of neurobiological correlates associated with resilience endophenotype may represent a critical first step in the identification of individuals with increased vulnerability to develop diseases, helping to identify brain signatures of resilience as biomarkers of vulnerability to stress-related diseases with implications for the development of training interventions, such as preconditioning or postconditioning effects associated to hormesis, which increases effective coping and management of stress conditions. In the brain, a resilient neuronal cell does not necessarily correspond to a healthy cell, as in the case of a cancerous phenotype being very resilient toward chemotherapy. In some tumor cells, high resilience mechanisms lead to a resistance to drugs despite being exposed to the same concentrations as their neighboring cells. Such changes can be long term, or even permanent, constituting cellular biological memories associated with beneficial outcomes, as in the context of cellular hormesis, where the concept of beneficial, particularly with respect to ischemia-reperfusion, where an initial stressor makes cells more resilient to subsequent stress to organs, the so-called preconditioning or alternatively after an adverse situation, a subsequent stressor "postconditioning" is used and developed, both experimentally and clinically (Smirnova et al., 2015). Long-term effects can be also detrimental and lead to adverse outcomes, especially when exposure is of limited duration, as in the case of mixture toxicities. The underlying strategy of hormesis is upregulation of adaptive mechanisms that results in the development of biological resilience. New resilient phenotype will conform to the quantitative and temporal features of the hormetic dose-time response relationship, often within a preconditioning context. While the amplitude of induced resilience is modest being about 30%–60% greater than the control group/background) at maximum, and the duration of resilience of limited duration, it appears possible to significantly extend the duration of the resilience in some models depending on preconditioning stimulation methods (Gidday, 2015). In the present review, we aimed to provide evidence for possible prevention and early intervention approaches targeting humans and animals and their correlation with environmental stressors to reduce the risk of brain disorders (i.e., neurodegenerative and neuropsychiatric disorders), fostering resilience mechanisms in order to maintain cellular homeostasis in response to stressors or adverse life experiences.

9.2 Regional specificity of brain resilience and vulnerability to stress

In recent years, neuroimaging techniques have become an increasingly important tool to study neural correlates of adaptive and nonadaptive behavior. Neuroimaging studies such as the functional magnetic resonance imaging and electroencephalography provided substantial evidence about both structure and function of brain supporting vulnerability or resilience to stress in different neurological disorders.

Resilience is associated with morphology changes of brain regions involved in cognitive and affective processes related to cortico-limbic system and plays an essential role in preserving mental function and physiological trajectories of brain network. In this context, higher levels of resilience are related to distinct morphological alterations in brain regions involved in executive control and emotional arousal networks, suggesting individuals with low resilience may have compromised cortico-limbic inhibition, making them more vulnerable to stress or trauma (Gupta et al., 2017).

Interestingly, the connectivity of these brain regions modulates resilience and vulnerability processes during stress or brain trauma (i.e., early life trauma, childhood maltreatment). Accordingly, these brain regions are implicated in storage of cognitive control, memory consolidation, neurogenesis, as well as emotion regulation. The brain regions' structural changes induced by chronic stress play a critical role in the pathophysiology of both neurodegenerative and neuropsychiatric disorders in animal and humans. In addition, alterations have been observed in cytoarchitecture of hippocampus and amygdala related to stress resilience and vulnerability in animal models of Posttraumatic Stress Disorder (PTSD).

In humans, several studies demonstrated that chronic stress or traumatic experiences in childhood maltreatment (i.e., parental neglect, early deprivation, physical, sexual and emotional abuse) are associated with morphological alterations in specific brain regions, such as increased atrophy of neurons in the hippocampus, prefrontal, and parietal cortex, involved in memory, selective attention, and executive functions (Gupta et al., 2017). Similarly, these negative events cause hypertrophy of neurons in the amygdala involved in emotional processing and arousal, fear conditioning, anxiety and aggression, as well as diminished striatal response to anticipated reward, thus, leading to increased vulnerability and development of neuropsychiatric and neurodegenerative illness. Moreover, animal studies have shown smaller amygdala volumes associated with lower levels of resilience, in keeping with neuroimaging findings demonstrating that amygdala volume is reduced in individuals who have been exposed to early adverse life events or maltreatment. For example, smaller amygdalae have been observed in individuals undergoing conditions of childhood poverty, as well as in adolescents having experiences of childhood maltreatment. Smaller amygdala volumes have been also observed in individuals exposed to childhood adversities such as physical abuse, neglect, or being raised in poor households and in PTSD populations compared to healthy controls. These studies indicate that impaired

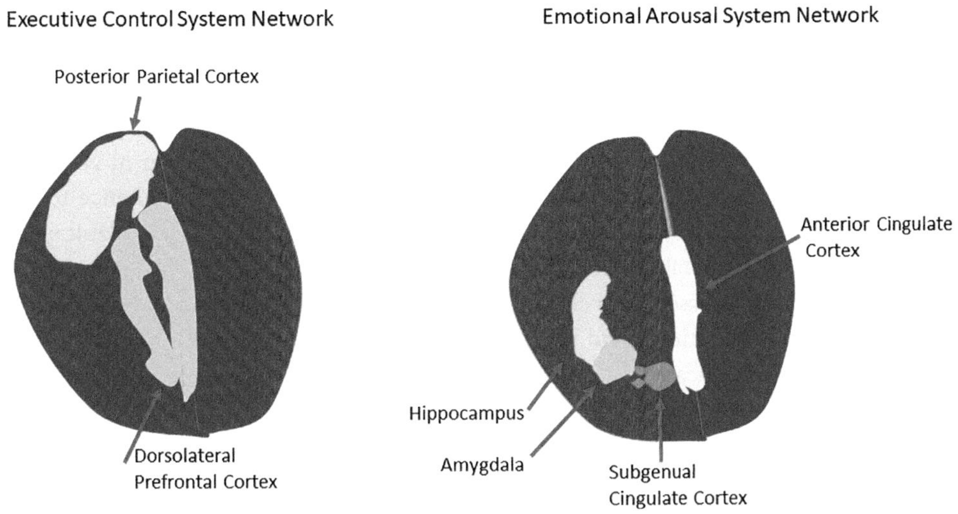

Executive Control System Network   Emotional Arousal System Network

Fig. 9.2 Executive control and emotional arousal function localization in the brain. Executive control network system: dorsal-lateral prefrontal cortex, posterior parietal cortex. Emotional arousal system network: anterior cingulate cortex, subgenual anterior cingulate cortex, amygdala and hippocampus.

executive control and emotional arousal networks in critical cortico-limbic structures, such as dorsal-lateral prefrontal cortex, posterior parietal cortex and, respectively, anterior cingulate cortex, anterior mid-cingulate cortex, subgenual anterior cingulate cortex, amygdala and hippocampus, which show inhibition in response to trauma, may play a critical role in the mediation of low resilience or vulnerability to disease (Fig. 9.2).

Moreover, a recent study demonstrated that glucocorticoid receptor stimulation by early maternal stress and long-term gene expression changes induced aberrant DNA methylation in prefrontal cortex of rats. A few neuroimaging studies have investigated the response to adversity as a "proxy" of resilience, reporting quite important resilience-related differences in brain structure. Retrieval of emotionally valenced words in females with histories of early abuse has been linked to decreased blood flow in the inferior parietal cortex. In an emotional Stroop task (https://en.wikipedia.org/wiki/Stroop_effect), there was decreased parietal cortex activity in females with histories of PTSD and abuse. The parietal cortex is a key region of the executive control network and is associated with inhibitory control, attention, working memory, planning, and response (Uddin et al., 2011). Therefore the findings are consistent with the hypothesis that high resilient individuals may be better able to engage the executive control network, including its role in inhibitory functions in relation to real or perceived challenge to their homeostasis.

The emerging notion on the links between neurogenesis and mental health supports the idea that improved resilience represent a neurogenic strategy to treat patients suffering from major depression, schizophrenia, and neurodegeneration. Neurogenesis plays a critical role in the synaptic plasticity of brain functions, such as olfactory discrimination, memory formation, and fear extinction. In this context, it has been reported that stressful or traumatic events induce glucocorticoid release and decrease adult hippocampal neurogenesis. Mice exposed to early life stress exhibited a reduction in amygdala/hippocampus-dependent fear memory because have reinforced stress resilience to cope future stressors and maintain a normal homeostatic state. Also, dysfunctional adult neurogenesis enhances vulnerability of the hippocampus and development of age-related neurodegenerative diseases, as well as neuropsychiatric diseases. Brain imaging data of the hippocampus in patients and stress-induced animal models with either depression or anxiety disorders indicated a remarkable reduction in region volume and dendritic spine numbers. By contrast, larger hippocampal volumes could be a biological marker of resilience, whereas loss of hippocampal neural plasticity (e.g., loss of glial cell and smaller neuronal cell nuclei) after chronic stress is a determinant factor to the pathophysiology of depression in vulnerable human and animal models (Park, 2019). Recently, several studies have correlated increased amygdala reactivity as a protective factor that promotes resilience to depression following early life stress. Thus lower depression symptoms were associated with higher connectivity between the amygdala and dorsal frontal networks in older adults. Consistent to this notion, repeated chronically acting stress induces a proinflammatory state by increasing the amygdala neuronal and microglial activation which triggers anxiety-like behaviors in rodents (Munshi et al., 2020). Taken together these data show that brain networks connectivity exist in the main brain regions, such as hippocampus, amygdala, prefrontal and parietal cortex in response to stress or trauma and that neural factors are involved to increase resilience or vulnerability after stressful events across the lifespan. A history of stress exposure can have a lasting impact on future stress reactivity in different brain regions. Finally, the elucidation on brain regions alterations will undoubtedly lead to more effective and better tolerated treatment approaches to enhance resilience and its use as a potential biomarker of healthier adulthood adaptations to childhood trauma.

9.3 Hydrogen sulfide: A resilient signaling molecule in brain disorders

Emerging interest has recently focused its attention on biological significance of hydrogen sulfide (H_2S) as a signaling mediator of resilience to modulate levels of multiple antioxidant pathways and ensure mental health during brain disorders, particularly, neurodegenerative and neuropsychiatric disorders. Accordingly, resilience against stress by redox-sensing molecules provides appropriate neuroprotection and repair systems and represents a first line of defense (in the form of, e.g., antioxidant enzymes and sulfur detoxifying molecules) as a system that can identify, repair, or excise damage to biomolecules or tissues caused by endogenous reactive species or exogenous toxicants and at last restore cellular redox homeostasis. Thus redox-resilient molecules against potentially harmful conditions and/or environmental threats require sensing, adaptation, defense, and repair systems working in concert to confer protection, stability, and offset damage. In this light, H_2S acts as a signaling mediator of resilience elicited through the modulation of cellular stress response pathways activated by oxidative stress and inflammatory processes. Oxidative stress and inflammation can induce the nuclear accumulation of Nrf2, a master regulator, which can upregulate downstream redox resilient genes thereby promoting the expression of antioxidant defense enzymes, including catalase, superoxide dismutase, heme oxygenase-1 (HO-1), and glutathione-S-transferase, thioredoxin (Trx), and Sirt1 (Calabrese et al., 2010). H_2S has been reported to activate the expression of Nrf-2, enhancing adaptive processes (Liu et al., 2012) H_2S is a gaseous biological mediator synthesized in mammalian through cystathionine β-synthase (CBS), cystathionine γ-lyase (CSE), glutamate-cysteine ligase, and 3-mercaptopyruvate sulfurtransferase. In the brain, CBS is the major H_2S-producing enzyme, present in both microglial and astrocyte cells. Instead, CSE generates H_2S in peripheral tissues, although both enzymes are present in the central nervous system as well as in peripheral tissues. Recently, H_2S has been considered the third endogenous gasotransmitter next to nitric oxide (NO) and carbon monoxide (CO) (Nagpure and Bian, 2015). For several times, H_2S has been considered a toxic and potentially lethal gas by scientific community. However, emerging studies have demonstrated its important physiological and pathophysiological roles as a gasotransmitter in the central nervous system (CNS). It plays a physiological role in a wide variety of cellular and organ functions and a protective role in multiple pathological conditions, displaying vasoactive, cytoprotective, antiinflammatory, antiapoptotic, antioxidant activities, and synaptic modulator. Synaptic plasticity is crucial for maintaining neuronal homeostasis in CNS. On the one hand, neuronal plasticity declines over time or under stressful conditions by excessive reactive oxidative species that can lead to progression of neuropsychiatric conditions (i.e., schizophrenia, bipolar disorder, major depressive disorder (MDD), and Alzheimer's disease (AD) and Parkinson (PD)) while on the other side, H_2S exerts a crucial role in maintaining synaptic plasticity for regulating health resilience and neuronal homeostasis. A revolutionary discovery has been provided by Kimura, in 1996, when revealed its role as an endogenous neuromodulator through the activation of N-methyl-D-aspartate receptor (NMDA) receptors. The latter modulate the long-term potentiation (LTP) and synaptic plasticity which is responsible for learning and memory function and therefore maintains synapse function. Moreover, NMDA receptors also lead to a sustained rise of neuronal cytosolic calcium ion, and interestingly, they are highly sensitive to oxidative stress. Specifically, in the CNS, H_2S enhances hippocampal LTP by modulating NMDA activity so regulates intracellular calcium concentration in brain cells (i.e., microglial, astrocyte, and neuron cells). Moreover, H_2S was found to be highly expressed in the hippocampus and cerebellum when compared with the cerebral cortex and brain stem. In addition, recent studies demonstrated that H_2S

modulates oxidative stress and neuroplasticity in in vivo models of cognitive dementia, depression, and anxiety as well as ischemic stroke (Chen et al., 2020). It is noteworthy that neurotrophic factors such as brain–derived neurotrophic factor (BDNF) represent a key modulator involved in synaptic plasticity and form memories and behavioral consolidation. Several studies suggested that H_2S might improve depressive and anxiety-related behaviors in nonstressed animal models, but its specific effects and molecular mechanisms on MDD models are still not very clear. To better understand the potential molecular mechanisms of action of H_2S, related to depressive-like behavior, recent in vitro and in vivo studies have indicated that intraperitoneally injection of the H_2S donor NaHS or inhaled H_2S exert neuroprotective effects by reversing the decrease in α-amino-3-hydroxy-5-methyl-4-isoxazolepropionic acid receptors, tropomyosin receptor kinase B (TrKB), and mTOR signaling pathways induced by chronic unpredictable mild stress. These finding are consistent with the ascertainment that H_2S has a pivotal role in neurotrophic signaling as antidepressant agent indirectly associated with synaptic protein synthesis or restoration of synaptic plasticity in depression-like behaviors. In fact, H_2S-induced hormetic responses have been reported in a broad range of cells, including astrocytes (200–400 µM—optimal hormetic concentration), endothelial cells (10–50 µM—hormetic range), mammary gland epithelia cells (10–200 µM—hormetic range), bone marrow stem cell (1 µM—optimal hormetic concentration), cardiomyocyte (10–50 µM—hormetic range), human lymphocytes (250–500 µM—hormetic range), human hepatocytes (50 µM—optimal hormetic concentration) (Shao et al., 2019), and human colon cancer cells (50–200 µM—hormetic concentration range). The optimal hormetic stimulation range in the earlier cited hormetic studies ranges from 1 µM (bone marrow stem cells) to 500 µM (lymphocytes). These concentrations are in general consistent with H_2S concentrations reported in the human, rat, and bovine brain of 50–160 µM (Wang et al., 2013) which suggests that hormetic responses commonly occur in humans and other mammalian species. In addition, H_2S attenuates endoplasmic reticulum stress (ER) and neuronal apoptosis in the hippocampus of homocysteine-treated rats via upregulating the BDNF-TrkB pathway. Based on these premises, H_2S has a significant impact on synapse signaling by interacting with neuronal receptors and thus maintains the neuronal function, excitability, as well as mental health. Under pathological condition, H_2S exerts antioxidant and antiinflammatory effects depending on its endogenous concentration. The concept of *hormesis* describes the ability of mild stresses to confer protection via activation of the *cellular stress response* pathways a universal defense redox reaction of cells to damage to cellular macromolecules such as DNA, proteins, and lipids. Accordingly, H_2S produces protective effects at lower concentrations and a variety of deleterious/cytotoxic effects at higher concentrations following a U-shaped dose-response curve. Since the concentration of H_2S can vary considerably both within and between tissues and under differing conditions, it is likely that H_2S could affect a complex array of biological responses that may challenge the capacity to make definitive biomedical and/or clinical interpretations or predictions. In addition, Li and Moore (Li, 2008) carried out insightful studies on the role of H_2S in health and disease with a claim that probed their biological conundrum as to how a molecule exhibits so very different effects on cells and could be due to different effects at different concentrations. The answer to this fundamental question lies in the intervention of the oral dose-response interaction which is independent of the biological model, tissue, and endpoint. In the brain, physiological levels of H_2S are associated with neuroprotection and neurotransmission. Recently, several studies have demonstrated that novel signaling molecules linked to polysulfide (H_2S_n) such as hydrogen persulfide and trisulfide (H_2S2 and H_2S_3) maintain normal neuronal transmission, vascular tone, protection, inflammatory regulation, and oxygen sensing. In this context, accumulating evidence has documented the significance role of H_2S_n and other persulfurated molecules elicited via S-sulfuration (S-sulfhydration) of specific cysteine residues of the target proteins to modify their activities in the CNS (Kimura, 2019). Moreover, it has been shown that polysulfides (H_2S_n) activate transient receptor potential ankyrin 1 (TRPA1) channels, facilitate the translocation of nuclear factor erythroid 2-related factor 2 (Nrf2) to the nucleus, and suppress the activity of phosphatase and tensin homolog (PTEN) by sulfurating (sulfhydrating) the target cysteine residues in the brain. In addition, H_2S_2 along with H_2S_3 shields neuronal cells from oxidative as well as carbonyl stresses through exerting reduced synthesis of glutathione, which is dependent on the Nrf2 pathway. Furthermore, H_2S impact on redox homeostasis and mitochondrial biogenesis through Nrf2 activity via Kelch-like ECH-associated protein (Keap1) persulfidation may also offer another explanation of therapeutic protection that H_2S confers against AD, given that NaHS enhances Nrf2 expression. Healthy adults are in sulfur balance with equilibrium between intake, transulfuration, and excretion. The conversion of methionine to cysteine via homocysteine is the only catabolic pathway of methionine. Specifically, sulfur is excreted via the kidney mainly as free sulfate, esterified sulfate, taurine, cysteine, and minor amounts of methionine, homocysteine, cystathionine, N-acetylcysteine, mercaptolactate, mercaptoacetate, thiosulfate, and thiocyanate. However, aberrant endogenous production and metabolism of H_2S is implicated in the pathogenesis of neurodegenerative disorders, including AD and PD as well as neuropsychiatric disorders (i.e., depression and anxiety) (Paul and Snyder, 2018). S-sulfhydration, a novel posttranslational modification, is emerging as a mechanism responsible for many biological effects mediated by H_2S during neurodegeneration. Consistent with this, H_2S sulfhydration plays a critical role in PD by activating the neuroprotective ubiquitin E3 ligase, parkin, responsible for clearance of toxic, misfolded proteins. Furthermore, parkin sulfhydration was found markedly depleted in the brains of patients with PD, suggesting that this loss may be pathologic and hydrogen sulfide donors may be therapeutic. For instance, H_2S inhibits inflammation of NF-B as well as increases D-serine-dependent synaptic plasticity in the hippocampus via S-sulfhydration. In addition, H_2S sulfhydrates protein thiol groups by transferring its sulfhydryl group to cysteine residue of targeted proteins. Intriguingly, H_2S also enhances the activity of KATP channels and CFTR Cl channels to suppress excessive excitation by stabilizing the membrane potential, while it suppresses the voltage-gated Ca^{2+} channels to decrease Ca^{2+} toxicity by enhancing the activity of H^+-ATPase (Mikami et al., 2011). In contrast, polysulfides (H_2S_n) facilitate the release of Nrf2 from Keap1/Nrf2 complex by sulfurating cysteine residues of Keap1, leading to its dissociation from Nrf2 which translocate into the nucleus and binds to the antioxidant response element (ARE) promoting antioxidant genes transcription, including γ-GCS (GCL), GSH, Hsp-70, Sirt-1, and HO-1. By these integrated mechanisms, H_2S and H_2S_n protect neurons from oxidative stress. Accordingly, as neuroprotectant H_2S and H2Sn protect neurons against oxidative stress by increasing the intracellular levels of GSH through enhancing the activity of cystine/glutamate antiporter, cysteine transporter as well as γ-glutamyl cysteine synthetase or glutamate cysteine ligase, a rate-limiting enzyme in the production of glutathione (GSH), to promote antioxidant defense (Kimura et al., 2019). Notably, GSH is a ubiquitous tripeptide containing cysteine, glutamate, and glycine with the amine group of cysteine forming a peptide bond with the carboxyl group of the side chain found in glutamate. It can exist alone in reduced forms as glutathione or in an oxidized dimer form also known as glutathione disulfide (GSSG). Glutathione biosynthesis is catalyzed by the enzyme γ-glutamylcysteine synthetase and glutathione synthase, while glutathione recovery from GSSG is catalyzed by GSSG reductase. Most studies have also demonstrated that H_2S can antagonize apoptosis through inhibiting ROS generation and thus promotes neuronal resilience in vitro and in vivo. Glutathione equally distributed throughout the brain occurs low in neuronal cells but high in astrocytes and oligodendrocytes, indicating that these glial cells may be the major source of glutathione generated from H_2S, which is also an indication of the presence of H_2S in astrocyte and microglial cells. Neuroinflammation involved neural immune crosstalk that activates immune cells, glial cells, and neurons. Several studies reported a possible role for H_2S as an antiinflammatory agent which improves spatial memory and cognitive functions in animal models of AD. Various exogenous H_2S donors (i.e., NaHS, GYY4137, and FW1256) exert healthy therapeutic effects in neurodegenerative disease models by targeting hallmark pathological events (e.g., amyloid-β production in AD and neuroinflammation in PD) (He et al., 2019). In this scenario, recent studies suggested that NaHS exhibits antinociceptive and antiinflammatory effects associated with the activation of the Nrf2/HO-1 pathway in the spinal cord of rats with neuropathic pain. Moreover, NaHS treatments in murine models of AD reduce oxidative stress through Nrf-2 and provided protection against homocysteine-induced cognitive dysfunction. Compelling evidence reported that NaHS attenuates lipopolysaccharide (LPS)-induced inflammation by inhibition of p38 mitogen–activated protein kinase (MAPK) and NF-κB signaling pathways in rodent and in rat microglia. Emerging studies have demonstrated the important physiological and pathophysiological roles of H_2S as a gasotransmitter for NLRP3 inflammasome related to neuroinflammation in the central nervous system. The NLRP3 (nucleotide-binding domain and leucine–rich repeat containing protein 3) inflammasome is a multiprotein complex that serves as a platform for caspase-1 activation and interleukin((IL)-1β maturation. NLRP3 mediates the inflammatory response to various exogenous and endogenous signals, including bacterial toxins, ATP, and ROS. Recently, in vitro and in vivo evidence has indicated that endogenous H_2S synthesis induces antiinflammatory and antioxidative effects against oxidative stress-induced ROS production and NLRP3 inflammasome-mediated inflammatory injury by suppressing NF-KB, the purinergic P2X7 receptor, STAT3 and cathepsin S as well as Nox4 signaling pathways (Lin et al., 2018). In addition, inhalation of H_2S protected against the PD-induced movement dysfunction and prevented neuronal apoptosis and microglia activation in the nigrostriatal region. Moreover, CBS overexpression has also been reported to protect against the 6 hydroxydopamine–induced model of PD. H_2S protects neurons from apoptosis and degeneration by exerting antiinflammatory effects and upregulating antioxidant enzymes. Additionally, treatment of H_2S in the Amyloid Precursor Protein (APP)/PS1 mouse model of AD has been reported to reduce cognitive impairment and mitigate oxidative stress. Moreover, a recent study reported that intraperitoneal administration of an H_2S donor (NaHS) downregulates β-secretase 1 (BACE1). The β-secretase processes APP and is responsible for Aβ generation. On the other hand, NaHS improved spatial memory impairment via regulation of the PI3K/Akt pathway in the APP/PS1 model of AD (He et al., 2016). Likewise, H_2S levels were found reduced in plasma of patients with AD compared to controls and H_2S levels correlated negatively with the severity of the disease. Depletion of H_2S and concomitant increase in homocysteine levels in AD has been attributed to a lack of S-adenosylmethionine (SAM), an allosteric activator of CBS. Recent finding indicated that H_2S makes significant neuroprotective impacts on methamphetamine neurotoxicity due to its antioxidant and antiinflammatory activities in hippocampal neurons. Aging, a progressive structural and functional decline, is considered to be a major risk factor underlying of neuroinflammation related to cognitive dementia. Importantly, aging influences the action and/or efficacy of many signaling molecules, including H_2S. Several lines of experimental evidence reported that exposure to low concentrations of exogenous H_2S donors prolongs the lifespan and delays the adverse health consequences of aging in *Caenorhabditis elegans* by an antioxidant effect suggesting a strong relationship between H_2S and aging. Moreover, a recent paper revealed a functional crosstalk between mTORC1 and H_2S signaling, two conserved pathways which play fundamental regulatory roles in aging of eukaryotic organisms. Within the context of aging and neuroinflammation, H_2S also exerts neuroprotective actions through the upregulation of redox resilient proteins which contributes to maintain cellular redox homeostasis during stressful conditions. Notably, a large body of evidence has demonstrated that Sirt-1, a nicotinamide(NAD)$^+$-dependent deacetylase of lysine residue of the target protein, ameliorates neurodegenerative disease, such as AD, PD, and this reveals that it delays senescence in the brain. Recent in vitro and in vivo evidence demonstrated that H_2S is neuroprotective against homocysteine-induced neurotoxicity and senescence through upregulation of Sirt-1 pathway (Wang et al., 2017). Reactive oxygen species (ROS) are constantly generated during neurodegeneration. Among various redox signals activated by ROS, thioredoxin (Trx) pathway is one of the extensively investigated signaling cascades that mediate oxidative cell proliferation, inflammation, senescence, and survival. It contains two redox-active cysteine residues that protect protein against unwarranted oxidant-mediated inter- or intra-molecular disulfide bond formation. The reductive activity of Trx is catalyzed by Trx reductase with NADPH as the electron donor, whereas the activity is reduced or lost after formation of disulfide bonds following exposure to oxidants. One of the well-documented endogenous Trx inhibitors is Trx-interacting protein (TXNIP), which interacts with Trx through the formation of the mixed disulfide bonds. Trx activity and its interaction with TXNIP or ASK1 are determined by its thiol function and involve dynamic sulfide exchanges. Given that H_2S modifies cysteine residues and possesses disulfide-bond reducing potential, it is reasonable to speculate that H_2S may regulate Trx signaling pathway through modification of its redox state. Moreover, a recent study reported that H_2S protected cells from oxidative cell injury through mechanisms involving its regulation on the redox state of thioredoxin and interferes with ASK1/P38 signaling pathway (Mao et al., 2019). Taken together the data earlier mentioned shed light on molecular mechanisms of action of H_2S signaling that at low concentrations, according to the hormesis concept, exerts neuroprotective effects through the modulation of multiple antioxidant redox pathways and could represent a promising approach to improve mental health and resilience in clinical setting during neurodegenerative and neuropsychiatric disorders in humans.

9.4 Plant polyphenols improve resilience and brain health via "Vitagenes"

Over the last few years, most efforts of the research focused on brain resilience, for neuroprotection, elicited by plant polyphenols through the activation of vitagene signaling pathway (Figs. 9.1 and 9.2). The latter is involved in preserving brain health and redox homeostasis in response to stress or trauma in major neuropsychiatric and neurodegenerative disorders (Trovato et al., 2016).

The resilience programs include also efficient functioning of maintenance and repair processes, so-called longevity assurance processes, which by enhancing endogenous cellular redox homeostatic mechanisms, integrate adaptive stress responses via Nrf-2-dependent vitagenes such as sirtuin, heme-oxygenase, HSP-70, and thioredoxin. These resilient genes represent a complex operational network to restore redox homeostasis (Calabrese et al., 2010) as well as epigenetic changes that leave a "memory or scar" as a consequence of the stress the cell has experienced. These memories might have long-term consequences, both positive (resilience) and negative (vulnerability), that contribute to chronic and delayed manifestations of hazard and, ultimately, disease. Notably, epigenetic changes in histone acetylation and expression of histone-modifying protective enzymes such as Sirt-1 and Sirt-2 in hippocampus, prefrontal cortex, and amygdala regions correlate with behavioral response and less resilience to chronic stress in rats. In addition, another recent study suggested that methylation level of HSP-70 promoter may reflect heat-stress-related epigenetic memory and may be useful in differentiating between individuals who are resilient or vulnerable to stress.

Emerging evidence suggests that dietary polyphenols are implicated in cognitive function and general brain health as well as in promoting neurogenesis and neuroplasticity and effectively counteract adverse mechanisms, such as oxidative stress, neuroinflammation, and epigenetic changes (Blaze et al., 2018). Interestingly, plant polyphenols are able to modify DNA methylation across human and mouse tissues following inflammation by promoting resilience and modulating synaptic plasticity against stress although this is just a beginning step in discovering of their therapeutic potential in mental health. Moreover, polyphenols can directly modulate specific cellular signaling pathways by induction of cAMP response element binding protein and subsequently by BDNF activation in psychiatric disorders promoting mental health. Modulation of vitagenes by plant polyphenols takes on a considerable importance in preserving cellular redox homeostasis and enhancing brain resilience in neurodegeneration. Within this context, both HO-1 and Hsp-70 have received great attention for their well-established antioxidant activity needed to maintain cell homeostasis, resilience, and longevity. Interestingly, a recent study suggested that stress compromises Hsp-70/Hsp-90 chaperone network and neuronal resilience in mammals and can cause age-dependent hippocampal neurodegeneration in rodents (Lackie et al., 2020). Moreover, several studies have shown the importance of HO-1 induction against oxidative and nitrosative stress. Excessive HO-1 upregulation may be deleterious for cells, due to accumulation of its by-products CO, iron, and the bilirubin precursor, biliverdin. Keeping in mind the concept of hormesis, it is plausible to consider the three products of HO-1 activity, due to their cytoprotective effects achieved at very low concentrations. Hormetic response is induced when physical, chemical, or biological threats exceed the cellular capacity for homeostasis. Emerging evidence shows that most chronic diseases are caused, rather than the initial injury, by the biological reaction to a given insult or to the agent of the injury itself. The initial components of this cascade evoke the release of ATP, ADP, Krebs cycle intermediates, oxygen, and reactive oxygen species, with involvement of purinergic signaling (Naviaux, 2019). The activities of plant polyphenols against neuroinflammation appear to target activated microglia resiliency and result in the reduction of proinflammatory factors induced by NRLP3 inflammasome. In particular, several data highlight the role played by phenolic components of extra virgin olive oil in counteracting amyloid aggregation and proteotoxicity, with a particular emphasis on the pathways involved in the onset and progression of AD and PD, including APP processing, Aβ peptide and tau aggregation, autophagy impairment, disruption of redox homeostasis, α-synuclein neurotoxicity, and neuroinflammation. In this context, extensive data indicate that oleuropein interferes with APP processing (Daccache et al., 2011), Aβ-amyloid aggregation, as well as tau protein, avoiding the growth of toxic Aβ oligomers both in vitro and in a mouse transgenic model of Aβ deposition. Additionally, it was recently discussed that the polyphenols oleuropein and hydroxytyrosol induce mild stress that prolong lifespan, improve stress resistance and healthspan of *C. elegans*. In accordance with these data, a recent study has reported that in rats, a diet supplemented with oleuropein displays strong protection against cognitive deterioration, decreases the apoptosis and oxidative stress levels, and prevents the spatial learning and memory impairments against morphine-induced neurotoxicity to the hippocampus (Shibani et al., 2019). It is noteworthy that hydroxytyrosol is present in the brain since it is a dopamine metabolite. Oxidative deamination of dopamine, catalyzed by monoamine oxidases, yields 3,4–dihydroxyphenylaldehyde (DOPA)L that can be oxidized by aldehyde dehydrogenase to 3,4-dihydroxyphenylacetic acid (DOPAC). Although DOPAC is the major metabolite of dopamine in the brain, a small portion of DOPAL may be reduced to hydroxytyrosol by aldehyde reductase and hydroxytyrosol can be converted to DOPAL by alcohol dehydrogenase. At the same time DOPAC can be transformed into hydroxytyrosol by DOPAC reductase. DOPAL is a highly reactive metabolite, suggesting that it might be a neurotoxic dopamine metabolite with a role in the pathogenesis of PD. Hydroxytyrosol ameliorates soluble oligomeric Aß$_{1-42}$ plus ibotenic acid-induced neurobehavioral dysfunction following intracerebroventricular injection of Aß$_{1-42}$ (Goldstein et al., 2016). Negatively altered spatial reference and working memories induced by Aß were rescued to a significant extent by hydroxytyrosol treatment, an effect associated with reduced activation of death-associated kinases, such as JNK and p38-MAPK, while concomitantly increasing survival-signaling pathways ERK-MAPK/RSK2, PI3K/Akt1, and JAK2/STAT3, as documented in hippocampal neurons. Recently, it has also been shown that oleuropein protects the hypothalamus from oxidative stress by improving mitochondrial function through activation of the Nrf-2-mediated signaling pathway. In conclusion, hydroxytyrosol is being considered with interest for possible use in pharmacological mitigation of neurodegenerative processes as a potent antioxidant and a novel small molecule that can induce the Nrf-2-ARE (antioxidant-responsive element) pathway.

Interestingly, plant polyphenols upregulate vitagene signaling pathway that represents a potential therapeutic target in the crosstalk of inflammatory response and oxidative stress in neurodegeneration. Accordingly, recent in vitro and in vivo studies have indicated that HT attenuates the NLRP3 inflammasome pathway by inhibiting proinflammatory IL-1β and IL-18 cytokine levels, oxidative stress, neuronal apoptosis, via activation of the Nrf-2/HO-1 signaling pathway and suppression of NF-κB

(Zrelli et al., 2015). Previous research revealed that hydroxytyrosol and oleuropein activate the signaling pathway from Sirt-1, a member of the vitagene family that utilizes NAD^+ as a substrate to catalyze the deacetylation of various substrates to restore adaptive homeostasis under stress conditions. By regulating cellular redox homeostasis, Sirt-1 plays a critical role in neuron survival, insulin sensitivity, mitochondrial biogenesis, neurogenesis, and inflammation. In the brain, Sirt-1 was shown to be essential for synaptic plasticity, cognitive functions, modulation of learning, and the preservation of memory processes that deteriorate during aging. Recent in vitro and in vivo studies suggest that hydroxytyrosol and oleuropein inhibit the inflammatory response through different pathways regulated by several members of the sirtuin family (e.g., Sirt-1, Sirt-2, Sirt-6). Increasing findings reveal mechanisms by which sedentary lifestyles accelerate brain aging, whereas lifestyle that includes intermittent bioenergetic challenges (i.e., caloric restriction, exercise, fasting and intellectual challenges, natural polyphenols) extends lifespan and foster healthy aging by brain resilience pathways. Recent longitudinal studies following large numbers of subjects, for example, find strong associations between exercise and brain health. Rodent models provide a useful tool to investigate features of, and mechanisms underlying, exercise-induced stress resilience (Arnold et al., 2020). A recent study suggested the beneficial effect of long-term physical exercise as a neuroprotective lifestyle and antiaging brain resilience process by the activation of Sirt-1–Sirt-3 axis in middle-aged active men. Furthermore, Sirt-1 pathway may also activate neuroprotective cascades involving proteostatic function and neurotrophic resilience mechanisms against AD in a mouse models. The thioredoxin system (Trx/TrxR) is an important thiol/disulfide redox controller ensuring the cellular redox homeostasis. In addition to regulation of expression, TrxR activity is also regulated posttranslationally by the TxNIP. Yet, a recent report showed that HT induces neuroprotection and cellular antioxidant defenses via activation of the Keap1-Nrf-2-TRXR1 pathway.

Polyphenols from plants (i.e., resveratrol, hydroxytyrosol, oleuropein, sulforaphane, curcumin, as well as gingko biloba) through the activation of adaptive responses induce brain resilience and neuroprotective defense against neurodegeneration. Increasing evidence suggests that plant polyphenols may exert healthy benefits acting in a hormetic-like manner through the modulation of adaptive stress–response pathways, such as vitagenes, making the hormesis concept fully applicable to the field of nutrition (Calabrese et al., 2018).

9.5 Conclusions and future perspectives

Resilience is a multifactorial construct that comprises physiological parameters, epigenetic modulators, and neurobiological markers. In this review, we have provided a brief overview of biological mechanisms underlying stress resilience and have explored how resilience changes throughout age. Emerging research has focused on biological resilience elicited by assumption of phytonutrients, particularly vitamins and plant polyphenols. Activation of stress-responsive mechanisms following moderate and chronic consumption of low doses of plant polyphenols, involving also H_2S signaling, induces *vitagenes* activating brain resilience mechanisms effective in the prevention of neuroinflammation and aging-associated cognitive decline, as well as neuropsychiatric disorder pathogenesis, thereby improving brain health-life and longevity in animals and humans. Epigenetic mechanisms are a key to understanding the effects of early stress in childhood, such as poverty, maltreatment, maternal social and nutritional deprivation, familial genetic, as well as sexual abuse. Accordingly, epigenome is a marker of resilience and brain health able to render individuals more resilient to mental disease across the generations. Thus medicine of the future must advocate the importance of clinical intervening early in life to prevent adverse early life experiences that have lifelong consequences for physical and mental health via targeted interventions not to reverse the condition rather "reprogram" brain function in healthier directions and promote resilience. In this light, the data mentioned before have highlighted the significance of preventive strategies to help operationalize resilience as a multifactorial determinant of brain healthy for translation into clinical settings: (i) psychobiological challenge tasks designed to evoke a resilient behavioral response, (ii) antioxidant therapy that enhance resilience, and (iii) neuromodulation strategies whose effects are characterized by neuroimaging of circuits that mediate resilience. Aging is one of the most challenging public health and it is considered as a "cellular danger response" to environmental stressors or injury leading to development of neurodegenerative disorders. Therefore identifying personalized biomarker signatures of resilience can help us characterize biologically vulnerable individuals (e.g., maltreated children) and on the other end, resilient individuals, such as centenarians, that represent the best model of longevity and healthy aging.

Further studies are, however, necessary to evaluate the real importance of neurobiological mechanisms impacting on target genes, as well as on epigenetic and mitochondrial pathways in specific brain regions, in order to elucidate resilience factors which promote brain health in response to stress, unraveling potential therapeutic interventions able to increase effectively resilience and, thus, stress management in vulnerable population.

References

Arnold MR, Greenwood BN, McArthur JA, Clark PJ, Fleshner M, Lowry CA. Effects of repeated voluntary or forced exercise on brainstem serotonergic systems in rats. Behav Brain Res 2020;378:112237.

Blaze J, Wang J, Ho L, Mendelev N, Haghighi F, Pasinetti GM. Polyphenolic compounds Alter stress-induced patterns of global DNA methylation in brain and blood. Mol Nutr Food Res 2018;62(8): e1700722.

Calabrese V, Cornelius C, Dinkova-Kostova AT, Calabrese EJ, Mattson MP. Cellular stress responses, the hormesis paradigm, and vitagenes: novel targets for therapeutic intervention in neurodegenerative disorders. Antioxid Redox Signal 2010;13(11):1763–811.

Calabrese V, Santoro A, Monti D, Crupi R, Di Paola R, Latteri S, Cuzzocrea S, Zappia M, Giordano J, Calabrese EJ, Franceschi C. Aging and Parkinson's Disease: Inflammaging, neuroinflammation and biological remodeling as key factors in pathogenesis. Free Radic Biol Med 2018;115:80–91.

Chen M, Pritchard C, Fortune D, Kodi P, Grados M. Hydrogen sulfide: a target to modulate oxidative stress and neuroplasticity for the treatment of pathological anxiety. Expert Rev Neurother 2020;20(1): 109–21.

Daccache A, Lion C, Sibille N, Gerard M, Slomianny C, Lippens G, Cotelle P. Oleuropein and derivatives from olives as tau aggregation inhibitors. Neurochem Int 2011;58:700–7.

Di Fabio A, Palazzeschi L. Hedonic and eudaimonic well-being: the role of resilience beyond fluid intelligence and personality traits. Front Psychol 2015;6:1367.

Gidday JM. Extending injury- and disease-resistant CNS phenotypes by repetitive epigenetic conditioning. Front Neurol 2015;6:42.

Goldstein DS, Jinsmaa Y, Sullivan P, Holmes C, Kopin IJ, Sharabi Y. 3,4-Dihydroxyphenylethanol (Hydroxytyrosol) mitigates the increase in spontaneous oxidation of dopamine during monoamine oxidase inhibition in PC12 cells. Neurochem Res 2016;41:2173–8.

Gupta A, Love A, Kilpatrick LA, Labus JS, Bhatt R, Chang L, Tillisch K, Naliboff B, Mayer EA. Morphological brain measures of cortico-limbic inhibition related to resilience. J Neurosci Res 2017;95(9): 1760–75.

He XL, Yan N, Chen XS, Qi YW, Yan Y, Cai Z. Hydrogen sulfide down-regulates BACE1 and PS1 via activating PI3K/Akt pathway in the brain of APP/PS1 transgenic mouse. Pharmacol Rep 2016;68:975–82.

He JT, Li H, Yang L, Mao CY. Role of hydrogen sulfide in cognitive deficits: evidences and mechanisms. Eur J Pharmacol 2019;849:146–53.

Kalisch R, Muller MB, Tuscher O. A conceptual framework for the neurobiological study of resilience. Behav Brain Sci 2015;38:e92.

Kimura H. Signaling by hydrogen sulfide (H_2S) and polysulfides (H_2S_n) in the central nervous system. Neurochem Int 2019;126:118–25.

Kimura Y, Shibuya N, Kimura H. Sulfite protects neurons from oxidative stress. Br J Pharmacol 2019;176(4):571–82.

Lackie RE, Razzaq AR, Farhan SMK, Qiu LR, Moshitzky G, Beraldo FH, Lopes MH, Maciejewski A, Gros R, Fan J, Choy WY, Greenberg DS, Martins VR, Duennwald ML, Lerch JP, Soreq H, Prado VF, Prado MAM. Modulation of hippocampal neuronal resilience during aging by the Hsp70/Hsp90 co-chaperone STI1. J Neurochem 2020;153:727–58.

Li, L.; Moore, P.K: Putative biological roles of hydrogen sulfide in health and disease: a breath of not so fresh air? Trends Pharmacol Sci 2008, 29, 84–90.

Lin Z, Altaf N, Li C, Chen M, Pan L, Wang D, Xie L, Zheng Y, Fu H, Han Y, Ji Y. Hydrogen sulfide attenuates oxidative stress-induced NLRP3 inflammasome activation via S-sulfhydrating c-Jun at Cys269 in macrophages. Biochim Biophys Acta Mol Basis Dis 2018;1864(9 Pt B):2890–900.

Liu W, Wang D, Liu K, Sun X. Nrf2 as a converging node for cellular signaling pathways of gasotransmitters. Med Hypotheses 2012;79:308–10.

Mao Z, Huang Y, Zhang Z, Yang X, Zhang X, Huang Y, Sawada N, Mitsui T, Takeda M, Yao J. Pharmacological levels of hydrogen sulfide inhibit oxidative cell injury through regulating the redox state of thioredoxin. Free Radic Biol Med 2019;134:190–9.

Mikami Y, Shibuya N, Kimura Y, Nagahara N, Yamada M, Kimura H. Hydrogen sulfide protects the retina from light-induced degeneration by the modulation of $Ca2+$ influx. J Biol Chem 2011;286: 39379–86.

Munshi S, Loh MK, Ferrara N, DeJoseph MR, Ritger A, Padival M, Record MJ, Urban JH, Rosenkranz JA. Repeated stress induces a pro-inflammatory state, increases amygdala neuronal and microglial activation, and causes anxiety in adult male rats. Brain Behav Immun 2020;84:180–99.

Nagpure BV, Bian JS. Brain, learning, and memory: role of H_2S in neurodegenerative diseases. Handb Exp Pharmacol 2015;230:193–215.

Naviaux RK. Metabolic features and regulation of the healing cycle-a new model for chronic disease pathogenesis and treatment. Mitochondrion 2019;46:278–97.

Park SC. Neurogenesis and antidepressant action. Cell Tissue Res 2019;377(1):95–106.

Paul BD, Snyder SH. Gasotransmitter hydrogen sulfide signaling in neuronal health and disease. Biochem Pharmacol 2018;149:101–9.

Shao Y, Chen Z, Wu L. Oxidative stress effects of soluble sulfide on human hepatocytes cell line L02. Int J Environ Res Public Health 2019;16:1662.

Shibani F, Sahamsizadeh A, Fatemi I, Allahtavakoli M, Hasanshahi J, Rahmani M, Azin M, Hassanipour M, Mozafari N, Kaeidi A. Effect of oleuropein on morphine-induced hippocampus neurotoxicity and memory impairments in rats. Naunyn Schmiedebergs Arch Pharmacol 2019;392(11):1383–91.

Smirnova L, Harris G, Leist M, Hartung T. Cellular resilience. ALTEX 2015;32(4):247–60.

Trovato A, Siracusa R, Di Paola R, Scuto M, Ontario ML, Bua O, Di Mauro P, Toscano MA, Petralia CCT, Maiolino L, Serra A, Cuzzocrea S, Calabrese V. Redox modulation of cellular stress response and lipoxin A4 expression by Hericium Erinaceus in rat brain: relevance to Alzheimer's disease pathogenesis. Immun Ageing 2016;13:23.

Uddin LQ, Supekar KS, Ryali S, Menon V. Dynamic reconfiguration of structural and functional connectivity across core neurocognitive brain networks with development. J Neurosci Off J Soc Neurosci 2011;31(50):18578–89.

Wang Z, Liu DX, Wang FW, Zhang Q, Du ZX, Zhan JM, Yuan QH, Ling EA, Hao AJ. L-cysteine promotes the proliferation and differentiation of neural stem cells via the CBS/H_2S pathway. Neuroscience 2013;237:106–17.

Wang CY, Zou W, Liang XY, et al. Hydrogen sulfide prevents homocysteine induced endoplasmic reticulum stress in PC12 cells by upregulating SIRT1. Mol Med Rep 2017;16:3587–93.

Zrelli H, Kusunoki M, Miyazaki H. Role of Hydroxytyrosol-dependent regulation of HO-1 expression in promoting wound healing of vascular endothelial cells via Nrf2 De novo synthesis and stabilization. Phytother Res 2015;29(7):1011–8.

CHAPTER 10

Different components of frailty in the aging subjects—The role of sarcopenia

Paolina Crocco, Serena Dato, Francesca Iannone, Giuseppe Passarino, and Giuseppina Rose
Department of Biology, Ecology and Earth Sciences, University of Calabria, Rende, Italy

10.1 Frailty definition and assessment

Frailty (from latin *fragilitas*) is an important but incompletely understood clinical concept in geriatrics, associated with adverse health outcomes, including increase in the risks of disability in older age, admission to long-term care, and increased mortality (Abate et al., 2007; Gordon et al., 2014). Looking to the literature from the last twenty years, there is no consensus for a comprehensive definition of frailty, but rather a general agreement that the frailty phenotype onsets as a consequence of a decline in several physiological systems (mainly affecting neuromuscular, neuroendocrine, and immunological systems) and reserve. The absence of an internationally agreed gold standard definition makes difficult the direct comparison between studies. Therefore, the prevalence rates in community-dwelling adults vary between studies, likely depending on the approach used to classify frailty (Collard et al., 2012; Espinoza et al., 2010; Santos-Eggimann et al., 2009). Furthermore, in many cases, we can find not a definition of frailty but instead a set of factors associated with frailty phenotype, like Cederholm's description of the physical phenotype of frailty: *"Encompasses state as exhaustion, weakness, and slowness"* (Cederholm, 2015).

Definitions of frailty can be grouped in "conceptual" and "operational," the last including the frailty assessments. From a conceptual point of view, some authors define frailty as a dynamic state between robustness and disability, associated with gradual losses, finally ending with death (Abellan van Kan et al., 2008; Alexa et al., 2013; Dent et al., 2016; Maxwell and Wang, 2017). The concept of deficit/decline was endorsed by Rockwood and Mitnitski (2011), who defined frailty as a state of accumulating deficits, where Fried et al. (2001) reported frailty as "a geriatric syndrome of decreased reserve and resistance to stressors resulting from cumulative declines across multiple physiological systems." The loss of organismal reserves recently inspired the Sunfrail Project (Longobucco et al., 2019), which defined frailty as a "physiological syndrome" characterized by reduction of functional reserves and resistance to "stressors" and "by gradual

☆ All the Authors, who are listed in alphabetical order, equally contributed to the manuscript.

Human Aging
https://doi.org/10.1016/B978-0-12-822569-1.00011-1

loss of energy, strength, endurance, and motor control." On the contrary, the definition of frailty which most of all embraces its complex nature is that of a multidimensional syndrome or multidomain phenotype, firstly proposed by Abellan van Kan et al. (2008) and further developed by many authors. The multidomain model, including physical/functional, nutritional, cognitive, sensory, and psychosocial domains of frailty, is important because it does not assume that frailty is homogenous across populations and is largely more accepted (Sezgin et al., 2019 and references therein). Fig. 10.1 outlines the four main frailty domains consensually included in multidimensional frailty definitions (Markle-Reid and Browne, 2003; Ernsth-Bravell and Mölstad, 2010; Mohandas et al. 2011; Maxwell and Wang, 2017; Panza et al. 2018).

Operational definitions of frailty imply the application of assessment tools based on a set of measures and/or a combination of deficits, in a multidimensional approach. Table 10.1 lists the most known frailty measures that include the largely applied frailty phenotype by Fried et al. (2001) and the frailty index by Rockwood and Mitnitski (2011), although the Tilburg Frailty Indicator (Gobbens et al., 2010a,b), the Groningen Frailty Indicator (Peters et al., 2012), the Frail Scale (Morley et al., 2012), and the comprehensive geriatric assessment (Wieland and Hirth, 2003) are also popular in literature. Common to all the frailty measures is the physical domain, while the cognitive domain is included in only 50% of frailty assessment tools, because there is still some uncertainty about the relationship between frailty, cognitive impairment, and dementia; so some authors exclude people severely cognitively compromised. It is also noticeable that the heterogeneity of frailty in different geographic areas makes it difficult to use standardized methods for measuring the quality of aging in different populations. Consequently, the necessity to carry out population-specific surveys to define tools that are able to highlight groups of subjects with homogeneous aging phenotypes within each population has emerged (Montesanto et al., 2010).

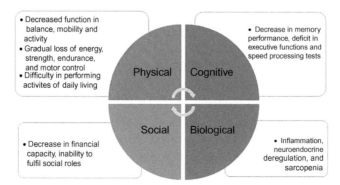

Fig. 10.1 Schematic description of the four main frailty domains, assessed and used to classify subjects for their frailty status.

Table 10.1 Frailty measures.

Frailty assessment tool	N° items	Domains	Subject's classification	Authors, Year
Frailty phenotype	5	Ph	Ordinal Scale: 0–5 3 levels (not frail, prefrail, frail); ≥ 3 frail	Fried et al., 2001
Frailty index (FI/ CSHA)	92	Ph, Ps, S	Continuous Scale: 0–1 Combination of tests and self-report	Mitnitski et al., 2002; Rockwood and Mitnitski, 2007; Mitniski et al., 2005; Widagdo et al., 2016; Abete et al., 2017
Clinical Frailty Scale—CSHA	70	Ph, Ps	Ordinal Scale: 1–7 7 levels (from robust to complete dependence)	Rockwood et al., 2005 Gregorevic et al., 2016
Edmonton Frail Scale—EFS	11	Ph, Ps, S	Ordinal Scale: 0–17; 5 levels (not frail, apparently vulnerable, mild, moderate, and severe frailty)	Rolfson et al., 2006; Fabrício-Wehbe et al., 2009
Tilburg Frailty Indicator	15	Ph, Ps, S	Dichotomous scale (frail-not frail). Range: 0–15, ≥5 frail	Gobbens et al., 2010a,b; Freitag et al., 2016; Dong et al., 2017
FiND—Frail nondisabled	5	Ph	Dichotomous scale (frail-not frail) Separates disability from frailty	Cesari et al., 2014a,b; Metzelthin et al., 2010; Daniels et al., 2010; Andreasen et al., 2015; Coelho et al., 2015; Faller et al., 2019
Groningen Frailty Indicator- GFI	15	Ph, Ps, S	Dichotomous scale (frail-not frail). Range: 0–15. ≥4 frail	Peters et al., 2012
Frail Scale	5	Ph	Ordinal Scale: 0–5 3 levels (not frail, prefrail, and frail) 0 Robust, 1 to 2 prefrail, ≥ 3 frail	Morley et al., 2012; Gardiner et al., 2015; Woo et al., 2015; Díaz de León González et al., 2016; Jung et al., 2016; Rosas-Carrasco et al., 2016; Aprahamian et al., 2017; Braun et al., 2018; Dong et al., 2018

Continued

Table 10.1 Frailty measures—cont'd

Frailty assessment tool	N° items	Domains	Subject's classification	Authors, Year
Comprehensive geriatric assessment		Ph, Ps, S, En	Continuous Scale: 0–1 Combination of tests and self-report	Wieland and Hirth, 2003
Comprehensive frailty assessment instrument— CFAI	23	Ph, Ps, S, En	Dichotomous scale (frail-not frail)	De Witte et al., 2013
Continuous Frailty Scale— CFS	5	Ph	Ordinal Scale: 3levels Range: 0–5, 0 Robust, 1–2 prefrail, $\geq$3 frail	Wu et al., 2018
Frailty score	4	Ph, Ps	3 clusters (not frail, prefrail, and frail) estimated by considering age and sex as covariates	Montesanto et al., 2010; Dato et al., 2012a,b

Scale: ordinal, continuous, dichotomous; Ps: Psychological/cognitive; Ph: Physical/functional; S: Social; En: Environmental.

10.2 Physical frailty and sarcopenia: two sides of the same coin

There is consensus among researchers that physical function plays a fundamental role in the determination and manifestation of frailty. In fact, many of the adverse outcomes of frailty are probably mediated by sarcopenia, which may be considered the biological substrate for the development of physical frailty.

The term "sarcopenia" [from the Greek sarx (flesh) and penia (loss)] was first used by Rosenberg in 1989 to characterize the age-associated decrease in skeletal muscle mass (Rosenberg, 1989). Nowadays, sarcopenia is considered as a geriatric syndrome characterized by the progressive and generalized loss of skeletal muscle mass, strength, and function. Besides to connect and support the skeletal system, skeletal muscle plays a crucial role in energy utilization and maintenance of metabolic health in humans. It is the largest reservoir of proteins and free amino acids in the body that can be used for energy production by various organs, accounting for approximately 20%–30% of an individual's resting metabolic rate (Zurlo et al., 1990), and for 20% of basal whole-body glucose uptake (Baron et al., 1988). It is also a key player in glycogen storage and lipid oxidation (Hearris et al., 2018). Therefore, the decrease of skeletal muscle performance has a marked influence on health with a profound negative impact on the individual quality of life, being associated with the loss of functional independence, increased risk of institutionalization and hospitalization, and increased incidence of adverse health outcomes (Cesari et al., 2014a,b). For all that, sarcopenia has recently been recognized as a specific disease by assignment of a single code within the International Classification of Disease (Cao and Morley, 2016).

The presence of comorbidities may impact the muscle mass/strength relationship. In fact, chronic diseases, including chronic liver diseases, diabetes, obesity, coronary heart diseases, hypertension, were demonstrated significant predictors of lower muscle strength (Lee et al., 2013; Batsis and Villareal, 2018; Mesinovic et al., 2019). Moreover, skeletal muscle mass loss is common in cancer patients and associated with poor clinical outcomes, including increased treatment-related toxicities and reduced survival (Peterson and Mozer, 2017). Often, the relationships between comorbidities and sarcopenia are bidirectional, which means that the existence of one condition may increase the risk of developing the other. Furthermore, the heterogeneity among older adults with respect to different pathological conditions may contribute to complicate the comparisons between different cohorts of sarcopenic patients.

The total number of people with sarcopenia is increasing. According to the current estimates, 5%–10% of subjects aged 60–70 years and 11%–50% of those over the age of 80 are facing with this disability, with higher prevalence in long-term and acute care settings (Shafiee et al., 2017; Mayhew et al., 2019; Shen et al., 2019). Differences in prevalence among studies are depending on the characteristics of the studied population, such as age, sex, and ethnicity. In Europe, the prevalence of sarcopenia may differ from 9.25% to 18% in subjects aged 65 years and older (Ethgen et al., 2017). Prevalence is also influenced by genetic background and environmental factors such as the level of physical activity, as well as by the different diagnostic criteria applied (Pagotto and Silveira, 2014; Marty et al., 2017; Tournadre et al., 2019). Regarding this last point, Table 10.2 reports diagnostic criteria and cutoff values used by different research groups for diagnosing sarcopenia. The most widely accepted criteria come from the European Working Group on Sarcopenia in Older People (EWGSOP), who developed diagnostic criteria, cutoff points for measured variables, and an operational definition of sarcopenia based on the presence of low muscle mass and either low muscular strength or low physical performance (Cruz-Jentoft et al., 2019).

10.3 Cellular and molecular mechanisms of Sarcopenia

Sarcopenia begins in the fourth decade of life with a rate of muscle loss of about 8.0% per decade until 70 years of age, accelerating up to 15% for each passing decade, with differences between sexes; muscle strength is lost at an even greater rate (Payette et al., 2003; Malafarina et al., 2012; Mitchell et al., 2012). Sarcopenia has a multifactorial origin. Potential contributing factors to the disease onset and progression and their consequences are schematized in Fig. 10.2 and briefly discussed in the following sections.

Table 10.2 Cutoff values used for sarcopenia diagnosis by different scientific groups.

Working groups	Sarcopenia definition	Muscle strength		Muscle mass		Physical performance	Reference
		Men	Women	Men	Women		
EWGSOP	Low muscle mass, muscle strength, and physical performance	HG < 27 kg	HG < 16 kg	DXA < 7.0 kg/m^2	DXA < 5.5 kg/m^2	GS ≤ 0.8 m/s (4-m course)	Cruz-Jentoft et al., 2019
AWGS	Low muscle mass, muscle strength, and physical performance	HG < 26 kg	HG < 18 kg	DXA < 7.0 cm^2/m^2	DXA < 5.4 cm^2/m^2	GS ≤ 0.8 m/s (6-m course)	Chen et al., 2014a
IWGS	Low muscle mass and physical performance	Inability to stand from chair unaided		DXA < 7.23 kg/m^2	DXA < 5.67 kg/m^2	GS < 1.0 m/s	Fielding et al., 2011
FNIH	Low muscle mass, muscle strength, and physical performance	HG < 26 kg Adj for BMI < 1.00	HG < 16 kg Adj for BMI < 0.56	DXA < 19.7 kg Adj for BMI < 0.789	DXA < 15 kg Adj for BMI < 0.512	GS ≤ 0.8 m/s	McLean et al., 2014

EWGSOP, European Working Group on Sarcopenia in Older People; AWGS, Asian Working Group for Sarcopenia; IWGS, International Working Group on sarcopenia; FNIH, Foundation for the National Institutes of Health; HG, Handgrip strength; Adj for BMI, Adjusted for body mass index; DXA, Dual-energy X-ray absorptiometry; GS, Gait speed.

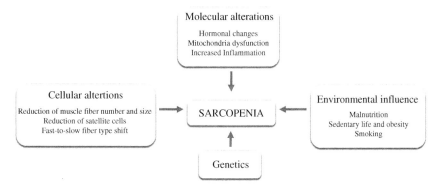

Fig. 10.2 Factors influencing the development and progression of sarcopenia.

10.3.1 Muscle structure and function changes

Human skeletal muscle fibers are grouped into three types: type I, type IIa, and type IIx fibers, which differ in contractile and metabolic properties. Type I fibers are slow-twitch fibers having a predominantly oxidative metabolism. Type IIa are fast-twitch fibers showing both oxidative and glycolytic metabolisms. Type IIx are also fast-twitch fibers, primarily relying upon glycolytic metabolism (Talbot and Maves, 2016; Ciciliot et al., 2013). Nilwik et al. (2013) provided evidence that age-related reduction in muscle mass is mainly due to smaller fast-twitch type II fiber size, with no substantial decline in muscle fiber numbers. Moreover, a fast-to-slow fiber-type shift, affecting mostly type IIx fibers, seems to characterize skeletal muscle aging (Miller et al., 2014; Miljkovic et al., 2015). The type I fibers are mainly responsible for endurance-type activities, while type II fibers are mainly responsible for higher intensity or highly fatiguing activities. The reduction in type II fibers may therefore result in a decline in muscle strength in the old adults. These fiber-specific changes are likely associated with the age-related remodeling of motor units that predominantly results in denervation of type II muscle fibers with collateral reinnervation of type I muscle fibers (Borzuola et al., 2020). Cycles of denervation-reinnervation result in a significant disruption of the neuromuscular junction components, finally leading to grouped fiber atrophy and muscle fiber loss (Hepple and Rice, 2016; Deschenes, 2011). As Fig. 10.2 shows, other factors such as decrease in myosin force and/or actin-myosin cross-bridge stability and defective excitation-contraction coupling may contribute to muscle weakness (Miljkovic et al., 2015; Qaisar et al., 2018).

A key role in the development of sarcopenia seems to be played by satellite cells, myogenic stem cells that reside in a niche between the basal lamina and the plasma membrane of muscle fibers. Under basal conditions, adult satellite cells remain quiescent but proliferate and differentiate in response to injury, becoming essential for muscle fiber regeneration, repair, and growth. The contribution of satellite cells to sarcopenia is based on the evidence that both the number and the function of muscle satellite cells decrease substantially with advancing age (Snijders et al., 2015; Verdijk et al., 2014; Always et al., 2014; Cosgrove et al., 2014; Sousa-Victor et al., 2014). A morphologic aspect of the aged skeletal muscle is the increased frequency of adipocyte or lipid deposition within muscle fibers (De Carvalho et al., 2019). A longitudinal study by Delmonico et al. (2009) reported an increase in muscular fat infiltration with aging, resulting in losses of strength and muscle quality. Fat accumulation can occur as intramuscular triglycerides (IMTG) deposition in lipid droplets, which has been suggested to contribute to the deterioration of skeletal muscle function in the oldest. Intramyocellular accumulation of lipids has been indeed linked with lower whole-muscle and single-fiber power (Choi et al., 2016), slower walking speed and chair stand time (Visser et al., 2002; Beavers et al., 2013), and higher risk of mobility limitation (St-Jean-Pelletier et al., 2017). Fat infiltration is one of the characteristic features of the interaction between sarcopenia and obesity in aging (De Stefano et al., 2015).

10.3.2 Mitochondrial dysfunction

Mitochondria integrate several cell signals including energy supply, ROS generation, and apoptosis, and for this reason, they are believed to have a central position in the maintenance of myocyte viability and myofiber plasticity. However, the exact mechanisms of how skeletal muscle aging affects mitochondria and, conversely, how aging mitochondria induce skeletal muscle dysfunction are still a challenging field of research.

Being a metabolically active tissue, skeletal muscle has a relatively high abundance of mitochondria that differ according to muscle fiber types: slow-twitch oxidative myofibers contain much less mitochondria than do fast-twitch glycolytic muscles myofibers. Two distinct mitochondrial subpopulations, subsarcolemmal mitochondria, located immediately beneath sarcolemma, and intermyofibrillar mitochondria, positioned deep into the myofibers between the myofibrils, are present in skeletal muscle. Both subpopulations, which differ both bioenergetically and structurally, communicate with each other likely to provide a conductive pathway of energy distribution among skeletal muscle fibers (Glancy et al., 2015).

Several studies showed that aged skeletal muscle typically exhibits fewer mitochondria (Peterson et al., 2012). These number variations are accompanied by perturbations in mitochondrial morphology and dynamics, mediated by the imbalance of mitochondria fusion and fission events. Some authors reported that skeletal muscle atrophy occurring with aging is associated with larger and more complex mitochondria (Leduc-Gaudet et al., 2015) while others report an aging-related increased mitochondrial fragmentation (Huang et al., 2010; Iqbal et al., 2013). It has also been seen that subsarcolemmal mitochondria show higher rates of fragmentation and degradation relative to the interfibrillar mitochondria subfraction (Wagatsuma et al., 2011). At the neuromuscular junction, mitochondria are numerically reduced and show signs of degeneration including formation of mega mitochondria due to multiple fusions between adjacent mitochondria (Garcia et al., 2013; Anagnostou and Hepple, 2020). Recent evidence also indicates that reduced mitophagy may cause the accumulation of mitochondrial dysfunctions in aging skeletal muscle (O'Leary et al., 2013; Gouspillou et al., 2014). In support of this, impairment of global autophagy, and therefore mitophagy, has been reported to lead to mitochondrial dysfunction and myofiber atrophy (Carnio et al., 2014).

Alterations in mitochondrial structure are strictly correlated with alterations in mitochondrial function. Plenty of studies showed that aged skeletal muscle shows increased mitochondrial reactive oxygen species (ROS) production, increased mitochondrial DNA (mtDNA), decreased activity of several mitochondrial enzymes and reduced mitochondrial respiratory chain efficiency, and increased susceptibility to permeability transition (reviewed in Peterson et al., 2012; Hepple, 2014; Picca et al., 2018).

Mitochondria are the major source of oxidants within cells. Through stimulation of redox-regulated processes, ROS mediate adaptations to contractile activity that includes an increase in mitochondrial biogenesis and in some catabolic processes. It has been shown that ROS production increases in old muscles both in the subsarcolemmal and intermyofibrillar pools of mitochondria (Chabi et al., 2008) This increase has been proposed to lead to attenuated activation of adaptive responses, playing a key role in age-related muscle dysfunction (Jackson and McArdle, 2011). Moreover, alteration in redox cross talk between muscle and neurons has been postulated to be associated with age-related denervation and impaired muscle reinnervation (McArdle et al., 2019). Besides to be a major source of oxidants within cells, mitochondria are primary target of oxidative stress. In aged skeletal muscle mtDNA mutations and deletions accumulate in fibers from, and these mutations are more frequent in muscles prone to sarcopenia (Bua et al., 2006). Interestingly, point mutations are often focused on the mtDNA replication control sites (Wang et al., 2001). This localization of mutation accumulation can reduce gene transcription and lead to a reduction of protein production, leading to alterations in the electron transport chain resulting in muscle fiber loss and apoptosis (Amara et al., 2007; Cheema et al., 2015). A recent study by Andreux et al. (2018) demonstrated a striking association of prefrailty status in older people with mitochondrial impairment in skeletal muscle. They found that prefrail subjects had significantly less abundant respiratory complexes I, IV, and V; lower activity of complexes I, II, and IV; and trend for a lower mtDNA/nuDNA that was accompanied by a significant downregulation of mitochondrial genes in muscle. A lower mitochondrial capacity and efficiency were both correlated with slower walking speed, a defining criterion for sarcopenia, within a group of older adults (Coen et al., 2013). Migliavacca et al. (2019) compared the genome-wide transcriptomic profiles of skeletal muscle biopsies from older men diagnosed with sarcopenia with age-matched controls. They found that regardless of their ethnicity, mitochondrial bioenergetic dysfunction was the strongest molecular signature of sarcopenia. Muscle from individuals with sarcopenia had fewer mitochondria, reduced mitochondrial respiratory complex expression and activity, and reduced levels of nicotinamide adenine dinucleotide (NAD^+). One of the consequences of mitochondrial dysfunction is the activation of apoptosis, a mechanism believed to represent a key driver of the onset and progression of muscle loss (Marzetti and Leeuwenburgh, 2006). Accordingly, it has been reported that mitochondrial apoptotic signaling correlates with reduced muscle volume and slower walking speed in older persons (Marzetti et al., 2012).

10.3.3 Anabolic resistance

Skeletal muscle mass is regulated by the balance between protein synthesis (MPS) and breakdown (MPB). MPS rates are largely controlled by responsiveness to anabolic stimuli, such as nutritional factors (dietary protein, amino acids, and insulin), and physical activity, while catabolism is influenced by multiple factors including inadequate nutrient intake, physical inactivity, inflammation, and illness. A negative net protein balance, arising from a reduction in MPS and/or increase in MPB, will result in a loss of skeletal muscle protein. However, studies focused on assessing rates of MPS and MPB in response to anabolic stimuli suggested that lower rates of MPS, associated with decreased expression and activation anabolic pathways components, are probably major contributors to the failure of muscle maintenance in the older adults (Cuthbertson et al., 2005; Wall et al., 2015). This age-related blunted MPS response to anabolic stimuli is known as anabolic resistance, whereby the "anabolic threshold" required to maximize anabolic pathways is increased (Haran et al., 2012; Dardevet et al., 2012). It has been for instance shown that the dose of dietary protein required to maximize MPS in older individuals is approximately double that of younger individuals (Moore et al., 2015). The role of physical activity in anabolic resistance of myofibrillar protein synthesis has also been investigated in several studies, showing that reduced physical activity, or period of bed rest, in older adults reduces muscle mass that is underpinned by a blunted MPS (Drummond et al., 2012; Breen et al., 2013; Wall et al., 2013). Factors underlying the age-related anabolic resistance include reduced postprandial amino acid delivery/availability, reduced uptake of amino acids into muscle, impairments in protein digestion and amino acid absorption, decrease in anabolic signaling proteins (reviewed in Burd et al., 2013). Alterations generated by comorbidities accompanying sarcopenia, such as insulin resistance, inflammation, and endocrine alterations, may play a role in muscle anabolic resistance. These factors will be tackled in subsequent sections of this chapter. At molecular level, the main signaling pathway known to control the rates of MPS consists of the insulin-like growth factor 1 (IGF1), the kinase Akt and its downstream effectors, the mammalian target of rapamycin (mTOR), and the transcription factor FoxO (Yoon, 2017; Hodson et al., 2019; Barclay et al., 2019). Presently, studies converge on mTORC1 dysregulation as a critical component in anabolic resistance able to remodel the skeletal muscle, being a fundamental site of integration for anabolic signals that stimulate MPS rates (Fry et al., 2011; Manifava et al., 2016). mTORC1 pathway plays an important role in stimulating postprandial/postexercise myofibrillar protein synthesis rates (Abou Sawan et al., 2018). On the contrary, several lines of evidence show that the activation of the mTORC1 and downstream targets is reduced after resistance exercise and protein ingestion in skeletal muscle of older with respect to younger subjects. Cuthbertson et al. (2005) have demonstrated that protein concentrations of mTORC1, and its downstream target p70S6K, differ between healthy young and older individuals, likely contributing to the reduced nutrient-sensing capacity of the muscle protein synthetic machinery in senescent muscle. Also, it has been reported that lower stimulation of protein synthesis after protein ingestion is associated with impaired skeletal muscle phosphorylation of p70S6K (Guillet et al., 2004), PKB, and S6K1 (Francaux et al., 2016). Similar results were observed in the older adults after acute resistance exercise (Kumar et al., 2009; Francaux et al., 2016).

10.3.4 Endocrine factors

Changes in the endocrine environment during aging contribute to the onset and progression of sarcopenia; sex hormones and growth factors act to regulate muscle tissue function and mass (Vitale et al., 2016; McKee and Morley, 2019). Testosterone is the main anabolic hormone for protein synthesis in skeletal muscle and also stimulates satellite cell activation and proliferation, thus influencing maintenance and growth of muscles. In several epidemiological studies, low levels of testosterone have been associated with reduced muscle strength and mass, as well as increased risk of falls (Baumgartner et al., 1999). Also, reduction in estrogen levels negatively influences the maintenance and growth of muscles (Messier et al., 2011). Dehydroepiandrosterone (DHEA), a precursor hormone that is converted into active androgens and estrogens, also declines with age leading to a loss in muscle mass and strength because of the reduced IGF-1 signaling (Valenti et al., 2004). IGF-1 is the primary mediator of muscle repair and growth, which stimulates satellite cell proliferation. IGF-1 also plays a crucial role in the processes of protein synthesis and growth hormone (GH) secretion (Barclay et al., 2019). Several lines of evidence support the role of the age–dependent decline in IGF-1 and GH levels in the pathogenesis of sarcopenia. Consistently, lower serum IGF-1 levels were associated with the risk of developing sarcopenia in a cross-sectional study (Volpato et al., 2014). Similarly, in a cohort of community-dwelling older adults, Tay et al. (2015) reported that IGF-1 was associated with sarcopenia. Moreover, low baseline serum IGF-1 levels correlated with a greater decrease in gait speed and appendicular lean mass in a cohort of older men (Gielen et al., 2015). Belonging to the same pathway, insulin plays an important role in the maintenance of a positive muscle protein balance. A recent systematic review and meta-analysis by Abdulla et al. (2016) concluded that insulin plays a permissive role in the stimulation of MPS in the presence of elevated AAs (Abdulla et al., 2016).

This function appears to be blunted in older people and in those with insulin resistance, resulting in a lower net protein balance and thus contributing to muscle loss in sarcopenia (Rasmussen et al., 2006). Consistently, longitudinal observational studies showed that insulin-resistant subjects have an accelerated loss of muscle mass over time, with respect to noninsulin-resistant individuals (Park et al., 2007; Park et al., 2009). Dysfunction with the hypothalamic-pituitary-adrenal (HPA) axis also occurs with aging, which leads to increased release of glucocorticoids. Among them, cortisol increases with aging leading to decreased anabolic processes and higher protein degradation (McKee and Morley, 2019). An increase in cortisol production has been reported in sarcopenic older persons compared with normal lean group (Waters et al., 2008). Vitamin D is a well-known regulator of bone metabolism, although modulation of muscle morphology, muscle strength, and physical performance has been reported (Halfon et al., 2015). Both observational and longitudinal studies found an association between low vitamin D levels and a decline in physical performance (Houston et al., 2007; Wicherts et al., 2007), as well as with increased risk for developing sarcopenia (Visser et al., 2003).

10.3.5 Inflammation

Aging is associated with low-grade chronic proinflammatory state that results from an imbalance between the inflammatory and anti-inflammatory networks. Recent years have seen increased interest in the potential role of inflammation in sarcopenia, due also to the fact the numerous inflammatory mediators, collectively known as "myokines," are synthesized and secreted by the muscle itself. These locally generated cytokines exert an autocrine function in regulating muscle metabolism, growth, and regeneration, as well as a paracrine/endocrine regulatory function on distant organs (Lee and Jun, 2019). One of the most abundant myokines is myostatin that acts as a negative regulator of muscle growth (White and LeBrasseur, 2014). Supportive evidence for a link between inflammation and sarcopenia comes from epidemiological data, which have shown that higher levels of proinflammatory factors such as interleukin (IL)-6, tumor necrosis factor (TNF)-α), and C-reactive protein (CRP) contribute to the loss of muscle mass and strength that accompanies aging (Visser et al., 2002; Schaap et al., 2009). Elevated IL-6 and TNF-α serum levels were found in older subjects with sarcopenia with respect to control group (Bian et al., 2017). Some studies also evaluated the measures of muscle mass, strength, and function in relation to inflammatory status in hospitalized and community-dwelling older subjects. For instance, Bautmans et al. (2005) proved that patients with inflammation had significantly worse muscle function and that reduced strength and fatigue resistance were significantly related to the levels of CRP, IL-6, and fibrinogen. Furthermore, a relationship was found between increased CRP and reduced grip strength (Norman et al., 2014) and reduced CRP and faster gait speed (Sousa et al., 2016). Similar associations between inflammation markers and muscle mass, strength, and physical performance were observed by Westbury et al. (2018). Possible mechanisms for the associations already mentioned may relate to the combined effects of the multiple age-related impairments caused by inflammation as, for instance, insulin dysregulation, hormonal alteration, endothelial dysfunction, and microvascular changes (Degens, 2010). Moreover, IL-6 and TNF-α at low levels have been shown to induce a transition from proliferation to differentiation of satellite cells; thus, in older individuals, the high levels of these cytokines may affect the maintenance of muscles (Langen et al., 2004; Kurosaka and Machida, 2013). Proinflammatory cytokines also impinge on muscle protein metabolism, with evidence showing that inflammation is associated with reduced rates of protein synthesis paralleled by enhanced protein breakdown (Jo et al., 2012). This likely occurs through different mechanisms, which include activation of the ubiquitin-proteasome system, suppression of the Akt/mTOR pathway, and downregulation of the anabolic effect of IGF-1 (Costamagna et al., 2015).

10.4 Genetic components of sarcopenia

Disentangling the genetic determinants of sarcopenia is not a simple challenge. Traditionally, researchers focused on loci associated with physical phenotypes characterizing sarcopenia, like low handgrip strength (Crocco et al., 2011; Dato et al., 2012a,b; Singh and Gasman, 2020), and decreased muscle mass (Westra et al., 2013; Cho et al., 2017a,b) like NUDT3, KLF5, and HLA-DQB1-AS1, or located in genes involved in mechanisms plausible acting in sarcopenia development, such as ATF6A or ATF6B (Jones et al., 2020), alpha-actinin 3 ACTN3 (Cho et al., 2017a,b). Other studies identified associations with physical performance of several variants in genes related to oxidative stress and mitochondrial functioning (Dato et al., 2014; Dato et al., 2015). Table 10.3 resumes the genes and SNPs related to phenotypes of sarcopenia in old people in different studies. This table is not exhaustive (a more detailed list may be found in Singh and Gasman, 2020), but it includes the genes mostly replicated in different populations, although contrasting findings were observed for some genes. For instance, ACE-rs1799752 was reported to be associated with muscle strength and volume in Northern Americans and to physical performance in Japanese (Charbonneau et al., 2008; Vigano et al., 2009; Yoshihara et al., 2009), but not in Danish or Caucasians of different origins (Frederiksen et al., 2003; Giaccaglia et al., 2008) or ACTN3-rs1815739, which was reported associated with muscle phenotypes by many but not all authors (McCauley et al., 2010). False-positive associations between SNPs and sarcopenia can also be experienced, due to the association with comorbidities and not with sarcopenia itself, but also with an indirect effect on the analyzed phenotype. Large genome-wide association studies (GWAS) for lean mass (LM) and HG were performed in different populations (Karasik et al., 2019; Zillikens et al., 2017; Willems et al., 2017; Tikkanen et al., 2018), sometime combining association studies with GTEX data (GTEx Consortium et al., 2017) and

Table 10.3 Summary of loci and SNPs related to phenotypes associated with sarcopenia and muscle phenotypes in old people in cross-sectional studies.

Locus	Biological role	Polymorphism	Phenotype	Reference
HLA-DQA1	HLA types	rs41268896	Sarcopenia	Jones et al., 2020
		rs29268645	Sarcopenia	Jones et al., 2020
HLA-DQB1-AS1	Antisense RNA 1	rs3129753	HG	Singh and Gasman, 2020
ACTN3	α-Actinin-3 protein	rsl815739[a]	Physical performance	Cho et al., 2017a,b
			Muscle strength in women	Walsh et al., 2008; Delmonico et al., 2007
			Muscle mass in women	Zempo et al., 2010
			Falling risk in women	Judson et al., 2011
ATF6B	Activating transcription factor 6 b; unfolded protein stress response	rs41268896	Sarcopenia	Jones et al., 2020
COL1A1	Type 1 collagen	rs1800012	Hand grip strength	Van Pottelbergh et al., 2001
ACVR2B	Activin A receptor type 2B	rs2268757	Muscle strength in women	Walsh et al., 2007
MSTN	Myostatin	rs1805086	Muscle strength	Corsi et al., 2002; Gonzalez-Freire et al., 2010
CNTF	Ciliary neurotrophic factor	rs1800169	Muscle strength	Arking et al., 2006
FTO	Fat mass and obesity-associated protein; alpha-ketoglutarate dependent dioxygenase	rs9939609	Sarcopenia	Khanal et al., 2020
IGF2	Insulin-like-growth factor 2	rs680	Hand grip strength	Sayer et al., 2002
ESR1	Estrogen receptor 1	rs4870044	Sarcopenia	Khanal et al., 2020
NOS3	Nitric oxide synthase 3	rsl799933	Sarcopenia	Khanal et al., 2020

Table 10.3 Summary of loci and SNPs related to phenotypes associated with sarcopenia and muscle phenotypes in old people in cross-sectional studies—cont'd

Locus	Biological role	Polymorphism	Phenotype	Reference
TRHR	Thyrotropin-releasing hormone	rs7832552	Sarcopenia	Khanal et al., 2020
KLF5	Kruppel-like factor 5; androgen-responsive protein	rs1028883	Lean mass	Singh and Gasman, 2020
NUDT3	Nudix hydrolase 3; homeostatic checkpoints	rs464553	Lean mass	Singh and Gasman, 2020
ST7L	Suppression of tumorigenicity 7 Like	rs1110043	Lean mass	Westra et al., 2013
RPS10	Ribosomal protein S10	rs464553	Lean mass	Westra et al., 2013
ACE	Angiotensin convening enzyme	rs1799752[a]	Muscle strength and volume	Charbonneau et al., 2008
VDR	Vitamin D receptor	rs1544410[a]	Muscle strength in women	Vandevyver et al., 1997
			Muscle strength in men	Bahat et al., 2010
			Falling risk	Onder et al., 2008
		rs2228570[a]	Muscle strength	Hopkinson et al., 2008
			Sarcopenia	Roth et al., 2004
		rs4516035	Handgrip strength	Bozsodi et al., 2016

[a]For these SNPs, contrasting results were observed. See the text for further information.

expression quantitative trait locus (eQTLs) linked to whole-body and appendicular lean mass (Westra et al., 2013). Recently, investigations were focused on older populations classified as sarcopenic/nonsarcopenic subjects or at different sarcopenia stages (Jones et al., 2020; Khanal et al., 2020). In this case, prevalence and association of SNPs with sarcopenia depends on sarcopenia definition itself, as well as on sex- and population-specificity. A clear example of such a great hurdle is documented by Khanal et al. (2020), who compared gene-phenotype associations obtained by applying different sarcopenia definitions. They found that when classifying their population by % SMM (skeletal muscle mass) definition, SNPs in FTO, ESR1, and NOS3 could be found. Based on the SMI (skeletal muscle mass index) definition, the association was found with

TRHR variability and in female only; noteworthy, by applying the EWGSOP definition, only four participants were defined as sarcopenic (Khanal et al., 2020).

10.5 Lifestyle risk factors for sarcopenia

Lifestyle habits are well-known risk factors of sarcopenia. Nutrition and physical activity, in particular, are so important factors that their deficiency according to many reports causes a failure of systemic homeostasis, especially if persisting for a long part of life.

10.5.1 Malnutrition

Physiological anorexia, decreased caloric intake, and weight loss are all related to aging and associated with increased mortality. The imbalance between energy intake and requirements is progressive in old people and can be determined by a reduction in appetite and impediments of eating like lack of teeth, cognitive disability, or lethargy. Risk factors for altered nutritional status include also sensory changes in the ability to taste food and adverse consequences of medications, depression. Weight loss can depend on malabsorption, mal-digestion, or obstruction of the gastrointestinal trait, dramatically influencing the effect of feeding. The effect of a poor nutritional regimen on sarcopenia may due to the fact that some nutritional factors, such as protein, vitamin D/calcium, and the vitamin D/acid-base balance of the diet, play an important role in maintaining muscle mass and, consequently, muscle strength and physical performance (Remelli et al., 2019). In the early stage of malnutrition, muscle succeeds in satisfying its protein and energy requirements using liver glycogen and body fat, obtained mobilizing protein stores in viscera. When these energy supplies are depleted, muscle proteins are broken down to provide amino acids for gluconeogenesis, thereby supplying energy to other parts of the body. The negative net balance between muscle protein synthesis and degradation decreases skeletal muscle mass. This inevitably triggers consequences also to the overall homeostasis of the organism, considering that skeletal muscle is increasingly recognized as one of the key regulators of energy and protein metabolism by way of metabolic crosstalk between body organs (Argilés et al., 2016). Malnutrition may lead to sarcopenia, but sarcopenia may exacerbate malnutrition due to its effects on swallowing, especially in hospitalized and institutionalized patients. The clinical presentation of both malnutrition and accelerated loss of lean body mass, strength, and/or functionality is a condition called Malnutrition-Sarcopenia Syndrome (MSS), which is used as a prognostic factor in the management of hospitalized older patients because correlated with mortality (Hu et al., 2017). The link between malnutrition and sarcopenia has been explored in several cross-sectional studies of community-dwelling older people, finding an association between a poor balanced diet in terms of reduced micro/macronutrients and the presence of sarcopenia at baseline (Beaudart et al., 2019 and reference therein). Important pieces of evidence came from the very few longitudinal studies, and in particular from the work of the GLISTEN (Gruppo Lavoro Italiano Sarcopenia—Trattamento e

Nutrizione) multicenter prospective study, which demonstrated that among participants diagnosed with malnutrition at baseline, the incidence of sarcopenia was higher, with 36.8% of them developing the syndrome throughout the four-year follow-up period, compared to 12.3% in the group of well-nourished participants (Martone et al., 2017). These results indicated that malnutrition or poor-balanced nutrition seems to be one of the risk factors for sarcopenia, but the onset of this condition is more than likely multifactorial with other factors, such as sedentary lifestyle and inflammatory biomarkers, which play an even important role in the development of sarcopenia.

10.5.2 Physical inactivity

Sedentary behavior, defined as time spent sitting or lying, has been shown to be a major modifiable risk factor for chronic disease, disablement, and frailty and has been found associated with reduced lean body mass and muscle mass and with an increased risk of sarcopenia (Chastin et al., 2012). Studies dealing with the effects of bed rest on skeletal muscle demonstrate the impact of sedentary behavior on muscle mass and metabolism: time spent sitting (>257 min/day for men and >330 min/day for women) or lying affects muscle physiology and accelerates sarcopenia (Hamer and Stamatakis, 2013). Moreover, prolonged sitting time is associated with metabolic syndrome (Gardiner et al., 2011), excessive body weight (Gómez-Cabello et al., 2012), and increased risk of mortality from all causes (Lee, 2016). At the cellular level, disuse-related changes in skeletal muscle mass commonly involve a decrease of fiber number and size (Narici et al., 2003) and reduction of the satellite cell number (Always et al., 2014), these last indispensable components of muscle regeneration. Moreover, it is physiologic that a lack of physical activity increases the resistance of muscle to anabolism, particularly the synthesis of proteins from amino acids (Dideriksen et al., 2013). In a vicious cycle, muscle weakness reduces the ability to perform daily functions and mobility and increases the risk of fall, in turn increasing or causing disability (Rezuş et al., 2020). In large population studies (Gennuso et al., 2013; León-Latre et al., 2014; Loprinzi, 2014), long periods spent in sedentary behavior were reported to be associated with elevated inflammatory state, measured by an increase in serum concentration of C-reactive protein and leukocyte count (Parsons et al., 2017). As stated previously, elevated inflammatory state contributes to the decrease of muscle mass, strength, power, and motor performance (Chen et al., 2014b). Moreover, a relationship between sedentary behavior and dysfunction of the immune system was proposed, with a reduced muscle contraction associate with an increase in muscle glucose level and decreased insulin sensitivity (Charansonney, 2011).

10.6 Management of sarcopenia

The management of sarcopenia includes a multidisciplinary approach, combining diet, exercise, and medications for improving the quality and quantity of skeletal muscle in older people. Potential therapies for sarcopenia can be divided into preventive measures, focused on the identification and modification of risk factors for disease development, and treatments, aimed to mitigate the progression of the disease after the onset.

10.6.1 Nutrition and physical activity

Nutritional screening and implementation of nutrition care plans should be considered to manage sarcopenia. In sarcopenic subjects older than 65 years, a modest increase of the recommended daily intake of proteins is currently proposed, from 0.8 g/kg/day to 1-1.2 g/kg/day (Deutz et al., 2014; Baum and Wolfe, 2015). This amount, as well as the modalities of protein intake throughout the day, can be modified with respect to the levels of physical exercise performed (Gryson et al., 2014). Important nutritional inputs for protein synthesis are essential amino acids (EAAs), whose supplement may benefit muscle mass and function (Cruz-Jentoft et al., 2014). Among EEAs, leucine is considered the primary nutritional regulator of muscle protein anabolism, a potent activator of the mammalian target of rapamycin (mTOR), a nutrient and energy-sensing signaling pathway affecting translation initiation and muscle protein synthesis (De Bandt, 2016). Therefore, it is recommended that older persons consume protein sources with higher proportions of EAAs (i.e., high-quality proteins), such as lean meat and other leucine-rich foods (e.g., soybeans, peanuts, cowpea, and lentils). Great interest has been devoted to antioxidant supplementation as a potential intervention in sarcopenia, since oxidative damage is considered to be one of the mechanisms leading to the loss of muscle mass and function (Kim et al., 2010). Thus, many studies have evaluated the effects of antioxidant dietary supplements like curcumin, resveratrol, a natural polyphenolic compound occurring in plants and food, or quercetin, a natural flavonoid, in limiting age-related muscle mass and performance decline. In mice, resveratrol, i.e., in combination with habitual exercise has been shown to improve mitochondrial function and avoid the aging-related decline in physical performance (Murase et al., 2009) or curcumin supplementation ameliorates exercise performance in rats increasing time of run to exhaustion compared to control animals (Sahin et al., 2016). However, the positive findings obtained in animals have not always been confirmed in human trials; controversial results were reported, probably due to the fact that oxidants (ROS/RNS, i.e., oxygen or nitrogen species) have a dual effect in skeletal muscle, because of their nature of signaling molecules: at low levels, they increase muscle force and adaptation to exercise, likely regulating intracellular signal transduction pathways directly or indirectly involved in skeletal muscle atrophy, motoneuronal degeneration, and impairment of muscle contractility, while at high levels, they lead to a decline of muscle performance (Damiano et al., 2019). Some other nutritional supplements, such as β-hydroxy β-methylbutyrate, creatine, and vitamin D, have been suggested to have an effect on muscle function, increasing muscle mass despite having no effects on muscle strength and physical performance (Deane et al., 2017). Together with resveratrol and green tea catechins, β-hydroxy-β-methylbutyrate has found able to regulate satellite cell function and therefore contributing to reducing sarcopenia or improving muscle mass after disuse, during rehabilitative loading in animal models of aging (Always et al., 2014). Although these compounds have not been rigorously tested in humans, the data from animal models prompt to carry out additional focused studies, to determine if these or other nutraceuticals can offset the muscle losses, or improve regeneration in sarcopenic muscles of older humans via improving satellite cell function. The combination of vitamin D supplementation with leucine-enriched whey protein was found to increase both postprandial protein synthesis and muscle mass in healthy older men (Chanet et al., 2017). Moreover, supplementation of omega-3 polyunsaturated fatty acids or monounsaturated fatty acids was found to improve protein anabolism, muscle mass, and muscle function aside from contributing to decrease insulin resistance and lipotoxicity and prevent fat mass (Gray and Mittendorfer, 2018). Exercise is the primary treatment option for sarcopenia, specifically load-bearing activities including resistance or strength training. Both kinds of training have been shown to be successful interventions in preventing sarcopenia: the first has been reported to positively influence the neuromuscular system as well as increase hormone concentrations and the rate of protein synthesis, whereas the second exercise has been demonstrated to increase the capability of skeletal muscle to synthesize proteins. Furthermore, with or without protein supplementation, physical activity diminishes lipotoxicity by increasing mitochondrial fatty-acid beta-oxidation in muscle cells (Bruce et al., 2006). Although physical exercise programs should be tailored with respect to disease stage and general health status of the patient, an optimal and regular training exercise program (3 times/week) should combine aerobic exercise (which improves cardiovascular function and endurance and decrease fat mass) with strength exercises (which increases muscle mass, strength, and function) (Phu et al., 2015). Thus, the screening of patients for impairment in their physical function and ADLs should be a routine part of healthcare visits for the older people. A combined approach of dietary supplements (proteins or EAAs) and exercise protein supplementation can be combined to prevent or recover the loss of muscle mass, with larger gains with respect to the diet and exercise alone (Tieland et al., 2012; Denison et al., 2015). Interactive effects of diet and exercise on physical function have been studied most extensively in relation to protein/amino acid supplementation in animal models (Deane et al., 2017) and were confirmed in humans (Dhillon and Hasni, 2017 and references therein). However, in a more recent meta-analysis, the findings were inconsistent among various populations (Papadopoulou, 2020 and reference therein).

10.6.2 Anabolic medications and pharmacological treatments

Actually, there are no pharmacological treatments approved by FDA (Food and Drug Administration) for the treatment of sarcopenia. Potential targets can address one or more of the multiple etiological factors of sarcopenia, as previously described, like the loss of regenerative capacity of muscle, the age-related changes in the expression of signaling molecules and in muscle physiology like denervation and mitochondrial dysfunction, or inflammation (Naranjo et al., 2017). Therapy with creatinine and myostatin has been demonstrated to regulate muscle growth, mass, and strength (Liu et al., 2007; Chilibeck et al., 2017). Growth hormone increases muscle protein synthesis and muscle mass but does not lead to gains in strength and function. Similarly, IGF1-based is also applied, although the lack of efficacy sometime experienced which may be due to local resistance to IGF-1 in aging muscle that results from inflammation and other age-related changes (Dhillon and Hasni, 2017). Anti-inflammatory drugs, steroid hormones, and growth factors do not target myogenesis or directly affect the restoration of tissue function, but can revert the functional decline of sarcopenia. However, in some case, as observed for testosterone, the modest positive effect on muscle strength and mass does not justify a dosage increase of these treatments, because of adverse effects, such as increased risk of prostate cancer and cardiovascular events (Wakabayashi and Sakuma, 2014). Clinical trials with metformin and angiotensin-converting enzyme inhibitors in combination with either physical activity or leucine supplementation, selective androgen receptor modulators, and vitamin D are under testing (Tournadre et al., 2019). New therapies for sarcopenia are under development. Regenerative medicine strategies try to restore skeletal muscle structure and function, by using exogenous delivery of stem/progenitor cells to repopulate the satellite cell pool and stimulate myogenesis (Naranjo et al., 2017). To date, little success has been achieved by exogenous stem cell delivery, because of difficulties in deliverability and in vitro expansion. Recently, evidence was found that changing the microenvironment of aged myogenic progenitor cells, through the use of ECM (extracellular matrix) biologic scaffolds, naturally derived or synthetic, can promote skeletal muscle regeneration (Barberi et al., 2013). By providing the appropriate tissue niche and directing the macrophage response, this method promises to facilitate the proliferative capacity of satellite cells and regeneration of skeletal muscle in the older adults (Wolf et al., 2015).

10.7 Conclusions and future perspectives

The number of 65 and older subjects in developed countries was increased dramatically over the last 50 years, accompanied by an increased prevalence of chronic diseases and disability level, putting an enormous burden on healthcare systems. Among the factors that increase the individual vulnerability and can predict health outcomes more than age, sarcopenia, the major component of the frailty syndrome in older adults, is often unrecognized and poorly managed in routine clinical practice, due to the lack of available uniform diagnostic testing and criteria. Appropriate treatments focused on the restorement of muscle structure and function to improve myogenesis or reverse the physiological decline in muscle loss are mandatory. The complex nature of sarcopenia implies that finding effective treatments strategies needs to be focused on a multidisciplinary approach that comprehends, among others, physical therapy, and nutritional and drug supplementation. A combination of these approaches can ameliorate the outcomes and improve the quality of life in patients suffering from sarcopenia.

References

Abate M, Di Iorio A, Di Renzo D, Paganelli R, Saggini R, Abate G. Frailty in the elderly: the physical dimension. Eura Medicophys 2007;43(3):407–15.

Abdulla H, Smith K, Atherton PJ, Idris I. Role of insulin in the regulation of human skeletal muscle protein synthesis and breakdown: A systematic review and meta-analysis. Diabetologia 2016;59:44–55.

Abellan van Kan G, Rolland YM, Morley JE, Vellas B. Frailty: toward a clinical definition. J Am Med Dir Assoc 2008;9(2):71–2.

Abete P, Basile C, Bulli G, Curcio F, Liguori I, Della-Morte D, Gargiulo G, Langellotto A, Testa G, Galizia G, Bonaduce D, Cacciatore F. The Italian version of the "frailty index" based on deficits in health: a validation study. Aging Clin Exp Res 2017;29(5):913–26.

Abou Sawan S, van Vliet S, Parel JT, Beals JW, Mazzulla M, West D, Philp A, Li Z, Paluska SA, Burd NA, Moore DR. Translocation and protein complex co-localization of mTOR is associated with postprandial myofibrillar protein synthesis at rest and after endurance exercise. Physiol Rep 2018;6(5)e13628.

Alexa ID, Ilie AC, Moroşanu A, Voica A. Approaching frailty as the new geriatric syndrome. Rev Med Chir Soc Med Nat Iasi 2013;117(3):680–5.

Always SE, Myers MJ, Mohamed JS. Regulation of Satellite Cell Function in Sarcopenia. Front Aging Neurosci 2014;6:246.

Amara CE, Shankland EG, Jubrias SA, Marcinek DJ, Kushmerick MJ, Conley KE. Mild mitochondrial uncoupling impacts cellular aging in human muscles in vivo. Proc Natl Acad Sci U S A 2007;104:1057–62.

Anagnostou ME, Hepple RT. Mitochondrial Mechanisms of Neuromuscular Junction Degeneration with Aging. Cells 2020;9(1):197.

Andreasen J, Lund H, Aadahl M, Gobbens RJ, Sorensen EE. Content validation of the Tilburg Frailty Indicator from the perspective of frail elderly. A qualitative explorative study. Arch Gerontol Geriatr 2015;61 (3):392–9.

Andreux PA, van Diemen MPJ, Heezen MR, Auwerx J, Rinsch C, Groeneveld GJ, Singh A. Mitochondrial function is impaired in the skeletal muscle of pre-frail elderly. Sci Rep 2018;8(1):8548 Erratum in: Sci Rep. 2019; 9(1),17821.

Aprahamian I, Cezar N, Izbicki R, Lin SM, Paulo D, Fattori A, Biella MM, Jacob Filho W, Yassuda MS. Screening for Frailty With the FRAIL Scale: A Comparison With the Phenotype Criteria. J Am Med Dir Assoc 2017;18(7):592–6.

Argilés JM, Campos N, Lopez-Pedrosa JM, Rueda R, Rodriguez-Mañas L. Skeletal Muscle Regulates Metabolism via Interorgan Crosstalk: Roles in Health and Disease. J Am Med Dir Assoc 2016;17 (9):789–96.

Arking DE, Fallin DM, Fried LP, Li T, Beamer BA, Xue QL, Chakravarti A, Walston J. Variation in the ciliary neurotrophic factor gene and muscle strength in older Caucasian women. J Am Geriatr Soc 2006;54(5):823–6.

Bahat G, Saka B, Erten N, Ozbek U, Coskunpinar E, Yildiz S, Sahinkaya T, Karan MA. BsmI polymorphism in the vitamin D receptor gene is associated with leg extensor muscle strength in elderly men. Aging Clin Exp Res 2010;22(3):198–205.

Barberi L, Scicchitano BM, De Rossi M, Bigot A, Duguez S, Wielgosik A, Stewart C, McPhee J, Conte M, Narici M, Franceschi C, Mouly V, Butler-Browne G, Musaro A. Age-dependent alteration in muscle regeneration: the critical role of tissue niche. Biogerontology 2013;14:273–92.

Barclay RD, Burd NA, Tyler C, Tillin NA, Mackenzie RW. The role of the IGF-1 signaling cascade in muscle protein synthesis and anabolic resistance in aging skeletal muscle. Front Nutr 2019;6:146.

Baron AD, Brechtel G, Wallace P, Edelman SV. Rates and tissue sites of non-insulin- and insulin-mediated glucose uptake in humans. Am J Physiol 1988;255(6Pt1):E769–74.

Batsis JA, Villareal DT. Sarcopenic obesity in older adults: aetiology, epidemiology and treatment strategies. Nat Rev Endocrinol 2018;14(9):513–37.

Baum JI, Wolfe RR. The link between dietary protein intake, skeletal muscle function and health in older adults. Healthcare (Basel) 2015;3(3):529–43.

Baumgartner RN, Waters DL, Gallagher D, Morley JE, Garry PJ. Predictors of skeletal muscle mass in elderly men and women. Mech Ageing Dev 1999;107(2):123–36.

Bautmans I, Njemini R, Lambert M, Demanet C, Mets T. Circulating acute phase mediators and skeletal muscle performance in hospitalized geriatric patients. J Gerontol: Ser A 2005;60(3):361–7.

Beaudart C, Locquet M, Touvier M, Reginster JY, Bruyère O. Association between dietary nutrient intake and sarcopenia in the SarcoPhAge study. Aging Clin Exp Res 2019;31:815–24.

Beavers KM, Beavers DP, Houston DK, Harris TB, Hue TF, Koster A, Newman AB, Simonsick EM, Studenski SA, Nicklas BJ, Kritchevsky SB. Associations between body composition and gait-speed decline: results from the Health, Aging, and Body Composition study. Am J Clin Nutr 2013;97 (3):552–60.

Bian AL, Hu HY, Rong YD, Wang J, Wang JX, Zhou XZ. A study on relationship between elderly sarcopenia and inflammatory factors IL-6 and TNF-α. Eur J Med Res 2017;22(1):25.

Borzuola R, Giombini A, Torre G, Campi S, Albo E, Bravi M, Borrione P, Fossati C, Macaluso A. Central and peripheral neuromuscular adaptations to ageing. J Clin Med 2020;9(3):741.

Bozsodi A, Boja S, Szilagyi A, Somhegyi A, Varga PP, Lazary A. Muscle strength is associated with vitamin D receptor gene variants. J Orthop Res 2016;34(11):2031–7.

Braun T, Gruneberg C, Thiel C. German translation, cross-cultural adaptation and diagnostic test accuracy of three frailty screening tools: PRISMA-7, FRAIL scale and Groningen Frailty Indicator. Z Gerontol Geriatr 2018;51(3):282–92.

Breen L, Stokes KA, Churchward-Venne TA, Moore DR, Baker SK, Smith K, Atherton PJ, Phillips SM. Two weeks of reduced activity decreases leg lean mass and induces "anabolic resistance" of myofibrillar protein synthesis in healthy elderly. J Clin Endocrinol Metab 2013;98:2604–12.

Bruce CR, Thrush AB, Mertz VA, Bezaire V, Chabowski A, Heigenhauser GJ, Dyck DJ. Endurance training in obese humans improves glucose tolerance and mitochondrial fatty acid oxidation and alters muscle lipid content. American journal of physiology. Endocrinol Metab 2006;291(1):E99–E107.

Bua E, Johnson J, Herbst A, Delong B, McKenzie D, Salamat S, Aiken JM. Mitochondrial DNA-deletion mutations accumulate intracellularly to detrimental levels in aged human skeletal muscle fibers. Am J Hum Genet 2006;79(3):469–80.

Burd NA, Gorissen SH, van Loon LJ. Anabolic resistance of muscle protein synthesis with aging. Exerc Sport Sci Rev 2013;41:169–73.

Cao L, Morley JE. Sarcopenia Is Recognized as an Independent Condition by an International Classification of Disease, Tenth Revision, Clinical Modification (ICD-10-CM) Code. J Am Med Dir Assoc 2016;17 (8):675–7.

Carnio S, LoVerso F, Baraibar MA, Longa E, Khan MM, Maffei M, Reischl M, Canepari M, Loefler S, Kern H, Blaauw B, Friguet B, Bottinelli R, Rudolf R, Sandri M. Autophagy impairment in muscle induces neuromuscular junction degeneration and precocious aging. Cell Rep 2014;8(5):1509–21.

Cederholm T. Overlaps between frailty and sarcopenia definitions. Nestle Nutr Inst Workshop Ser 2015;83:65–9.

Cesari M, Demougeot L, Boccalon H, Guyonnet S, Abellan Van Kan G, Vellas B, Andrieu S. A self-reported screening tool for detecting community-dwelling older persons with frailty syndrome in the absence of mobility disability: the FiND questionnaire. PLoS One 2014a;9(7)e101745.

Cesari M, Landi F, Vellas B, Bernabei R, Marzetti E. Sarcopenia and Physical Frailty: Two Sides of the Same Coin. Front Aging Neurosci 2014b;6:192.

Chabi B, Ljubicic V, Menzies KJ, Huang JH, Saleem A, Hood DA. Mitochondrial function and apoptotic susceptibility in aging skeletal muscle. Aging Cell 2008;7:2–12.

Chanet A, Verlaan S, Salles J, Giraudet C, Patrac V, Pidou V, Pouyet C, Hafnaoui N, Blot A, Cano N, Farigon N, Bongers A, Jourdan M, Luiking Y, Walrand S, Boirie Y. Supplementing breakfast with a vitamin D and leucine-enriched whey protein medical nutrition drink enhances postprandial muscle protein synthesis and muscle mass in healthy older men. J Nutr 2017;147(12):2262–71.

Charansonney OL. Physical activity and aging: a life-long story. Discov Med 2011;12(64):177–85.

Charbonneau DE, Hanson ED, Ludlow AT, Delmonico MJ, Hurley BF, Roth SM. ACE genotype and the muscle hypertrophic and strength responses to strength training. Med Sci Sports Exerc 2008;40:677–83.

Chastin SF, Ferriolli E, Stephens NA, Fearon KC, Greig C. Relationship between sedentary behaviour, physical activity, muscle quality and body composition in healthy older adults. Age Ageing 2012;41 (1):111–4.

Cheema N, Herbst A, McKenzie D, Aiken JM. Apoptosis and necrosis mediate skeletal muscle fiber loss in age-induced mitochondrial enzymatic abnormalities. Aging Cell 2015;14:1085–93.

Chen LK, Liu LK, Woo J, Assantachai P, Auyeung TW, Bahyah KS, Chou MY, Chen LY, Hsu PS, Krairit O, Lee JS, Lee WJ, Lee Y, Liang CK, Limpawattana P, Lin CS, Peng LN, Satake S, Suzuki T, Won CW, et al. Sarcopenia in Asia: consensus report of the Asian Working Group for Sarcopenia. J Am Med Dir Assoc 2014a;15(2):95–101.

Chen X, Mao G, Leng SX. Review Frailty syndrome: an overview. Clin Interv Aging 2014b;9:433–4.

Chilibeck P, Kaviani M, Candow D, Zello GA. Effect of creatine supplementation during resistance training on lean tissue mass and muscular strength in older adults: a meta-analysis. Open Access J Sports Med 2017;8:213–26.

Cho J, Lee I, Kang H. ACTN3 Gene and Susceptibility to Sarcopenia and Osteoporotic Status in Older Korean Adults. Biomed Res Int 2017a;2017:4239648.

Cho KM, Park H, Oh DY, Kim TY, Lee KH, Han SW, Im SA, Kim TY, Bang YJ. Skeletal muscle depletion predicts survival of patients with advanced biliary tract cancer undergoing palliative chemotherapy. Oncotarget 2017b;8(45):79441–52.

Choi SJ, Files DC, Zhang T, Wang ZM, Messi ML, Gregory H, Stone J, Lyles MF, Dhar S, Marsh AP, Nicklas BJ, Delbono O. Intramyocellular lipid and impaired myofiber contraction in normal weight and obese older adults. J Gerontol A Biol Sci Med Sci 2016;71(4):557–64.

Ciciliot S, Rossi AC, Dyar KA, Blaauw B, Schiaffino S. Muscle type and fiber type specificity in muscle wasting. Int J Biochem Cell Biol 2013;45(10):2191–9.

Coelho T, Paúl C, Gobbens RJ, Fernandes L. Determinants of frailty: the added value of assessing medication. Front Aging Neurosci 2015;7:56.

Coen PM, Jubrias SA, Distefano G, Amati F, Mackey DC, Glynn NW, Manini TM, Wohlgemuth SE, Leeuwenburgh C, Cummings SR, Newman AB, Ferrucci L, Toledo FG, Shankland E, Conley KE, Goodpaster BH. Skeletal muscle mitochondrial energetics are associated with maximal aerobic capacity and walking speed in older adults. J Gerontol A Biol Sci Med Sci 2013;68(4):447–55.

Collard RM, Boter H, Schoevers RA, Oude Voshaar RC. Prevalence of frailty in community-dwelling older persons: a systematic review. J Am Geriatr Soc 2012;60(8):1487–92.

Corsi AM, Ferrucci L, Gozzini A, Tanini A, Brandi ML. Myostatin polymorphisms and age-related sarcopenia in the Italian population. J Am Geriatr Soc 2002;50(8):1463.

Cosgrove BD, Gilbert PM, Porpiglia E, Mourkioti F, Lee SP, Corbel SY, Llewellyn ME, Delp SL, Blau HM. Rejuvenation of the muscle stem cell population restores strength to injured aged muscles. Nat Med 2014;20:255–64.

Costamagna D, Costelli P, Sampaolesi M, Penna F. Role of inflammation in muscle homeostasis and myogenesis. Mediators Inflamm 2015;2015:805172.

Crocco P, Montesanto A, Passarino G, Rose G. A common polymorphism in the UCP3 promoter influences hand grip strength in elderly people. Biogerontology 2011;12(3):265–71.

Cruz-Jentoft AJ, Landi F, Schneider SM, Zúñiga C, Arai H, Boirie Y, Chen LK, Fielding RA, Martin FC, Michel JP, Sieber C, Stout JR, Studenski SA, Vellas B, Woo J, Zamboni M, Cederholm T. Prevalence of and interventions for sarcopenia in ageing adults: a systematic review. Report of the International Sarcopenia Initiative (EWGSOP and IWGS). Age Ageing 2014;43(6):748–59.

Cruz-Jentoft AJ, Bahat G, Bauer J, Boirie Y, Bruyère O, Cederholm T, Cooper C, Landi F, Rolland YS, Avan A, Schneider SM, Sieber CC, Topinkova E, Vandewoude M, Visser M, Zamboni M, Writing Group for the European Working Group on Sarcopenia in Older People 2 (EWGSOP2), and the Extended Group for EWGSOP2. Sarcopenia: revised European consensus on definition and diagnosis. Age Ageing 2019;48(1):16–31.

Cuthbertson D, Smith K, Babraj J, Leese G, Waddell T, Atherton P, Wackerhage H, Taylor PM, Rennie MJ. Anabolic signaling deficits underlie amino acid resistance of wasting, aging muscle. FASEB J 2005;19(3):422–4.

Damiano S, Muscariello E, La Rosa G, Di Maro M, Mondola P, Santillo M. Dual role of reactive oxygen species in muscle function: can antioxidant dietary supplements counteract age-related sarcopenia? Int J Mol Sci 2019;20(15):3815.

Daniels R, Metzelthin S, van Rossum E, de Witte L, van den Heuvel W. Interventions to prevent disability in frail community-dwelling older persons: an overview. Eur J Ageing 2010;7(1):37–55.

Dardevet D, Rémond D, Peyron MA, Papet I, Savary-Auzeloux I, Mosoni L. Muscle wasting and resistance of muscle anabolism: the "anabolic threshold concept" for adapted nutritional strategies during sarcopenia. Sci World J 2012;2012:269531.

Dato S, Montesanto A, Lagani V, Jeune B, Christensen K, Passarino G. Frailty phenotypes in the elderly based on cluster analysis: a longitudinal study of two Danish cohorts. Evidence for a genetic influence on frailty. Age (Dordr) 2012a;34(3):571–82.

Dato S, Soerensen M, Montesanto A, Lagani V, Passarino G, Christensen K, Christiansen L. UCP3 polymorphisms, hand grip performance and survival at old age: association analysis in two Danish middle aged and elderly cohorts. Mech Ageing Dev 2012b;133(8):530–7.

Dato S, Soerensen M, Lagani V, Montesanto A, Passarino G, Christensen K, Tan Q, Christiansen L. Contribution of genetic polymorphisms on functional status at very old age: a gene-based analysis of 38 genes (311 SNPs) in the oxidative stress pathway. Exp Gerontol 2014;52:23–9.

Dato S, De Rango F, Crocco P, Passarino G, Rose G. Antioxidants and quality of aging: further evidences for a major role of TXNRD1 gene variability on physical performance at old age. Oxid Med Cell Longev 2015;2015:926067.

De Bandt JP. Leucine and mammalian target of rapamycin-dependent activation of muscle protein synthesis in aging. J Nutr 2016;146(12):2616S–2624S.

De Carvalho FG, Justice JN, Freitas EC, Kershaw EE, Sparks LM. Adipose tissue quality in aging: how structural and functional aspects of adipose tissue impact skeletal muscle quality. Nutrients 2019;11(11):2553.

De Stefano F, Zambon S, Giacometti L, Sergi G, Corti MC, Manzato E, Busetto L. Obesity, muscular strength, muscle composition and physical performance in an elderly population. J Nutr Health Aging 2015;19(7):785–91.

De Witte N, Gobbens R, De Donder L, Dury S, Buffel T, Schols J, Verté D. The comprehensive frailty assessment instrument: development, validity and reliability. Geriatric nursing (New York, NY) 2013;34(4):274–81.

Deane CS, Wilkinson DJ, Phillips BE, Smith K, Etheridge T, Atherton PJ. "Nutraceuticals" in relation to human skeletal muscle and exercise. Am J Physiol Endocrinol Metab 2017;312(4):E282–99.

Degens H. The role of systemic inflammation in age-related muscle weakness and wasting. Scand J Med Sci Sports 2010;20:28–38.

Delmonico MJ, Kostek MC, Doldo NA, Hand BD, Walsh S, Conway JM, Carignan CR, Roth SM, Hurley BF. Alpha-actinin-3 (ACTN3) R577X polymorphism influences knee extensor peak power response to strength training in older men and women. J Gerontol A Biol Sci Med Sci 2007;62(2):206–12. https://doi.org/10.1093/gerona/62.2.206.

Delmonico MJ, Harris TB, Visser M, Park SW, Conroy MB, Velasquez-Mieyer P, Boudreau R, Manini TM, Nevitt M, Newman AB, Goodpaster BH, Health, Aging, and Body. Longitudinal study of muscle strength, quality, and adipose tissue infiltration. Am J Clin Nutr 2009;90(6):1579–85.

Denison HJ, Cooper C, Sayer AA, Robinson SM. Prevention and optimal management of sarcopenia: a review of combined exercise and nutrition interventions to improve muscle outcomes in older people. Clin Interv Aging 2015;10:859–69.

Dent E, Kowal P, Hoogendijk EO. Frailty measurement in research and clinical practice: A review. Eur J Intern Med 2016;31:3–10.

Deschenes MR. Motor unit and neuromuscular junction remodelling with aging. Curr Aging Sci 2011;4:209–20.

Deutz NE, Bauer JM, Barazzoni R, Biolo G, Boirie Y, Bosy-Westphal A, Cederholm T, Cruz-Jentoft A, Krznarić Z, Nair KS, Singer P, Teta D, Tipton K, Calder PC. Protein intake and exercise for optimal muscle function with aging: recommendations from the ESPEN Expert Group. Clin Nutrit (Edinburgh, Scotland) 2014;33(6):929–36.

Dhillon RJ, Hasni S. Pathogenesis and management of sarcopenia. Clin Geriatr Med 2017;33(1):17–26.

Díaz de León González E, Gutiérrez Hermosillo H, Martinez Beltran JA, Chavez JH, Palacios Corona R, Salinas Garza DP, Rodriguez Quintanilla KA. Validation of the FRAIL scale in Mexican elderly: results from the Mexican Health and Aging Study. Aging Clin Exp Res 2016;28(5):901–8. https://doi.org/10.1007/s40520-015-0497-y.

Dideriksen K, Reitelseder S, Holm L. Influence of amino acids, dietary protein, and physical activity on muscle mass development in humans. Nutrients 2013;5(3):852–76.

Dong L, Liu N, Tian X, Qiao X, Gobbens R, Kane RL, Wang C. Reliability and validity of the Tilburg Frailty Indicator (TFI) among Chinese community-dwelling older people. Arch Gerontol Geriatr 2017;73:21–8.

Dong L, Qiao X, Tian X, Liu N, Jin Y, Si H, Wang C. Cross-cultural adaptation and validation of the FRAIL scale in chinese community-dwelling older adults. J Am Med Dir Assoc 2018;19(1):12–7.

Drummond MJ, Dickinson JM, Fry CS, Walker DK, Gundermann DM, Reidy PT, Timmerman KL, Markofski MM, Paddon-Jones D, Rasmussen BB, Volpi E. Bed rest impairs skeletal muscle amino acid transporter expression, mTORC1 signaling, and protein synthesis in response to essential amino acids in older adults. AmJ Physiol Endocrinol Metab 2012;302:E1113–22.

Ernsth-Bravell M, Mölstad S. Easy-to-use definition of frailty for guiding care decisions in elderly individuals: probability or utopia? Aging Health 2010;6(6):697–9.

Espinoza SE, Jung I, Hazuda H. Lower frailty incidence in older Mexican Americans than in older European Americans: the San Antonio Longitudinal Study of Aging. J Am Geriatr Soc 2010;58(11):2142–8.

Ethgen O, Beaudart C, Buckinx F, Bruyère O, Reginster JY. The future prevalence of sarcopenia in Europe: a claim for public health action. Calcif Tissue Int 2017;100(3):229–34.

Fabrício-Wehbe SC, Schiaveto FV, Vendrusculo TR, Haas VJ, Dantas RA, Rodrigues RA. Cross-cultural adaptation and validity of the 'Edmonton Frail Scale—EFS' in a Brazilian elderly sample. Rev Lat Am Enfermagem 2009;17(6):1043–9.

Faller JW, Pereira DDN, de Souza S, Nampo FK, Orlandi FS, Matumoto S. Instruments for the detection of frailty syndrome in older adults: A systematic review. PLoS One 2019;14(4)e0216166.

Fielding RA, Vellas B, Evans WJ, Bhasin S, Morley JE, Newman AB, Abellan van Kan G, Andrieu S, Bauer J, Breuille D, Cederholm T, Chandler J, De Meynard C, Donini L, Harris T, Kannt A, Keime Guibert F, Onder G, Papanicolaou D, Rolland Y, et al. Sarcopenia: an undiagnosed condition in older adults. Current consensus definition: prevalence, etiology, and consequences. International working group on sarcopenia. J Am Med Dir Assoc 2011;12(4):249–56.

Francaux M, Demeulder B, Naslain D, Fortin R, Lutz O, Caty G, Deldicque L. Aging reduces the activation of the mTORC1 pathway after resistance exercise and protein intake in human skeletal muscle: potential role of REDD1 and impaired anabolic sensitivity. Nutrients 2016;8(1):47.

Frederiksen H, Bathum L, Worm C, Christensen K, Puggaard L. ACE genotype and physical training effects: a randomized study among elderly Danes. Aging Clin Exp Res 2003;15(4):284–91.

Freitag S, Schmidt S, Gobbens RJ. Tilburg frailty indicator. German translation and psychometric testing. Z Gerontol Geriatr 2016;49(2):86–93.

Fried LP, Tangen CM, Walston J, Newman AB, Hirsch C, Gottdiener J, Seeman T, Tracy R, Kop WJ, Burke G, McBurnie MA. Cardiovascular Health Study Collaborative Research Group. Frailty in older adults: evidence for a phenotype. J Gerontol A Biol Sci Med Sci 2001;56(3):M146–56.

Fry CS, Drummond MJ, Glynn EL, Dickinson JM, Gundermann DM, Timmerman KL, Walker DK, Dhanani S, Volpi E, Rasmussen BB. Aging impairs contraction-induced human skeletal muscle mTORC1 signaling and protein synthesis. Skeletal Muscle 2011;1(1):11.

Garcia ML, Fernandez A, Solas MT. Mitochondria, motor neurons and aging. J Neurol Sci 2013;330:18–26.

Gardiner PA, Healy GN, Eakin EG, Clark BK, Dunstan DW, Shaw JE, Zimmet PZ, Owen N. Associations between television viewing time and overall sitting time with the metabolic syndrome in older men and women: the Australian Diabetes, Obesity and Lifestyle study. J Am Geriatr Soc 2011;59(5):788–96.

Gardiner PA, Mishra GD, Dobson AJ. Validity and responsiveness of the FRAIL scale in a longitudinal cohort study of older Australian women. J Am Med Dir Assoc 2015;16(9):781–3.

Gennuso KP, Gangnon RE, Matthews CE, Thraen-Borowski KM, Colbert LH. Sedentary behavior, physical activity, and markers of health in older adults. Med Sci Sports Exerc 2013;45(8):1493–500.

Giaccaglia V, Nicklas B, Kritchevsky S, Mychalecky J, Messier S, Bleecker E, Pahor M. Interaction between angiotensin converting enzyme insertion/deletion genotype and exercise training on knee extensor strength in older individuals. Int J Sports Med 2008;29(1):40–4.

Gielen E, O'Neill TW, Pye SR, Adams JE, Wu FC, Laurent MR, Claessens F, Ward KA, Boonen S, Bouillon R, Vanderschueren D, Verschueren S. Endocrine determinants of incident sarcopenia in middle-aged and elderly European men. J Cachexia Sarcopenia Muscle 2015;6(3):242–52.

Glancy B, Hartnell LM, Malide D, Yu ZX, Combs CA, Connelly PS, Subramaniam S, Balaban RS. Mitochondrial reticulum for cellular energy distribution in muscle. Nature 2015;523(7562):617–20.

Gobbens RJ, van Assen MA, Luijkx KG, Wijnen-Sponselee MT, Schols JM. Determinants of frailty. J Am Med Dir Assoc 2010a;11(5):356–64.

Gobbens RJ, van Assen MA, Luijkx KG, Wijnen-Sponselee MT, Schols JM. The Tilburg Frailty Indicator: psychometric properties. J Am Med Dir Assoc 2010b;11(5):344–55.

Gómez-Cabello A, Pedrero-Chamizo R, Olivares PR, Hernández-Perera R, Rodríguez-Marroyo JA, Mata E, Aznar S, Villa JG, Espino-Torón L, Gusi N, González-Gross M, Casajús JA, Ara I, Vicente-Rodríguez G, EXERNET Study Group. Sitting time increases the overweight and obesity risk independently of walking time in elderly people from Spain. Maturitas 2012;73(4):337–43.

Gonzalez-Freire M, Rodriguez-Romo G, Santiago C, Bustamante-Ara N, Yvert T, Gomez-Gallego F, Serra Rexach JA, Ruiz JR, Lucia A. The K153R variant in the myostatin gene and sarcopenia at the end of the human lifespan. Age (Dordr) 2010;32(3):405–9.

Gordon AL, Masud T, Gladman JRF. Now that we have a definition for physical frailty, what shape should frailty medicine take? Age Ageing 2014;43(1):8–9.

Gouspillou G, Sgarioto N, Kapchinsky S, Purves-Smith F, Norris B, Pion CH, Barbat-Artigas S, Lemieux F, Taivassalo T, Morais JA, Aubertin-Leheudre M, Hepple RT. Increased sensitivity to mitochondrial permeability transition and myonuclear translocation of endonuclease G in atrophied muscle of physically active older humans. FASEB J 2014;28:1621–33.

Gray SR, Mittendorfer B. Fish oil-derived n-3 polyunsaturated fatty acids for the prevention and treatment of sarcopenia. Curr Opin Clin Nutr Metab Care 2018;21:104–9.

Gregorevic KJ, Hubbard RE, Lim WK, Katz B. The clinical frailty scale predicts functional decline and mortality when used by junior medical staff: a prospective cohort study. BMC Geriatr 2016;16:117.

Gryson C, Ratel S, Rance M, Penando S, Bonhomme C, Le Ruyet P, Duclos M, Boirie Y, Walrand S. Four-month course of soluble milk proteins interacts with exercise to improve muscle strength and delay fatigue in elderly participants. J Am Med Dir Assoc 2014;15(12) 958.e1–958.e9589.

GTEx Consortium, Laboratory, Data Analysis & Coordinating Center (LDACC)—Analysis Working Group, Statistical Methods groups—Analysis Working Group. Genetic effects on gene expression across human tissues. Nature 2017;550(7675):204–13.

Guillet C, Prod'homme M, Balage M, Gachon P, Giraudet C, Morin L, Grizard J, Boirie Y. Impaired anabolic response of muscle protein synthesis is associated with S6K1 dysregulation in elderly humans. FASEB J 2004;18(13):1586–7.

Halfon M, Phan O, Teta D. Vitamin D: a review on its effects on muscle strength, the risk of fall, and frailty. Biomed Res Int 2015;2015:953241.

Hamer M, Stamatakis E. Screen-based sedentary behavior, physical activity, and muscle strength in the English longitudinal study of ageing. PLoS One 2013;8(6)e66222.

Haran PH, Rivas DA, Fielding RA. Role and potential mechanisms of anabolic resistance in sarcopenia. J Cachexia Sarcopenia Muscle 2012;3(1):57–62.

Hearris MA, Hammond KM, Fell JM, Morton JP. Regulation of muscle glycogen metabolism during exercise: implications for endurance performance and training adaptations. Nutrients 2018;10(3):298.

Hepple RT. Mitochondrial involvement and impact in aging skeletal muscle. Front Aging Neurosci 2014;6:211.

Hepple RT, Rice CL. Innervation and neuromuscular control in ageing skeletal muscle. J Physiol 2016;594 (8):1965–78.

Hodson N, West D, Philp A, Burd NA, Moore DR. Molecular regulation of human skeletal muscle protein synthesis in response to exercise and nutrients: a compass for overcoming age-related anabolic resistance. American journal of physiology. Cell Physiol 2019;317(6):C1061–78.

Hopkinson NS, Li KW, Kehoe A, Humphries SE, Roughton M, Moxham J, Montgomery H, Polkey MI. Vitamin D receptor genotypes influence quadriceps strength in chronic obstructive pulmonary disease. Am J Clin Nutr 2008;87(2):385–90.

Houston DK, Cesari M, Ferrucci L, Cherubini A, Maggio D, Bartali B, Johnson MA, Schwartz GG, Kritchevsky SB. Association between vitamin D status and physical performance: the InCHIANTI study. J Gerontol A Biol Sci Med Sci 2007;62(4):440–6.

Hu X, Zhang L, Wang H, Hao Q, Dong B, Yang M. Malnutrition-sarcopenia syndrome predicts mortality in hospitalized older patients. Sci Rep 2017;7(1):3171.

Huang JH, Joseph AM, Ljubicic V, Iqbal S, Hood DA. Effect of age on the processing and import of matrix-destined mitochondrial proteins in skeletal muscle. J Gerontol A Biol Sci Med Sci 2010;65:138–46.

Iqbal S, Ostojic O, Singh K, Joseph AM, Hood DA. Expression of mitochondrial fission and fusion regulatory proteins in skeletal muscle during chronic use and disuse. Muscle Nerve 2013;48:963–70.

Jackson MJ, McArdle A. Age-related changes in skeletal muscle reactive oxygen species generation and adaptive responses to reactive oxygen species. J Physiol 2011;589(Pt 9):2139–45.

Jo E, Lee SR, Park BS, Kim JS. Potential mechanisms underlying the role of chronic inflammation in age-related muscle wasting. Aging Clin Exp Res 2012;24(5):412–22.

Jones G, Pilling LC, Kuo CL, Kuchel G, Ferrucci L, Melzer D. Sarcopenia and Variation in the Human Leukocyte Antigen Complex. J Gerontol: Ser A 2020;75(2):301–8.

Judson RN, Wackerhage H, Hughes A, Mavroeidi A, Barr RJ, Macdonald HM, Ratkevicius A, Reid DM, Hocking LJ. The functional ACTN3 577X variant increases the risk of falling in older females: results from two large independent cohort studies. J Gerontol A Biol Sci Med Sci 2011;66(1):130–5.

Jung HW, Yoo HJ, Park SY, Kim SW, Choi JY, Yoon SJ, Kim CH, Kim KI. The Korean version of the FRAIL scale: clinical feasibility and validity of assessing the frailty status of Korean elderly. Korean J Intern Med 2016;31(3):594–600.

Karasik D, Zillikens MC, Hsu YH, Aghdassi A, Akesson K, Amin N, Barroso I, Bennett DA, Bertram L, Bochud M, Borecki IB, Broer L, Buchman AS, Byberg L, Campbell H, Campos-Obando N, Cauley JA, Cawthon PM, Chambers JC, Chen Z, et al. Disentangling the genetics of lean mass. Am J Clin Nutr 2019;109(2):276–87.

Khanal P, He L, Stebbings G, Onambele-Pearson GL, Degens H, Williams A, Thomis M, Morse CI. Prevalence and association of single nucleotide polymorphisms with sarcopenia in older women depends on definition. Sci Rep 2020;10(1):2913.

Kim JS, Wilson JM, Lee SR. Dietary implications on mechanisms of sarcopenia: Roles of protein, amino acids and antioxidants. J Nutr Biochem 2010;21:1–13.

Kumar V, Selby A, Rankin D, Patel R, Atherton P, Hildebrandt W, Williams J, Smith K, Seynnes O, Hiscock N, Rennie MJ. Age-related differences in the dose-response relationship of muscle protein synthesis to resistance exercise in young and old men. J Physiol 2009;587(1):211–7.

Kurosaka M, Machida S. Interleukin-6-induced satellite cell proliferation is regulated by induction of the JAK2/STAT3 signalling pathway through cyclin D1 targeting. Cell Prolif 2013;46:365–73.

Langen RC, Van Der Velden JL, Schols AM, Kelders MC, Wouters EF, Janssen-Heininger YM. Tumor necrosis factor-alpha inhibits myogenic differentiation through MyoD protein destabilization. FASEB J 2004;18:227–37.

Leduc-Gaudet JP, Picard M, St-Jean Pelletier F, Sgarioto N, Auger MJ, Vallée J, Robitaille R, St-Pierre DH, Gouspillou G. Mitochondrial morphology is altered in atrophied skeletal muscle of aged mice. Oncotarget 2015;6(20):17923–37.

Lee PH. Examining non-linear associations between accelerometer-measured physical activity, sedentary behavior, and all-cause mortality using segmented cox regression. Front Physiol 2016;7:272.

Lee JH, Jun HS. Role of myokines in regulating skeletal muscle mass and function. Front Physiol 2019;10:42.

Lee WJ, Liu LK, Peng LN, Lin MH, Chen LK, ILAS Research Group. Comparisons of sarcopenia defined by IWGS and EWGSOP criteria among older people: results from the I-Lan longitudinal aging study. J Am Med Dir Assoc 2013;14(7) 528.e1-7.

León-Latre M, Moreno-Franco B, Andrés-Esteban EM, Ledesma M, Laclaustra M, Alcalde V, Peñalvo JL, Ordovás JM, Casasnovas JA, Aragon Workers' Health Study investigators. Sedentary lifestyle and its relation to cardiovascular risk factors, insulin resistance and inflammatory profile. Rev Esp Cardiol (Engl Ed) 2014;67(6):449–55.

Liu H, Bravata DM, Olkin I, Nayak S, Roberts B, Garber AM, Hoffman AR. Systematic review: the safety and efficacy of growth hormone in the healthy elderly. Ann Intern Med 2007;146(2):104–15.

Longobucco Y, Benedetti C, Tagliaferri S, Angileri VV, Adorni E, Pessina M, Zerbinati L, Cicala L, Pelà G, Giacomini V, Barbolini M, Lauretani F, Maggio MG. Proactive interception and care of Frailty and Multimorbidity in older persons: the experience of the European Innovation Partnership on Active and Healthy Ageing and the response of Parma Local Health Trust and Lab through European Projects. Acta bio-medica: Atenei Parmensis 2019;90(2):364–74.

Loprinzi PD. Accelerometer-determined sedentary and physical activity estimates among older adults with diabetes: considerations by demographic and comorbidity characteristics. J Aging Phys Act 2014;22 (3):432–40.

Malafarina V, Uriz-Otano F, Iniesta R, Gil-Guerrero L. Sarcopenia in the elderly: diagnosis, physiopathology and treatment. Maturitas 2012;71:109–14.

Manifava M, Smith M, Rotondo S, Walker S, Niewczas I, Zoncu R, Clark J, Ktistakis NT. Dynamics of mTORC1 activation in response to amino acids. Elife 2016;5:e19960.

Markle-Reid M, Browne G. Conceptualizations of frailty in relation to older adults. J Adv Nurs 2003;44 (1):58–68. https://doi.org/10.1046/j.1365-2648.2003.02767.x.

Martone AM, Bianchi L, Abete P, Bellelli G, Bo M, Cherubini A, Corica F, Di Bari M, Maggio M, Manca GM, Marzetti E, Rizzo MR, Rossi A, Volpato S, Landi F. The incidence of sarcopenia among hospitalized older patients: results from the Glisten study. J Cachexia Sarcopenia Muscle 2017;8 (6):907–14.

Marty E, Liu Y, Samuel A, Or O, Lane J. A review of sarcopenia: Enhancing awareness of an increasingly prevalent disease. Bone 2017;105:276–86.

Marzetti E, Leeuwenburgh C. Skeletal muscle apoptosis, sarcopenia and frailty at old age. Exp Gerontol 2006;41:1234–8.

Marzetti E, Lees HA, Manini TM, Buford TW, Aranda Jr. JM, Calvani R, Capuani G, Marsiske M, Lott DJ, Vandenborne K, Bernabei R, Pahor M, Leeuwenburgh C, Wohlgemuth SE. Skeletal muscle apoptotic signaling predicts thigh muscle volume and gait speed in community-dwelling older persons: an exploratory study. PLoS One 2012;7(2)e32829.

Maxwell CA, Wang J. Understanding frailty: a nurse's guide. Nurs Clin North Am 2017;52(3):349–61. https://doi.org/10.1016/j.cnur.2017.04.003.

Mayhew AJ, Amog K, Phillips S, Parise G, McNicholas PD, de Souza RJ, Thabane L, Raina P. The prevalence of sarcopenia in community-dwelling older adults, an exploration of differences between studies and within definitions: a systematic review and meta-analyses. Age Ageing 2019;48(1):48–56.

McArdle A, Pollock N, Staunton CA, Jackson MJ. Aberrant redox signalling and stress response in age-related muscle decline: Role in inter- and intra-cellular signalling. Free Radic Biol Med 2019;132:50–7.

McCauley T, Mastana SS, Folland JP. ACE I/D and ACTN3 R/X polymorphisms and muscle function and muscularity of older Caucasian men. Eur J Appl Physiol 2010;109(2):269–77.

McKee A, Morley JE. Hormones and sarcopenia. Curr Opin Endocrine Metab Res 2019;9:34–9.

McLean RR, Shardell MD, Alley DE, Cawthon PM, Fragala MS, Harris TB, Kenny AM, Peters KW, Ferrucci L, Guralnik JM, Kritchevsky SB, Kiel DP, Vassileva MT, Xue QL, Perera S, Studenski SA, Dam TT. Criteria for clinically relevant weakness and low lean mass and their longitudinal association with incident mobility impairment and mortality: the foundation for the National Institutes of Health (FNIH) sarcopenia project. J Gerontol A Biol Sci Med Sci 2014;69(5):576–83.

Mesinovic J, Zengin A, De Courten B, Ebeling PR, Scott D. Sarcopenia and type 2 diabetes mellitus: a bidirectional relationship. Diabetes Metab Syndr Obes 2019;12:1057–72.

Messier V, Rabasa-Lhoret R, Barbat-Artigas S, Elisha B, Karelis AD, Aubertin-Leheudre M. Menopause and sarcopenia: a potential role for sex hormones. Maturitas 2011;68(4):331–6.

Metzelthin SF, Daniëls R, van Rossum E, de Witte L, van den Heuvel WJ, Kempen GI. The psychometric properties of three self-report screening instruments for identifying frail older people in the community. BMC Public Health 2010;10:176.

Migliavacca E, Tay S, Patel HP, Sonntag T, Civiletto G, McFarlane C, Forrester T, Barton SJ, Leow MK, Antoun E, Charpagne A, Seng Chong Y, Descombes P, Feng L, Francis-Emmanuel P, Garratt ES, Giner MP, Green CO, Karaz S, Kothandaraman N, et al. Mitochondrial oxidative capacity and NAD + biosynthesis are reduced in human sarcopenia across ethnicities. Nat Commun 2019;10(1):5808.1.

Miljkovic N, Lim JY, Miljkovic I, Frontera WR. Aging of skeletal muscle fibers. Ann Rehabil Med 2015;39 (2):155–62.

Miller MS, Callahan DM, Toth MJ. Skeletal muscle myofilament adaptations to aging, disease, and disuse and their effects on whole muscle performance in older adult humans. Front Physiol 2014;5:369.

Mitchell WK, Williams J, Atherton P, Larvin M, Lund J, Narici M. Sarcopenia, dynapenia, and the impact of advancing age on human skeletal muscle size and strength; a quantitative review. Front Physiol 2012;3:260.

Mitniski A, Song X, Skoog I, Broe GA, Cox JL, Grunfeld E, Rockwood K. Relative fitness and frailty of elderly men and women in developed countries and their relationship with mortality. J Am Geriatr Soc 2005;53:2184–9.

Mitnitski AB, Mogilner AJ, MacKnight C, Rockwood K. The mortality rate as a function of accumulated deficits in a frailty index. Mech Ageing Dev 2002;123(11):1457–60.

Mohandas A, Reifsnyder J, Jacobs M, Fox T. Current and future directions in frailty research. Popul Health Manag 2011;14(6):277–83. https://doi.org/10.1089/pop.2010.0066.

Montesanto A, Lagani V, Martino C, Dato S, De Rango F, Berardelli M, Corsonello A, Mazzei B, Mari V, Lattanzio F, Conforti D, Passarino G. A novel, population-specific approach to define frailty. Age (Dordr) 2010;32(3):385–95.

Moore DR, Churchward-Venne TA, Witard O, Breen L, Burd NA, Tipton KD, Phillips SM. Protein ingestion to stimulate myofibrillar protein synthesis requires greater relative protein intakes in healthy older versus younger men. J Gerontol A Biol Sci Med Sci 2015;70(1):57–62.

Morley JE, Malmstrom TK, Miller DK. A simple frailty questionnaire (FRAIL) predicts outcomes in middle aged African Americans. J Nutr Health Aging 2012;16(7):601–8.

Murase T, Haramizu S, Ota N, Hase T. Suppression of the aging associated decline in physical performance by a combination of resveratrol intake and habitual exercise in senescence accelerated mice. Biogerontology 2009;10:423–34.

Naranjo JD, Dziki JL, Badylak SF. Regenerative medicine approaches for age-related muscle loss and sarcopenia: a mini-review. Gerontology 2017;63(6):580–9.

Narici MV, Maganaris CN, Reeves ND, Capodaglio P. Effect of aging on human muscle architecture. J Appl Physiol 2003;95:2229–34.

Nilwik R, Snijders T, Leenders M, Groen BB, van Kranenburg J, Verdijk LB, van Loon LJ. The decline in skeletal muscle mass with aging is mainly attributed to a reduction in type II muscle fiber size. Exp Gerontol 2013;48(5):492–8.

Norman K, Stobäus N, Kulka K, Schulzke J. Effect of inflammation on handgrip strength in the non-critically ill is independent from age, gender and body composition. Eur J Clin Nutr 2014;68(2):155–8.

O'Leary MF, Vainshtein A, Iqbal S, Ostojic O, Hood DA. Adaptive plasticity of autophagic proteins to denervation in aging skeletal muscle. Am J Physiol Cell Physiol 2013;304:C422–30.

Onder G, Capoluongo E, Danese P, Settanni S, Russo A, Concolino P, Bernabei R, Landi F. Vitamin D receptor polymorphisms and falls among older adults living in the community: results from the ilSIRENTE study. J Bone Miner Res: Off J Am Soc Bone Miner Res 2008;23(7):1031–6.

Pagotto V, Silveira EA. Methods, diagnostic criteria, cutoff points, and prevalence of sarcopenia among older people. Scientific World Journal 2014;2014:231312.

Panza F, Lozupone M, Solfrizzi V, Sardone R, Dibello V, Di Lena L, D'Urso F, Stallone R, Petruzzi M, Giannelli G, Quaranta N, Bellomo A, Greco A, Daniele A, Seripa D, Logroscino G. Different cognitive frailty models and health- and cognitive-related outcomes in older age: from epidemiology to prevention. J Alzheimers Dis 2018;62(3):993–1012. https://doi.org/10.3233/JAD-170963.

Papadopoulou SK. Sarcopenia: A Contemporary Health Problem among Older Adult Populations. Nutrients 2020;12(5):1293.

Park SW, Goodpaster BH, Strotmeyer ES, Kuller LH, Broudeau R, Kammerer C, de Rekeneire N, Harris TB, Schwartz AV, Tylavsky FA, Cho YW, Newman AB, Health, Aging, and Body Composition Study. Accelerated loss of skeletal muscle strength in older adults with type 2 diabetes: the health, aging, and body composition study. Diabetes Care 2007;30(6):1507–12.

Park SW, Goodpaster BH, Lee JS, Kuller LH, Boudreau R, de Rekeneire N, Harris TB, Kritchevsky S, Tylavsky FA, Nevitt M, Cho YW, Newman AB, Health, Aging, and Body Composition Study. Excessive loss of skeletal muscle mass in older adults with type 2 diabetes. Diabetes Care 2009;32 (11):1993–7.

Parsons TJ, Sartini C, Welsh P, Sattar N, Ash S, Lennon LT, Wannamethee SG, Lee IM, Whincup PH, Jefferis BJ. Physical activity, sedentary behavior, and inflammatory and hemostatic markers in men. Med Sci Sports Exerc 2017;49(3):459–65.

Payette H, Roubenoff R, Jacques PF, Dinarello CA, Wilson PW, Abad LW, Harris T. Insulin-like growth factor-1 and interleukin 6 predict sarcopenia in very old community-living men and women: the Framingham Heart Study. J Am Geriatr Soc 2003;51(9):1237–43.

Peters LL, Boter H, Buskens E, Slaets JP. Measurement properties of the Groningen Frailty Indicator in home-dwelling and institutionalized elderly people. J Am Med Dir Assoc 2012;13(6):546–51.

Peterson SJ, Mozer M. Differentiating sarcopenia and cachexia among patients with cancer. Nutr Clin Pract 2017;32(1):30–9.

Peterson CM, Johannsen DL, Ravussin E. Skeletal muscle mitochondria and aging: A review. J Aging Res 2012;2012:19482.

Phu S, Boersma D, Duque G. Exercise and sarcopenia. J Clin Densitom 2015;18(4):488–92.

Picca A, Calvani R, Bossola M, Allocca E, Menghi A, Pesce V, Lezza AMS, Bernabei R, Landi F, Marzetti E. Update on mitochondria and muscle aging: all wrong roads lead to sarcopenia. Biol Chem 2018;399 (5):421–36.

Qaisar R, Bhaskaran S, Premkumar P, Ranjit R, Natarajan KS, Ahn B, Riddle K, Claflin DR, Richardson A, Brooks SV, Van Remmen H. Oxidative stress-induced dysregulation of excitation-contraction coupling contributes to muscle weakness. J Cachexia Sarcopenia Muscle 2018;9(5):1003–17.

Rasmussen BB, Fujita S, Wolfe RR, Mittendorfer B, Roy M, Rowe VL, Volpi E. Insulin resistance of muscle protein metabolism in aging. FASEB J 2006;20(6):768–9.

Remelli F, Vitali A, Zurlo A, Volpato S. Vitamin D deficiency and sarcopenia in older persons. Nutrients 2019;11(12):2861.

Rezuş E, Burlui A, Cardoneanu A, Rezuş C, Codreanu C, Pârvu M, Rusu Zota G, Tamba BI. Inactivity and skeletal muscle metabolism: a vicious cycle in old age. Int J Mol Sci 2020;21(2):592.

Rockwood K, Mitnitski A. Frailty in relation to the accumulation of deficits. J Gerontol A Biol Sci Med Sci 2007;62A:722–7.

Rockwood K, Mitnitski A. Frailty defined by deficit accumulation and geriatric medicine defined by frailty. Clin Geriatr Med 2011;27(1):17–26.

Rockwood K, Song X, MacKnight C, Bergman H, Hogan DB, McDowell I, Mitnitski A. A global clinical measure of fitness and frailty in elderly people. CMAJ 2005;173(5):489–95.

Rolfson DB, Majumdar SR, Tsuyuki RT, Tahir A, Rockwood K. Validity and reliability of the Edmonton Frail Scale. Age Ageing 2006;35(5):526–9.

Rosas-Carrasco O, Cruz-Arenas E, Parra-Rodríguez L, García-González AI, Contreras-González LH, Szlejf C. Cross-cultural adaptation and validation of the FRAIL scale to assess frailty in Mexican adults. J Am Med Dir Assoc 2016;17(12):1094–8. https://doi.org/10.1016/j.jamda.2016.07.008.

Rosenberg I. Summary comments: epidemiological and methodological problems in determining nutritional status of older persons. Am J Clin Nutr 1989;50:1231–3.

Roth DE, Soto G, Arenas F, Bautista CT, Ortiz J, Rodriguez R, Cabrera L, Gilman RH. Association between vitamin D receptor gene polymorphisms and response to treatment of pulmonary tuberculosis. J Infect Dis 2004;190(5):920–7.

Sahin K, Pala R, Tuzcu M, Ozdemir O, Orhan C, Sahin N, Juturu V. Curcumin prevents muscle damage by regulating NF-kappaB and Nrf2 pathways and improves performance: An in vivo model. J Inflamm Res 2016;9:147–54.

Santos-Eggimann B, Cuénoud P, Spagnoli J, Junod J. Prevalence of frailty in middle-aged and older community-dwelling Europeans living in 10 countries. J Gerontol A Biol Sci Med Sci 2009;64 (6):675–81.

Sayer AA, Syddall H, O'Dell SD, Chen XH, Briggs PJ, Briggs R, Day IN, Cooper C. Polymorphism of the IGF2 gene, birth weight and grip strength in adult men. Age Ageing 2002;31(6):468–70.

Schaap LA, Pluijm SM, Deeg DJ, Harris TB, Kritchevsky SB, Newman AB, Colbert LH, Pahor M, Rubin SM, Tylavsky FA, Visser M, Health ABC Study. Higher inflammatory marker levels in older persons: associations with 5-year change in muscle mass and muscle strength. J Gerontol A Biol Sci Med Sci 2009;64(11):1183–9.

Sezgin D, O'Donovan M, Cornally N, Liew A, O'Caoimh R. Defining frailty for healthcare practice and research: A qualitative systematic review with thematic analysis. Int J Nurs Stud 2019;92:16–26.

Shafiee G, Keshtkar A, Soltani A, Ahadi Z, Larijani B, Heshmat R. Prevalence of sarcopenia in the world: a systematic review and meta-analysis of general population studies. J Diabetes Metab Disord 2017;16:21.

Shen Y, Chen J, Chen X, Hou L, Lin X, Yang M. Prevalence and associated factors of sarcopenia in nursing home residents: a systematic review and meta-analysis. J Am Med Dir Assoc 2019;20(1):5–13.

Singh AN, Gasman B. Disentangling the genetics of sarcopenia: prioritization of NUDT3 and KLF5 as genes for lean mass & HLA-DQB1-AS1 for hand grip strength with the associated enhancing SNPs & a scoring system. BMC Med Genet 2020;21(1):40.

Snijders T, Nederveen JP, McKay BR, Joanisse S, Verdijk LB, van Loon LJ, Parise G. Satellite cells in human skeletal muscle plasticity. Front Physiol 2015;6:283.

Sousa AC, Zunzunegui MV, Li A, Phillips SP, Guralnik JM, Guerra RO. Association between C-reactive protein and physical performance in older populations: results from the International Mobility in Aging Study (IMIAS). Age Ageing 2016;45(2):274–80.

Sousa-Victor P, Gutarra S, García-Prat L, Rodriguez-Ubreva J, Ortet L, Ruiz-Bonilla V, Jardí M, Ballestar E, González S, Serrano AL, Perdiguero E, Muñoz-Cánoves P. Geriatric muscle stem cells switch reversible quiescence into senescence. Nature 2014;506:316–21.

St-Jean-Pelletier F, Pion CH, Leduc-Gaudet JP, Sgarioto N, Zovilé I, Barbat-Artigas S, Reynaud O, Alkaterji F, Lemieux FC, Grenon A, Gaudreau P, Hepple RT, Chevalier S, Belanger M, Morais JA, Aubertin-Leheudre M, Gouspillou G. The impact of ageing, physical activity, and pre-frailty on skeletal muscle phenotype, mitochondrial content, and intramyocellular lipids in men. J Cachexia Sarcopenia Muscle 2017;8(2):213–28.

Talbot J, Maves L. Skeletal muscle fiber type: using insights from muscle developmental biology to dissect targets for susceptibility and resistance to muscle disease. Wiley Interdiscip Rev Dev Biol 2016;5 (4):518–34.

Tay L, Ding YY, Leung BP, Ismail NH, Yeo A, Yew S, Tay KS, Tan CH, Chong MS. Sex-specific differences in risk factors for sarcopenia amongst community-dwelling older adults. Age (Dordr) 2015;37 (6):121.

Tieland M, Dirks ML, van der Zwaluw N, Verdijk LB, van de Rest O, de Groot LC, van Loon LJ. Protein supplementation increases muscle mass gain during prolonged resistance-type exercise training in frail elderly people: a randomized, double-blind, placebo-controlled trial. J Am Med Dir Assoc 2012;13 (8):713–9.

Tikkanen E, Gustafsson S, Amar D, Shcherbina A, Waggott D, Ashley EA, Ingelsson E. Biological insights into muscular strength: genetic findings in the UK Biobank. Sci Rep 2018;8(1):6451.

Tournadre A, Vial G, Capel F, Soubrier M, Boirie Y. Sarcopenia. Joint Bone Spine 2019;86(3):309–14.

Valenti G, Denti L, Maggio M, Ceda G, Volpato S, Bandinelli S, Ceresini G, Cappola A, Guralnik JM, Ferrucci L. Effect of DHEAS on skeletal muscle over the life span: the InCHIANTI study. J Gerontol A Biol Sci Med Sci 2004;59(5):466–72.

Van Pottelbergh I, Goemaere S, Nuytinck L, De Paepe A, Kaufman JM. Association of the type I collagen alpha1 Sp1 polymorphism, bone density and upper limb muscle strength in community-dwelling elderly men. Osteoporosis Int 2001;12(10):895–901.

Vandevyver C, Wylin T, Cassiman JJ, Raus J, Geusens P. Influence of the vitamin D receptor gene alleles on bone mineral density in postmenopausal and osteoporotic women. J Bone Miner Res 1997;12(2):241–7.

Verdijk LB, Snijders T, Drost M, Delhaas T, Kadi F, van Loon LJ. Satellite cells in human skeletal muscle; from birth to old age. Age (Dordr) 2014;36:545–7.

Vigano A, Trutschnigg B, Kilgour RD, Hamel N, Hornby L, Lucar E, Foulkes W, Tremblay ML, Morais JA. Relationship between angiotensin-converting enzyme gene polymorphism and body composition, functional performance, and blood biomarkers in advanced cancer patients. Clin Cancer Res 2009;15(7):2442–7.

Visser M, Kritchevsky SB, Goodpaster BH, Newman AB, Nevitt M, Stamm E, Harris TB. Leg muscle mass and composition in relation to lower extremity performance in men and women aged 70 to 79: the health, aging and body composition study. J Am Geriatr Soc 2002;50(5):897–904.

Visser M, Deeg DJ, Lips P, Longitudinal Aging Study Amsterdam. Low vitamin D and high parathyroid hormone levels as determinants of loss of muscle strength and muscle mass (sarcopenia): the Longitudinal Aging Study Amsterdam. J Clin Endocrinol Metab 2003;88(12):5766–72.

Vitale G, Cesari M, Mari D. Aging of the endocrine system and its potential impact on sarcopenia. Eur J Intern Med 2016;35:10–5.

Volpato S, Bianchi L, Cherubini A, Landi F, Maggio M, Savino E, Bandinelli S, Ceda GP, Guralnik JM, Zuliani G, Ferrucci L. Prevalence and clinical correlates of sarcopenia in community-dwelling older people: application of the EWGSOP definition and diagnostic algorithm. J Gerontol A Biol Sci Med Sci 2014;69(4):438–46.

Wagatsuma A, Kotake N, Mabuchi K, Yamada S. Expression of nuclear-encoded genes involved in mitochondrial biogenesis and dynamics in experimentally denervated muscle. J Physiol Biochem 2011;67:359–70.

Wakabayashi H, Sakuma K. Comprehensive approach to sarcopenia treatment. Curr Clin Pharmacol 2014;9 (2):171–80.

Wall BT, Snijders T, Senden JMG, Ottenbros CLP, Gijsen AP, Verdijk LB, van Loon LJC. Disuse impairs the muscle protein synthetic response to protein ingestion in healthy men. J Clin Endocrinol Metab 2013;98:4872–81.

Wall BT, Gorissen SH, Pennings B, Koopman R, Groen BBL, Verdijk LB, van Loon LJC. Aging is accompanied by a blunted muscle protein synthetic response to protein ingestion. PLoS One 2015;10: e0140903.

Walsh S, Metter EJ, Ferrucci L, Roth SM. Activin-type II receptor B (ACVR2B) and follistatin haplotype associations with muscle mass and strength in humans. J Appl Physiol 2007;102(6):2142–8.

Walsh S, Liu D, Metter EJ, Ferrucci L, Roth SM. ACTN3 genotype is associated with muscle phenotypes in women across the adult age span. J Appl Physiol 2008;105(5):1486–91.

Wang Y, Michikawa Y, Mallidis C, Bai Y, Woodhouse L, Yarasheski KE, Miller CA, Askanas V, Engel WK, Bhasin S, Attardi G. Muscle-specific mutations accumulate with aging in critical human mtDNA control sites for replication. Proc Natl Acad Sci U S A 2001;98:4022–7.

Waters DL, Qualls CR, Dorin RI, Veldhuis JD, Baumgartner RN. Altered growth hormone, cortisol, and leptin secretion in healthy elderly persons with sarcopenia and mixed body composition phenotypes. J Gerontol A Biol Sci Med Sci 2008;63(5):536–41.

Westbury LD, Fuggle NR, Syddall HE, Duggal NA, Shaw SC, Maslin K, Dennison EM, Lord JM, Cooper C. Relationships between markers of inflammation and muscle mass, strength and function: findings from the Hertfordshire Cohort study. Calcif Tissue Int 2018;102(3):287–95.

Westra HJ, Peters MJ, Esko T, Yaghootkar H, Schurmann C, Kettunen J, Christiansen MW, Fairfax BP, Schramm K, Powell JE, Zhernakova A, Zhernakova DV, Veldink JH, Van den Berg LH, Karjalainen J, Withoff S, Uitterlinden AG, Hofman A, Rivadeneira F, Hoen P, et al. Systematic identification of trans eQTLs as putative drivers of known disease associations. Nat Genet 2013;45 (10):1238–43.

White TA, LeBrasseur NK. Myostatin and sarcopenia: opportunities and challenges—a mini-review. Gerontology 2014;60:289–93.

Wicherts IS, van Schoor NM, Boeke AJ, Visser M, Deeg DJ, Smit J, Knol DL, Lips P. Vitamin D status predicts physical performance and its decline in older persons. J Clin Endocrinol Metab 2007;92 (6):2058–65.

Widagdo IS, Pratt N, Russell M, Roughead EE. Construct validity of four frailty measures in an older Australian population: a rasch analysis. J Frailty Aging 2016;5(2):78–81.

Wieland D, Hirth V. Comprehensive geriatric assessment. Cancer Control 2003;10(6):454–62.

Willems SM, Wright DJ, Day FR, Trajanoska K, Joshi PK, Morris JA, Matteini AM, Garton FC, Grarup N, Oskolkov N, Thalamuthu A, Mangino M, Liu J, Demirkan A, Lek M, Xu L, Wang G, Oldmeadow C, Gaulton KJ, Lotta LA, et al. Large-scale GWAS identifies multiple loci for hand grip strength providing biological insights into muscular fitness. Nat Commun 2017;8:16015.

Wolf MT, Dearth CL, Sonnenberg SB, Loboa EG, Badylak SF. Naturally derived and synthetic scaffolds for skeletal muscle reconstruction. Adv Drug Deliv Rev 2015;84:208–21.

Woo J, Yu R, Wong M, Yeung F, Wong M, Lum C. Frailty screening in the community using the FRAIL scale. J Am Med Dir Assoc 2015;16(5):412–9.

Wu C, Geldhof GJ, Xue QL, Kim DH, Newman AB, Odden MC. Development, construct validity, and predictive validity of a continuous frailty scale: results from 2 large US cohorts. Am J Epidemiol 2018;187 (8):1752–62.

Yoon MS. mTOR as a key regulator in maintaining skeletal muscle mass. Front Physiol 2017;8:788.

Yoshihara A, Tobina T, Yamaga T, Ayabe M, Yoshitake Y, Kimura Y, Shimada M, Nishimuta M, Nakagawa N, Ohashi M, Hanada N, Tanaka H, Kiyonaga A, Miyazaki H. Physical function is weakly associated with angiotensin-converting enzyme gene I/D polymorphism in elderly Japanese subjects. Gerontology 2009;55(4):387–92.

Zempo H, Tanabe K, Murakami H, Iemitsu M, Maeda S, Kuno S. ACTN3 polymorphism affects thigh muscle area. Int J Sports Med 2010;31(2):138–42.

Zillikens MC, Demissie S, Hsu YH, Yerges-Armstrong LM, Chou WC, Stolk L, Livshits G, Broer L, Johnson T, Koller DL, Kutalik Z, Luan J, Malkin I, Ried JS, Smith AV, Thorleifsson G, Vandenput L, Hua Zhao J, Zhang W, Aghdassi A, et al. Large meta-analysis of genome-wide association studies identifies five loci for lean body mass. Nat Commun 2017;8(1):80.

Zurlo F, Larson K, Bogardus C, Ravussin E. Skeletal muscle metabolism is a major determinant of resting energy expenditure. J Clin Invest 1990;86(5):1423–7.

CHAPTER 11

Hormones in aging

Andrea Sansone[a] and Francesco Romanelli[b]

[a]Chair of Endocrinology and Medical Sexology (ENDOSEX), Department of Systems Medicine, University of Rome Tor Vergata, Roma, Italy
[b]Department of Experimental Medicine, Section of Medical Pathophysiology, Food Science and Endocrinology, Sapienza University of Rome, Rome, Italy

11.1 Introduction

Aging is a natural phenomenon, resulting in several changes in human physiology. There is solid evidence supporting the notion that all organs and systems are affected by aging, although at a different pace; neurological function, physical strength, and bone density are a clear example of this. Endocrine function is similarly affected by aging, resulting in completely different hormonal levels at different stages of life; while changes occurring through puberty are more clinically evident, endocrine function is similarly changed in old age. Some features are different between males and females, suggesting a gender-based approach to diagnosis and treatment of aging-related endocrine dysfunction. Failing to recognize any pathological endocrine changes might result in worse quality of life and general health; at the same time, providing unnecessary treatments in the hopes of finding a "rejuvenation elixir" might have undesirable, and possibly harmful, side effects. Up-to-date and evidence-based knowledge of endocrine aging-related pathophysiology is therefore necessary to provide adequate patient care and to support healthy aging.

11.2 Endocrine physiology: The role of the pituitary gland and hypothalamus

The hypothalamus-pituitary complex is the master regulator of endocrine function in the body. This function is expressed through a finely regulated system involving negative feedback mechanisms: as a rough example, the pituitary releases TSH (thyroid stimulating hormone) upon stimulation by the hypothalamic TRH (thyrotropin releasing hormone), and secretion of TSH is suppressed by levels of thyroid hormones through different feedback loops (Lechan and Fekete, 2004; Prummel et al., 2004). Similar mechanisms are well described for the hypothalamic-pituitary-adrenal axis (HPA, featuring hypothalamic corticotropin-releasing hormone, CRH and pituitary adrenocorticotropic hormone, ACTH) and the hypothalamic-pituitary-gonadal axis (HPG, featuring hypothalamic gonadotropin-releasing hormone, GnRH and pituitary follicle-stimulating hormone, FSH and luteinizing hormone, LH).

Human Aging
https://doi.org/10.1016/B978-0-12-822569-1.00007-X

As the master regulator of endocrine function, any disturbance affecting the hypothalamus-pituitary complex is likely resulting in impaired endocrine function. In these regards, aging is a "disruptor" of pituitary physiology; however, whether this is strictly depending on the effects of aging, rather than on the interaction of aging with endocrine disruptors (Pamphlett et al., 2019), obesity (Cavadas et al., 2016; Rudman et al., 1981), or calorie intake (Cavadas et al., 2016), is an open question.

11.3 Gonadal function in aging

11.3.1 Menopause

Age-related hypogonadism is prominently featured in any discussion on the topic of aging-related endocrine disorders. Indeed, there is no example which is as clear-cut as menopause in order to describe to what extent aging might affect endocrine function. Oocyte number and quality decline progressively throughout adult life, with severely decreased rates of fertility from the fourth decade of life (Broekmans et al., 2009). As the number of follicles declines, concentrations of the two hormones secreted by granulosa cells, namely anti-Müllerian hormone (AMH) and inhibin B, follow a similar trend (Broer et al., 2011; Hale et al., 2007). It should therefore be unsurprising that AMH levels are considered a good readout for ovarian reserve (Steiner et al., 2017). Depletion of follicular reserve is not immediately noticeable: in fact, the onset of menopausal transition usually occurs around 46 years of age (van den Beld et al., 2018), and then progresses through anovulatory bleeding and oligomenorrhea, ultimately ending with menopause around age 51 (range 40–60) years of age (van den Beld et al., 2018). The decline in follicle count is mirrored by additional endocrine manifestations involving sexual hormones: as estrogens and inhibins get progressively lower, LH and FSH pulses amplitude increases, whereas the usual preovulatory peaks are gradually lost. As estradiol is mainly produced by the follicles (Hall, 2015), the decreasing amount of developing follicles leads to reduced levels of estrogens. During menopausal transition, FSH gradually increases, despite persistence of menstrual cycles; this is mainly due to the declining concentrations of estradiol and inhibin B. In fact, estradiol decreases GnRH secretion via negative feedback and decreases the gonadotrope response to GnRH at the pituitary level, whereas inhibin B similarly inhibits FSH secretion from the pituitary. When estradiol and inhibin B levels decrease, GnRH pulses occur more frequently and lead to higher serum levels of FSH (and LH, to a lesser extent) (Gill et al., 2002). Following menopause, levels of gonadotropin decline, despite being still higher than premenopause, being possibly sustained by conversion of other hormones to estradiol—mainly testosterone, secreted by the adrenal and by theca cells, and androstenedione/estrone.

It should be clear that menopause is not a "on-off" situation, but yet a progressing condition ultimately resulting in complete ovarian failure; a staging system consisting of seven different phases, from the early reproductive phase to menopausal transition to postmenopause, has been proposed (Soules et al., 2001).

11.3.2 "Andropause," male late-onset hypogonadism

Compared to the female side, the effects of aging on the male gonads are less dramatic (Basaria, 2013). Spermatogenesis is maintained throughout all life of a man, although some minor changes occur in sperm quality (Sartorius and Nieschlag, 2009). However, the endocrine function of the testis is progressively impaired with aging, resulting in a form of LOH which has often been inappropriately called "andropause." LOH is a syndrome which is defined by biochemical evidence of low serum testosterone associated with several signs and symptoms: several guidelines on the topic have been published in the last decades, with varying thresholds used for definition of "low serum testosterone" (Giagulli et al., 2020; Kwong et al., 2019). Signs and symptoms of hypogonadism include sexual dysfunction (Sansone et al., 2017), loss of libido and muscle strength, as well as reduced mineral bone density and depressed mood. Different thresholds have been proposed by International Societies investigating androgen deficiency: while no definite consensus has been reached so far, most guidelines, including those released in 2020 by the European Academy of Andrology (Corona et al., 2020a) and by the European Association of Urology (Salonia et al., 2020), as well as by the International Society for the Study of Aging Male in 2015 (Lunenfeld et al., 2015), agree on using 12 nmoL/L (350 ng/dL) as a threshold for suspecting hypogonadism. Clinical investigation is of utmost importance in the management of LOH; indeed, total testosterone should never be considered as a motive for treatment initiation in the absence of symptoms of LOH, and similarly patients with serum testosterone levels above 12 nmol/L who still complain about symptoms and signs suggestive of testosterone deficiency should be investigated more in detail, possibly by calculating free and bioavailable testosterone.

Several mechanisms lead to the development of LOH, Leydig cell numbers decrease with aging and response to LH is similarly impaired, and the small increase in pituitary LH release is not adequate to compensate. Additionally, possibly also due to obesity, Sex Hormone Binding Globulin levels increase with aging, therefore further reducing the amount of free and bioavailable testosterone (Basaria, 2013). Obesity also has negative bearing on androgen status; as testosterone is aromatized to estradiol in the adipose tissue, obese subjects are prone to have higher serum estradiol, which acts on the hypothalamic-pituitary-gonadal axis by exerting negative feedback.

Treatment of androgen deficiency should focus on the removal of the initiating factor, if identifiable: a clear-cut example comes from hyperprolactinemia, a condition known to downregulate the hypothalamic-pituitary-gonadal axis and frequently associated with use of selected medication, severe hypothyroidism, or pituitary neoplasms (prolactinomas) (Casanueva et al., 2006). In such patients, treatment should be focused on treating hyperprolactinemia rather than on artificially increasing testosterone levels (Casanueva et al., 2006). Similarly, changing lifestyles, such as smoking cessation (Corona et al., 2020b), being physically active (Di Luigi et al., 2012), and following an adequate diet (Corona et al., 2013) might improve testosterone levels, hence sparing the need to start administration of exogenous testosterone. In all cases, increasing serum testosterone levels is associated with improvements in regards to signs and symptoms of androgen deficiency; testosterone replacement therapy (TRT) could lead to faster improvement (Saad et al., 2011), but should be prescribed with caution, and only following adequate investigation (Sansone et al., 2017). Additionally, it should be considered that while effects on mood, muscle strength, and sexual function begin shortly after treatment initiation, improvements on bone density, insulin metabolism, and erectile function might take several months to be clinically noticeable (Lunenfeld et al., 2015). Therefore counseling is necessary in order to adequately manage patients' expectations in regards to the efficacy of treatment. Likewise, specialist consultation is necessary to reduce the potential side effects associated with TRT: due to the potential effects on prostate tumor growth and on hemopoiesis, TRT is contraindicated in patients with history of prostate cancer and/or with high hematocrit (Sansone et al., 2017). More recently, a small number of studies have reported a significant risk for cardiovascular health in subjects undergoing TRT; however, to the present date, a growing body of evidence is actually supporting the notion that treatment of hypogonadism would have beneficial, rather than harmful, effects on cardiovascular and metabolic health (Corona et al., 2019; Rastrelli et al., 2018).

11.4 Growth hormone and aging

Growth hormone (GH) has a variety of pleiotropic effects, among which bone growth is only the most known (Prader, 1967); in fact, glucose and lipid metabolism, as well as protein synthesis and calcium retention, are all influenced by GH. GH deficiency has well-known effects when occurring before the closure of the epiphyseal plate, although its effects in the adult are more nuanced and only partially investigated. GH action is mediated by the insulin-like growth factor 1 (IGF-1), a protein secreted by the liver upon GH stimulation.

Somatopause, i.e., the gradual decline in the secretion of GH, is a common condition in aging, and a common trait even for other species (Di Somma et al., 2011). GH secretion decreases significantly since the third decade of life, with a ~15% decrease for every decade. During aging, both IGF-1 and GH concentrations decline, suggesting that the somatopause acts directly on the pituitary, rather than through increased peripheral resistance.

Symptoms of "normal" aging closely mirror those of adult growth hormone deficiency (GHD), although to a lesser extent: indeed, features of GHD include lower muscle and bone mass, increased visceral fat, and reduced cardiac output. However, the physiologic decline in GH levels associated with aging should not be mistaken for adult GHD, and treatment should not be suggested lightly. Decline of GH levels in aging is possibly acting as a beneficial, rather than harmful, condition: evidence from studies in humans and animal models has suggested that GH deficiency could be associated with longer lifespan, as well as better preserved cognitive function in old people (Frater et al., 2018; Junnila et al., 2013; Nashiro et al., 2017). It should also be considered that administration of GH has several known side effects, such as arthralgia and hyperglycemia, which might be particularly troublesome in old patients, who are by definition more at risk of developing adverse events associated with GH treatment (Molitch et al., 2011); additionally, some concerns over the possibility that GH might stimulate tumor cell growth have been raised, although data in these regards mostly comes from animal models (Clayton et al., 2011). Therefore the efficacy and safety of GH administration for somatopause are still questionable.

11.5 Adrenal function in aging

11.5.1 Glucocorticoids

HPA is finely regulated in order to provide the necessary adjustments to stressors, whether endogenous or exogenous, hence maintaining homeostasis. Hypothalamic secretion of CRH acts on the pituitary by stimulating release of ACTH. This fine regulation is preserved in aging, although some changes occur in the feedback mechanisms in both directions, with both decreased ACTH responsivity to cortisol and decreased cortisol release upon ACTH stimulation. However, no clinically relevant decline in cortisol levels occurs during aging (Parker et al., 2000). Clinical interpretation of laboratory results is extremely relevant in the context of adrenal aging, as cortisol levels—already wildly fluctuating in healthy, young subjects—show larger variations in old subjects (Bergendahl et al., 2000), as well as higher average 24-h concentrations (Van Cauter et al., 1996), possibly associated with altered circadian rhythm (Sherman et al., 1985). Alterations in circadian rhythms are in turn associated with increased risks for the development of cognitive impairment (Ennis et al., 2017) and increased body fat (Purnell et al., 2004).

11.5.2 Adrenal androgens

Adrenal androgens are similarly affected by aging, with dehydroepiandrosterone (DHEA) and its sulfate derivative (DHEAS) being the most prevalently secreted hormone. DHEAS markedly declines in aging, with concentrations in old people which are up to 20%–30% of peak values in younger subjects (Ravaglia et al., 1996). Several roles have been attributed to DHEA and DHEAS, including immune-modulating, cardioprotective, and antiobesity properties (Ohlsson et al., 2015; Yen and Laughlin, 1998); it is largely assumed that most of these effects are actually indirect, being mediated by the conversion to estradiol and testosterone, although some binding sites specific for DHEAS have been identified.

Lower serum DHEAS levels have been associated with higher risk of cardiovascular events and mortality in old people (Ohlsson et al., 2015; Yen and Laughlin, 1998). However, treatment with DHEA, while resulting in increased concentrations of DHEA, DHEAS, and sex steroids, has not produced any clinically significant effect (Elraiyah et al., 2014). Therefore no clear indication in regards to the actual beneficial effects of DHEA administration is currently available.

11.5.3 Mineralocorticoids and aging

No significant evidence has been collected so far in regards to changes in mineralocorticoids levels during aging. This can partially be explained by the different regulation mechanisms: while glucocorticoids and androgens are secreted upon ACTH stimulation, regulation of mineralocorticoids is independent of the HPA axis and only affected by the renin-angiotensin system. However, new evidence is suggesting that changes in the mineralocorticoid receptor might occur during aging, possibly affecting vascular function and providing a target for research in the context of cardiovascular aging (Gorini et al., 2019).

11.6 Thyroid function in aging

Aging has an effect on the thyroid gland as well, with structural and functional changes being found in old patients. Histology specimens from older patients show a decreased number of follicles as well as lower thyroglobulin contents (Gesing et al., 2012; Tabatabaie and Surks, 2013); this could possibly suggest why thyroid disease, although common in the general population with seemingly increasing prevalence (Duntas, 2018), could have higher incidence in old people, with hypothyroidism being the most prevalent condition. However, evidence in regards to the changes occurring in thyroid function during aging is somewhat controversial; while there is reason to believe that direct age-related effects exist, there is no doubt that indirect effects on the hypothalamic-pituitary-thyroid axis might result in impaired thyroid status. An age-related decrease in serum TSH levels has been found in healthy subjects, possibly due to blunted TSH secretion by the pituitary (Cuttelod et al., 1974; Van Coevorden et al., 1989), independently of presence of autoantibodies; at the same time, TSH concentrations might actually be reduced in older, apparently euthyroid patients, although this case is suggestive of endogenous or exogenous thyrotoxicosis (Chahal and Drake, 2007. Additionally, the effects of TSH of FT4 to FT3 conversion are largely believed to be affected by aging; indeed, as TSH levels increase, the FT3-to-FT4 ratio is similarly increasing in subjects up to 40 years of age, but this enhancement does not occur in older subjects, hence suggesting a decrease in thyroxine to triiodothyronine conversion (Strich et al., 2016). It's also worth noticing that several factors, such as low-grade inflammation and concomitant disease (Bremner et al., 2012), might influence hormone metabolism, resulting in different FT4-to-FT3 ratios.

Prevalence of autoimmune thyroid disease and autonomous nodules is increasing with age (Surks et al., 2004), possibly increasing the incidence of both hypo- and hyperthyroidism in aging subjects. Whether these changes actually prevent or contribute to the pathogenesis of other age-related conditions, such as atherosclerosis, osteoporosis, or coronary heart disease, is still an unresolved question (Chahal and Drake, 2007; van den Beld et al., 2018). Hyperthyroidism, even in its subclinical form, is associated with an increased risk of atrial fibrillation, osteoporosis fractures, and cognitive decline (Rieben et al., 2016; Segna et al., 2018; Selmer et al., 2014); on the other hand, subclinical hypothyroidism is associated with lower mortality (Atzmon et al., 2009; Selmer et al., 2012), and slightly hampered thyroid function has been linked to reduced frailty and better physical function (Virgini et al., 2015). Therefore there seems to be quite good reason for delaying levothyroxine treatment to old patients with moderately elevated TSH levels.

Prevalence of thyroid nodules seems to be linearly associated with aging (Kwong et al., 2015). Reports from almost 30 years ago already suggested that overall more than 50% of subjects above 65 years of age had nodules in their thyroid: to the present date, an even higher prevalence in subjects over 60 years old, almost 80% in women and 74% in men, has been demonstrated (Guth et al., 2009). Some bias in these regards might be suspected due to the ever-increasing use of imaging techniques, such as magnetic resonance and computed tomography, which might lead to increased diagnosis of otherwise unnoticeable thyroid nodules (i.e., without any clinical bearing). Independently of the patients' age, or—rather—*considering* the patients' age, investigation of thyroid nodules is of paramount importance. As previously described, thyrotoxicosis resulting from autonomous ("hot") nodules is associated with a significant decline in quality of life and overall health. In regards to the risk of malignancy, while it is true that the overall prevalence of thyroid nodules increases with age, it would seem that old patients have lower likelihood of developing thyroid cancer compared to younger subjects. Age, however, is included in most staging systems: a recent report suggests that worse prognosis is not related to age per se, but rather to the accumulated mutations and changes in immune competence which are ultimately mirrored by changes in the endocrine system (Ylli et al., 2018). In these regards, higher TSH levels—just as occurring in subclinical hypothyroidism—might stimulate tumor cell growth, and lead to worse clinical outcomes due to several genetic mutations accrued over many years (Haymart, 2009).

11.7 Conclusions and future perspective

Endocrine function plays a pivotal role in many phases of human life and development, from intrauterine development to the transition through puberty and adulthood. Evidence in regards to the effects of endocrine changes in healthy and unhealthy aging, on the other hand, is still confusing. While some benefits have been clearly proven for treatment of endocrine deficiencies, such as testosterone replacement therapy in male hypogonadism, there is still no clear evidence concerning the benefits of supplementation for other conditions, such as growth hormone deficiency in older age. There is reason to suspect that these changes, while seemingly hazardous for human health, might actually be beneficial; therefore, future studies should be aimed at investigating to which extent treatment should actually be suggested.

References

Atzmon G, Barzilai N, Surks MI, Gabriely I. Genetic predisposition to elevated serum thyrotropin is associated with exceptional longevity. J Clin Endocrinol Metabol 2009;94(12):4768–75. https://doi.org/10.1210/jc.2009-0808.

Basaria S. Reproductive aging in men. Endocrinol Metab Clin North Am 2013;42(2):255–70. https://doi.org/10.1016/j.ecl.2013.02.012.

Bergendahl M, Iranmanesh A, Mulligan T, Veldhuis JD. Impact of age on cortisol secretory dynamics basally and as driven by nutrient-withdrawal stress. J Clin Endocrinol Metabol 2000;85(6):2203–14. https://doi.org/10.1210/jc.85.6.2203.

Bremner AP, Feddema P, Leedman PJ, Brown SJ, Beilby JP, Lim EM, Wilson SG, O'Leary PC, Walsh JP. Age-related changes in thyroid function: a longitudinal study of a community-based cohort. J Clin Endocrinol Metabol 2012;97(5):1554–62. https://doi.org/10.1210/jc.2011-3020.

Broekmans FJ, Soules MR, Fauser BC. Ovarian aging: mechanisms and clinical consequences. Endocr Rev 2009;30(5):465–93. https://doi.org/10.1210/er.2009-0006.

Broer SL, Eijkemans MJC, Scheffer GJ, Van Rooij IAJ, De Vet A, Themmen APN, Laven JSE, De Jong FH, Te Velde ER, Fauser BC, Broekmans FJM. Anti-Müllerian hormone predicts menopause: a long-term follow-up study in normoovulatory women. J Clin Endocrinol Metabol 2011;96(8):2532–9. https://doi.org/10.1210/jc.2010-2776.

Casanueva FF, Molitch ME, Schlechte JA, Abs R, Bonert V, Bronstein MD, Brue T, Cappabianca P, Colao A, Fahlbusch R, Fideleff H, Hadani M, Kelly P, Kleinberg D, Laws E, Marek J, Scanlon M, Sobrinho LG, Wass JAH, Giustina A. Guidelines of the pituitary society for the diagnosis and management of prolactinomas. Clin Endocrinol (Oxf) 2006;65(2):265–73. https://doi.org/10.1111/j.1365-2265.2006.02562.x.

Cavadas C, Aveleira CA, Souza GFP, Velloso LA. The pathophysiology of defective proteostasis in the hypothalamus—from obesity to ageing. Nat Rev Endocrinol 2016;12(12):723–33. https://doi.org/10.1038/nrendo.2016.107.

Chahal HS, Drake WM. The endocrine system and ageing. J Pathol 2007;211(2):173–80. https://doi.org/10.1002/path.2110.

Clayton PE, Banerjee I, Murray PG, Renehan AG. Growth hormone, the insulin-like growth factor axis, insulin and cancer risk. Nat Rev Endocrinol 2011;7(1):11–24. https://doi.org/10.1038/nrendo.2010.171.

Corona G, Rastrelli G, Monami M, Saad F, Luconi M, Lucchese M, Facchiano E, Sforza A, Forti G, Mannucci E, Maggi M. Body weight loss reverts obesity-associated hypogonadotropic hypogonadism: a systematic review and meta-analysis. Eur J Endocrinol 2013;168(6):829–43. https://doi.org/10.1530/EJE-12-0955.

Corona G, Rastrelli G, Guaraldi F, Tortorici G, Reismann Y, Sforza A, Maggi M. An update on heart disease risk associated with testosterone boosting medications. Expert Opin Drug Saf 2019;18(4):321–32. https://doi.org/10.1080/14740338.2019.1607290.

Corona G, Goulis DG, Huhtaniemi I, Zitzmann M, Toppari J, Forti G, Vanderschueren D, Wu FC. European Academy of Andrology (EAA) guidelines on investigation, treatment and monitoring of functional hypogonadism in males: endorsing organization: European Society of Endocrinology. Andrology 2020a;8(5):970–87. https://doi.org/10.1111/andr.12770.

Corona G, Sansone A, Pallotti F, Ferlin A, Pivonello R, Isidori A, et al. People smoke for nicotine, but lose sexual and reproductive health for tar: a narrative review on the effect of cigarette smoking on male sexuality and reproduction. J Endocrinol Invest 2020b;43(10):1391–408. https://doi.org/10.1007/s40618-020-01257-x.

Cuttelod S, Lemarchand-Béraud T, Magnenat P, Perret C, Poli S, Vannotti A. Effect of age and role of kidneys and liver on thyrotropin turnover in man. Metabolism 1974;23(2):101–13. https://doi.org/10.1016/0026-0495(74)90107-3.

Di Luigi L, Romanelli F, Sgrò P, Lenzi A. Andrological aspects of physical exercise and sport medicine. Endocrine 2012;42(2):278–84. https://doi.org/10.1007/s12020-012-9655-6.

Di Somma C, Brunelli V, Savanelli MC, Scarano E, Savastano S, Lombardi G, Colao A. Somatopause: state of the art. Minerva Endocrinol 2011;36(3):243–55.

Duntas LH. Thyroid function in aging: a discerning approach. Rejuvenation Res 2018;21(1):22–8. https://doi.org/10.1089/rej.2017.1991.

Elraiyah T, Sonbol MB, Wang Z, Khairalseed T, Asi N, Undavalli C, Nabhan M, Altayar O, Prokop L, Montori VM, Murad MH. The benefits and harms of systemic dehydroepiandrosterone (DHEA) in postmenopausal women with normal adrenal function: a systematic review and meta-analysis. J Clin Endocrinol Metabol 2014;99(10):3536–42. https://doi.org/10.1210/jc.2014-2261.

Ennis GE, An Y, Resnick SM, Ferrucci L, O'Brien RJ, Moffat SD. Long-term cortisol measures predict Alzheimer disease risk. Neurology 2017;88(4):371–8. https://doi.org/10.1212/WNL.0000000000003537.

Frater J, Lie D, Bartlett P, McGrath JJ. Insulin-like growth factor 1 (IGF-1) as a marker of cognitive decline in normal ageing: a review. Ageing Res Rev 2018;42:14–27. https://doi.org/10.1016/j.arr.2017.12.002.

Gesing A, Lewiński A, Karbownik-Lewińska M. The thyroid gland and the process of aging; what is new? Thyroid Res 2012;5(1):16. https://doi.org/10.1186/1756-6614-5-16.

Giagulli VA, Castellana M, Lisco G, Triggiani V. Critical evaluation of different available guidelines for late-onset hypogonadism. Andrology 2020. https://doi.org/10.1111/andr.12850.

Gill S, Lavoie HB, Bo-Abbas Y, Hall JE. Negative feedback effects of gonadal steroids are preserved with aging in postmenopausal women. J Clin Endocrinol Metabol 2002;87(5):2297–302. https://doi.org/10.1210/jcem.87.5.8510.

Gorini S, Kim SK, Infante M, Mammi C, La Vignera S, Fabbri A, Jaffe IZ, Caprio M. Role of aldosterone and mineralocorticoid receptor in cardiovascular aging. Front Endocrinol 2019;10. https://doi.org/10.3389/fendo.2019.00584.

Guth S, Theune U, Aberle J, Galach A, Bamberger CM. Very high prevalence of thyroid nodules detected by high frequency (13 MHz) ultrasound examination. Eur J Clin Invest 2009;39(8):699–706. https://doi.org/10.1111/j.1365-2362.2009.02162.x.

Hale GE, Zhao X, Hughes CL, Burger HG, Robertson DM, Fraser IS. Endocrine features of menstrual cycles in middle and late reproductive age and the menopausal transition classified according to the Staging of Reproductive Aging Workshop (STRAW) staging system. J Clin Endocrinol Metabol 2007;92(8):3060–7. https://doi.org/10.1210/jc.2007-0066.

Hall JE. Endocrinology of the menopause. Endocrinol Metab Clin North Am 2015;44(3):485–96. https://doi.org/10.1016/j.ecl.2015.05.010.

Haymart MR. Understanding the relationship between age and thyroid cancer. Oncologist 2009;14(3):216–21. https://doi.org/10.1634/theoncologist.2008-0194.

Junnila RK, List EO, Berryman DE, Murrey JW, Kopchick JJ. The GH/IGF-1 axis in ageing and longevity. Nat Rev Endocrinol 2013;9(6):366–76. https://doi.org/10.1038/nrendo.2013.67.

Kwong N, Medici M, Angell TE, Liu X, Marqusee E, Cibas ES, Krane JF, Barletta JA, Kim MI, Larsen PR, Alexander EK. The influence of patient age on thyroid nodule formation, multinodularity, and thyroid cancer risk. J Clin Endocrinol Metabol 2015;100(12):4434–40. https://doi.org/10.1210/jc.2015-3100.

Kwong JCC, Krakowsky Y, Grober E. Testosterone deficiency: a review and comparison of current guidelines. J Sex Med 2019;16(6):812–20. https://doi.org/10.1016/j.jsxm.2019.03.262.

Lechan RM, Fekete C. Feedback regulation of thyrotropin-releasing hormone (TRH): mechanisms for the non-thyroidal illness syndrome. J Endocrinol Invest 2004;27(6):105–19.

Lunenfeld B, Mskhalaya G, Zitzmann M, Arver S, Kalinchenko S, Tishova Y, Morgentaler A. Recommendations on the diagnosis, treatment and monitoring of hypogonadism in men. Aging Male 2015;18 (1):5–15. https://doi.org/10.3109/13685538.2015.1004049.

Molitch ME, Clemmons DR, Malozowski S, Merriam GR, Vance ML. Evaluation and treatment of adult growth hormone deficiency: an endocrine society clinical practice guideline. J Clin Endocrinol Metabol 2011;96(6):1587–609. https://doi.org/10.1210/jc.2011-0179.

Nashiro K, Guevara-Aguirre J, Braskie MN, Hafzalla GW, Velasco R, Balasubramanian P, Wei M, Thompson PM, Mather M, Nelson MD, Guevara A, Teran E, Longo VD. Brain structure and function associated with younger adults in growth hormone receptor-deficient humans. J Neurosci 2017;37 (7):1696–707. https://doi.org/10.1523/JNEUROSCI.1929-16.2016.

Ohlsson C, Vandenput L, Tivesten. DHEA and mortality: what is the nature of the association? J Steroid Biochem Mol Biol 2015;145:248–53. https://doi.org/10.1016/j.jsbmb.2014.03.006.

Pamphlett R, Jew SK, Doble PA, Bishop DP. Elemental analysis of aging human pituitary glands implicates mercury as a contributor to the somatopause. Front Endocrinol 2019;10. https://doi.org/10.3389/fendo.2019.00419.

Parker CR, Slayden SM, Azziz R, Crabbe SL, Hines GA, Boots LR, Bae S. Effects of aging on adrenal function in the human: responsiveness and sensitivity of adrenal androgens and cortisol to adrenocorticotropin in premenopausal and postmenopausal women. J Clin Endocrinol Metabol 2000;85(1):48–54. https://doi.org/10.1210/jc.85.1.48.

Prader A. Dwarfism, hypopituitarism, and growth hormone. Arch Dis Child 1967;42(223):225–7. https://doi.org/10.1136/adc.42.223.225.

Prummel MF, Brokken LJS, Wiersinga WM. Ultra short-loop feedback control of thyrotropin secretion. Thyroid 2004;14(10):825–9. https://doi.org/10.1089/thy.2004.14.825.

Purnell JQ, Brandon DD, Isabelle LM, Loriaux DL, Samuels MH. Association of 24-hour cortisol production rates, cortisol-binding globulin, and plasma-free cortisol levels with body composition, leptin levels, and aging in adult men and women. J Clin Endocrinol Metabol 2004;89(1):281–7. https://doi.org/10.1210/jc.2003-030440.

Rastrelli G, Dicuio M, Reismann Y, Sforza A, Maggi M, Corona G. Cardiovascular impact of testosterone therapy for hypogonadism. Expert Rev Cardiovasc Ther 2018;16(9):617–25. https://doi.org/10.1080/14779072.2018.1510314.

Ravaglia G, Forti P, Maioli F, Boschi F, Bernardi M, Pratelli L, Pizzoferrato A, Gasbarrini G. The relationship of Dehydroepiandrosterone Sulfate (DHEAS) to endocrine-metabolic parameters and functional status in the oldest-old. Results from an Italian study on healthy free-living over-ninety-year-olds. J Clin Endocrinol Metabol 1996;81(3):1173–8. https://doi.org/10.1210/jcem.81.3.8772596.

Rieben C, Segna D, Da Costa BR, Collet TH, Chaker L, Aubert CE, Baumgartner C, Almeida OP, Hogervorst E, Trompet S, Masaki K, Mooijaart SP, Gussekloo J, Peeters RP, Bauer DC, Aujesky D, Rodondi N. Subclinical thyroid dysfunction and the risk of cognitive decline: a meta-analysis of prospective cohort studies. J Clin Endocrinol Metabol 2016;101(12):4945–54. https://doi.org/10.1210/jc.2016-2129.

Rudman D, Kutner MH, Rogers CM, Lubin MF, Fleming GA, Bain RP. Impaired growth hormone secretion in the adult population. Relation to age and adiposity. J Clin Investig 1981;67(5):1361–9. https://doi.org/10.1172/JCI110164.

Saad F, Aversa A, Isidori AM, Zafalon L, Zitzmann M, Gooren L. Onset of effects of testosterone treatment and time span until maximum effects are achieved. Eur J Endocrinol 2011;165(5):675–85. https://doi.org/10.1530/EJE-11-0221.

Salonia A, Bettocchi C, Carvalho G, Corona G, Jones TH, Kadioğlu A, et al. European association of urology guidelines, 2020 edition [Internet]. Vol. presented at the EAU Annual Congress Amsterdam 2020. European Association of Urology Guidelines Office; 2020. Available from: https://uroweb.org/guideline/sexual-and-reproductive-health-2/.

Sansone A, Sansone M, Lenzi A, Romanelli F. Testosterone replacement therapy: the Emperor's new clothes. Rejuvenation Res 2017;20(1):9–14. https://doi.org/10.1089/rej.2016.1818.

Sartorius GA, Nieschlag E. Paternal age and reproduction. Hum Reprod Update 2009;16(1):65–79. https://doi.org/10.1093/humupd/dmp027.

Segna D, Bauer DC, Feller M, Schneider C, Fink HA, Aubert CE, Collet TH, da Costa BR, Fischer K, Peeters RP, Cappola AR, Blum MR, van Dorland HA, Robbins J, Naylor K, Eastell R, Uitterlinden AG, Rivadeneira Ramirez F, … Gogakos A. Association between subclinical thyroid dysfunction and change in bone mineral density in prospective cohorts. J Intern Med 2018;283(1):56–72. https://doi.org/10.1111/joim.12688.

Selmer C, Olesen JB, Hansen ML, Lindhardsen J, Olsen AM, Madsen JC, et al. The spectrum of thyroid disease and risk of new onset atrial fibrillation: a large population cohort study. BMJ 2012;345: e7895https://doi.org/10.1136/bmj.e7895.

Selmer C, Olesen JB, Hansen ML, Von Kappelgaard LM, Madsen JC, Hansen PR, Pedersen OD, Faber J, Torp-Pedersen C, Gislason GH. Subclinical and overt thyroid dysfunction and risk of all-cause mortality and cardiovascular events: a large population study. J Clin Endocrinol Metabol 2014;99(7):2372–82. https://doi.org/10.1210/jc.2013-4184.

Sherman B, Wysham C, Pfohl B. Age-related changes in the circadian rhythm of plasma cortisol in man. J Clin Endocrinol Metabol 1985;61(3):439–43. https://doi.org/10.1210/jcem-61-3-439.

Soules MR, Sherman S, Parrott E, Rebar R, Santoro N, Utian W, Woods N. Executive summary: stages of reproductive aging workshop (STRAW). In: Fertility and sterility. vol. 76. 2001. p. 874–8. https://doi.org/10.1016/S0015-0282(01)02909-0 Issue 5.

Steiner AZ, Pritchard D, Stanczyk FZ, Kesner JS, Meadows JW, Herring AH, Baird DD. Association between biomarkers of ovarian reserve and infertility among older women of reproductive age. JAMA 2017;318(14):1367–76. https://doi.org/10.1001/jama.2017.14588.

Strich D, Karavani G, Edri S, Gillis D. TSH enhancement of FT4 to FT3 conversion is age dependent. Eur J Endocrinol 2016;175(1):49–54. https://doi.org/10.1530/EJE-16-0007.

Surks MI, Ortiz E, Daniels GH, Sawin CT, Col NF, Cobin RH, Franklyn JA, Hershman JM, Burman KD, Denke MA, Gorman C, Cooper RS, Weissman NJ. Subclinical thyroid disease: scientific review and guidelines for diagnosis and management. JAMA 2004;291(2):228–38. https://doi.org/10.1001/jama.291.2.228.

Tabatabaie V, Surks MI. The aging thyroid. Curr Opin Endocrinol Diabetes Obes 2013;20(5):455–9. https://doi.org/10.1097/01.med.0000433055.99570.52.

Van Cauter E, Leproult R, Kupfer DJ. Effects of gender and age on the levels and circadian rhythmicity of plasma cortisol. J Clin Endocrinol Metabol 1996;81(7):2468–73. https://doi.org/10.1210/jc.81.7.2468.

Van Coevorden A, Laurent E, Decoster C, Kerkhofs M, Neve P, Van Cauter E, Mockel J. Decreased basal and stimulated thyrotropin secretion in healthy elderly men. J Clin Endocrinol Metabol 1989;69(1):177–85. https://doi.org/10.1210/jcem-69-1-177.

van den Beld AW, Kaufman J-M, Zillikens MC, Lamberts SWJ, Egan JM, van der Lely AJ. The physiology of endocrine systems with ageing. Lancet Diabetes Endocrinol 2018;647–58. https://doi.org/10.1016/s2213-8587(18)30026-3.

Virgini VS, Rodondi N, Cawthon PM, Harrison SL, Hoffman AR, Orwoll ES, Ensrud KE, Bauer DC. Subclinical thyroid dysfunction and frailty among older men. J Clin Endocrinol Metabol 2015;100(12):4524–32. https://doi.org/10.1210/jc.2015-3191.

Yen SSC, Laughlin GA. Aging and the adrenal cortex. In: Experimental gerontology. vol. 33. Elsevier Inc; 1998. p. 897–910. https://doi.org/10.1016/S0531-5565(98)00046-1 Issues 7–8.

Ylli D, Burman KD, Van Nostrand D, Wartofsky L. Eliminating the age cutoff in staging of differentiated thyroid cancer: the safest road? J Clin Endocrinol Metabol 2018;103(5):1813–7. https://doi.org/10.1210/jc.2017-02725.

CHAPTER 12

Chronobiology and chrononutrition: Relevance for aging

Damiano Galimberti and Giuseppe Mazzola
Italian Association of Anti-Ageing Physicians, Milano, Italy

12.1 Introduction

Chrononutrition is the science that investigates the correlations between mealtimes, metabolism, physiology, and circadian rhythm. Nevertheless, according to another interpretative key, the term also outlines a diet in harmony with biological biorhythms and, in this sense, it can be considered as a branch of chronobiology, which studies biorhythms and therefore the periodic oscillations of parameters and functions biological of living organisms.

Biorhythms are classified, based on their duration, into three types: (i) those that take place within 24 h (usually from 1 ms to a few hours) such as heart rate and respiratory rate; they are called ultradian; (ii) those that last an entire day, such as the sleep-wake cycle; they are called circadian; those that last for more than a day (usually from 1 week to 1 year)—they are called infradian, and the menstrual cycle is part of it.

The set of biological biorhythms, especially circadian ones, constitutes what is commonly known as the biological clock. Among the most studied biorhythms there is the sleep-wake cycle, directly connected to the light-dark cycle which in turn is a consequence of the rotation of the earth's axis, the motion that for millions of years has accompanied evolutionary processes, those of natural selection and the adaptation of all organisms to life on earth.

Adaptation to the light-dark cycle is an ancestral heritage, so deeply rooted in life that most organisms perceive the intensity and frequency of light intensity by using conserved molecular mechanisms and specific cell structures, which coordinate cell types of the whole organism. Consequently, each cell in the body can capture the light-dark alternation through neuroendocrine signals originating from the signaling of the accommodation facilities themselves. Consequently, the same alignment of the sleep-wake cycle with the light-dark cycle is directly associated with the increase in fitness of many species, i.e., the relative value most used in evolutionary biology to estimate the adaptation of an organism or species to a given ecosystem and to another organism of the same species (Peijun et al., 2013).

Human Aging
https://doi.org/10.1016/B978-0-12-822569-1.00006-8

12.2 Biorhythms

Biological clocks are conserved cellular pathways, adaptable to environmental changes. There are biological clocks associated with intrinsic circadian biorhythms in all organs and somatic cells, which regulate the expression of a group of genes called clock-controlled genes (CCG) (Takahashi, 2017). A fundamental characteristic of circadian clocks is that they are self-sufficient and they do not need external signals to be generated. However, some environmental signs have proven able to influence them deeply. These are called zeitgebers (from the German "time donor"), and the main ones are light, temperature, and nutrition (Eckel-Mahan and Sassone-Corsi, 2013; Gaucher et al., 2018).

In *Homo sapiens* circadian biorhythms influence the secretion of different hormonal molecules such as melatonin (which varies according to the light-dark cycle), growth hormone (which reaches the maximum levels in the early hours of the night), cortisol (maximum early in the morning on waking), but also other biological parameters such as body temperature, which in the dark drops by about 1°C and urinary excretion of sodium and potassium. As a result, many hormones are influenced by circadian rhythms and therefore have different physiological blood levels at different times of the day.

Biorhythms have the physiological meaning of anticipating homeostatic mechanisms, intervening just before a recurring stimulus occurs. In advanced biology this mechanism saves a significant amount of energy; think of the peristaltic movements and the synthesis of digestive juices, which begin just before mealtime, optimizing the entire digestive process, which for many herbivorous species constitutes a significant source of energy expenditure. Biorhythms are endogenous, spontaneous, and tend to be independent of many environmental factors; nevertheless, some external stimuli can influence these cycles so much as to modify the configuration of the biological clock itself, by moving the "hands" forward or backward. The most important external stimuli in this regard are the light-dark cycle, the ambient temperature, the time and composition of the meal, social stimuli, and physical activity (Duffy and Czeisler, 2009; Takahashi, 2017).

The duration of the cycle also has heterogeneous evidence: according to some authors, the circadian cycle is slightly longer than 24h; in one study, it was around 25h. In a clinical study conducted on isolated subjects, which had been imposed an automated light-dark cycle of 8 plus 16h (24h), the study group followed a circadian cycle of about 25h when, always in the same environment, it was given them the possibility of independently establishing the light-dark alternation (Mills et al., 1974). On the contrary, a more rigorous study concluded that the circadian pacemaker of *Homo sapiens* oscillates with a period of 24/18h in a constant way (Dijk et al., 1999).

The external light can modify the configuration of the circadian biorhythms; in fact, when traveling in areas with a large difference in time zone compared to the departure time (for example Rome-Sydney, +9h), the misalignment between the external light-dark cycle and the subject's biological clock can frequently cause the symptoms of jet lag, a clinical condition resulting from a homeostatic disorder associated with gastrointestinal and sleep disturbances, reduced wakefulness and attention, migraine, confusion, and a feeling of general malaise. The symptoms disappear following the adaptation of the organism to the new light signals associated with the external time zone, demonstrating that the light stimuli are able to change the configuration of the biological clock progressively, until complete alignment with the external clock (Herxheimer, 2014).

Other evidence of how light stimuli are associated with changes in circadian biorhythms associated with them derives from studies conducted on workers undergoing permanent or rotating night shifts. Exposure to low intensity light during night work shifts inhibits the adaptation mechanisms of these subjects to the new schedule, even after years, leading to sleep restriction and a deficit of melatonin secretion, which decreases the inhibitory stimuli of the increase of cortisol levels at night, and an increased risk of type 2 diabetes, insulin resistance, and obesity (Sookoian et al., 2008; Zvonic et al., 2007), all situations favoring unsuccessful aging. Nevertheless, in a particular study, it was found that by exposing subjects to very bright light (Adamovich et al., 2014; La Fleur et al., 1999) cyclically, if workers at the end of the shift reached the house wearing sunglasses (which decreased by 85% l intensity of light stimulation) and they slept in conditions of complete darkness for 8 h, then there was a complete adaptation to the new schedule with a new configuration of the melatonin secretion biorhythms and a new decrease in body temperature in the period of daytime darkness (Boivin and James, 2002). Although to date there is no direct evidence that these measures can reduce the risk factors set out before, surely these measures could be an interesting starting point for further analysis in this regard. In some experimental studies conducted on astronauts, to improve their ability to adapt to the new light-dark cycle, the subjects were exposed to a light of 10,000 lx in the prelaunch phase (Stewart et al., 1995).

In addition to light, sleep duration is also able to change the biological clock and numerous metabolisms, including appetite regulation and food intake. The decrease in hours of night sleep can facilitate the onset of glucose intolerance and insulin resistance, as well as alter the production of ghrelin and leptin, both phenomena associated with a significant increase in the release of ghrelin (hormone implicated in the induction of hunger) and a reduction in plasma leptin (hormone implicated in the onset of satiety).

Sleep inhibits the release of ghrelin and its acylation, which is associated with its activation; precisely these metabolic effects could favor the establishment of the correct fast during the night's rest (Spiegel et al., 2011). All of these sleep-induced dysmetabolic phenomena can induce an increase in hunger and carbohydrate cravings upon awakening, both factors associated with an increase in daily calorie intake, BMI, risk of obesity, and type 2 diabetes (Knutson et al., 2006; Lauderdale et al., 2009; Shahrad et al., 2004; Spiegel et al., 2005; Vorona et al., 2005).

These experimental outcomes, in a western lifestyle context, are of extreme importance, since several epidemiological studies have found significant and evident associations between the percentage of Americans who sleep less than 7 h, which increased from 15.6% of 1969 to 37.1% in 2002, and the increase in obesity and over-weight in that population, which is deeply increased (Chaput et al., 2008).

In addition, these figures are a cause for concern, considering that a good percentage of Americans today sleep only 5 or 6 h a day and that in the past 40 years, the duration of sleep reported by Americans has decreased by an average of 1.5–2.0 h, highlighting a sta-tistically significant and substantial sleep debt (Jean-Louis et al., 2000; Kripke et al., 1979).

The effect of sleep restriction should be feared in the richest country including Italy, in fact similar epidemiological evidence exists in Europe too. For example, a Spanish research from 2000 showed that people who sleep only 6 h a night have an increased risk of obesity (Vioque et al., 2000). The biological and molecular mechanisms underlying the effect of sleep restriction are not yet clear; however, starting from the fact that mel-atonin and cortisol have profiles other than those considered physiological, the first with evening increase and night peak and the second with awakening peak and minimum eve-ning, and that numerous evidences on primates and rodents have shown that melatonin inhibits directly, through the melatonin receptor(MT)1 present in the adrenal gland, the action of adrenocorticotropic hormone (ACTH) in mediating the release of cortisol.

(Richter et al., 2008; Torres-Farfan et al., 2003), it can be assumed that sleep depri-vation or even chronic sleep fragmentation (for example caused by obstructive dyspnea nocturnal) reduces the release and signaling of melatonin, increasing the sympathetic activity, and consequently the production of cortisol. Ultimately, this would result in an increase in visceral fat, a decrease in growth hormone signaling (the main hormone related to the increase in muscle mass), and consequently increasing the risk of patholog-ical conditions such as insulin resistance and obesity, both sarcopenic and nonsarcopenic (Spiegel et al., 2005). Furthermore, alteration of the release and signaling of melatonin is likely to promote oxidative distress in a marked way, being melatonin one of the most important and effective antioxidants in the body (Reiter et al., 2018). Consequently, it would not be surprising if some future studies will show an association between the increase in sleep deficiency and the increased risk of incurring neoplastic and/or patho-logical phenomena associated with alterations in the redox state, such as the cytokine storm, or accelerated skin and brain aging.

12.3 Central oscillator and peripheral oscillators

Oscillations are an important type of cellular signaling characterized by the periodic change of the system over time. Oscillations can occur generally by positive feedback loops (positive loops), alone or in combination with negative feedback (Kholodenko, 2006). The molecular structures that originate oscillations are called oscillators. Fungal and animal cells share circadian oscillators of conserved structure, consisting of feedback mechanisms, transcription-translation feedback loops (TTFL), composed of two parts: i) a positive circuit with a heterodimer complex that acts as activator, promoting the transcription of the associated genes; ii) one or more components that make up a negative circuit that inhibit the activity of the first circuit (Kholodenko, 2006; Marshall, 1995; Murphy et al., 2002).

Experimental evidence has allowed us to establish that the circadian rhythm originates in the hypothalamic suprachiasmatic nucleus (SCN) located in the anterior hypothalamus. This nucleus would therefore behave as a central oscillator, that is, as a biological pacemaker, capable of organizing the biological rhythm and coordinating it with other peripheral oscillators and therefore, indirectly, to all the cells of the body. Although the biorhythm originating from the SCN is independent, it can be synchronized with the light-dark cycle, which is the main input signal capable of modifying the configuration of the biological clock, acting as a synchronization signal (technically defined zeitgeber) of the circadian rhythm. In other words, the hands of the main clock (master clock) or that of the hypothalamic suprachiasmatic nucleus are moved and set by the external light.

Many circadian processes are therefore regulated at the level of the SCN and are influenced by light; among the main ones we mention pressure, body temperature, cardiovascular efficiency, muscle strength, hormone levels, and cognitive abilities. These effects can be so great that they could explain the diurnal variation of some cardiovascular events, including changes in blood pressure, activity of the autonomic nervous system and of the renin-angiotensin axis, coagulation, vascular tone, and intracellular metabolism of cardiomyocytes (Potter et al., 2016).

In mouse models, the ablation of the SCN determines the loss of the circadian rhythm (Stephan and Zucker, 1972). Furthermore, there are numerous other similar evidences, which demonstrate how the removal of the ventromedial hypothalamic nuclei, which are located near the regulator nucleus of the circadian rhythm and are directly related to it, due to obesity (King, 2006), while that of the nuclei lateral hypothalamus, which express neuropeptides with orexigenic activity, and connected with the SCN, induce anorexia (Rovere-Jovene et al., 2016).

Specialized retina ganglion cells, are able to perceive the luminous intensity (luminance) through the presence of specific chromophore not involved in the complex mechanism of image acquisition and projection, melanopsin and cryptochrome, the first of which belongs to the rhodopsin family and has a sensitivity peak at 480 nm, a wavelength between blue and the beginning of green spectrum (Dacey et al., 2005), while the second, which is preserved in animals and plants, is more sensitive in the blue light. The retinal ganglion cells are anatomically and functionally connected with the main structures connected with the regulation of the biological clock; in fact they project into the SCN through the retino-hypothalamic tract, with glutamatergic transmission, which in turn projects into different areas of the central nervous system and into epiphysis (Moore, 1995). The latter structure, in conditions of absence of light signals, begins to synthesize and release melatonin into the bloodstream starting from serotonin, in a process that reaches its maximum secretion peak between 2 and 4 in the morning, promoting the sleep and drop in body temperature. Furthermore, some evidence suggests that melatonin also has a peripheral action, as its receptors coupled to G-proteins (MT1, MT2) are also present in numerous other districts, such as vascular tissue, gastrointestinal tract, liver, pancreas, adipose tissue, and skin. Finally, in addition to MT1 and MT2, there may also be a third receptor (MT3), not yet identified in humans, a homologue of the quinone reductase receptor of the hamster and the rabbit, and another nuclear receptor consisting of the retinoid orphan receptors(ROR)/retinoid Z receptors dimer (Adamczyk-Sowa et al., 2012; Slominski et al., 2012). Experimental evidence indicates that melatonin signal inhibition is associated with an increased risk of impaired glucose homeostasis and the onset of type 2 diabetes (Peschke et al., 2007), as well as a polymorphic intronic variant (rs10830963) in the MT2 receptor, expressed in pancreatic β cells, it is associated with a significant increase in receptor expression and with reduced insulin secretion (Laakso et al., 2009); in addition to a decrease in weight loss during low-calorie and high-protein diets, particularly in female subjects (Goni et al., 2014).

The SCN, modulating the activity of the hypothalamic–pituitary–adrenal axis, exerts an inhibitory effect on the secretion of the corticoid release factor, which is normally regulated in turn by the signal of vasopressin, produced at the level of the hypothalamic paraventricular nucleus. That induces a consequent effect on the release of ACTH from the adeno-hypophysis and thus regulating glucocorticoid production levels (Kalsbeek et al., 1996).

The SCN releases other local humoral factors that modulate the activity of the hypothalamus (Li et al., 2012), markedly impacting on behavior and locomotor activity, body temperature, and energy metabolism in animal models (Kramer et al., 2001; Kraves and Weitz, 2006). Furthermore, there are several direct experimental evidences which demonstrate that this modulates the activity of other hypothalamic nuclei, whose axons project both into the pituitary portal system, producing the factors that release or inhibit the release of adeno-hypophyseal hormones, and in the neurohypophysis (supraoptic and paraventricular). It is precisely thanks to these neurohormonal regulatory mechanisms that the activity of the SCN is able to synchronize with that of other specialized cells called peripheral oscillators (Mendoza and Challet, 2009; Yamazaki et al., 2000), which have been identified in almost all organs and which are directly involved in the circadian biological clock.

The peripheral oscillators present in the liver, intestine, and adipose tissue (Gómez-Santos et al., 2009; Zvonic et al., 2006) could play a central action in the regulation of metabolic processes. Nonetheless, it should be emphasized that peripheral oscillators have their own intrinsic rhythm and continue to operate even after the ablation of SCN in animal models (Husse et al., 2015; Welsh et al., 2009).

The systemic levels of insulin, glucagon, and adrenaline can vary considerably following the activity of the SCN (Challet, 2015). In particular, the circadian biorhythm associated with the use of glucose is directly regulated by the activity of the SCN through sympathetic pathways connected with the liver (Cailotto et al., 2005).

The activation of the central oscillator induces an increase in insulin sensitivity mainly by increasing muscle glucose absorption and hepatic glucose production (La Fleur et al., 1999, 2001). In addition, the SCN also exercises direct control over cardiac metabolism (Buijs and Kalsbeek, 2001; Plano et al., 2017). Peripheral clocks are strongly influenced by cellular metabolism: numerous metabolic signals such as NADH / NAD +, NADPH/NADP +, ATP/AMP, oxygen and reactive oxygen species (ROS), carbon monoxide, or glucose directly influence the transcriptional activity of circadian proteins (Aguilar-Arnal and Sassone-Corsi, 2013; Plano et al., 2017). Furthermore, the circadian clock modulates specific elements of the cell cycle, such as the inhibitor WEE1 kinase (Gaucher et al., 2018; Hirayama et al., 2005; Matsuo et al., 2003). It is important to underline that the regulation of body weight involves the integration of different hypothalamic areas, including SCN, the arcuate nucleus, the ventromedial hypothalamic nucleus, and the paraventricular nucleus, whose interactions control appetite and food intake, fat deposition, and energy expenditure (Buijs et al., 2016; Morton et al., 2006; Plano et al., 2017).

Another important peripheral oscillator is the FEO (Food Entrainable Oscillator), a circadian regulation system capable of triggering the anticipatory processes of the meal and capable of synchronizing with the time and composition of them, provided that there is certain regularity. There is experimental evidence that suggests that the FEO is responsible for the circadian secretion of ghrelin and gastrointestinal motility in animal models, stimulating the search for food in the animal 2h before the meal (Laermans et al., 2015).

According to current scientific evidence, FEO is located in the stomach cells that release ghrelin and express the period proteins PER1 and PER2 (Lesauter et al., 2009), but some authors hypothesize its multiple localization in the intestine (Davidson, 2006). PER2 protein is a central component of FEO whose alteration is associated with various pathological conditions such as obesity, cancer, gastrointestinal pathologies, metabolic syndrome, and sleep disturbances (Garaulet et al., 2010b). Central oscillator and peripheral oscillators work synchronously and correlatively in regulating the metabolism, modulating the sensitivity of the synapses of the autonomic nervous system, regulating the transcription of at least 20%–30% of the cell transcriptome subject to variations associated with environmental stimuli such as brightness, the length of the day (the photoperiod), temperature, and even the time and the composition of the meal. The chronic desynchronization of peripheral and central oscillators, which takes on syndromic or pathological connotations, is called chronodisruption and is associated with various risk conditions, as we will see later.

12.4 Clock-controlled genes

The biological event known as "oscillations," generated by the central and peripheral oscillators, regulate the transcription of the CCG, whose expression controls many other genes, greatly influencing the entire metabolism cellular, with a rhythm of about 24 h (Li and Zhang, 2015).

Circadian biorhythm is based on 6 molecular motifs called TTFLs. A close examination of the mechanisms and interactions underlying these motifs could make this discussion unnecessarily complex; nevertheless, for example loop 3 allows PPAR-α, a member of peroxisomal activation receptor family, to generate positive feedback by activating the transcription of the gene Brain and Muscle Aryl-hydrocarbon receptor nuclear translocator-Like 1(BMAL1), while loop 4 allows nicotinamide phosphoribosyl transferase (NAMPT) to induce the formation of negative feedback that increases NAD levels and modulates deacetylating activity (Hurley et al., 2016; Takahashi et al., 2009; Ye et al., 2014).

As we have previously stated, the expression of these genes is finely controlled by oscillatory mechanisms, which are guided by feedback systems, TTFL, and by epigenetic modifications originating both from regulatory processes independent of environmental signals than by the peripheral oscillators themselves (Prasai et al., 2008). In mammals two of these genes play an important action on gene transcription, Circadian Locomotor Output Cycles Kaput (CLOCK) and BMAL1 which are expressed in the first phase of biorhythm (phase light), while they are not in the second phase (dark phase). The two phases last 12 h each.

In the light phase, the protein products deriving from the transcription of CLOCK and BMAL1, dimer in the cytosol and move to the nucleus, binding to specific response elements called *E-Box*; CLOCK acetylates BMAL1 and together activate the transcription of genes PER1, PER2, PER3, the Cryptochrome genes (CRY1, CRY2), the Reverse Erythroblastosis Virus a (REV-ERB-α) gene and RORα, and other CCG genes associated with circadian biorhythms and behavior. While CLOCK is transcribed continuously, BMAL1 presents a rhythmic transcription, which is regulated by the REV-ERB-α antagonist effect for the binding of RORα with CLOCK, necessary for the induction of the transcription of BMAL 1.

The dark phase is characterized by the suppression of the expression of CLOCK and BMAL1, guided by the inhibitory effect exerted by the PER–CRY dimer. The CRY protein, which acts independently from light, dimers with CRY in the cytosol and together the two products move to the nucleus, where they inhibit the transcription of BMAL1 and CLOCK. That, at least if present in the nuclear site beyond a certain concentration threshold, inhibits the expression of PER, CRY, and REV-ERB-α through a negative feedback mechanism. This event constitutes the main signal that induces the resumption of the transcription of CLOCK and BMAL1 and the new phase of light. The half-life of the PER–CRY dimer is controlled by casein kinase phosphorylase, which acts on PER by marking it for ubiquitination and degradation, and protein kinase A(PKA), which phosphorylates CRY by stimulating proteasomal degradation (Angelousi et al., 2018; Trott and Menet, 2018).

The circadian biological clock and the cell cycle are structurally related; in fact, the metabolism of nucleotides, lipids, and the DNA replication is directly related to the cellular availability of energy nutrients and is also regulated by mechanisms dependent on the biological clock (Dang et al., 2017; de Winter et al., 2014; Kaplon et al., 2015). Circadian mechanism not only would synchronize energy metabolism, cell cycle, and circadian clock, but could help to separate over time those cellular processes that otherwise are difficult to separate in space and limit the response to the external environment of the cell. In yeast, oxidative metabolism, capable of generating ROS, and DNA replication are coordinated with each other and with precise timing (Burnetti et al., 2016; Chen and McKnight, 2007; Tu et al., 2005). Indeed, some single-cell organisms have a cell division directly controlled by circadian processes (Gaucher et al., 2018; Miyagishima et al., 2014). In multicellular organisms these mechanisms are much more complex and still to be defined, but for some extent similar. Obviously, more research is still needed to clarify these mechanisms. In mammals, some evidence suggests that the biological clock can modulate the rhythmicity and mitotic progression of epidermis, hair follicles, and intestinal stem cells (Janich et al., 2011; Karpowicz et al., 2013; Plikus et al., 2013; Takahashi et al., 2012; Weger et al., 2017), in addition to the repairing of liver damage, intestinal damage, and skin damage (Bouchard-Cannon et al., 2013; Gaucher et al., 2018; Karpowicz et al., 2013; Kowalska et al., 2013; Stanger, 2015). The proteins of the CCG genes interact directly with the protein complexes of cell checkpoints to mediate the response to DNA damage (Gaucher et al., 2018; Sancar et al., 2010).

As the biological clock responds differentially during cell cycle progression, the cell cycle may affect the biological clock. For example, P53 directly binds the PER2 promoter to a response element to P53, temporarily repressing the CLOCK-mediated transcription of BMAL1 (Miki, 2013). The CCG genes might induce the elimination of specific elements of the biological clock and the cell cycle, including phosphorylation, acetylation, ubiquitination, and SUMOylation (Gaucher et al., 2018). However, it is now known that there are numerous multilevel connections between the cell cycle, the circadian biorhythms, the sleep-wake cycle, the light-dark cycle, and the cellular energy metabolism, as well as with the tissue and basal ones.

It is interesting to note that the activity of adenylate cyclase and, more generally, the AMP/ATP ratio, could represent the link between the cell's energy status and the circadian rhythm, given that PKA is directly activated influenced by deacetylations induced by the sirtuin-1 (Sirt-1), an NAD+ dependent deacetylase whose activity could therefore be influenced by cellular redox status (Merbitz-Zahradnik and Wolf, 2015), as well as by the availability of NAD+. Some studies suggest that Sirt1 can not only influence the activity of CLOCK-BMAL1, but can activate the Peroxisome proliferator-activated receptor-γ coactivator (PGC)-1α with the same structural modifications, which in turn would bind the promoter of BMAL1 and activate the transcription of the enzymes Sirt-1 and NAMPT (the latter responsible for the synthesis of NAD +). It is interesting to note that the transcription of the Sirt-1 and NAMPT enzymes and that of the RORα and PGC-1α factors is under the gene control of the CLOCK-BMAL1 dimer. That suggests the existence of feedback mechanisms such that the sirtuins and the BMAL1 pathways influence each other. It seems that the levels of Sirt-1 decrease with age and that the consequence is a dysfunction of the circadian rhythm; as a result, some authors hypothesize that the use of some nutraceuticals, as resveratrol, active on Sirt-1 may maintain the functionality of the cell watch (Belden and Dunlap, 2013), exerting a beneficial effect on successful aging and longevity.

Some in vivo studies reported that surgical removal of the epiphysis and the administration of its extracts induced significant changes in glycemia and glucose use in many tissues (Diaz and Blazquez, 1986; Mellado et al., 1989; Rodríguez et al., 1989). In fact, the lack of melatonin can cause insulin resistance (Cipolla-Neto et al., 1998; Nogueira et al., 2011), because this hormone controls pancreatic function (indirectly regulates the synthesis of glucagon and insulin) (Anhê et al., 2004; Bähr et al., 2011; Picinato et al., 2002), induce other metabolic effects on adipose tissue, hormone synthesis, thyroid glands activity, and modulate the energy metabolism (Brydon et al., 2001; Cardoso Alonso-Vale et al., 2004, 2006). It even seems that melatonin can prevent obesity in animal models, probably through its effect on adipocytes and/or on thyroid hormones (Plano et al., 2017; Tan et al., 2011).

Melatonin can modulate the leptin secretion (Szewczyk-Golec et al., 2015) but the effect of the photoperiod on leptin modulation in food intake is not clear. Plasma levels of leptin and adiponectin respond differently on fat mass amount and photoperiodic changes of the adipose tissue (Ahima and Flier, 2000). However, in animal models a significant increase in fat mobilization is related to leptin signaling, but it was not associated with a contextual change in the expression levels of anorectic or anabolic peptides leptin correlated in the arcuate nucleus. Neuropeptide Y, pro-opiomelanocortin (Reddy et al., 1999), as well as hormones stimulated by the activity of orexin and melanin (Ebling, 2015; Parsons et al., 2015) are probably not involved in that, and the experimental evidences suggest probably the main hormone linked to satiety and hunger photoperiodic related mechanism is leptin itself (Plano et al., 2017).

Other hormones are implicated in the synchronization of zeitgebers with energy metabolism, like the thyroid hormone. Pars tuberalis melatonin receptors regulate the photoperiodic synthesis and secretion of thyroid stimulating hormone (TSH), and during highly dilated photoperiods, it happens an increase in TSH and diosidase 2 levels. That leads to the conversion of thyroxine (T4) to triiodothyronine (T3), increasing the energy expenditure. On the contrary, shortened photoperiods induce decreases in TSH and promote the activity of diosidase 3, which convert T4 to inverse T3 and diiodothyronine, decreasing the basal metabolic rate and increasing the food intake and the fat deposits (Ebling, 2015; Plano et al., 2017).

Data on organ-specific circadian metabolism are still limited. Nonetheless, many studies have shown that the liver oscillator is strongly associated with feeding and fasting cycles and the hepatic function itself is probably regulated according to circadian rhythms (Akhtar et al., 2002; Panda et al., 2002; Storch et al., 2002; Zhang et al., 2014). The liver is involved in some pathways circadian rhythm correlated as the expression of glucose transporters, the glucagon receptor, and enzymes related to lipid homeostasis that can be modified by variations of the central oscillator (Adamovich et al., 2014). In in vivo models, Bmal1 deletion in the liver reduced basal blood glucose (Lamia et al., 2008), insulin/glucagon secretion and production (Boden et al., 1996; Ruiter et al., 2003), and even glucose uptake (Bolli et al., 1984; La Fleur et al., 2001). Many CCG and CCG-linked genes have specific functions in the context of glucose metabolism; for example, the cryptochromes regulate liver gluconeogenesis increasing the transcription of gluconeogenic genes regulated by CREB-binding protein; again, the overexpression of Cry1 in the liver influences glucose levels in model animals (Brenner et al., 2010).

The modulatory signals of the central oscillator seem to have a significant effect on the circadian regulation of glucose metabolism in the liver, as well as lipid metabolism. In fact, multiple signaling pathways converge on phosphoenolpyruvate carboxykinase or pyruvate kinase modulation, affecting glucose metabolism (Plano et al., 2017; Reinke and Asher, 2016).

The biological clock is also implicated in the regeneration of epithelia. In fact, despite the cellular complexity of the skin, the populations of skin stem cells are well defined, experimentally traceable, and have remarkably different circadian cell proliferation cycles. There is a correlation even between circadian clock and hair's time cycle. During the telogen phase and in the transition from telogen to anagen, circadian genes are expressed robustly in the progenitors of the stem cells of the hair (Lin et al., 2009), in contrast to the fragmented expression in the stem cells of the bulb (Janich et al., 2011; Plikus et al., 2013).

Moreover, the regulation of the immune system reveals that it is correlated with circadian clock (Scheiermann et al., 2013). The number of circulating leukocytes peaks during the day in rodents and during the night in humans. Furthermore, in tissues, the circadian variation in the expression of adhesion molecules of endothelial cells and chemokines and their receptors mediates the recruitment of leukocytes in a time-dependent manner. Those studies on several epithelial organs are confirmed, which have shown that susceptibility to infections depends on daytime. Studies have already linked the circadian clock to controlling DNA damage and UV-B skin-induced cancers. It has been observed a relationship between CLOCK and psoriasis (Plikus et al., 2015).

12.5 Biological clock modulation and chronodisruption

Chronodisruption is the phenomenon of desynchronization of peripheral oscillators from the central one. Some causes of this phenomenon are night work and jet lag, including social jet lag, i.e., the discrepancy in a person's sleep pattern between the weekday and the weekend which, if continued chronically, have been associated with an increased risk of metabolic syndrome, obesity, and cardiovascular disease (Parsons et al., 2015), all pathologies associated with unsuccessful aging.

Different environmental stimuli influence the biological clock; in particular, the most relevant factors for intensity and frequency of exposure are: (i) light and temperature, which can act directly on the central clock; (ii) fats and other nutrient molecules, which seem to modify the synchronization of the central oscillator with those peripheral. Following is an exhaustive illustration of the main literature evidence regarding these stimuli.

Evidence on animal models has led many researchers to speculate that chronic exposure to artificial light sources in the evening could help induce an increase in the risk of obesity and metabolic disorders. In one study the exposure of mice to 12 h of light and constant light for 4 consecutive days causes an increase in food intake, a reduction in energy expenditure, and an acute increase in body weight (Hamaguchi et al., 2018). The light reaches the SCN nucleus through the retinal hypothalamic tract. The fibers of the tract carry glutamatergic signals which activate the synaptic N-methyl-D-aspartate receptors present in the cell body of the SCN neurons. This phenomenon promotes the activation of the MAPkinase pathway and the phosphorylation of CREB, which, moved to the nucleus, would induce the transcription of the PER genes (Lee et al., 2010).

Pulses of light given at the beginning of the night phase cause an activation delay, while the light stimuli if present at the end of the night phase induce an acceleration of the rhythm. Animal models show that the acute exposure to light stimuli during the night induces an increase in the corticosterone peak (Ishida et al., 2005; Mohawk et al., 2007) through nervous stimuli that induce a rapid increase in the expression of PER1 and other genes involved in the synthesis of corticosterone (Ishida et al., 2005), as well as an increase in PER 1,2 in the liver (Cailotto et al., 2009). The evidence published in literature since today suggests an important role of the light in the biological clock independent synchronization of the peripheral oscillators of kidney, liver, heart, and adrenal glands, induced by one of the hypothalamic retinal nerve pathways. In mouse model, continuous light exposure protocols induce high increases in fat mass, alterations in energy expenditure, insulin resistance, dysregulations of carbohydrate metabolism (Fonken et al., 2010; Hamaguchi et al., 2015; Han and Mallampalli, 1950; Meijer et al., 2013), changes in circulating levels of cholesterol, melatonin, glucose, and corticosterone (Dauchy et al., 2010; Vauzour et al., 2010; Wideman and Murphy, 2009). Furthermore, 6 h of night light every 2 days induce an increase in weight, fat mass, triglycerides, and adipocyte size, without simultaneously increasing food consumption (Plano et al., 2017).

Stimulations with 200 lx (low intensity) during the night can modulate the melatonin blood levels (Boivin et al., 1996; Brainard et al., 1988), while exposures at the same time of day with a > 300 lx stimulation are associated with weightier and higher BMI index in older subjects. Moreover, these subjects show an increase in triglycerides and low-density lipoproteins (LDL) (Obayashi et al., 2013). The onset of time-destruction phenomena following constant exposure to night light induces an alteration of the energy expenditure, a chronic reduction of leptin, an increase in basal and postprandial glycemia insulin resistance, and a total inversion of the daily profile of cortisol, a marker of major risk of high blood pressure and of a significant reduction in sleep efficiency and quality (Leproult et al., 2014; Marti et al., 2016). The data also show an association between night work and low high-density lipoproteins levels (Guo et al., 2015; Pellettieri and Alvarado, 2007) stronger in females than in males (Chen et al., 2016; Guo et al., 2015). Moreover, it has been shown that in male night worker the risk of obesity is not significantly associated with this lifestyle (Haus et al., 2016). In summary, the data published in the literature suggest that dysfunctional eating behavior and exposure to night light can lead to negative and significant alteration in carbohydrate metabolism, related to a reduced functionality of pancreatic β cells and an increase of insulin resistance (Plano et al., 2017).

Regarding the absence of light, two days of darkness in one animal model study increased circadian gene expression and stimulated the activity of fat catabolism enzymes in peripheral metabolic tissues. Other tested animal showed that 7 days of darkness induce a reduction in consumption of food and water, in body weight, in glycemia, and an increase in the energy expenditure (Zhang et al., 2006). Despite these results, the evidence concerning darkness circadian effect on animal models is controversial and heterogeneous (Plano et al., 2017).

Experimental evidence indicates that the temperature promotes the transcription of the Heat Shock Factor 1, which, by binding to the promoters of Heat Shock Proteins (HSP), increases the transcription of HSP90 and HSP70, cytosolic factors normally linked to the inactive receptor of glucocorticoids, which in turn are able to bind to glucocorticoid response element present in the PER gene promoter. HSP90 and HSP70 form a heterodimer with this receptor and act as the molecular chaperones necessary to allow posttranslation maturation of the same. Temperature changes can exert a modulatory effect on the expression of the PER genes, by increasing the HSP (Buhr et al., 2010).

Glucocorticoids could induce transcription of the PER and BMAL1 genes and according to some authors they would constitute the zeitgeber responsible for synchronizing peripheral tissues and transcriptome expression with the rhythm of the SCN (Reddy et al., 2007); on the contrary, a study recently concluded that glucose administered to rat fibroblasts reduces the expression of the PER1 and PER2 genes (Hirota et al., 2002).

Fats influence the biological clock, through the peroxisomal activation receptors PPAR-γ and PPAR-α, which have been seen to be potentially able to promote the expression of BMAL1. The intake of high-fat foods and the continued consumption of snacks between meals could affect the correct synchronization of peripheral oscillators with the central one (Engin, 2017).

High salt intake could stimulate an acceleration of the dark phase, also inducing the transcription of the glucose transporters in the intestine, which in turn would increase the acute absorption of glucose (Oike et al., 2010).

Caffeine can prolong the circadian rhythm (Oike et al., 2011), probably by inhibiting phosphodiesterases and inducing an increase in the cAMP/PKA signal, which would induce CREB phosphorylation. According to some studies, a careful use of coffee could facilitate the restoration of the correct synchronization to the new light-dark cycle, in the cases of subjects affected by jet lag (Piérard et al., 2001).

Resveratrol is able, like the calorie restriction, to increase the transcription of BMAL1, PER1, and PER2 via SIRT-1/PGC-1α (Oike and Kobori, 2008), thus exerting a beneficial action on successful aging and longevity. It would increase the expression of PER1 and PER2 (Yagita and Okamura, 2000).

Many studies show that mealtimes are a relevant factor for the health of the body and for the synchronization of peripheral tissues with the SCN. In mouse models, hyperlipidic ad libitum diets were associated with obesity, while simple food deprivation at specific times of the same diet was associated with a significant reduction in the incidence of obesity, hyperinsulinemia, hepatic steatosis, and inflammation, suggesting that the meal time shifting can have an effect on the health of magnitude comparable to its composition (Ellisman et al., 2012). Furthermore, mice with nocturnal habits, fed only during the day (an experimental model studied to mimic the conditions to which night workers are subjected), had a significant weight increase compared to other animals with nocturnal habits and fed at night (Arble et al., 2009). Another study concluded that mouse model deprived of the first meal of the day (breakfast) showed significant alterations of liver metabolism, while those deprived of the last meal (dinner) did not suffer from these alterations and had a significantly lower body weight (Wu et al., 2011). This evidence suggests that breakfast has an important effect on the correct alignment of peripheral oscillators with the SCN, those present in the liver, while dinner can influence lipid metabolism and fat accumulation. Ultimately, the authors of a mouse study observed that a first meal made using foods with high glycemic index and amino acids significantly induces the expression of endoplasmic reticulum proteins, by means of an insulin-dependent mechanism (Bandín et al., 2015; Garaulet et al., 2010a). These and other evidences suggest that insulin is a powerful signal of synchronization of liver oscillators and adipose tissue, having an important effect on cardiovascular, dysmetabolic risk, and therefore on successful aging and longevity (Hashimoto et al., 2012). Some recent evidence corroborated the argument that modulating the body temperature and the glucocorticoids levels and limiting food intake during the dark phase in animals completely reverses the expression of CCG genes in the liver, kidneys, heart, and pancreas, but does not affect the function of the SCN (Damiola et al., 2000; Plano et al., 2017). Besides, animals induced desynchronization of the white adipose tissue peripheral oscillators, show significant alterations of genes involved in energy metabolism and in the synthesis of adipokines (but not in the CCG genes), and the abolition of the daily peak of corticosterone in these animals made the CCG genes of the peripheral clock arrhythmic (Plano et al., 2017; Su et al., 2016).

Clinical studies have shown that subjects associated with dietary regimens characterized by the consumption of the main meal of the day in the evening (late eaters) have higher levels of insulinemia, an altered glucose tolerance, an altered cortisol profile, and a tendency to resistance to weight loss compared to those who eat the main meal in the morning (early eaters) (Bandín et al., 2015; Garaulet et al., 2010a). In addition, clinical studies have shown that skipping breakfast correlates with greater weight in adolescence and an increased risk of obesity in adulthood, risk factor for unsuccessful aging. These same studies conclude that a "healthy" breakfast should consist of complex carbohydrates and a good fiber intake, deriving respectively from regular and appropriate consumption of whole grains and fruit (Timlin et al., 2008; Ma et al., 2003). Furthermore, these foods would also be correlated with a decrease in the sense of hunger and greater stability of the glycemic profile. Other studies show that subjects suffering from night eating syndrome have a significantly higher BMI [92]. Many authors speculate that these alterations in mealtimes would be associated with a lower production of leptin, a hormone subjected directly to circadian control, giving rise to a lower satiating effect (Stern et al., 2014; Scheer et al., 2009). Furthermore, diet-induced thermogenesis would be subject to circadian variations, greater in the morning and lesser at night (Romon et al., 1993).

Exercise can be another factor capable of exerting an effect on the biological clock. In fact, physical activity can have different effects on the organism based on the time of day, exerting a resulting effect of parasympathetic activation during the morning and an ultimate analysis effect of sympathetic activation during the evening, therefore probably inducing significant changes in the cholesterol profile (Yamanaka et al., 2015). The wheel motion has been proposed as a potential therapeutic tool for time-destruction models, because some in vivo studies have shown that it has a strong feedback signal of the circadian system, increasing the amplitude of the biorhythm linked to the central oscillator of the SCN (van Oosterhout et al., 2012), and promoting the synchronization of the adrenal and hepatic oscillators with the cardiometabolic activity (Plano et al., 2017; Schroeder et al., 2012).

Alterations in the expression profile of clock genes are also of important in determining an increased risk of chronodestruction. Members of the PPAR family, implicated in lipid and glucose metabolism, are subject to some circadian regulation. BMAL1 has a regulatory effect on PPAR-α expression levels and at the same time PPAR-γ and PPAR-α exert a regulatory effect on the transcription of BMAL1 itself. In addition, PPAR-α regulates the production of essential elements for lipid and cholesterol and cholesterol metabolism, such as Sterol Regulatory Element Binding Protein, Fatty Acid Synthase, and 3-hydroxy-3-methyl-glutaryl-coenzyme A reductase (Chen and Yang, 2014), whose concentrations are directly related to those of PPAR-α during the day.

Finally, recent experimental evidence suggests that the enzymes involved in mitochondrial b-oxidation also undergo circadian variations in their expression levels, as well as those of carbohydrate metabolism (Chen and Yang, 2014).

Some gene variations induced experimentally in models can interfere with the speed of the clock and with the synchronization of the central oscillator with the peripheral ones. It is known that mice mutant homozygous for the CLOCK gene completely lose the circadian rhythm and have an altered feeding rhythm. This leads to hyperphagia, obesity, metabolic syndrome, hyperleptinemia, hyperlipidemia, hepatic steatosis, hyperglycemia, and hypoinsulinemia (Eckel et al., 2005). Similarly, the deletion of the BMAL1 gene in adipose tissue is associated with a significant and marked increase in the incidence of obesity in mice (Griffin et al., 2012). This allows to hypothesize the existence of genetic variations in the CCG genes, capable of having a significant effect on the circadian clock and on the energy metabolism, as also suggested by some clinical studies in the literature. In humans, particular haplotypic combinations in the CLOCK gene were associated with an increased risk of metabolic syndrome (Scott et al., 2008) and obesity (Sookoian et al., 2008), as well as genetic variations in the PER2 and REV-ERB-α genes were significantly correlated with obesity (Garaulet et al., 2010a).

The evidence reported suggests that all the factors capable of inducing a desynchronization between the SCN and peripheral oscillators can determine a condition of chronodestruction which, if prolonged over time, increases the risk of metabolic pathologies, such as metabolic syndrome, l obesity, and type 2 diabetes (Fig. 12.1), as previously stated risk factors for unsuccessful aging.

12.6 Diet, circadian rhythm, aging, and longevity

12.6.1 Distribution of macronutrients throughout the day

Epidemiological evidence shows that an isocaloric and hyperlipidic diet would induce obesity, also due to the "phase shift" of the circadian rhythm, even more when food intake is stimulated in the resting phase (Oosterman et al., 2015). More generally, the poor quality of meals, in 24h, can alter the natural synchronism between SCN and peripheral clock. The transition from a diet high in carbohydrates and low in fiber to a diet high in content fiber and low carbohydrate content reduces salivary cortisol levels and positively modulates the expression of PER genes, confirming the interaction between nutrients and peripheral oscillators. Other studies support the idea that fasting, during the day, in the activity phase, involves a greater increase in triglyceride and LDL levels, compared to subjects who consume the same calories, but distributed throughout the day. Furthermore, the common Western diet is rich in animal proteins, saturated fats, and simple sugars and constitutes a powerful negative zeitgeber associated with a reduction in successful aging and longevity. On the contrary, a Mediterranean-style diet is a powerful positive zeitgeber, as well as for the essential compounds also for the functional

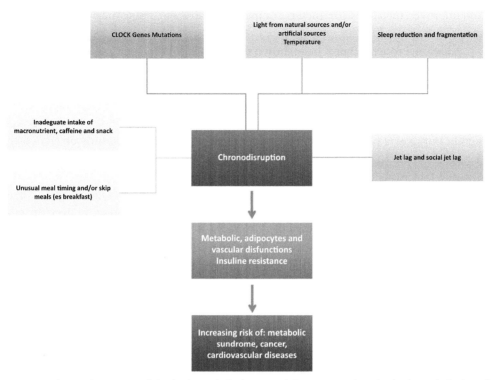

Fig. 12.1 Chronodisruption of the biological clock. Several factors can alter the biological clock and desynchronize the SCN from peripheral oscillators: mutations in the CLOCK, BMAL1, CRY, and PER genes; artificial light, especially in the evening and at night, which can interrupt sleep; temperature changes; changes in the sleep-wake cycle, such as those caused by night work or by factors that reduce or fragment sleep; the chronic jet lag, to which, for example, travelers by profession or hostesses and aircraft pilots are subjected; fats contained in food, active ingredients such as caffeine, taking food at unusual times (for example at night) or skipping breakfast and having abundant dinners. All these factors lead to alterations in the metabolism which lead to an increased risk of metabolic, cardiovascular, and cancer pathologies.

effects of the individual nutrients, including polyphenols, pro-anthocyanidins (which regulate CCGs such as Bmal1 and NAMPT), polyamines (which regulate the functionality of the PER2–CRY1 complex), monounsaturated and polysaturated fatty acids (which modulate SIRT-6, retinoid x receptor (RXR), PPAR-γ, and PPAR-α) and phytosterols (which modulate the signaling of vitamin A and vitamin D, modulating the activity of RXRs) (Heyde and Oster, 2019; Hood and Amir, 2017; Monk, 2010).

A study compared metabolic parameters in a group of mice taking one single large meal at the beginning of the active (night) phase and mice that had a small additional meal at the end of the active phase. Although the mice in each group consumed an equal amount of food in the 24 h, the mice with two meals showed reduced body weight gain and improved metabolic parameters compared to those of a single meal or freely fed animals (Fuse et al., 2012).

Additional rodent studies have shown that early night fasting increases body weight, while late night fasting reduces weight gain (Jakubowicz et al., 2012, 2013). In a study on weight loss, the comparison between two groups with isocaloric diet revealed, in the group that had the largest breakfast and the smallest dinner, the greatest improvement in metabolic parameters and in another study, it was shown that the early meals (morning) significantly lower blood lipid levels. Furthermore, in a large cohort study on adolescents it was observed that breakfast consumption is inversely associated with weight gain (Jakubowicz et al., 2012, 2013).

Several studies have shown the correlation between short sleep duration (<5 h) or advanced sleep (midpoint of sleep > 5:30 AM) and the consumption of dinners/the consumption of more calories late in the evening with a significantly risk higher in obesity and diabetes (Dashti et al., 2015; Hamaguchi et al., 2018; Hsieh et al., 2011; Shan et al., 2015). Furthermore, the nocturnal consumption syndrome characterized by a delayed diet over time is positively associated with a high BMI (Colles et al., 2007).

12.6.2 Meals frequency

Individuals isolated and in the absence of external stimuli organize their diet according to the 3 + 2 scheme, characterized by three main meals and two snacks (Aschoff, 1979). As early as 1964 it was observed that individuals who habitually consumed three or less meals had a greater risk of obesity than those who followed the earlier scheme (Fabry, 1964). Furthermore, it has been shown that in male subjects, the frequency of meals is inversely correlated with the increase in body weight (Drummond et al., 1998). Furthermore, in adolescent subjects the consumption of a hearty breakfast and a light dinner is associated with a lower BMI (Summerbell et al., 1996). A recent study concluded that skipping breakfast is associated with an increase in BMI and waist circumference, as well as eating a dinner less than 3 h before bedtime was significantly associated with a higher value of BMI (Yasuhiko et al., 2014). A clinical study conducted on obese women shows how these subjects tended to spontaneously consume about six meals during the day, compared to five of the nonobese, as well as showing a tendency to eat more abundant meals in the evening (Bertéus Forslund et al., 2002). However, an increase in meal frequency reduces the peak in glucose and insulin, especially in combination with a diet rich in fiber and this can have a protective effect on the risk of obesity and metabolic syndrome (Lundin et al., 2004). Consequently, it is underlined how attention should be paid to avoiding the consumption of meals after the evening hours or at least three hours before going to bed. Furthermore, regularity in the consumption of meals is also important, which is associated with a greater postmeal thermogenic effect, a better lipid profile, and a lower insulinemic peak (Farshchi et al., 2005).

As reported in the previous paragraphs, the consumption of meals at delayed times affects the circadian biorhythm associated with the regulation of glycemia and the PER genes activity in the adipose tissue, suggesting that the timing of meals can have both a positive and negative effect on the correct alignment of circadian biorhythms in night worker (Wehrens et al., 2017). A 16-year longitudinal study reported a 27% higher risk of coronary heart disease in men who skipped breakfast and a 55% higher risk in men who were late night eater. No association was observed between food frequency and coronary heart disease risk, suggesting that mealtimes may have a greater effect on cardiovascular risk (Cahill et al., 2013). A recent study has shown that over 50% of adults distribute their daily calorie intake over 15 h or more and reducing the duration of daily food consumption in 10–11 h for 16 weeks, without other nutritional modification like reducing calories or changing the food choices, induced weight loss and improved the sleep quality of the subjects (Gill and Panda, 2015). On the whole, animal and clinical studies suggest a strong correlation between the altered exposure to light at night, and a dysfunctional glucocorticoid profile and probably light intensity may increase or decrease this alteration (Plano et al., 2017).

12.6.3 Caloric restriction

Moreover, postprandial glycemia is greatly affected by the biological clock because the circadian variation of cortisol and other glucocorticoids modulates the glucose tolerance, especially in the evening and in the night hours (Basu et al., 2012; Morris et al., 2016; Yuen et al., 2013). Furthermore, an excessive caloric intake occurring when the organism is not able to manage it (like store it, mobilize it, and generate it) is an independent risk factor for insulin resistance conditions, metabolic syndrome, and obesity (Bo et al., 2014; Morgan and Newland, 1999). In humans and in other mammals, chronodestruction and other more moderate circadian alterations can deeply alter the endocrine signaling associated with the metabolic management of glucose, fatty acids, or ketone bodies, inducing an increase in postprandial glycemia, due to the desynchronization of the pancreatic oscillators involved in glucose management (Plano et al., 2017). In the night workers, eating at night behavior reduces the total daily energy expenditure (McHill et al., 2014). Therefore the evidences suggest that changes in glucose uptake are correlated with chronic changes in circadian biorhythms. Reduced postprandial glucose tolerance, chronic dysregulation in the absorption and use of glucose may lead to an imbalance between the use of carbohydrates and triglycerides that can lead to increase in BMI, dyslipidemia, and hypertriglyceridemia. These alterations have also been correlated with the negative energy balance imposed by sleep debt (Eckel et al., 2015; McHill et al., 2014), and this dysfunctional behavior (eating or being awake when the circadian clock promotes sleep) is a risk factor for metabolic syndrome, weight gain, and obesity (Plano et al., 2017).

Several studies have suggested that both in animal models and humans, dietary intervention can prevent or decrease various age-related diseases, by positively regulating aging process through the modulation of nutrient-sensing pathways such as the insulin/insulin-like growth factor (IGF)-1, the mechanistic target of rapamycin (mTOR), and the sirtuin pathways (Aiello et al., 2017).

An example of dietary intervention is caloric restriction (CR), the only known strategy for increasing successful aging and longevity in almost all living organisms. Nonetheless, the data published regarding the long-term food deprivation in human population are conflicting and heterogeneous and sometimes it has been associated with negative effects (Bahijri et al., 2013; Charmandari et al., 2005; Golbidi et al., 2017).

To discuss CR in this chapter is appropriate because it entrains the central clock in the SCN (Mendoza et al., 2005). It also upregulates clock genes in several tissues (Swindell, 2008). Thus CR affects the clock and metabolism both centrally and peripherally. As previously stated, the Sirt-1 protein deacetylase forms a complex with the clock acetylase. This complex regulates the activity of clock-controlled genes through chromatin remodeling (Nakahata et al., 2008) and by acetylating/deacetylating Bmal1 and Per2 (Asher et al., 2008). SIRT-1 is also an accessory protein to the core clock machinery (Asher et al., 2008).

However, CR is not simply fasting, but a diet model characterized by a strong decrease in the caloric intake without providing insufficient micronutrients and essential compounds. Its role in human aging is difficult to ascertain because human lifespan makes long-term investigations and there are no universally accepted biomarkers to measure the rate of human aging, i.e., not chronological age, but biological age. However, some epidemiologic and short-term human studies support CR-related health benefits (Fontana et al., 2010). Moreover, in 2004, a long-term epidemiologic study linked CR to human longevity. The authors followed 1915 healthy non-smokers, aged 45–68, for 36 years. This study suggested that maintaining low energy intake in midlife determines lowest late-life mortality risk. Indeed, a weak trend toward lower all-cause mortality was reported in healthy never-smoking Japanese-American men whose caloric intake had been reduced by 15% with respect to the cohort average (Willcox et al., 2004). Moreover, a controlled randomized study on nonobese individuals reported that a two year 25% CR is feasible for humans and provides health benefits, such as decrease of inflammatory markers and cardiometabolic risk factors (Redman et al., 2014).

Despite the positive effects, calorie restriction may be difficult to maintain over time, also with the help of a health professional and, without adequate compliance to the dietary prescription, it probably increases the risk of malnutrition. Some authors suggest the application of intermittent abstinence from food and caloric beverages, as an alternative to the classic caloric restriction, which could reduce the risk of malnutrition. Intermittent abstinence for at least 12 h (IF) or 48–120 h (PF) improves health effects and protect from age-related disabilities. In humans, IF (e.g., consumption of approximately 500 kcal/day for 2 days a week) has beneficial effects on insulin, blood glucose levels, C-reactive protein, and blood pressure (Harvie et al., 2011). In mice, PF cycles, lasting 2 or more days, but separated by a week of normal diet, have emerged as a highly effective strategy to reduce circulating IGF-1 and glucose levels, and, consequently, to downregulate mTOR pathway. In humans, IF and PF help to reduce obesity, hypertension, asthma, and rheumatoid arthritis (Lee and Longo, 2016; Longo and Mattson, 2014).

Another alternative to caloric restriction is the regular use of caloric restriction mimetic that promote the deacetylation of certain cellular proteins (Lakin-Thomas, 2019) through the decrease of acyl-coA reserves, inhibition of acetyltransferases, and stimulation of the deacetylases. Those that in in vivo studies have proven more positive effects on successful aging and longevity are hydroxycitrate and other acyl CoA depleting agents, metabolic intermediates of the synthesis of NAD, anacardic acid, curcumin, polyphenols such as epigallocatechins, garcinol, spermidine, rapamycin, and resveratrol (Madeo et al., 2019).

12.7 Meals composition for successful aging

In order to promote successful aging and longevity, the following recommendations must be ensured: a proper diet model compliance, a diet regime with the correct daily meal composition, an optimal modulation of the light-dark cycle, and an efficient sleep in terms of quality and quantity. All the following indications must always be implemented in the context of a diet conforming to the FAO and OMS recommendation. In addition, all these indications are however to be harmonized to the clinical characteristics of the subject.

The carbohydrates intake in the morning is necessary to reduce cortisol levels and avoid the onset of those dysfunctional glucocorticoids level. It is well known that glucose tolerance is higher in the morning than in the afternoon and evening (Charmandari et al., 2005). In healthy subjects, the period before awakening is characterized by a maximum peak of ACTH and cortisol, to compensate the hypoglycemic stimulus due to night fasting. This hormonal structure induces a considerable hepatic gluconeogenetic stimulus, as well as an increase in the Cahill cycle, i.e., the series of reactions in which amino groups and carbons from muscle are transported to the liver, and the mobilization of muscle energy reserves. In animal models, this leads to a stimulation of carbohydrate intake and the search for food. On the contrary, the intake of food, and of carbohydrate sources, reduces cortisol levels. Chronically high levels of glucocorticoids lead to an increase in the levels of circulating free fatty acids, directly related to a greater accumulation of fat, especially at central and visceral level (as occurs in people with Cushing's syndrome) and can promote insulin resistance. Moreover, in these conditions there is a clear stimulation of the differentiation processes of the preadipocytes in the adipose tissue. In the muscles, an excessive protein catabolism supports the gluconeogenesis and contributes to increase the risk of insulin resistance. Finally, glucocorticoids can directly inhibit insulin release, once again increasing the risk of insulin resistance (Dimitriadis et al., 1997; Lee et al., 2014).

Therefore accordingly of the previous considerations and considering that breakfast is the first meal of the day that interrupts the overnight fast, it is recommend to eat a breakfast that provides 20%–25% of the daily calories. The breakfast should be mostly based on sources of low glycemic index carbohydrates and fiber, both to reduce cortisol and the ghrelin signaling. The reasonably more suitable foods are therefore whole grains, fruit, together with milk, yogurt, and/or eggs, according to the personal preferences. During breakfast, attention should be paid to the salt levels introduced, since in the morning the aldosterone levels are maximum and accordingly even the reabsorption of water and sodium. It is therefore advisable to pay attention to the consumption of prepackaged foods, which are often added with significant quantities of sucrose and sodium chloride. It is recommended to check the values of simple carbohydrates contained in yogurt and the sodium amount added in biscuits, in rusks, and other similar cereal-based foods.

The midmorning snack (which should also be based on whole grains and/or fruit) can help stabilize blood sugar and therefore avoid excessive calorie intake in lunch.

Lunch should provide 40%–45% of daily calories, especially considering that in the afternoon there is the greatest energy expenditure from physical activity of the day. It is recommended the consumption of whole grains and vegetables. In some subjects, especially those who are overweight and heavily overweight, it is advisable to consume the finely chopped vegetables before the main meal since one of the main inhibitory stimuli of the release of ghrelin, and subsequently it is recommended to finely cut the vegetables (Koliaki, 2010). Fats sources must be consumed, for example, in the form of extra virgin olive oil on vegetables.

Clinical studies have shown that an increase in plasma glucose reduces the release of hypothalamic growth hormone (GH)-releasing hormone (GRH) and therefore of GH, while hypoglycemia stimulates the release of GRH and GH; in addition, the release of GH is inhibited by the effect of insulin and cortisol. In general, a meal rich in carbohydrates reduces the production of GH, while a meal rich in proteins and above all of arginine stimulates their synthesis. At this regards, it has to be considered that GH is responsible for the production of IGF-1, and high levels of IGF-1 downregulate FOXO3A signaling. FOXO3A signaling is responsible for the suppression of proinflammatory mediators and induction of antioxidant enzymes, so favoring successful aging and longevity (Aiello et al., 2017). The oxidation of lipids is greater in the evening and higher than that of carbohydrates (Sensi and Capani, 1987), also considering that the nocturnal hormonal profile is particularly favorable for the consumption of fats, the intake of optimal sources of fats at dinner is certainly advisable.

The afternoon snack should be about 5% of daily calories intake, while dinner should be light (about 30% daily calories intake) and provide on good protein and/or fat sources, since during the evening hours, in order to better sustain the physiological increases in the levels of thyroid hormones (which have a peak after midnight) and the production of GH which has a maximum around midnight (however, see earlier). Dinner should be composed of meat, fish, eggs, legumes, and dairy products to be consumed together with vegetables. The continuous consumption of snack based on fatty foods between meals must be avoided, because fats, in such conditions, alter the cellular clock (Kohsaka and Bass, 2007).

Today, recommendations claim that in older subjects it is necessary to increase the protein intake from 0.8 g/kg of body weight to at least 0.95 g/kg of body weight, mainly for preventing sarcopenia and sarcopenic obesity. In vivo studies suggest that a low-caloric diet and a low intake of animal proteins would significantly reduce blood levels of IGF-1 (Fontana et al., 2008), responsible for inhibition of mTOR, and downregulation of FOXO3A signaling (Aiello et al., 2017). In this regard, further clinical studies and human longitudinal studies are needed, although some evidence on humans is already present in the literature, such as the CALERIE Study; the interviewees aged between 50 and 65 years, who reported a high protein intake, showed an increase in IGF-1 levels, a 75% increase in the risk of general mortality, and a 3–4 times increase in the risk of cancer mortality, in accordance with the results of experiments on mouse models. Other studies show that a low protein intake is associated with a reduction of IGF-1, cancer, and general mortality in the populations under 65 years, but not in the older population (Ravussin et al., 2015; Rickman et al., 2011).

In all the main meal of the day, it is recommended to consume season vegetables, due to the maximum mineral and vitamin content, using a spoon for adding the oil, in order to avoid the accidental consumption of not safe quantities of fats. The oil and the lemon juice should be added immediately before eating, preventing oxygen and light from further reducing the content of sensitive compounds and functional nutrients. The chrononutrition studies do not indicate specific quantities of nutrients to be consumed during the day and in the absence of such indications, it is necessary to follow the international guidelines on nutrition drawn up by WHO and OMS.

12.8 Conclusion and future perspectives

In the end, we can conclude that the optimal diet model for promoting successful aging and longevity and health is the one capable of satisfying at least the points later considering data from centenarian studies.

Epidemiological Italian studies on populations with a high incidence of centenarians have concluded that these subjects had the following common denominators: high intake of fruit, legumes, and vegetables; outdoor movement; fewer calories intake than recommendation; prevalence of local products and gastronomy; and at the end, correct circadian distribution of meals, which includes a hearty breakfast, light snacks, and frugal dinner (Franceschi et al., 2018; Sonya et al., 2012).

Guarantee over time the quality of the food introduced, inspired by the food choices typical of the Mediterranean model with oriental influences.

An optimal modulation of the light–dark cycle and an efficient sleep in terms of quality and quantity.

Adopt a calorie intake that, depending on the proposed diet, provides a quantity of calories that is between a range of 15% and 20% around the basal metabolic rate in subjects with a low level of physical activity. From data published in the literature, it appears that a sustainable calorie restriction regimen for humans may start from a calorie intake 15%–20% lower than the basal metabolic rate, since this calorie intake has been shown to be safe in human studies. Furthermore, significant effects of calorie restriction (10%–15% lower than the basal metabolism) are demonstrated even in humans and adopting a balanced diet with that caloric intake is considered effective and safe (Rickman et al., 2011).

Provide a protein content of around 0.95–1 g/kg body weight. For older people, see earlier. The animal proteins should be limited.

But what should be the distribution and timing of meals and related nutrients in the day? These two parameters really matter and to what extent.

Both a low-calorie diet and the regular use of functional and mimetic compounds could be the nutritional basis for successful aging and longevity, obviously only together with all the other measures that characterize a correct lifestyle, but in this context, there remain some questions: Which low-calorie level is sustainable over time and for how long? Furthermore, which food proposal is closest to our customs? Which longitudinal human studies can be a reference model of positive biology for the general population?

References

Adamczyk-Sowa, Sowa, Zwirska-Korczala K, Pierzchala, et al. Labeled [^{3}H]—thymidine incorporation in the DNA of 3T3-L1 preadipocytes due to MT$_2$- and not MT$_3$- melatonin receptor. J Physiol Pharmacol 2012;65(1):152–66.

Adamovich Y, Rousso-Noori L, Zwighaft Z, Neufeld-Cohen A, Golik M, Kraut-Cohen J, Wang M, Han X, Asher G. Circadian clocks and feeding time regulate the oscillations and levels of hepatic triglycerides. Cell Metab 2014;19(2):319–30. https://doi.org/10.1016/j.cmet.2013.12.016.

Aguilar-Arnal L, Sassone-Corsi P. The circadian epigenome: how metabolism talks to chromatin remodeling. Curr Opin Cell Biol 2013;25(2):170–6. https://doi.org/10.1016/j.ceb.2013.01.003.

Ahima RS, Flier JS. Adipose tissue as an endocrine organ. Trends Endocrinol Metab 2000;11(8):327–32. https://doi.org/10.1016/S1043-2760(00)00301-5.

Aiello A, Accardi G, Candore G, Gambino CM, Mirisola M, Taormina G, Virruso C, Caruso C. Nutrient sensing pathways as therapeutic targets for healthy ageing. Expert Opin Ther Targets 2017;21(4):371–80. https://doi.org/10.1080/14728222.2017.1294684.

Akhtar RA, Reddy AB, Maywood ES, Clayton JD, King VM, Smith AG, Gant TW, Hastings MH, Kyriacou CP. Circadian cycling of the mouse liver transcriptome, as revealed by cDNA microarray, is driven by the suprachiasmatic nucleus. Curr Biol 2002;12(7):540–50. https://doi.org/10.1016/S0960-9822(02)00759-5.

Angelousi A, Kassi E, Nasiri-Ansari N, Weickert MO, Randeva H, Kaltsas G. Clock genes alterations and endocrine disorders. Eur J Clin Invest 2018;48(6)https://doi.org/10.1111/eci.12927.

Anhê GF, Caperuto LC, Pereira-Da-Silva M, Souza LC, Hirata AE, Velloso LA, Cipolla-Neto J, Carvalho CRO. In vivo activation of insulin receptor tyrosine kinase by melatonin in the rat hypothalamus. J Neurochem 2004;90(3):559–66. https://doi.org/10.1111/j.1471-4159.2004.02514.x.

Arble DM, Bass J, Laposky AD, Vitaterna MH, Turek FW. Circadian timing of food intake contributes to weight gain. Obesity 2009;17(11):2100–2. https://doi.org/10.1038/oby.2009.264.

Aschoff J. Circadian rhythms: influences of internal and external factors on the period measured in constant conditions. Z Tierpsychol 1979;49(3):225–49. https://doi.org/10.1111/j.1439-0310.1979.tb00290.x.

Asher G, Gatfield D, Stratmann M, Reinke H, Dibner C, Kreppel F, Mostoslavsky R, Alt FW, Schibler U. SIRT1 regulates circadian clock gene expression through PER2 deacetylation. Cell 2008;134(2):317–28. https://doi.org/10.1016/j.cell.2008.06.050.

Bahijri S, Borai A, Ajabnoor G, Abdul Khaliq A, AlQassas I, Al-Shehri D, Chrousos G. Relative metabolic stability, but disrupted circadian cortisol secretion during the fasting month of Ramadan. PLoS One 2013;8(4)https://doi.org/10.1371/journal.pone.0060917.

Bähr I, Mühlbauer E, Schucht H, Peschke E. Melatonin stimulates glucagon secretion in vitro and in vivo. J Pineal Res 2011;50(3):336–44. https://doi.org/10.1111/j.1600-079X.2010.00848.x.

Bandín C, Scheer FAJL, Luque AJ, Ávila-Gandiá V, Zamora S, Madrid JA, Gómez-Abellán P, Garaulet M. Meal timing affects glucose tolerance, substrate oxidation and circadian-related variables: a randomized, crossover trial. Int J Obes (Lond) 2015;39(5):828–33. https://doi.org/10.1038/ijo.2014.182.

Basu A, Man CD, Nandy DK, Levine JA, Bharucha AE, Rizza RA, Basu R, Carter RE, Cobelli C, Kudva YC. Diurnal pattern to insulin secretion and insulin action in healthy individuals. Diabetes 2012;61(11):2691–700. https://doi.org/10.2337/db11-1478.

Belden WJ, Dunlap JC. XAging well with a little wine and a good clock. Cell 2013;153(7):1421. https://doi.org/10.1016/j.cell.2013.05.055.

Bertéus Forslund H, Lindroos AK, Sjöström L, Lissner L. Meal patterns and obesity in Swedish women—a simple instrument describing usual meal types, frequency and temporal distribution. Eur J Clin Nutr 2002;56(8):740–7. https://doi.org/10.1038/sj.ejcn.1601387.

Bo S, Musso G, Beccuti G, Fadda M, Fedele D, Gambino R, Gentile L, Durazzo M, Ghigo E, Cassader M. Consuming more of daily caloric intake at dinner predisposes to obesity. A 6-year population-based prospective cohort study. PLoS One 2014;9(9)https://doi.org/10.1371/journal.pone.0108467.

Boden G, Ruiz J, Urbain JL, Chen X. Evidence for a circadian rhythm of insulin secretion. Am J Physiol Endocrinol Metab 1996;E246–52. https://doi.org/10.1152/ajpendo.1996.271.2.E246.

Boivin DB, James FO. Circadian adaptation to night-shift work by judicious light and darkness exposure. J Biol Rhythms 2002;17(6):556–67. https://doi.org/10.1177/0748730402238238.

Boivin DB, Duffy JF, Kronauer RE, Czeisler CA. Dose-response relationships for resetting of human circadian clock by light. Nature 1996;379(6565):540–2. https://doi.org/10.1038/379540a0.

Bolli G, de Feo P, Perriello G, De Cosmo S, Compagnucci P, Santeusanio F, Brunetti P, Unger RH. Mechanisms of glucagon secretion during insulin-induced hypoglycemia in man. Role of the beta cell and arterial hyperinsulinemia. J Clin Investig 1984;73(4):917–22. https://doi.org/10.1172/JCI111315.

Bouchard-Cannon P, Mendoza-Viveros L, Yuen A, Kærn M, Cheng HYM. The circadian molecular clock regulates adult hippocampal neurogenesis by controlling the timing of cell-cycle entry and exit. Cell Rep 2013;5(4):961–73. https://doi.org/10.1016/j.celrep.2013.10.037.

Brainard GC, Lewy AJ, Menaker M, Fredrickson RH, Miller LS, Weleber RG, Cassone V, Hudson D. Dose-response relationship between light irradiance and the suppression of plasma melatonin in human volunteers. Brain Res 1988;454(1–2):212–8. https://doi.org/10.1016/0006-8993(88)90820-7.

Brenner DA, Montminy M, Kay SA, Pongsawakul PY, Liu AC, Hirota T, Nusinow DA, Sun X, Landais S, Kodama Y. Cryptochrome mediates circadian regulation of cAMP signaling and hepatic gluconeogenesis. Nat Med 2010;16(10):1152–6. https://doi.org/10.1038/nm.2214.

Brydon L, Petit L, Delagrange P, Strosberg AD, Jockers R. Functional expression of MT2 (Mel 1b) melatonin receptors in human PAZ6 adipocytes. Endocrinology 2001;142(10):4264–71. https://doi.org/10.1210/endo.142.10.8423.

Buhr ED, Yoo SH, Takahashi JS. Temperature as a universal resetting cue for mammalian circadian oscillators. Science 2010;330(6002):379–85. https://doi.org/10.1126/science.1195262.

Buijs RM, Kalsbeek A. Hypothalamic integration of central and peripheral clocks. Nat Rev Neurosci 2001;2(7):521–6. https://doi.org/10.1038/35081582.

Buijs FN, León-Mercado L, Guzmán-Ruiz M, Guerrero-Vargas NN, Romo-Nava F, Buijs RM. The circadian system: a regulatory feedback network of periphery and brain. Physiology 2016;31(3):170–81. https://doi.org/10.1152/physiol.00037.2015.

Burnetti AJ, Aydin M, Buchler NE. Cell cycle start is coupled to entry into the yeast metabolic cycle across diverse strains and growth rates. Mol Biol Cell 2016;27(1):64–74. https://doi.org/10.1091/mbc.E15-07-0454.

Cahill LE, Chiuve SE, Mekary RA, Jensen MK, Flint AJ, Hu FB, Rimm EB. Prospective study of breakfast eating and incident coronary heart disease in a cohort of male US health professionals. Circulation 2013;128(4):337–43. https://doi.org/10.1161/CIRCULATIONAHA.113.001474.

Cailotto C, La Fleur SE, Van Heijningen C, Wortel J, Kalsbeek A, Feenstra M, Pévet P, Buijs RM. The suprachiasmatic nucleus controls the daily variation of plasma glucose via the autonomic output to the liver: are the clock genes involved? Eur J Neurosci 2005;22(10):2531–40. https://doi.org/10.1111/j.1460-9568.2005.04439.x.

Cailotto C, Lei J, van der Vliet J, van Heijningen C, van Eden CG, Kalsbeek A, Pévet P, Buijs RM. Effects of nocturnal light on (clock) gene expression in peripheral organs: a role for the autonomic innervation of the liver. PLoS One 2009;4(5)https://doi.org/10.1371/journal.pone.000565.

Cardoso Alonso-Vale MI, Forato Anhê G, Das Neves Borges-Silva C, Andreotti S, Barnabé Peres S, Cipolla-Neto J, Bessa Lima F. Pinealectomy alters adipose tissue adaptability to fasting in rats. Metab Clin Exp 2004;53(4):500–6. https://doi.org/10.1016/j.metabol.2003.11.009.

Cardoso Alonso-Vale MI, Andreotti S, Borges-Silva CDN, Mukai PY, Cipolla-Neto J, Lima FB. Intermittent and rhythmic exposure to melatonin in primary cultured adipocytes enhances the insulin and dexamethasone effects on leptin expression. J Pineal Res 2006;41(1):28–34. https://doi.org/10.1111/j.1600-079X.2006.00328.x.

Challet E. Keeping circadian time with hormones. Diabetes Obes Metab 2015;17(1):76–83. https://doi.org/10.1111/dom.12516.

Chaput JP, Després JP, Bouchard C, Tremblay A. The association between sleep duration and weight gain in adults: a 6-year prospective study from the Quebec Family Study. Sleep 2008;31(4):517–23. https://doi.org/10.1093/sleep/31.4.517.

Charmandari E, Tsigos C, Chrousos G. Endocrinology of the stress response. Annu Rev Physiol 2005;67:259–84. https://doi.org/10.1146/annurev.physiol.67.040403.120816.

Chen Z, McKnight SL. A conserved DNA damage response pathway responsible for coupling the cell division cycle to the circadian and metabolic cycles. Cell Cycle 2007;6(23):2906–12. https://doi.org/10.4161/cc.6.23.5041.

Chen L, Yang G. PPARs integrate the mammalian clock and energy metabolism. PPAR Res 2014;2014: https://doi.org/10.1155/2014/653017.

Chen M, Yu S, Yu B, Yang Y, Jin Z, Duan W, Zhao G, Zhai M, Liu L, Yi D. Berberine attenuates myocardial ischemia/reperfusion injury by reducing oxidative stress and inflammation response: role of silent information regulator 1. Oxid Med Cell Longev 2016;2016:https://doi.org/10.1155/2016/1689602.

Cipolla-Neto J, Machado UF, Bartol I, Seraphim PM, Sumida DH, Moraes SMF, Hell NS, Okamoto MM, Saad MJA, Carvalho CRO. Pinealectomy causes glucose intolerance and decreases adipose cell responsiveness to insulin in rats. Am J Physiol Endocrinol Metab 1998;275(6):E934–41. https://doi.org/10.1152/ajpendo.1998.275.6.e934.

Colles SL, Dixon JB, O'Brien PE. Night eating syndrome and nocturnal snacking: Association with obesity, binge eating and psychological distress. Int J Obes (Lond) 2007;31(11):1722–30. https://doi.org/10.1038/sj.ijo.0803664.

Dacey DM, Liao HW, Peterson BB, Robinson FR, Smith VC, Pokorny J, Yau KW, Gamlin PD. Melanopsin-expressing ganglion cells in primate retina signal colour and irradiance and project to the LGN. Nature 2005;433(7027):749–54. https://doi.org/10.1038/nature03387.

Damiola, F., Le Minli, N., Preitner, N., Kornmann, B., Fleury-Olela, F., & Schibler, U. (2000). Restricted feeding uncouples circadian oscillators in peripheral tissues from the central pacemaker in the suprachiasmatic nucleus. Genes Dev, 14 (23), 2950–2961. doi:https://doi.org/10.1101/gad.183500

Dang CV, Hogenesch JB, Weljie AM, ER U, Sianati B, Sengupta A, Anafi RC, Kavakli IH, Sancar A, Baur JA. Clock regulation of metabolites reveals coupling between transcription and metabolism. Cell Metab 2017;25(4)https://doi.org/10.1016/j.cmet.2017.03.019 961–974.e4.

Dashti HS, Scheer FAJL, Jacques PF, Lamon-Fava S, Ordovás JM. Short sleep duration and dietary intake: epidemiologic evidence, mechanisms, and health implications. Adv Nutr 2015;6(6):648–59. https://doi.org/10.3945/an.115.008623.

Dauchy RT, Dauchy EM, Tirrell RP, Hill CR, Davidson LK, Greene MW, Tirrell PC, Wu J, Sauer LA, Blask DE. Dark-phase light contamination disrupts circadian rhythms in plasma measures of endocrine physiology and metabolism in rats. Comp Med 2010;60(5):348–56.

Davidson AJ. Search for the feeding-entrainable circadian oscillator: a complex proposition. Am J Physiol Regul Integr Comp Physiol 2006;290(6):R1524–6. https://doi.org/10.1152/ajpregu.00073.2006.

de Winter L, Schepers LW, Cuaresma M, Barbosa MJ, Martens DE, Wijffels RH. Circadian rhythms in the cell cycle and biomass composition of Neochloris oleoabundans under nitrogen limitation. J Biotechnol 2014;187:25–33. https://doi.org/10.1016/j.jbiotec.2014.07.016.

Diaz B, Blazquez E. Effect of pinealectomy on plasma glucose, insulin and glucagon levels in the rat. Horm Metab Res 1986;18(4):225–9. https://doi.org/10.1055/s-2007-1012279.

Dijk DJ, Kronauer RE, Shanahan TL, Brown EN, Mitchell JF, Rimmer DW, Ronda JM, Silva EJ, Allan JS, Emens JS. Stability, precision, and near-24-hour period of the human circadian pacemaker. Science 1999;284(5423):2177–81. https://doi.org/10.1126/science.284.5423.2177.

Dimitriadis G, Leighton B, Parry-Billings M, Sasson S, Young M, Krause U, Bevan S, Piva T, Wegener G, Newsholme EA. Effects of glucocorticoid excess on the sensitivity of glucose transport and metabolism to insulin in rat skeletal muscle. Biochem J 1997;321(3):707–12. https://doi.org/10.1042/bj3210707.

Drummond SE, Crombie NE, Cursiter MC, Kirk TR. Evidence that eating frequency is inversely related to body weight status in male, but not female, non-obese adults reporting valid dietary intakes. Int J Obes (Lond) 1998;22(2):105–12. https://doi.org/10.1038/sj.ijo.0800552.

Duffy JF, Czeisler CA. Effect of light on human circadian physiology. Sleep Med Clin 2009;4(2):165–77. https://doi.org/10.1016/j.jsmc.2009.01.004.

Ebling FJP. Hypothalamic control of seasonal changes in food intake and body weight. Front Neuroendocrinol 2015;37:97–107. https://doi.org/10.1016/j.yfrne.2014.10.003.

Eckel RH, Takahashi JS, Bass J, Lin E, Ivanova G, McDearmon E, Laposky A, Losee-Olson S, Easton A, Jensen DR. Obesity and metabolic syndrome in circadian clock mutant nice. Science 2005;308 (5724):1043–5. https://doi.org/10.1126/science.1108750.

Eckel RH, Depner CM, Perreault L, Markwald RR, Smith MR, McHill AW, Higgins J, Melanson EL, Wright KP. Morning circadian misalignment during short sleep duration impacts insulin sensitivity. Curr Biol 2015;25(22):3004–10. https://doi.org/10.1016/j.cub.2015.10.011.

Eckel-Mahan K, Sassone-Corsi P. Metabolism and the circadian clock converge. Physiol Rev 2013;93 (1):107–35. https://doi.org/10.1152/physrev.00016.2012.

Ellisman MH, Panda S, Zarrinpar A, DiTacchio L, Bushong EA, Gill S, Leblanc M, Chaix A, Joens M, Fitzpatrick JAJ. Time-restricted feeding without reducing caloric intake prevents metabolic diseases in mice fed a high-fat diet. Cell Metab 2012;15(6):848–60. https://doi.org/10.1016/j.cmet.2012.04.019.

Engin A. Circadian rhythms in diet-induced obesity. In: Advances in experimental medicine and biology. vol. 960. Springer New York LLC; 2017. p. 19–52. https://doi.org/10.1007/978-3-319-48382-5_2.

Fabry P. The frequency of meals. its relation to overweight, hypercholesterolaemia, and decreased glucose-tolerance. Lancet 1964;2(7360):614–5.

Farshchi HR, Taylor MA, Macdonald IA. Beneficial metabolic effects of regular meal frequency on dietary thermogenesis, insulin sensitivity, and fasting lipid profiles in healthy obese women. Am J Clin Nutr 2005;81(1):16–24. https://doi.org/10.1093/ajcn/81.1.16.

Fonken LK, Workman JL, Walton JC, Weil ZM, Morris JS, Haim A, Nelson RJ. Light at night increases body mass by shifting the time of food intake. Proc Natl Acad Sci U S A 2010;107(43):18664–9. https://doi.org/10.1073/pnas.1008734107.

Fontana L, Weiss EP, Villareal DT, Klein S, Holloszy JO. Long-term effects of calorie or protein restriction on serum IGF-1 and IGFBP-3 concentration in humans. Aging Cell 2008;7(5):681–7. https://doi.org/10.1111/j.1474-9726.2008.00417.x.

Fontana L, Partridge L, Longo VD. Extending healthy life span-from yeast to humans. Science 2010;328 (5976):321–6. https://doi.org/10.1126/science.1172539.

Franceschi C, Ostan R, Santoro A. Nutrition and inflammation: are centenarians similar to individuals on calorie-restricted diets? Annu Rev Nutr 2018;38:329–56. https://doi.org/10.1146/annurev-nutr-082117-051637.

Fuse Y, Hirao A, Kuroda H, Otsuka M, Tahara Y, Shibata S. Differential roles of breakfast only (one meal per day) and a bigger breakfast with a small dinner (two meals per day) in mice fed a high-fat diet with regard to induced obesity and lipid metabolism. J Circadian Rhythms 2012;10:https://doi.org/10.1186/1740-3391-10-4.

Garaulet M, Corbalán-Tutau MD, Madrid JA, Baraza JC, Parnell LD, Lee YC, Ordovas JM. PERIOD2 variants are associated with abdominal obesity, psycho-behavioral factors, and attrition in the dietary treatment of obesity. J Am Diet Assoc 2010a;110(6):917–21. https://doi.org/10.1016/j.jada.2010.03.017.

Garaulet M, Ordovás JM, Madrid JA. The chronobiology, etiology and pathophysiology of obesity. Int J Obes (Lond) 2010b;34(12):1667–83. https://doi.org/10.1038/ijo.2010.11.

Gaucher J, Montellier E, Sassone-Corsi P. Molecular cogs: interplay between circadian clock and cell cycle. Trends Cell Biol 2018;28(5):368–79. https://doi.org/10.1016/j.tcb.2018.01.006.

Gill S, Panda S. A smartphone app reveals erratic diurnal eating patterns in humans that can be modulated for health benefits. Cell Metab 2015;22(5):789–98. https://doi.org/10.1016/j.cmet.2015.09.005.

Golbidi S, Daiber A, Korac B, Li H, Essop MF, Laher I. Health benefits of fasting and caloric restriction. Curr Diab Rep 2017;17(12)https://doi.org/10.1007/s11892-017-0951-7.

Gómez-Santos C, Gómez-Abellán P, Madrid JA, Hernández-Morante JJ, Lujan JA, Ordovas JM, Garaulet M. Circadian rhythm of clock genes in human adipose explants. Obesity 2009;17(8):1481–5. https://doi.org/10.1038/oby.2009.164.

Goni L, Cuervo M, Milagro FI, Martínez JA. Gene-gene interplay and gene-diet interactions involving the MTNR1B rs 10830963 variant with body weight loss. Lifestyle Genom 2014;7(4–6):232–42. https://doi.org/10.1159/000380951.

Griffin JL, Wang F, Lawson JA, Fitzgerald GA, Grant G, Reyes TM, Bradfield CA, Vaughan CH, Eiden M, Masoodi M. Obesity in mice with adipocyte-specific deletion of clock component Arntl. Nat Med 2012;18(12):1768–77. https://doi.org/10.1038/nm.2979.

Guo Y, Rong Y, Huang X, Lai H, Luo X, Zhang Z, Liu Y, He M, Wu T, Chen W. Shift work and the relationship with metabolic syndrome in chinese aged workers. PLoS One 2015;10(3)https://doi.org/10.1371/journal.pone.0120632.

Hamaguchi Y, Tahara Y, Hitosugi M, Shibata S. Impairment of circadian rhythms in peripheral clocks by constant light is partially reversed by scheduled feeding or exercise. J Biol Rhythms 2015;30(6):533–42. https://doi.org/10.1177/0748730415609727.

Hamaguchi M, Tanaka M, Asano M, Yamazaki M, Oda Y, Fukui M, Fukuda T, Majima S, Matsumoto S, Senmaru T. Late-night-dinner is associated with poor glycemic control in people with type 2 diabetes: the KAMOGAWA-DM cohort study. Endocr J 2018;65(4):395–402. https://doi.org/10.1507/endocrj.EJ17-0414.

Han, Mallampalli R. The acute respiratory distress syndrome: from mechanism to translation. J Immunol 1950;194:855–60. https://doi.org/10.4049/jimmunol.1402513.

Harvie MN, Pegington M, Mattson MP, Frystyk J, Dillon B, Evans G, Cuzick J, Jebb SA, Martin B, Cutler RG, Son TG, Maudsley S, Carlson OD, Egan JM, Flyvbjerg A, Howell A. The effects of intermittent or continuous energy restriction on weight loss and metabolic disease risk markers: a randomized trial in young overweight women. Int J Obes (Lond) 2011;35(5):714–27. https://doi.org/10.1038/ijo.2010.171.

Hashimoto S, Oda H, Haruma T, Okubo S, Kataoka Y, Kobayashi S, Ikegami K, Laurent T, Kojima T, Noutomi K. Real-time monitoring in three-dimensional hepatocytes reveals that insulin acts as a synchronizer for liver clock. Sci Rep 2012;2:https://doi.org/10.1038/srep00439.

Haus E, Reinberg A, Mauvieux B, Le Floc'h N, Sackett-Lundeen L, Touitou Y. Risk of obesity in male shift workers: A chronophysiological approach. Chronobiol Int 2016;33(8):1018–36. https://doi.org/10.3109/07420528.2016.1167079.

Herxheimer A. Jet lag. BMJ Clin Evid 2014;2303.

Heyde I, Oster H. Differentiating external zeitgeber impact on peripheral circadian clock resetting. Sci Rep 2019;9(1)https://doi.org/10.1038/s41598-019-56323-z.

Hirayama J, Cardone L, Doi M, Sassone-Corsi P. Common pathways in circadian and cell cycle clocks: Light-dependent activation of Fos/AP-1 in zebrafish controls CRY-1a and WEE-1. Proc Natl Acad Sci U S A 2005;102(29):10194–9. https://doi.org/10.1073/pnas.0502610102.

Hirota T, Okano T, Kokame K, Shirotani-Ikejima H, Miyata T, Fukada Y. Glucose down-regulates Per 1 and Per 2 mRNA levels and induces circadian gene expression in cultured rat-1 fibroblasts. J Biol Chem 2002;277(46):44244–51. https://doi.org/10.1074/jbc.M206233200.

Hood S, Amir S. The aging clock: circadian rhythms and later life. J Clin Investig 2017;127(2):437–46. https://doi.org/10.1172/JCI90328.

Hsieh SD, Muto T, Murase T, Tsuji H, Arase Y. Association of short sleep duration with obesity, diabetes, fatty liver and behavioral factors in Japanese men. Intern Med 2011;50(21):2499–502. https://doi.org/10.2169/internalmedicine.50.5844.

Hurley JM, Loros JJ, Dunlap JC. Circadian oscillators: around the transcription–translation feedback loop and on to output. Trends Biochem Sci 2016;41(10):834–46. https://doi.org/10.1016/j.tibs.2016.07.009.

Husse J, Eichele G, Oster H. Synchronization of the mammalian circadian timing system: Light can control peripheral clocks independently of the SCN clock: alternate routes of entrainment optimize the alignment of the body's circadian clock network with external time. Bioessays 2015;37(10):1119–28. https://doi.org/10.1002/bies.201500026.

Ishida A, Mutoh T, Ueyama T, Bando H, Masubuchi S, Nakahara D, Tsujimoto G, Okamura H. Light activates the adrenal gland: timing of gene expression and glucocorticoid release. Cell Metab 2005;2(5):297–307. https://doi.org/10.1016/j.cmet.2005.09.009.

Jakubowicz D, Froy O, Wainstein J, Boaz M. Meal timing and composition influence ghrelin levels, appetite scores and weight loss maintenance in overweight and obese adults. In: Steroids. vol. 77. 2012. p. 323–31. https://doi.org/10.1016/j.steroids.2011.12.006 Issue 4.

Jakubowicz D, Barnea M, Wainstein J, Froy O. High Caloric intake at breakfast vs. dinner differentially influences weight loss of overweight and obese women. Obesity 2013;21(12):2504–12. https://doi.org/10.1002/oby.20460.

Janich P, Pascual G, Merlos-Suárez A, Batlle E, Ripperger J, Albrecht U, Cheng HYM, Obrietan K, Di Croce L, Benitah SA. The circadian molecular clock creates epidermal stem cell heterogeneity. Nature 2011;480(7376):209–14. https://doi.org/10.1038/nature10649.

Jean-Louis G, Kripke DF, Ancoli-Israel S. Sleep and quality of well-being. Sleep 2000;23(8):1115–21. https://doi.org/10.1093/sleep/23.8.1k.

Kalsbeek A, Heerikhuize J, Wortel J, Buijs R. A diurnal rhythm of stimulatory input to the hypothalamo-pituitary–adrenal system as revealed by timed intrahypothalamic administration of the vasopressin V1 antagonist. J Neurosci Off J Soc Neurosci 1996;16:5555–65. https://doi.org/10.1523/JNEUROSCI.16-17-05555.1996.

Kaplon J, van Dam L, Peeper D. Two-way communication between the metabolic and cell cycle machineries: the molecular basis. Cell Cycle 2015;14(13):2022–32. https://doi.org/10.1080/15384101.2015.1044172.

Karpowicz P, Zhang Y, Hogenesch JB, Emery P, Perrimon N. The circadian clock gates the intestinal stem cell regenerative state. Cell Rep 2013;3(4):996–1004. https://doi.org/10.1016/j.celrep.2013.03.016.

Kholodenko BN. Cell-signalling dynamics in time and space. Nat Rev Mol Cell Biol 2006;7(3):165–76. https://doi.org/10.1038/nrm1838.

King BM. The rise, fall, and resurrection of the ventromedial hypothalamus in the regulation of feeding behavior and body weight. Physiol Behav 2006;87(2):221–44. https://doi.org/10.1016/j.physbeh.2005.10.007.

Knutson KL, Ryden AM, Mander BA, Van Cauter E. Role of sleep duration and quality in the risk and severity of type 2 diabetes mellitus. Arch Intern Med 2006;166(16):1768–74. https://doi.org/10.1001/archinte.166.16.1768.

Kohsaka A, Bass J. A sense of time: how molecular clocks organize metabolism. Trends Endocrinol Metab 2007;18(1):4–11. https://doi.org/10.1016/j.tem.2006.11.005.

Koliaki C. The effect of ingested macronutrients on postprandial ghrelin response: a critical review of existing literature data. Int J Pept 2010;710852. https://doi.org/10.1155/2010/710852.

Kowalska E, Ripperger JA, Hoegger DC, Bruegger P, Buch T, Birchler T, Mueller A, Albrecht U, Contaldo C, Brown SA. NONO couples the circadian clock to the cell cycle. Proc Natl Acad Sci U S A 2013;110(5):1592–9. https://doi.org/10.1073/pnas.1213317110.

Kramer A, Yang FC, Snodgrass P, Li X, Scammell TE, Davis FC, Weitz CJ. Regulation of daily locomotor activity and sleep by hypothalamic EGF receptor signaling. Science 2001;294(5551):2511–5. https://doi.org/10.1126/science.1067716.

Kraves S, Weitz CJ. A role for cardiotrophin-like cytokine in the circadian control of mammalian locomotor activity. Nat Neurosci 2006;9(2):212–9. https://doi.org/10.1038/nn1633.

Kripke DF, Simons RN, Garfinkel L, Hammond EC. Short and long sleep and sleeping pills: is increased mortality associated? Arch Gen Psychiatry 1979;36(1):103–16. https://doi.org/10.1001/archpsyc.1979.01780010109014.

La Fleur SE, Kalsbeek A, Wortel J, Buijs RM. A suprachiasmatic nucleus generated rhythm in basal glucose concentrations. J Neuroendocrinol 1999;11(8):643–52. https://doi.org/10.1046/j.1365-2826.1999.00373.x.

La Fleur SE, Kalsbeek A, Wortel J, Fekkes ML, Buijs RM. A daily rhythm in glucose tolerance: A role for the suprachiasmatic nucleus. Diabetes 2001;50(6):1237–43. https://doi.org/10.2337/diabetes.50.6.1237.

Laakso M, Marchetti P, Watanabe RM, Mulder H, Groop L, Boehnke M, Altshuler D, Sundler F, Eriksson JG, Jackson AU. Common variant in MTNR1B associated with increased risk of type 2 diabetes and impaired early insulin secretion. Nat Genet 2009;41(1):82–8. https://doi.org/10.1038/ng.288.

Laermans J, Vancleef L, Tack J, Depoortere I. Role of the clock gene Bmal 1 and the gastric ghrelin-secreting cell in the circadian regulation of the ghrelin-GOAT system. Sci Rep 2015;5:https://doi.org/10.1038/srep16748.

Lakin-Thomas P. Circadian rhythms, metabolic oscillators, and the target of rapamycin (TOR) pathway: the Neurospora connection. Curr Genet 2019;65(2):339–49. https://doi.org/10.1007/s00294-018-0897-6.

Lamia KA, Storch KF, Weitz CJ. Physiological significance of a peripheral tissue circadian clock. Proc Natl Acad Sci U S A 2008;105(39):15172–7. https://doi.org/10.1073/pnas.0806717105.

Lauderdale DS, Knutson KL, Rathouz PJ, Yan LL, Hulley SB, Liu K. Cross-sectional and longitudinal associations between objectively measured sleep duration and body mass index: the CARDIA sleep study. Am J Epidemiol 2009;805–13. https://doi.org/10.1093/aje/kwp230.

Lee C, Longo V. Dietary restriction with and without caloric restriction for healthy aging. F1000Res 2016;5. https://doi.org/10.12688/f1000research.7136.1.

Lee B, Li A, Hansen KF, Cao R, Yoon JH, Obrietan K. CREB influences timing and entrainment of the SCN circadian clock. J Biol Rhythms 2010;25(6):410–20. https://doi.org/10.1177/0748730410381229.

Lee MJ, Pramyothin P, Karastergiou K, Fried SK. Deconstructing the roles of glucocorticoids in adipose tissue biology and the development of central obesity. BBA-Mol Basis Dis 2014;1842(3):473–81. https://doi.org/10.1016/j.bbadis.2013.05.029.

Leproult R, Holmbäck U, Van Cauter E. Circadian misalignment augments markers of insulin resistance and inflammation, independently of sleep loss. Diabetes 2014;63(6):1860–9. https://doi.org/10.2337/db13-1546.

Lesauter J, Hoque N, Weintraub M, Pfaff DW, Silver R. Stomach ghrelin-secreting cells as food-entrainable circadian clocks. Proc Natl Acad Sci U S A 2009;106(32):13582–7. https://doi.org/10.1073/pnas.0906426106.

Li S, Zhang L. Circadian control of global transcription. Biomed Res Int 2015;2015:https://doi.org/10.1155/2015/187809.

Li JD, Hu WP, Zhou QY. The circadian output signals from the suprachiasmatic nuclei. In: Progress in brain research. vol. 199. Elsevier B.V; 2012. p. 119–27. https://doi.org/10.1016/B978-0-444-59427-3.00028-9.

Lin KK, Kumar V, Geyfman M, Chudova D, Ihler AT, Smyth P, Paus R, Takahashi JS, Andersen B. Circadian clock genes contribute to the regulation of hair follicle cycling. PLoS Genet 2009;5(7)https://doi.org/10.1371/journal.pgen.1000573.

Longo VD, Mattson MP. Fasting: molecular mechanisms and clinical applications. Cell Metab 2014;19(2):181–92. https://doi.org/10.1016/j.cmet.2013.12.008.

Lundin EA, Zhang JX, Lairon D, Tidehag P, Åman P, Adlercreutz H, Hallmans G. Effects of meal frequency and high-fibre rye-bread diet on glucose and lipid metabolism and ileal excretion of energy and sterols in ileostomy subjects. Eur J Clin Nutr 2004;58(10):1410–9. https://doi.org/10.1038/sj.ejcn.1601985.

Ma Y, Bertone ER, Stanek EJ, Reed GW, Hebert JR, Cohen NL, Merriam PA, Ockene IS. Association between eating patterns and obesity in a free-living US adult population. Am J Epidemiol 2003;158 (1):85–92. https://doi.org/10.1093/aje/kwg117.

Madeo F, Carmona-Gutierrez D, Hofer SJ, Kroemer G. Caloric restriction mimetics against age-associated disease: targets, mechanisms, and therapeutic potential. Cell Metab 2019;29(3):592–610. https://doi.org/10.1016/j.cmet.2019.01.018.

Marshall CJ. Specificity of receptor tyrosine kinase signaling: transient versus sustained extracellular signal-regulated kinase activation. Cell 1995;80(2):179–85. https://doi.org/10.1016/0092-8674(95)90401-8.

Marti AR, Meerlo P, Grønli J, van Hasselt SJ, Mrdalj J, Pallesen S, Pedersen TT, Henriksen TEG, Skrede S. Shift in food intake and changes in metabolic regulation and gene expression during simulated night-shiftwork: a rat model. Nutrients 2016;8(11)https://doi.org/10.3390/nu8110712.

Matsuo T, Yamaguchi S, Mitsui S, Emi A, Shimoda F, Okamura H. Control mechanism of the circadian clock for timing of cell division in vivo. Science 2003;302(5643):255–9. https://doi.org/10.1126/science.1086271.

McHill AW, Melanson EL, Higgins J, Connick E, Moehlman TM, Stothard ER, Wright KP. Impact of circadian misalignment on energy metabolism during simulated nightshift work. Proc Natl Acad Sci U S A 2014;111(48):17302–7. https://doi.org/10.1073/pnas.1412021111.

Meijer JH, Van Den Berg SAA, Houben T, Van Klinken JB, Van Den Berg R, Pronk ACM, Havekes LM, Romijn JA, Van Dijk KW, Biermasz NR. Detrimental effects of constant light exposure and high-fat diet on circadian energy metabolism and insulin sensitivity. FASEB J 2013;27(4):1721–32. https://doi.org/10.1096/fj.12-210898.

Mellado C, Rodríguez V, de Diego JG, Alvarez E, Blázquez E. Effect of pinealectomy and of diabetes on liver insulin and glucagon receptor concentrations in the rat. J Pineal Res 1989;6(4):295–306. https://doi.org/10.1111/j.1600-079X.1989.tb00425.x.

Mendoza J, Challet E. Brain clocks: from the suprachiasmatic nuclei to a cerebral network. Neuroscientist 2009;15(5):477–88. https://doi.org/10.1177/1073858408327808.

Mendoza J, Graff C, Dardente H, Pevet P, Challet E. Feeding cues alter clock gene oscillations and photic responses in the suprachiasmatic nuclei of mice exposed to a light/dark cycle. J Neurosci 2005;25 (6):1514–22. https://doi.org/10.1523/JNEUROSCI.4397-04.2005.

Merbitz-Zahradnik T, Wolf E. How is the inner circadian clock controlled by interactive clock proteins?: structural analysis of clock proteins elucidates their physiological role. FEBS Lett 2015;589(14):1516–29. https://doi.org/10.1016/j.febslet.2015.05.024.

Miki T. Regulates period 2 expression and the circadian clock. Nat Commun 2013;4:2444. https://doi.org/10.1038/ncomms3444.

Mills JN, Minors DS, Waterhouse JM. The circadian rhythms of human subjects without timepieces or indication of the alternation of day and night. J Physiol 1974;240(3):567–94. https://doi.org/10.1113/jphysiol.1974.sp010623.

Miyagishima SY, Fujiwara T, Sumiya N, Hirooka S, Nakano A, Kabeya Y, Nakamura M. Translation-independent circadian control of the cell cycle in a unicellular photosynthetic eukaryote. Nat Commun 2014;5:https://doi.org/10.1038/ncomms4807.

Mohawk JA, Pargament JM, Lee TM. Circadian dependence of corticosterone release to light exposure in the rat. Physiol Behav 2007;92(5):800–6. https://doi.org/10.1016/j.physbeh.2007.06.009.

Monk TH. Enhancing circadian zeitgebers. Sleep 2010;33(4):421–2. https://doi.org/10.1093/sleep/33.4.421.

Moore RY. Neural control of the pineal gland. Behav Brain Res 1995;125–30. https://doi.org/10.1016/0166-4328(96)00083-6.

Morgan K, Newland L. Global maps for GIS. Environ Sci Pollut Res 1999;6(1):59–60. https://doi.org/10.1007/BF02987122.

Morris CJ, Purvis TE, Mistretta J, Scheer FAJL. Effects of the internal circadian system and circadian misalignment on glucose tolerance in chronic shift workers. J Clin Endocrinol Metabol 2016;101 (3):1066–74. https://doi.org/10.1210/jc.2015-3924.

Morton GJ, Cummings DE, Baskin DG, Barsh GS, Schwartz MW. Central nervous system control of food intake and body weight. Nature 2006;443(7109):289–95. https://doi.org/10.1038/nature05026.

Murphy LO, Smith S, Chen RH, Fingar DC, Blenis J. Molecular, interpretation of ERK signal duration by immediate early gene products. Nat Cell Biol 2002;4(8):556–64. https://doi.org/10.1038/ncb822.

Nakahata Y, Kaluzova M, Grimaldi B, Sahar S, Hirayama J, Chen D, Guarente LP, Sassone-Corsi P. The NAD+-dependent deacetylase SIRT1 modulates CLOCK-mediated chromatin remodeling and circadian control. Cell 2008;134(2):329–40. https://doi.org/10.1016/j.cell.2008.07.002.

Nogueira TC, Lellis-Santos C, Jesus DS, Taneda M, Rodrigues SC, Amaral FG, Lopes AMS, Cipolla-Neto J, Bordin S, Anhê GF. Absence of melatonin induces night-time hepatic insulin resistance and increased gluconeogenesis due to stimulation of nocturnal unfolded protein response. Endocrinology 2011;152(4):1253–63. https://doi.org/10.1210/en.2010-1088.

Obayashi K, Saeki K, Iwamoto J, Ikada Y, Kurumatani N. Exposure to light at night and risk of depression in the elderly. J Affect Disord 2013;151(1):331–6. https://doi.org/10.1016/j.jad.2013.06.018.

Oike H, Kobori M. Resveratrol regulates circadian clock genes in Rat-1 fibroblast cells. Biosci Biotechnol Biochem 2008;72(11):3038–40. https://doi.org/10.1271/bbb.80426.

Oike H, Nagai K, Fukushima T, Ishida N, Kobori M. High-salt diet advances molecular circadian rhythms in mouse peripheral tissues. Biochem Biophys Res Commun 2010;402(1):7–13. https://doi.org/10.1016/j.bbrc.2010.09.072.

Oike H, Kobori M, Suzuki T, Ishida N. Caffeine lengthens circadian rhythms in mice. Biochem Biophys Res Commun 2011;410(3):654–8. https://doi.org/10.1016/j.bbrc.2011.06.049.

Oosterman JE, Kalsbeek A, La Fleur SE, Belsham DD. Impact of nutrients on circadian rhythmicity. Am J Physiol Regul Integr Comp Physiol 2015;308(5):R337–50. https://doi.org/10.1152/ajpregu.00322.2014.

Panda S, Antoch MP, Miller BH, Su AI, Schook AB, Straume M, Schultz PG, Kay SA, Takahashi JS, Hogenesch JB. Coordinated transcription of key pathways in the mouse by the circadian clock. Cell 2002;109(3):307–20. https://doi.org/10.1016/S0092-8674(02)00722-5.

Parsons MJ, Moffitt TE, Gregory AM, Goldman-Mellor S, Nolan PM, Poulton R, Caspi A. Social jetlag, obesity and metabolic disorder: investigation in a cohort study. Int J Obes (Lond) 2015;39(5):842–8. https://doi.org/10.1038/ijo.2014.201.

Peijun M, Mark AW, Hirschie JC. An evolutionary fitness enhancement conferred by the circadian system in cyanobacteria. Chaos, Solitons Fractals 2013;65–74. https://doi.org/10.1016/j.chaos.2012.11.006.

Pellettieri J, Alvarado AS. Cell turnover and adult tissue homeostasis: from humans to planarians. Annu Rev Genet 2007;41:83–105. https://doi.org/10.1146/annurev.genet.41.110306.130244.

Peschke E, Stumpf I, Bazwinsky I, Litvak L, Dralle H, Mühlbauer E. Melatonin and type 2 diabetes—a possible link? J Pineal Res 2007;42(4):350–8. https://doi.org/10.1111/j.1600-079X.2007.00426.x.

Picinato MC, Haber EP, Carpinelli AR, Cipolla-Neto J. Daily rhythm of glucose-induced insulin secretion by isolated islets from intact and pinealectomized rat. J Pineal Res 2002;33(3):172–7. https://doi.org/10.1034/j.1600-079X.2002.02925.x.

Piérard C, Beaumont M, Enslen M, Chauffard F, Tan DX, Reiter RJ, Fontan A, French J, Coste O, Lagarde D. Resynchronization of hormonal rhythms after an eastbound flight in humans: effects of slow-release caffeine and melatonin. Eur J Appl Physiol 2001;85(1–2):144–50. https://doi.org/10.1007/s004210100418.

Plano SA, Casiraghi LP, Moro PG, Paladino N, Golombek DA, Chiesa JJ. Circadian and metabolic effects of light: implications in weight homeostasis and health. Front Neurol 2017;8:https://doi.org/10.3389/fneur.2017.00558.

Plikus MV, Vollmers C, De La Cruz D, Chaix A, Ramos R, Panda S, Chuong CM. Local circadian clock gates cell cycle progression of transient amplifying cells during regenerative hair cycling. Proc Natl Acad Sci U S A 2013;110(23):E2106–15. https://doi.org/10.1073/pnas.1215935110.

Plikus MV, Van Spyk EN, Pham K, Geyfman M, Kumar V, Takahashi JS, Andersen B. The circadian clock in skin: Implications for adult stem cells, tissue regeneration, cancer, aging, and immunity. J Biol Rhythms 2015;30(3):163–82. https://doi.org/10.1177/0748730414563537.

Potter GDM, Skene DJ, Arendt J, Cade JE, Grant PJ, Hardie LJ. Circadian rhythm and sleep disruption: causes, metabolic consequences, and countermeasures. Endocr Rev 2016;37(6):584–608. https://doi.org/10.1210/er.2016-1083.

Prasai MJ, George JT, Scott EM. Molecular clocks, type 2 diabetes and cardiovascular disease. Diab Vasc Dis Res 2008;5(2):89–95. https://doi.org/10.3132/dvdr.2008.015.

Ravussin E, Redman LM, Rochon J, Das SK, Fontana L, Kraus WE, Romashkan S, Williamson DA, Meydani SN, Villareal DT, Smith SR, Stein RI, Scott TM, Stewart TM, Saltzman E, Klein S, Bhapkar M, Martin CK, Gilhooly CH, CALERIE Study Group. A 2-year randomized controlled trial of human caloric restriction: feasibility and effects on predictors of health span and longevity. J Gerontol A Biol Sci Med Sci 2015;70(9):1097–104. https://doi.org/10.1093/gerona/glv057.

Reddy AB, Cronin AS, Ford H, Ebling FJP. Seasonal regulation of food intake and body weight in the male Siberian hamster: studies of hypothalamic orexin (hypocretin), neuropeptide Y (NPY) and pro-opiomelanocortin (POMC). Eur J Neurosci 1999;11(9):3255–64. https://doi.org/10.1046/j.1460-9568.1999.00746.x.

Reddy AB, Maywood ES, Karp NA, King VM, Inoue Y, Gonzalez FJ, Lilley KS, Kyriacou CP, Hastings MH. Glucocorticoid signaling synchronizes the liver circadian transcriptome. Hepatology 2007;45(6):1478–88. https://doi.org/10.1002/hep.21571.

Redman LM, Kraus WE, Bhapkar M, Das SK, Racette SB, Martin CK, Fontana L, Wong WW, Roberts SB, Ravussin E. Energy requirements in nonobese men and women: Results from CALERIE. Am J Clin Nutr 2014;99(1):71–8. https://doi.org/10.3945/ajcn.113.065631.

Reinke H, Asher G. Circadian clock control of liver metabolic functions. Gastroenterology 2016;150 (3):574–80. https://doi.org/10.1053/j.gastro.2015.11.043.

Reiter RJ, Tan DX, Rosales-Corral S, Galano A, Zhou XJ, Xu B. Mitochondria: central organelles for melatonins antioxidant and anti-aging actions. Molecules 2018;23(2)https://doi.org/10.3390/molecules23020509.

Richter HG, Torres-Farfan C, Garcia-Sesnich J, Abarzua-Catalan L, Henriquez MG, Alvarez-Felmer M, Gaete F, Rehren GE, Seron-Ferre M. Rhythmic expression of functional MT1 melatonin receptors in the rat adrenal gland. Endocrinology 2008;149(3):995–1003. https://doi.org/10.1210/en.2007-1009.

Rickman AD, Williamson DA, Martin CK, Gilhooly CH, Stein RI, Bales CW, Roberts S, Das SK. The CALERIE study: design and methods of an innovative 25% caloric restriction intervention. Contemp Clin Trials 2011;32(6):874–81. https://doi.org/10.1016/j.cct.2011.07.002.

Rodríguez V, Mellado C, Alvarez E, De Diego JG, Blázquez E. Effect of pinealectomy on liver insulin and glucagon receptor concentrations in the rat. J Pineal Res 1989;6(1):77–88. https://doi.org/10.1111/j.1600-079X.1989.tb00405.x.

Romon M, Edme JL, Boulenguez C, Lescroart JL, Frimat P. Circadian variation of diet-induced thermogenesis. Am J Clin Nutr 1993;57(4):476–80. https://doi.org/10.1093/ajcn/57.4.476.

Rovere-Jovene C, Tolle V, Viltart O, Epelbaum J, Hanachi M, Fetissov S, Godart N, Melchior JC, Ramoz N. New insights in anorexia nervosa. Front Neurosci 2016;10:https://doi.org/10.3389/fnins.2016.00256.

Ruiter M, La Fleur SE, Van Heijningen C, Van der Vliet J, Kalsbeek A, Buijs RM. The daily rhythm in plasma glucagon concentrations in the rat is modulated by the biological clock and by feeding behavior. Diabetes 2003;52(7):1709–15. https://doi.org/10.2337/diabetes.52.7.1709.

Sancar A, Lindsey-Boltz LA, Kang TH, Reardon JT, Lee JH, Ozturk N. Circadian clock control of the cellular response to DNA damage. FEBS Lett 2010;584(12):2618–25. https://doi.org/10.1016/j.febslet.2010.03.017.

Scheer FAJL, Hilton MF, Mantzoros CS, Shea SA. Adverse metabolic and cardiovascular consequences of circadian misalignment. Proc Natl Acad Sci U S A 2009;106(11):4453–8. https://doi.org/10.1073/pnas.0808180106.

Scheiermann C, Kunisaki Y, Frenette PS. Circadian control of the immune system. Nat Rev Immunol 2013;13(3):190–8. https://doi.org/10.1038/nri3386.

Schroeder AM, Truong D, Loh DH, Jordan MC, Roos KP, Colwell CS. Voluntary scheduled exercise alters diurnal rhythms of behaviour, physiology and gene expression in wild-type and vasoactive intestinal peptide-deficient mice. J Physiol 2012;590(23):6213–26. https://doi.org/10.1113/jphysiol.2012.233676.

Scott EM, Carter AM, Grant PJ. Association between polymorphisms in the clock gene, obesity and the metabolic syndrome in man. Int J Obes (Lond) 2008;32(4):658–62. https://doi.org/10.1038/sj.ijo.0803778.

Sensi S, Capani F. Chronobiological aspects of weight loss in obesity: effects of different meal timing regimens. Chronobiol Int 1987;4(2):251–61. https://doi.org/10.3109/07420528709078532.

Shahrad T, Ling L, Diane A, Terry Y, Emmanuel M, Philippe F. Short sleep duration is associated with reduced leptin, elevated ghrelin, and increased body mass index. PLoS Med 2004;e62. https://doi.org/10.1371/journal.pmed.0010062.

Shan Z, Ma H, Xie M, Yan P, Guo Y, Bao W, Rong Y, Jackson CL, Hu FB, Liu L. Sleep duration and risk of type 2 diabetes: a meta-analysis of prospective studies. Diabetes Care 2015;38(3):529–37. https://doi.org/10.2337/dc14-2073.

Slominski RM, Reiter RJ, Schlabritz-Loutsevitch N, Ostrom RS, Slominski AT. Melatonin membrane receptors in peripheral tissues: distribution and functions. Mol Cell Endocrinol 2012;351(2):152–66. https://doi.org/10.1016/j.mce.2012.01.004.

Sonya V, Claudia R, Calogero C. Centenarians and diet: what they eat in the Western part of Sicily. Immun Ageing 2012;https://doi.org/10.1186/1742-4933-9-10.

Sookoian S, Gemma C, Gianotti TF, Burgueño A, Castaño G, Pirola CJ. Genetic variants of clock transcription factor are associated with individual susceptibility to obesity. Am J Clin Nutr 2008;87(6):1606–15. https://doi.org/10.1093/ajcn/87.6.1606.

Spiegel K, Knutson K, Leproult R, Tasali E, Van Cauter E. Sleep loss: a novel risk factor for insulin resistance and type 2 diabetes. J Appl Physiol 2005;99(5):2008–19. https://doi.org/10.1152/japplphysiol.00660.2005.

Spiegel K, Tasali E, Leproult R, Scherberg N, Van Cauter E. Twenty-four-hour profiles of acylated and total ghrelin: relationship with glucose levels and impact of time of day and sleep. J Clin Endocrinol Metabol 2011;96(2):486–93. https://doi.org/10.1210/jc.2010-1978.

Stanger BZ. Cellular homeostasis and repair in the mammalian liver. Annu Rev Physiol 2015;77:179–200. https://doi.org/10.1146/annurev-physiol-021113-170255.

Stephan FK, Zucker I. Circadian rhythms in drinking behavior and locomotor activity of rats are eliminated by hypothalamic lesions. Proc Natl Acad Sci U S A 1972;69(6):1583–6. https://doi.org/10.1073/pnas.69.6.1583.

Stern JH, Grant AS, Thomson CA, Tinker L, Hale L, Brennan KM, Woods NF, Chen Z. Short sleep duration is associated with decreased serum leptin, increased energy intake and decreased diet quality in postmenopausal women. Obesity 2014;E55–61. https://doi.org/10.1002/oby.20683.

Stewart KT, Hayes BC, Eastman CI. Light treatment for NASA shiftworkers. Chronobiol Int 1995;12(2):141–51. https://doi.org/10.3109/07420529509064509.

Storch KF, Lipan O, Leykin I, Viswanathan N, Davis FC, Wong WH, Weitz CJ. Extensive and divergent circadian gene expression in liver and heart. Nature 2002;417(6884):78–83. https://doi.org/10.1038/nature744.

Su Y, Foppen E, Zhang Z, Fliers E, Kalsbeek A. Effects of 6-meals-a-day feeding and 6-meals-a-day feeding combined with adrenalectomy on daily gene expression rhythms in rat epididymal white adipose tissue. Genes Cells 2016;21(1):6–24. https://doi.org/10.1111/gtc.12315.

Summerbell CD, Moody RC, Shanks J, Stock MJ, Geissler C. Relationship between feeding pattern and body mass index in 220 free-living people in four age groups. Eur J Clin Nutr 1996;50(8):513–9.

Swindell WR. Comparative analysis of microarray data identifies common responses to caloric restriction among mouse tissues. Mech Ageing Dev 2008;129(3):138–53. https://doi.org/10.1016/j.mad.2007.11.003.

Szewczyk-Golec K, Woźniak A, Reiter RJ. Inter-relationships of the chronobiotic, melatonin, with leptin and adiponectin: Implications for obesity. J Pineal Res 2015;59(3):277–91. https://doi.org/10.1111/jpi.12257.

Takahashi JS. Transcriptional architecture of the mammalian circadian clock. Nat Rev Genet 2017;18(3):164–79. https://doi.org/10.1038/nrg.2016.150.

Takahashi JS, Imai SI, Bass J, Abrassart D, Kobayashi Y, Marcheva B, Hong HK, Chong JL, Buhr ED, Lee C. Circadian clock feedback cycle through NAMPT-mediated NAD+ biosynthesis. Science 2009;324(5927):651–4. https://doi.org/10.1126/science.1171641.

Takahashi JS, Andersen B, Liu Q, Ruiz R, Gordon W, Espitia F, Cam E, Millar SE, Smyth P, Ihler A. Brain and muscle Arnt-like protein-1 (BMAL1) controls circadian cell proliferation and susceptibility to UVB-induced DNA damage in the epidermis. Proc Natl Acad Sci U S A 2012;109(29):11758–63. https://doi.org/10.1073/pnas.1209592109.

Tan DX, Manchester LC, Fuentes-Broto L, Paredes SD, Reiter RJ. Significance and application of melatonin in the regulation of brown adipose tissue metabolism: relation to human obesity. Obes Rev 2011;12 (3):167–88. https://doi.org/10.1111/j.1467-789X.2010.00756.x.

Timlin MT, Pereira MA, Story M, Neumark-Sztainer D. Breakfast eating and weight change in a 5-year prospective analysis of adolescents: Project EAT (Eating Among Teens). Pediatrics 2008;121 (3):638–45.

Torres-Farfan C, Richter HG, Rojas-García P, Vergara M, Forcelledo ML, Valladares LE, Torrealba F, Valenzuela GJ, Serón-Ferré M. Mt1 melatonin receptor in the primate adrenal gland: Inhibition of adrenocorticotropin-stimulated cortisol production by melatonin. J Clin Endocrinol Metabol 2003;88 (1):450–8. https://doi.org/10.1210/jc.2002-021048.

Trott AJ, Menet JS. Regulation of circadian clock transcriptional output by CLOCK:BMAL1. PLoS Genet 2018;14(1) https://doi.org/10.1371/journal.pgen.1007156.

Tu BP, Kudlicki A, Rowicka M, McKnight SL. Cell biology: logic of the yeast metabolic cycle: temporal compartmentalization of cellular processes. Science 2005;310(5751):1152–8. https://doi.org/10.1126/science.1120499.

van Oosterhout F, Lucassen EA, Houben T, vander Leest HT, Antle MC, Meijer JH. Amplitude of the SCN clock enhanced by the behavioral activity rhythm. PLoS One 2012;7(6)https://doi.org/10.1371/journal.pone.0039693.

Vauzour D, Rodriguez-Mateos A, Corona G, Oruna-Concha MJ, Spencer JPE. Polyphenols and human health: prevention of disease and mechanisms of action. Nutrients 2010;2(11):1106–31. https://doi.org/10.3390/nu2111106.

Vioque J, Torres-Cantero A, Quiles J. Time spent watching television, sleep duration and obesity in adults living in Valencia, Spain. Int J Obes Relat Metab Disord 2000;24:1683–8. https://doi.org/10.1038/sj.ijo.0801434.

Vorona RD, Winn MP, Babineau TW, Eng BP, Feldman HR, Ware JC. Overweight and obese patients in a primary care population report less sleep than patients with a normal body mass index. Arch Intern Med 2005;165(1):25–30. https://doi.org/10.1001/archinte.165.1.25.

Weger M, Diotel N, Dorsemans AC, Dickmeis T, Weger BD. Stem cells and the circadian clock. Dev Biol 2017;431(2):111–23. https://doi.org/10.1016/j.ydbio.2017.09.012.

Wehrens SMT, Christou S, Isherwood C, Middleton B, Gibbs MA, Archer SN, Skene DJ, Johnston JD. Meal timing regulates the human circadian system. Curr Biol 2017;27(12)https://doi.org/10.1016/j.cub.2017.04.059 1768–1775.e3.

Welsh DK, Takahashi JS, Kay SA. Suprachiasmatic nucleus: cell autonomy and network properties. Annu Rev Physiol 2009;72:551–77. https://doi.org/10.1146/annurev-physiol-021909-135919.

Wideman CH, Murphy HM. Constant light induces alterations in melatonin levels, food intake, feed efficiency, visceral adiposity, and circadian rhythms in rats. Nutr Neurosci 2009;12(5):233–40. https://doi.org/10.1179/147683009X423436.

Willcox BJ, Yano K, Chen R, Willcox DC, Rodriguez BL, Masaki KH, Donlon T, Tanaka B, Curb JD. How much should we eat? The association between energy intake and mortality in a 36-year follow-up study of Japanese-American men. J Gerontol Ser A Biol Sci Med Sci 2004;59(8):789–95. https://doi.org/10.1093/gerona/59.8.b789.

Wu T, Sun L, Zhuge F, Guo X, Zhao Z, Tang R, Chen Q, Chen L, Kato H, Fu Z. Differential roles of breakfast and supper in rats of a daily three-meal schedule upon circadian regulation and physiology. Chronobiol Int 2011;28(10):890–903. https://doi.org/10.3109/07420528.2011.622599.

Yagita K, Okamura H. Forskolin induces circadian gene expression of rPer1, rPer2 and dbp in mammalian rat-1 fibroblasts. FEBS Lett 2000;465(1):79–82. https://doi.org/10.1016/S0014-5793(99)01724-X.

Yamanaka Y, Hashimoto S, Takasu NN, Tanahashi Y, Nishide SY, Honma S, Honma KI. Morning and evening physical exercise differentially regulate the autonomic nervous system during nocturnal sleep in humans. Am J Physiol Regul Integr Comp Physiol 2015;309(9):R1112–R 1121. https://doi.org/10.1152/ajpregu.00127.2015.

Yamazaki S, Numano R, Abe M, Hida A, Takahashi RI, Ueda M, Block GD, Sakaki Y, Menaker M, Tei H. Resetting central and peripheral circadian oscillators in transgenic rats. Science 2000;288(5466):682–5. https://doi.org/10.1126/science.288.5466.682.

Yasuhiko A, Isao S, Ikuyo H, Kana Y, Kotatsu M, Kanako Y, Tatsuhiro M, Tadahiro K, Takeshi T, Taro K. Skipping breakfast is correlated with obesity. J Rural Med 2014;51–8. https://doi.org/10.2185/jrm.2887.

Ye R, Selby CP, Chiou YY, Ozkan-Dagliyan I, Gaddameedhi S, Sancar A. Dual modes of CLOCK: BMAL1 inhibition mediated by Cryptochrome and period proteins in the mammalian circadian clock. Genes Dev 2014;28(18):1989–98. https://doi.org/10.1101/gad.249417.114.

Yuen KCJ, Chong LE, Riddle MC. Influence of glucocorticoids and growth hormone on insulin sensitivity in humans. Diabet Med 2013;30(6):651–63. https://doi.org/10.1111/dme.12184.

Zhang J, Kaasik K, Blackburn MR, Cheng CL. Constant darkness is a circadian metabolic signal in mammals. Nature 2006;439(7074):340–3. https://doi.org/10.1038/nature04368.

Zhang R, Lahens NF, Ballance HI, Hughes ME, Hogenesch JB. A circadian gene expression atlas in mammals: Implications for biology and medicine. Proc Natl Acad Sci U S A 2014;111(45):16219–24. https://doi.org/10.1073/pnas.1408886111.

Zvonic S, Ptitsyn AA, Conrad SA, Scott LK, Floyd ZE, Kilroy G, Wu X, Goh BC, Mynatt RL, Gimble JM. Characterization of peripheral circadian clocks in adipose tissues. Diabetes 2006;55(4):962–70. https://doi.org/10.2337/diabetes.55.04.06.db05-0873.

Zvonic S, Floyd ZE, Mynatt RL, Gimble JM. Circadian rhythms and the regulation of metabolic tissue function and energy homeostasis. Obesity 2007;15(3):539–43. https://doi.org/10.1038/oby.2007.544.

CHAPTER 13

Nutraceutical approach to age-related diseases—The clinical evidence on cognitive decline

Arrrgo F.G. Cicero[a,b] and Alessandro Colletti[b,c]
[a]Medical and Surgery Sciences Department, Alma Mater Studiorum University of Bologna, Bologna, Italy
[b]Italian Nutraceutical Society (SINut), Bologna, Italy
[c]Pharmacology Department, University of Turin, Turin, Italy

13.1 Introduction

13.1.1 Chronic age-related diseases and cognitive impairment

Most of the epidemiologic studies report a protective association between Mediterranean diet (MD) adherence, cognitive impairment, and brain health. Data from clinical trials supporting these observational findings are also emerging (Gardener and Caunca, 2018). In particular, it is well known that MD is characterized by many foods that could be "functional food" and/or "nutraceuticals" that contribute also to the brain health and the prevention of dementia (Cicero et al., 2018).

Dementia is a pathological condition considered to be a health- and social-care priority for many high-income countries focusing a large part of their economic resources on prevention (Schaller et al., 2015). Evidence indicates that the prevalence of dementia is rapidly expanding also in the low- and middle-income sectors. From a global perspective, the prevalence of dementia increases exponentially with age, doubling about every 5 years from ages in North America, Latin America, and Asia-Pacific, whereas in Australia, Western Europe, and South, East, and Southeast Asia, the incidence is slightly lower (Prince et al., 2013).

Senile dementia is characterized by the impairment of many cognitive abilities, with particular regard to reasoning, memory, perceptual speed, and language. There is evidence that many genetic, nutritional, and metabolic risk factors are able to promote the cognitive impairment development in this group of patients (Flirski and Sobow, 2005; Zanchetti et al., 2014). Obviously, there is still an issue of the causality of these factors, which requires further investigation (Banach et al., 2017). However, with the exclusion of the genetic impact, all the other possible determinants represent viable and potentially effective targets for nonpharmacological therapeutic interventions. In fact, not only there are several drugs available to treat dementia, but only a few of these

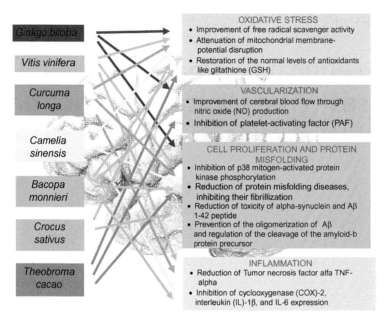

Fig. 13.1 Botanical extracts active on brain protection and main putative mechanisms of actions.

are supported by being classified as a strong evidence-based medicine (EBM) and they are mostly not well tolerated (O'Brien et al., 2017). This suggests the importance of lifestyle strategies as effective alternative treatments to prevent and treat neurodegeneration and vascular complications (Polidori and Schulz, 2014). For example, because of its antioxidant-rich and cardioprotective dietary pattern, the MD seems to delay the progression of mild cognitive impairment and Alzheimer's disease (Psaltopoulou et al., 2013; Singh et al., 2014). It might be similarly plausible that antioxidant-rich foods are able to afford protection from cognitive decline and major neurodegenerative diseases (Von Bernhardi and Eugenín, 2012). To date, several botanicals and phytochemicals have been tested in vitro because of their suggested neuroprotective properties (Fig. 13.1); however, a relatively low number of them have been shown to have some activity in humans affected by cognitive impairment or decline (Mecocci et al., 2014).

In this context, this chapter will summarize the available evidence supporting the use in clinical practice of some botanicals and phytochemicals with confirmed activity on human central nervous system and demonstrated effects in modulating cognitive decline.

13.2 Data selection

A systematic search strategy was developed to identify metaanalyses and randomized clinical trials (RCTs) from January 1970 to March 2020, in both MEDLINE (National Library of Medicine, Bethesda, MD) and the Cochrane Register of Controlled Trials

(The Cochrane Collaboration, Oxford, UK). The terms "botanicals," "dietary supplements, " "herbal drug," "phytochemical," "cognitive impairment," "Alzheimer's disease," "clinical trial," and "humans" were incorporated into an electronic search strategy. Two authors then independently reviewed all of the citations retrieved from the electronic search in order to identify potentially relevant articles for this review and determine their eligibility. Quality assessment of each article was performed evaluating each study's aim, case and control definitions, inclusion and exclusion criteria, sample section and analysis, and statistical definition of significant differential expression. Bibliographies of all identified studies, and review articles were reviewed looking for additional papers of interest. A preference has been given to botanicals or phytochemicals with newly and strongly demonstrated effects in humans (Table 13.1).

13.3 The state of the art

13.3.1 *Ginkgo biloba*

Ginkgo biloba (Gb) is one of the most widely used natural compounds for the prevention and treatment of Alzheimer's disease (AD). Most available RCTs were carried out using the EGb761 *Ginkgo biloba* extract, which is produced from the Gb leaves by a validated production process and contains pharmacologically active compounds proanthocyanidins, other than flavonol glycosides and terpenoids, within a narrow range of 22%–27% and 5%–7%, respectively (Biber, 2003). Mechanisms to explain Gb's potential procognitive and/or its neuroprotective effects are mostly attributed to an increase in cerebral blood flow (CBF). Gb extract inhibits platelet-activating factor (PAF) and enhances nitric oxide (NO) production in vessels, with subsequent beneficial effects on peripheral and cerebral blood flow (Hirsch et al., 2017). Additionally, Gb produces modification of the monoamine neurotransmitter systems—it displays a free radical scavenger activity and has neuroprotective and antiapoptotic properties (Brondino et al., 2013; Schneider, 2012). In animal models of Alzheimer's disease, Gb extract inhibits Aβ 1–42-induced hippocampal neuron dysfunction and death, Aβ-induced pathological behavior, and Aβ aggregation and also enhances neurogenesis (Suliman et al., 2016). Also, results obtained from a recent study of the effects of EGb761 on hippocampal neuronal injury and carbonyl stress of aging rats provided more clarity regarding the antiaging mechanism of protein carbonylation. EGb761 was found to be an effective agent against D-gal-induced hippocampal neuronal loss due to its antioxidant as well as carbonyl stress characteristics. The findings of this study also confirm the carbonylation hypothesis regarding high levels of 4-hydroxynonenal (4-HNE) being a strong contributor to age-related neurodegenerative disorders (Li et al., 2019). Trials looking for dementia prevention in participants with no cognitive impairment at baseline have administered EGb761 for 4–6 years in doses of 240 mg/day. No decrement of dementia incidence was reported in treated participants in the GuidAge trial (2854 participants) (Vellas et al., 2012), nor the Ginkgo Evaluation of

Table 13.1 Botanicals and phytochemicals with known activity on the central nervous system and clinically relevant effects in modulating cognitive decline in humans.

Botanicals	Active compounds	Effective dosage	Beneficial proprieties involved in cognitive decline modulating	Clinically relevant effects involved in cognitive decline modulating
Ginkgo biloba (EGb761)	Flavonol glycosides, proanthocyanidins, terpenoids	>200 mg/day for at least 5 months	Neuro protective and antiapoptotic properties	↑CBF, ↑MMSE, and ↑ADL (in combination with conventional medicine)
Vitis vinifera	Trans-resveratrol	≥75 mg/day for almost 6 weeks (+quercetin 320 mg/day to increase resveratrol's bioavailability)	Antioxidant activity and antiinflammatory effects	↑CBF, ↑FMD, ↑CVR (high dose)
Camellia sinensis	Epigallocatechin-3-gallate, ʟ-theanine	Not detectable	Antioxidant activity, antiinflammatory activity, blood-brain barrier crossing ability, and antistress effects	Not detectable
Theobroma cacao	Cocoa flavanols	≥494 mg/day for a most 8 weeks	Antioxidant activity and neuroprotective propriety	↑FMD, ↑insulin sensitivity, ↑CBF, ↑neurovascular coupling ↑TMT-A, ↑TMT-B, ↑VFT
Bacopa monnieri	Bacosides A and B, alkaloids, saponins	200 mg/kg/day for almost 3 months	Antiinflammatory, antioxidant, metal chelation, amyloid, and cholinergic effects	↑TMT-B, ↓CRT
Crocus sativus	Crocin	30 mg/day for 1 year	Antidepressant and antiinflammatory effects, and improves learning and memory	↑ADAS-cog, ↑SCIRS, ↑FAST
Curcuma longa	Curcumin	Unknown	Antioxidant, antiinflammatory, and amyloid disaggregating activities	Not yet available

ADAS-cog, Alzheimer's disease assessment scale–cognitive subscale; *ADL*, activity of daily living; *CBF*, cerebral blood flow; *CVR*, cerebral vasodilator responsiveness; *FMD*, flow-mediated dilatation; *MMSE*, Mini-Mental State Examination; *SCIRS*, Severe Cognitive Impairment Rating Scale; *FAST*, functional assessment staging; *TMT-A*, Trail Making Test A; *TMT-B*, Trail Making Test B; *VFT*, verbal fluency test.

Memory trial (3069 participants) (DeKosky et al., 2008). In a metaanalysis of eight trials of noncognitively impaired participants treated for up to 13 weeks with Gb, no cognitive enhancement benefits were found (Canter and Ernst, 2007). In the case of patients with dementia or cognitive impairment, a recent metaanalysis of 21 trials with 2608 patients concluded that Gb in combination with conventional medicine was superior in improving Mini-Mental State Examination (MMSE) scores at 24 weeks for patients with Alzheimer's disease (mean difference [MD] 2.39, 95% confidence interval [CI]: 1.28–3.50, $P < 0.0001$) and mild cognitive impairment (MD 1.90, 95% CI: 1.41–2.39, $P < 0.00001$). This was also seen in Activity of daily living (ADL) scores at 24 weeks for Alzheimer's disease (MD -3.72, 95% CI: -5.68 to -1.76, $P = 0.0002$) when compared with conventional medicine alone (Brondino et al., 2013). Adverse events were mild and mainly involving the gastrointestinal tract (Yang et al., 2016). These results have been confirmed by the most recent metaanalysis, concluding that Gb extract has potentially beneficial effects for people with dementia when it is administered at doses greater than 200 mg/day for at least 5 months (Yuan et al., 2017). Moreover, treatment with EGb761 may improve behavioral performance in patients with mild to moderate dementia (vascular dementia or AD) and therefore assist with improving the well-being of their caregivers (Ihl, 2013).

In conclusion, Gb is potentially beneficial for the improvement of cognitive function, activities of daily living, and global clinical assessment in patients living with mild cognitive impairment or Alzheimer's disease and there is also evidence that it could alleviate some neuropsychiatric symptoms in dementia.

13.3.2 *Vitis vinifera*

Resveratrol (3,5,4′-trihydroxystilbene) belongs to a family of polyphenolic compounds known as stilbenes, which are particularly concentrated in grapes and red wine (Shi et al., 2014; Sahebkar et al., 2015). It is also one of the phytoalexins, a group of low molecular mass substances produced by plants in response to many exogenous stimuli, such as UV radiation, chemical stressors, and particularly microbial infections (Cvejic et al., 2010). The present compound exists in two isomeric forms. The *trans*-isomer occurs in the berry skins of most grape cultivars, and its synthesis is stimulated by UV light, injury, and fungal infection. The *cis*-isomer is formed by UV irradiation of the *trans*-isomer and is generally absent or only slightly detectable in grapes even though it is produced during vinification (Moreno et al., 2008). Most research on resveratrol concerns the *trans*-isomer, which is the most stable compound (Trela and Waterhouse, 1996).

Cognitive deficits were demonstrated to correlate with higher reactive oxygen species (ROS) levels and nitrogen species. In effect, the oxidative stress seems to precede the senile plaques formation (Wahlster et al., 2013). In this regard, resveratrol exerts potent antioxidant activity that could be useful in preventing neurodegeneration in Alzheimer's disease (Kim et al., 2010). In fact, it scavenges free radicals, protects neurons and microglia (Zhuang et al., 2003; Candelario-Jalil et al., 2007), and attenuates the Aβ-induced accumulation of intracellular ROS (Jang and Surh, 2003). The treatment of a murine HT22 hippocampal cell line with resveratrol was demonstrated to attenuate ROS production and mitochondrial membrane-potential disruption, also restoring the normal levels of glutathione (GSH) depleted by the Aβ1-42 (Kwon et al., 2010).

Resveratrol may also attenuate Aβ-induced intracellular ROS accumulation (Koukoulitsa et al., 2016), by inducing upregulation of cellular antioxidants (i.e., glutathione) and gene expression of phase 2 enzymes, protecting against oxidative and electrophilic injury (Cao and Li, 2004), and strengthening the HO-1 pathway (Kwon et al., 2011). Chronic administration of resveratrol also significantly reduces increased levels of malondialdehyde in rats (Sharma and Gupta, 2002; Kumar et al., 2007) and contrasts brain inflammation, which is another important pathological hallmark of Alzheimer's disease contributing to neuronal damage and enhancing Aβ formation (Sadigh-Eteghad et al., 2016). In fact, this compound interferes with the neuroinflammatory process (Venigalla et al., 2016) through the suppression of astrocytes and microglia activation (Wang et al., 2002; Bi et al., 2005) and the inhibition of p38 mitogen-activated protein kinase phosphorylation and nuclear factor (NF)-κB activation. This ultimately reduces tumor necrosis factor-alpha (TNF-α) and NO production (Cheng et al., 2015). This effect also blocks cyclooxygenase (COX)-2 and *inducible NO synthase* (iNOS) expression (Rahman et al., 2006). Resveratrol treatment was also shown to reverse Aβ-induced iNOS overexpression (Huang et al., 2011) and exert antiinflammatory effects through the inhibition of TNF-α, interleukin (IL)-1β, and IL-6 expression (Yao et al., 2015), as well as signal transducer and activator of transcription 1 (STAT1) and STAT3 phosphorylation (Capiralla et al., 2012).

The neuroprotective effect of resveratrol has been tested in some randomized, double-blind, placebo-controlled clinical trials (Lee et al., 2017). Among others, Kennedy et al. (2010) evaluated the effect of a single oral dose of resveratrol on mental function and cerebral blood flow (CBF) in the frontal cortex of young, healthy humans. Enrolled subjects received three single-dose treatments: placebo, 250 mg *trans*-resveratrol, and 500 mg *trans*-resveratrol. Following administration, cognitive function did not enhance acutely, but resveratrol significantly increased CBF in a dose-dependent manner. The result was magnified by the contemporaneous administration of piperine 20 mg (probably because of a direct effect of piperine on CBF, since piperine coadministration was demonstrated not to affect resveratrol plasma level) (Wightman et al., 2014). A further crossover clinical trial has been designed to evaluate the effects of resveratrol on circulatory function and cognitive performance in obese adults (Wong et al., 2013). Participants were randomized to consume a capsule containing either 75 mg of resveratrol or a color-matched placebo daily for 6 weeks. Then, participants were crossed over to an alternate dose for another 6 weeks. Then, after an hour following the assessments in week 6 and 12, participants consumed a single additional dose of the supplement. The primary objective was to measure the degree of change in vasodilator function assessed by flow-mediated dilatation (FMD) in the brachial artery. Resveratrol supplementation was found to be well tolerated and induced a 23% increase in FMD compared with placebo. Moreover, a single dose of resveratrol (75 mg) following chronic resveratrol supplementation resulted in an acute FMD response 35% greater than the placebo. However, attention and concentration were unaffected by chronic resveratrol supplementation (Lee et al., 2017). Witte et al. (2014) carried out a study aimed at assessing the ability of resveratrol, given over 26 weeks, to enhance cognitive performance. Overweight older adults were randomly divided into an active treated group, receiving 200 mg/day of resveratrol in a formulation with quercetin 320 mg (Serban et al., 2016) in order to increase its bioavailability, and a control group, placebo treated. Volume, microstructure, and functional connectivity of the hippocampus were explored using magnetic resonance imaging. Resveratrol supplementation induced retention of memory and improved the functional connectivity between the hippocampus and frontal, parietal, and occipital areas, compared to the placebo (Kennedy et al., 2010). The changes in resting-state functional connectivity networks of the hippocampus after resveratrol intake were linked with behavioral improvements. Additionally, glucose metabolism was improved, which may account for some of the beneficial effects of resveratrol on neuronal function. These data have been recently confirmed in another trial by Köbe et al. (2017) A long-term (52-week), larger ($n = 119$) trial of resveratrol in individuals with mild to moderate AD was also conducted (Turner et al., 2015). Participants were recruited and randomly divided into placebo and resveratrol (500 mg orally once daily with dose escalation by 500 mg increments every 13 weeks, ending with 1000 mg twice daily) groups, respectively. The levels of $A\beta_{40}$ in the cerebrospinal fluid and plasma declined more in the placebo group than in the resveratrol-treated group with a significant difference at week 52 (note that $A\beta_{40}$ levels decline as dementia advances). The trial also confirmed that resveratrol was safe and well tolerated. Recently, Wong et al. (2016) carried out a balanced crossover clinical trial to evaluate the effects of resveratrol supplementation on cerebrovascular function in type 2 diabetes adult patients. Participants were randomly allocated to receive placebo or resveratrol at doses of 75, 150, or 300 mg, taken as a single dose during four intervention visits that took place at 7-day intervals over 4 weeks. The main objective was to determine the most effective dose of resveratrol to improve the cerebral vasodilator responsiveness (CVR) to hypercapnia in the middle cerebral artery, using transcranial Doppler ultrasound. Resveratrol consumption significantly increased CVR in the middle cerebral artery at all tested doses, and the maximum improvement was observed with the lowest dose. In a more recent 14-week study carried out on postmenopausal women, the improved CVR was also associated with improvement in cognitive function scores (Evans et al., 2017), which further confirms the potential usefulness of resveratrol in subjects with cognitive decline.

13.3.3 *Camelia sinensis*

Daily tea intake might be associated with a significant reduction in the risk to develop cognitive disorders (odds ratio [OR]$=0.65$, 95% CI: 0.58–0.73) (Ma et al., 2016). Epigallocatechin-3-gallate (EGCG; $C_{22}H_{18}O_{11}$) is the main polyphenol compound extracted from *Camelia sinensis* (tea plant) and binds directly to a large number of proteins that are involved in protein misfolding diseases, inhibiting their fibrillization (Chakrawarti et al., 2016). This hydrophilic catechin is also a potent antioxidant, capable of capturing and forming a complex between iron ions and copper and might, in this way, neutralize catalytic effects that these ions have on the production of ROS. It also exhibits antiinflammatory activity and has the ability to cross the blood-brain barrier (Mandel et al., 2006; Serban et al., 2015). In turn, it has become clear from many studies using models of neurodegenerative disorders that EGCG protects against oxidative stress insults and decreases cellular and animal neuronal death. Furthermore, it promotes the formation of aggregates that are spherically stable. These spherical aggregates, which are not considered cytotoxic, have a low content of β-sheet fibrils and do not catalyze the formation of fibrils (Singh et al., 2016). This property of EGCG allows the reduction of toxicity of alpha-synuclein and Aβ1-42 peptides (Ehrnhoefer et al., 2008). Studies carried out in transgenic mouse model of AD analyzed the effects of EGCG on cleavage of the amyloid-beta protein precursor (AβPP) and the reduction of cerebral amyloidosis through the use of this catechin and found that EGCG reduces the generation of Aβ in neuroblastoma N2alpha cells transfected with Swedish mutant human AβPP and in neurons in primary culture of Swedish mutant mice overexpressing AβPP (Rezai-Zadeh et al., 2008; Bieschke et al., 2010). On the contrary, when dealing with transgenic AβPPsw transgenic mice overexpressing Aβ, it was found that after being exposed to catechins, there was a decrease in the levels of Aβ and plaques associated with promotion of nonamyloidogenic alpha-secretase proteolytic pathway (Singh et al., 2016). Despite impressive preclinical literature, sufficient clinical data are not available to support the use of EGCG as a cognitive function improver in clinical practice. Thus, the protective effect of green tea on human cognitive performance is likely related to other green tea compounds.

A recent study was conducted on female Swiss mice who were given a normal diet or a hypercaloric diet (CD) over 8 weeks, and, concomitantly, receiving oral doses of white, green, red, or black teas (1% dose) or water. The mice subjected to CD showed weight gain, body fat accumulation, increased glucose, cholesterol, and triglycerides, associated with recognition memory deficits and increased reactive species (RS) levels and acetylcholinesterase (AChE) activity in the hippocampus. A significant reduction in AChE activity and partially reduced fat accumulation was observed in all the mice receiving teas. Green and red teas reduced memory deficit. White, green, and black teas reduced RS levels, whereas only green and black tea reduced plasma triglyceride levels (Soares et al., 2019).

L-Theanine ($C_7H_{14}N_2O_3$) is a major amino acid found in green tea, able to enhance cognitive functions and to have a positive effect on relaxation, emotional status, and quality sleep (Aguiar and Borowski, 2013; Türközü and Şanlier, 2017). It is well known that L-theanine can facilitate the longer-lasting processes responsible for supporting attention across the time frame of a difficult task (Gomez-Ramirez et al., 2009). Its mechanism of action might be based on the increase of brain serotonin, dopamine, and *gamma-aminobutyric acid* (GABA) levels (Nathan et al., 2006). Its antistress effect is probably due to the inhibition of cortical neuron excitation (Kimura et al., 2007). Changes in brain alpha oscillatory activity using electroencephalography (EEG) have been studied as correlates of L-theanine potential antistress effects, using resting-state recordings and cognitive-enhancing effects with task-related recordings (Juneja et al., 1999). However, the positive effects of L-theanine in humans seem to be magnified when it is associated with caffeine (Dietz and Dekker, 2017). In fact, caffeine can block all four subtypes of adenosine receptors (A1, A2A, A2B, and A3) with most of its actions mediated by high-affinity A1 and A2A receptors. The antagonistic effect of caffeine on adenosine receptors may induce upregulation of those receptors, improve the functioning of the blood-brain barrier, and, through this mechanism, protect against AD (Chen et al., 2010). In fact, in humans, the chronic consumption of caffeine is related to a significantly lower risk of developing cognitive impairment/decline (0.84, 95% CI, 0.72–0.99, $I2 = 42.6\%$] (Santos et al., 2010; Wierzejska, 2017).

13.3.4 *Theobroma cacao*

In recent years, great attention has been given to the potential positive cognitive effects of flavanols contained in *Theobroma cacao* (Grassi et al., 2016). It has been supposed that these flavanols act directly on neurons and improve brain blood flow (Nehlig, 2013).

To date, several flavonoid-binding sites on neurons have been reported: adenosine, GABAA, δ-opioid, nicotinic, TrkB, estrogen, and testosterone receptors (Williams and Spencer, 2012). However, it is still unclear whether any of these molecules are involved in mediating flavanol-induced stimulation of neuronal activity, which could account for increased blood flow in specific regions. Furthermore, it is difficult to establish the cause-effect relationship between increased neuronal activity and increased regional specific blood flow (Dietz and Dekker, 2017). However, receptor-binding flavanols and their metabolites will trigger the activation of various downstream kinases, including multiple members of both the MAP kinase and PI3 kinase pathways, as well as inducing the opening of agonist-activated ion channels leading to increased activity (Luh et al., 2000). Moreover, cocoa extracts seem to be useful in preventing the oligomerization of Aβ by restoring the long-term potentiation response reduced by oligomeric Aβ (Wang et al., 2014).

Cocoa flavanols affect both insulin sensitivity and endothelial vascular function through reciprocal mechanisms involving NO (Hooper et al., 2012; Kim et al., 2006) and similar mechanisms may occur in the cerebrovascular system. Short-term interventions with flavanol-rich cocoa in older volunteers reported improvements in cerebral blood flow and neurovascular coupling (Sorond et al., 2013), which reflects the coordination between neuronal activity, hemodynamic factors, and cellular interactions (Girouard and Iadecola, 2006).

In a recently published double-blind parallel-arm study carried out on 90 old volunteers with mild cognitive impairment randomized to receive a drink containing ≈ 990 mg (high flavanols), ≈ 520 mg (intermediate flavanols), or ≈ 45 mg (low flavanols) of cocoa flavanols once daily for 8 weeks, cognitive function was assessed by MMSE, Trail Making Test A (TMT-A) and B (TMT-B), and verbal fluency test (VFT). At the end of the follow-up period, there were no significant differences in MMSE ($P = 0.13$) among the three treatment groups. The time required to complete TMT-A and TMT-B was significantly ($P < 0.05$) lower in subjects treated with high flavanols (38.1 ± 10.9 and 104.1 ± 28.7 s, respectively) and intermediate flavanols (40.2 ± 11.3 and 115.9 ± 28.3 s, respectively) in comparison with those assigned to low flavanols (52.6 ± 17.9 and 139.2 ± 43.0 s, respectively). Similarly, the VFT score was significantly ($P < 0.05$) better in subjects assigned to high flavanols in comparison with those assigned to low flavanols (27.5 ± 6.7 vs 22.3 ± 8.1 words per 60 s). Insulin resistance, blood pressure, and lipid peroxidation also decreased among subjects in the high-flavanol and intermediate-flavanol groups. Changes in insulin resistance explained $\approx 40\%$ of composite z score variability through the study period ($P < 0.0001$) (Desideri et al., 2012).

A high-flavanol intervention was demonstrated to enhance dentate gyrus function, a region in the hippocampal formation that decreases in function with increasing age and is therefore believed to be a possible cause of age-related memory decline (Brickman et al., 2014). In a recent trial carried out on healthy older adults, CBF was measured by arterial spin labeling functional magnetic resonance imaging pre- and postconsumption of low (23 mg) or high (494 mg) 330 mL equicaloric flavanol drinks assessed for caffeine, theobromine, taste, and appearance, according to a randomized counterbalanced crossover double-blind design (Chen et al., 2010). Significant increases in regional perfusion across the brain were observed following consumption of the high-flavanol drink relative to the low-flavanol drink, particularly in the anterior cingulate cortex and the central opercular cortex of the parietal lobe. In the opinion of the authors, this could be associated with improvement in cognitive performance (Lamport et al., 2015).

13.3.5 *Bacopa monnieri*

Bacopa monnieri has been used for centuries in Ayurvedic medicine, as a neural tonic, sedative, antiepileptic, memory, and learning enhancer, as it contains several active phytochemical constituents including bacosides A and B, alkaloids, and saponins (Sivaramakrishna et al., 2005). Its principal mechanisms of action are related to antiinflammatory, antioxidant, metal chelation, amyloid, and cholinergic effects (Stough et al., 2015). Moreover, recent findings support the Bacopa's ability to inhibit catechol-O-methyl transferase, prolyl endopeptidase, and poly (ADP-ribose) polymerase (Dethe et al., 2016). It also exerts an antagonistic effect on serotonin $_6$ and $_{2A}$ (5-HT$_6$ and 5-HT$_{2A}$) receptors, which are commonly recognized as influencing several neurological pathways associated with memory and learning disorders and age-associated memory impairment (Dethe et al., 2016). Rastogi et al. also suggest that a long-term (200 mg/ kg orally per day for 3 months) bacoside administration can lead to a significant decrease in proinflammatory cytokines (IL-1beta, TNF-alpha, but not interferon-gamma), a significant induction in iNOS expression, and a significant reduction in total nitrite and lipofuscin content in the cortex (Rastogi et al., 2012). A metaanalysis of nine RCTs involving 437 subjects showed improved cognition by improved TMT-B test (-17.9 ms; 95% CI -24.6 to -11.2; $P < 0.001$) and decreased choice reaction time (CRT) (10.6 ms; 95% CI -12.1 to -9.2; $P < 0.001$). The tested extracts were overall well tolerated (Kongkeaw et al., 2014).

Finally, a recent study was designed to evaluate the prevention potential of Bacopa (BM) supplementation on colchicine-induced inflammation and Aβ production. Dementia was induced by a single intracerebroventricular injection of colchicine (15 μg/5 μL), whereas BM extract was administered orally (50 mg/kg body weight, daily) for 15 days. The study reveals that BM reverses colchicine-induced dementia through antiinflammatory and antioxidant action, suggesting that it may be an effective therapeutic intervention to ameliorate the progression of AD (Saini et al., 2019).

13.3.6 *Crocus sativus*

Crocus sativus L. is a perennial herb member of the Iridaceae family and saffron is obtained from its dark-red stigmas. The main chemical compound of saffron is crocin ($C_{44}H_{64}O_{24}$); it is known to possess antidepressant and antiinflammatory effects and to improve learning and memory (Schmidt et al., 2007). Crocin may combat the cognitive deficits caused by neurotoxic agents like streptozocin (Naghizadeh et al., 2013). Its main mechanism of action is based on the inhibition of acetyl-cholinesterase activity and on its antioxidant properties (Pitsikas, 2015). Evidence from in vitro and in vivo research indicates that saffron also has potential anticarcinogenic (cancer-suppressing), antimutagenic (mutation-preventing), antioxidant, and antidepressant characteristics (Bathaie and Mousavi, 2010).

Even if the majority of available clinical trials have tested the effects of saffron on mood (Hausenblas et al., 2015), some trials have also confirmed its effects on cognitive function. An early trial was carried out on 46 subjects with moderate AD symptoms, showing that 16-week supplementation with 30 mg saffron per day was able to significantly improve the AD assessment scale-cognitive subscale (ADAS-cog) when compared to the placebo (Akhondzadeh et al., 2010). The same research group then demonstrated in 68 patients that 1-year supplementation with 30 mg saffron per day was also able to improve the Severe Cognitive Impairment Rating Scale (SCIRS) and functional assessment staging (FAST), similarly to 20 mg in moderate to severe Alzheimer's disease (Farokhnia et al., 2014). More recently, similar results have been achieved in a different research group that consists of patients affected by mild cognitive impairment (Hausenblas et al., 2015). In all available trials, the incidence of adverse events was similar to that of the placebo (Tsolaki et al., 2016).

13.3.7 *Curcuma longa*

Curcumin ($C_{21}H_{20}O_6$) is the active component isolated from the grass rhizome *Curcuma longa*. It is a hydrophobic polyphenol with low bioavailability in humans (Sahebkar et al., 2017). Among the characteristics that have been identified for this compound, there are antioxidant, antiinflammatory, and amyloid disaggregating effects (Ullah et al., 2017; Ganjali et al., 2017). It also decreases inflammation induced by Aβ and moderately inhibits gamma-secretase and acetylcholinesterase activities (Lim et al., 2001; Yang et al., 2005). Even if several authors have reported results surpassing expectations in animal models, clinical studies using curcumin in humans have described conflicting findings, reporting positive effects in two studies out of five (Mazzanti and Di Giacomo, 2016). This discrepancy could be partly related to the low bioavailability or the low doses of the tested product. However, besides the published research, a number of studies on curcumin supplementation in healthy older people or in patients with mild cognitive impairment or Alzheimer's disease are still underway or completed, but results are not yet available.

13.4 Conclusions and future perspectives

A large selection of botanicals with memory-enhancing properties in humans also exerts other positive effects on health, which could contribute to the perceived improvement of cognitive function, independently from a specific action on memory-related mechanisms (for instance, they could exhibit simultaneous mood-improving effects). Reasons that justify the partially discrepant results of different trials carried out with the same botanical species are the different dosages tested, the different levels of quality of the extracts, and the different study designs. In particular, the products are sometimes underdosed in order to reduce costs or because the product would require administration of more doses in 1 day. However, by using 603 methodologies derived from pharmaceutical and drug industry research, which include the follow-up of therapeutic responses with validated tests and with the help of Alzheimer's disease biomarkers, new and more reliable results are now becoming available (Mattsson et al., 2015). One of the possible perspectives is to associate more botanicals with different mechanisms of action in order to improve their efficacy. For instance, in a recent trial from our group, we tested the short-term effect of a dietary supplement containing *Bacopa monnieri* dry extract 320 mg, L-theanine 100 mg, and *Crocus sativus* 30 mg, in young and old patients. These patients exhibited MMSE at baseline between 20 and 27, and self-perceived cognitive decline without known diagnosis of cognitive decline nor dementia. After 8 weeks of treatment, test scores on the enrolled subjects' cognitive decline, perceived stress, and depression were significantly improved (Cicero et al., 2017). Future studies addressing whether long-term dietary intake of single or combined botanicals can reduce the severity and incidence of neurodegenerative and other age-related diseases appear critical and crucial for the future.

References

Aguiar S, Borowski T. Neuropharmacological review of the nootropic herb *Bacopa monnieri*. Rejuvenation Res 2013;16:313–26. https://doi.org/10.1089/rej.2013.1431.

Akhondzadeh S, Sabet MS, Harirchian MH, Togha M, Cheraghmakani H, Razeghi S, et al. Saffron in the treatment of patients with mild to moderate Alzheimer's disease: a 16-week, randomized and placebo-controlled trial. J Clin Pharm Ther 2010;35:581–8. https://doi.org/10.1111/j.1365-2710.2009.01133.x.

Banach M, Rizzo M, Nikolic D, Howard G, Howard V, Mikhailidis D. Intensive LDL-cholesterol lowering therapy and neurocognitive function. Pharmacol Ther 2017;170:181–91.

Bathaie SZ, Mousavi SZ. New applications and mechanisms of action of saffron and its important ingredients. Crit Rev Food Sci Nutr 2010;50:761–86. https://doi.org/10.1080/10408390902773003.

Bi XL, Yang JY, Dong YX, Wang JM, Cui YH, Ikeshima T, et al. Resveratrol inhibits nitric oxide and TNF-α production by lipopolysaccharide-activated microglia. Int Immunopharmacol 2005;5:185–93.

Biber A. Pharmacokinetics of *Ginkgo biloba* extracts. Pharmacopsychiatry 2003;36:S32–7.

Bieschke J, Russ J, Friedrich RP, Ehrnhoefer DE, Wobst H, Neugebauer K, et al. EGCG remodels mature alpha-synuclein and amyloid-beta fibrils and reduces cellular toxicity. Proc Natl Acad Sci U S A 2010;107:7710–5.

Brickman AM, Khan UA, Provenzano FA, Yeung LK, Suzuki W, Schroeter H, et al. Enhancing dentate gyrus function with dietary flavanols improves cognition in older adults. Nat Neurosci 2014;17:1798–803. https://doi.org/10.1038/nn.3850.

Brondino N, De Silvestri A, Re S, Lanati N, Thiemann P, Verna A, Emanuele E, et al. A systematic review and meta-analysis of *Ginkgo biloba* in neuropsychiatric disorders: from ancient tradition to modern-day medicine. Evid Based Complement Alternat Med 2013;2013:915691. https://doi.org/10.1155/2013/915691.

Candelario-Jalil E, de Oliveira AC, Graf S, Bhatia HS, Hull M, Munoz E, Fiebich BL. Resveratrol potently reduces prostaglandin E2 production and free radical formation in lipopolysaccharide-activated primary rat microglia. J Neuroinflammation 2007;4:25.

Canter PH, Ernst E. *Ginkgo biloba* is not a smart drug: an updated systematic review of randomised clinical trials testing the nootropic effects of *G. biloba* extracts in healthy people. Hum Psychopharmacol 2007;22:265–78.

Cao Z, Li Y. Potent induction of cellular antioxidants and phase 2 enzymes by resveratrol in cardiomyocytes: protection against oxidative and electrophilic injury. Eur J Pharmacol 2004;489:39–48.

Capiralla H, Vingtdeux V, Zhao H, Sankowski R, Al-Abed Y, Davies P, et al. Resveratrol mitigates lipopolysaccharide- and Aβ-mediated microglial inflammation by inhibiting the TLR4/NF-κB/STAT signaling cascade. J Neurochem 2012;120:461–72.

Chakrawarti L, Agrawal R, Dang S, Gupta S, Gabrani R. Therapeutic effects of EGCG: a patent review. Expert Opin Ther Pat 2016;26:907–16. https://doi.org/10.1080/13543776.2016.1203419.

Chen X, Ghribi O, Geiger JD. Caffeine protects against disruptions of the blood-brain barrier in animal models of Alzheimer's and Parkinson's diseases. J Alzheimers Dis 2010;20:S127–41.

Cheng X, Wang Q, Li N, Zhao H. Effects of resveratrol on hippocampal astrocytes and expression of TNF-α in Alzheimer's disease model rate. Wei Sheng Yan Jiu 2015;44:610–4.

Cicero AF, Bove M, Colletti A, Rizzo M, Fogacci F, Giovannini M, et al. Short-term impact of a combined nutraceutical on cognitive function, perceived stress and depression in young elderly with cognitive impairment: a pilot, double-blind, randomized clinical trial. J Prev Alzh Dis 2017;4:12–5. https://doi.org/10.14283/jpad.2016.10.

Cicero AF, Fogacci F, Banach M. Botanicals and phytochemicals active on cognitive decline: the clinical evidence. Pharmacol Res 2018;130:204–12. https://doi.org/10.1016/j.phrs.2017.12.029.

Cvejic JM, Djekic SV, Petrovic AV, Atanackovic MT, Jovic SM, Brceski ID, et al. Determination of trans- and cis-resveratrol in Serbian commercial wines. J Chromatogr Sci 2010;48:229–34.

DeKosky ST, Williamson JD, Fitzpatrick AL, Kronmal RA, Ives DG, Saxton JA, et al. *Ginkgo biloba* for prevention of dementia: a randomized controlled trial. JAMA 2008;300:2253–62.

Desideri G, Kwik-Uribe C, Grassi D, Necozione S, Ghiadoni L, Mastroiacovo D, et al. Benefits in cognitive function, blood pressure, and insulin resistance through cocoa flavanol consumption in elderly subjects with mild cognitive impairment: the Cocoa, Cognition, and Aging (CoCoA) study. Hypertension 2012;60:794–801.

Dethe S, Deepak M, Agarwal A. Elucidation of molecular mechanism(s) of cognition enhancing activity of Bacomind(®): a standardized extract of *Bacopa monnieri*. Pharmacogn Mag 2016;12:S482–7.

Dietz C, Dekker M. Effect of green tea phytochemicals on mood and cognition. Curr Pharm Des 2017;23:2876–905. https://doi.org/10.2174/1381612823666170105151800.

Ehrnhoefer DE, Bieschke J, Boeddrich A, Herbst M, Masino L, Lurz R, et al. EGCG redirects amyloidogenic polypeptides into unstructured, off pathway oligomers. Nat Struct Mol Biol 2008;15:558–66.

Evans HM, Howe PR, Wong RH. Effects of resveratrol on cognitive performance, mood and cerebrovascular function in post-menopausal women; a 14-week randomised placebo-controlled intervention trial. Nutrients 2017;9. https://doi.org/10.3390/nu9010027. pii: E27.

Farokhnia M, ShafieeSabet M, Iranpour N, Gougol A, Yekehtaz H, Alimardani R, et al. Comparing the efficacy and safety of Crocus sativus L. with memantine in patients with moderate to severe Alzheimer's disease: a double-blind randomized clinical trial. Hum Psychopharmacol 2014;29:351–9. https://doi.org/10.1002/hup.2412.

Flirski M, Sobow T. Biochemical markers and risk factors of Alzheimer's disease. Curr Alzheimer Res 2005;2:47–64.

Ganjali S, Blesso CN, Banach M, Pirro M, Majeed M, Sahebkar A. Effects of curcumin on HDL functionality. Pharmacol Res 2017;119:208–18. https://doi.org/10.1016/j.phrs.2017.02.008.

Gardener H, Caunca M. Mediterranean diet in preventing neurodegenerative diseases. Curr Nutr Rep 2018;7:10–20. https://doi.org/10.1007/s13668-018-0222-5.

Girouard H, Iadecola C. Neurovascular coupling in the normal brain and in hypertension, stroke, and Alzheimer disease. J Appl Physiol 2006;100:328–35.

Gomez-Ramirez M, Kelly SP, Montesi JL, Foxe JJ. The effects of L-theanine on alpha-band oscillatory brain activity during a visuo-spatial attention task. Brain Topogr 2009;22:44–51. https://doi.org/10.1007/s10548-008-0068-z.

Grassi D, Ferri C, Desideri G. Brain protection and cognitive function: cocoa flavonoids as nutraceuticals. Curr Pharm Des 2016;22:145–51.

Hausenblas HA, Heekin K, Mutchie HL, Anton S. A systematic review of randomized controlled trials examining the effectiveness of saffron (Crocus sativus L.) on psychological and behavioral outcomes. J Integr Med 2015;13:231–40. https://doi.org/10.1016/S2095-4964(15)60176-5.

Hirsch GE, Viecili PR, de Almeida AS, Nascimento S, Porto FG, Otero J, et al. Natural products with antiplatelet action. Curr Pharm Des 2017;23:1228–46. https://doi.org/10.2174/1381612823666161123151611.

Hooper L, Kay C, Abdelhamid A, Kroon PA, Cohn JS, Rimm EB, et al. Effects of chocolate, cocoa, and flavan-3-ols on cardiovascular health: a systematic review and meta-analysis of randomized trials. Am J Clin Nutr 2012;95:740–51.

Huang TC, Lu KT, Wo YY, Wu YJ, Yang YL. Resveratrol protects rats from Aβ-induced neurotoxicity by the reduction of iNOS expression and lipid peroxidation. PLoS One 2011;6:e29102.

Ihl R. Effects of *Ginkgo biloba* extract EGb 761 (R) in dementia with neuropsychiatric features: review of recently completed randomised, controlled trials. Int J Psychiatry Clin Pract 2013;17:S8–14.

Jang JH, Surh YJ. Protective effect of resveratrol on β-amyloid-induced oxidative PC12 cell death. Free Radic Biol Med 2003;34:1100–10.

Juneja LR, Chu DC, Okubo T, Nagato Y, Yokogoshi H. L-Theanine—a unique amino acid of green tea and its relaxation effect in humans. Trends Food Sci Technol 1999;10:199–204.

Kennedy D, Wightman EL, Reay JL, Lietz G, Okello EJ, Wilde A, et al. Effects of resveratrol on cerebral blood flow variables and cognitive performance in humans: a double-blind, placebo-controlled, crossover investigation. Am J Clin Nutr 2010;91:1590–7.

Kim JA, Montagnani M, Koh KK, Quon MJ. Reciprocal relationships between insulin resistance and endothelial dysfunction: molecular and pathophysiological mechanisms. Circulation 2006;113:1888–904.

Kim J, Lee HJ, Lee KW. Naturally occurring phytochemicals for the prevention of Alzheimer's disease. J Neurochem 2010;112:1415–30.

Kimura K, Ozeki M, Juneja LR, Ohira H. L-Theanine reduces psychological and physiological stress responses. Biol Psychol 2007;74:39–45.

Köbe T, Witte AV, Schnelle A, Tesky VA, Pantel J, Schuchardt JP, et al. Impact of resveratrol on glucose control, hippocampal structure and connectivity, and memory performance in patients with mild cognitive impairment. Front Neurosci 2017;11:105. https://doi.org/10.3389/fnins.2017.00105.

Kongkeaw C, Dilokthornsakul P, Thanarangsarit P, Limpeanchob N, Norman SC. Meta-analysis of randomized controlled trials on cognitive effects of *Bacopa monnieri* extract. J Ethnopharmacol 2014;151:528–35. https://doi.org/10.1016/j.jep.2013.11.008.

Koukoulitsa C, Villalonga-Barber C, Csonka R, Alexi X, Leonis G, Dellis D, et al. Biological and computational evaluation of resveratrol inhibitors against Alzheimer's disease. J Enzyme Inhib Med Chem 2016;31:67–77.

Kumar A, Naidu PS, Seghal N, Padi SS. Neuroprotective effects of resveratrol against intracerebroventricular colchicine-induced cognitive impairment and oxidative stress in rats. Pharmacology 2007;79:17–26.

Kwon KJ, Kim HJ, Shin CY, Han SH. Melatonin potentiates the neuroprotective properties of resveratrol against beta amyloid-induced neurodegeneration by modulating AMP activated protein kinase pathways. J Clin Neurol 2010;6:127–37.

Kwon KJ, Kim JN, Kim MK, Lee J, Ignarro LJ, Kim HJ, et al. Melatonin synergistically increases resveratrol induced heme oxygenase-1 expression through the inhibition of ubiquitin-dependent proteasome pathway: a possible role in neuroprotection. J Pineal Res 2011;50:110–23.

Lamport DJ, Pal D, Moutsiana C, Field DT, Williams CM, Spencer JP, Butler LT. The effect of flavanol-rich cocoa on cerebral perfusion in healthy older adults during conscious resting state: a placebo controlled, crossover, acute trial. Psychopharmacology 2015;232(17):3227–34. https://doi.org/10.1007/s00213-015-3972-4.

Lee J, Torosyan N, Silverman DH. Examining the impact of grape consumption on brain metabolism and cognitive function in patients with mild decline in cognition: a double-blinded placebo controlled pilot study. Exp Gerontol 2017;87:121–8. https://doi.org/10.1016/j.exger.2016.10.004.

Li J, Zhang YC, Chen G. Effect of *Ginkgo biloba* extract EGb761 on hippocampal neuronal injury and carbonyl stress of D-gal-induced aging rats. Evid Based Complement Alternat Med 2019;2019:1–12.

Lim GP, Chu T, Yang F, Beech W, Frautschy SA, Cole GM. The curry spice curcumin reduces oxidative damage and amyloid pathology in an Alzheimer transgenic mouse. J Neurosci 2001;21:8370–7.

Luh WM, Wong EC, Bandettini PA, Ward BD, Hyde JS. Comparison of simultaneously measured perfusion and BOLD signal increases during brain activation with T(1)-based tissue identification. Magn Reson Med 2000;44:137–43.

Ma QP, Huang C, Cui QY, Yang DJ, Sun K, Chen X, Li XH. Meta-analysis of the association between tea intake and the risk of cognitive disorders. PLoS One 2016;11. https://doi.org/10.1371/journal.pone.0165861.

Mandel S, Amit T, Reznichenko L, Weinreb O, Youdim MB. Green tea catechins as brain-permeable, natural iron chelators-antioxidants for the treatment of neurodegenerative disorders. Mol Nutr Food Res 2006;50:229–34.

Mattsson N, Carrillo MC, Dean RA, Devous Sr. MD, Nikolcheva T, Pesini P, et al. Revolutionizing Alzheimer's disease and clinical trials through biomarkers. Alzheimers Dement 2015;1:412–9. https://doi.org/10.1016/j.dadm.2015.09.001.

Mazzanti G, Di Giacomo S. Curcumin and resveratrol in the management of cognitive disorders: what is the clinical evidence? Molecules 2016;21. https://doi.org/10.3390/molecules21091243. pii: E1243.

Mecocci P, Tinarelli C, Schulz RJ, Polidori MC. Nutraceuticals in cognitive impairment and Alzheimer's disease. Front Pharmacol 2014;5:147. https://doi.org/10.3389/fphar.2014.00147.

Moreno M, Castro E, Falqué E. Evolution of trans- and cis-resveratrol content in red grapes (*Vitis vinifera L.* cv Menciá, Albarello and Merenzao) during ripening. Eur Food Res Technol 2008;227:667–74.

Naghizadeh B, Mansouri MT, Ghorbanzadeh B, Farbood Y, Sarkaki A. Protective effects of oral crocin against intracerebroventricular streptozotocin-induced spatial memory deficit and oxidative stress in rats. Phytomedicine 2013;20:537–42. https://doi.org/10.1016/j.phymed.2012.12.019.

Nathan PJ, Lu K, Gray M, Oliver C. The neuropharmacology of L-theanine (N-ethyl-L-glutamine): a possible neuroprotective and cognitive enhancing agent. J Herb Pharmacother 2006;6:21–30.

Nehlig A. The neuroprotective effects of cocoa flavanol and its influence on cognitive performance. Br J Clin Pharmacol 2013;75:716–27. https://doi.org/10.1111/j.1365-2125.2012.04378.x.

O'Brien JT, Holmes C, Jones M, Jones R, Livingston G, McKeith I, et al. Clinical practice with anti-dementia drugs: a revised (third) consensus statement from the British Association for Psychopharmacology. J Psychopharmacol 2017;31:147–68. https://doi.org/10.1177/0269881116680924.

Pitsikas N. The effect of *Crocus sativus* L. and its constituents on memory: basic studies and clinical applications. Evid Based Complement Alternat Med 2015;2015:926284. https://doi.org/10.1155/2015/926284.

Polidori MC, Schulz RJ. Nutritional contributions to dementia prevention: main issues on antioxidant micronutrients. Genes Nutr 2014;9:382. https://doi.org/10.1007/s12263-013-0382-2.

Prince M, Bryce R, Albanese E, Wimo A, Ribeiro W, Ferri CP. The global prevalence of dementia: a systematic review and metaanalysis. Alzheimers Dement 2013;9:63–75. https://doi.org/10.1016/j.jalz.2012.11.007.

Psaltopoulou T, Sergentanis TN, Panagiotakos DB, Sergentanis IN, Kosti R, Scarmeas N. Mediterranean diet, stroke, cognitive impairment, and depression: a meta-analysis. Ann Neurol 2013;74:580–91. https://doi.org/10.1002/ana.23944.

Rahman I, Biswas SK, Kirkham PA. Regulation of inflammation and redox signaling by dietary polyphenols. Biochem Pharmacol 2006;72:1439–52.

Rastogi M, Ojha R, Prabu PC, Devi DP, Agrawal A, Dubey GP. Amelioration of age associated neuroinflammation on long term bacosides treatment. Neurochem Res 2012;37:869–74. https://doi.org/10.1007/s11064-011-0681-1.

Rezai-Zadeh K, Arendash GW, Hou H, Fernandez F, Jensen M, Runfeldt M, et al. Green tea epigallocatechin-3-gallate (EGCG) reduces beta-amyloid mediated cognitive impairment and modulates tau pathology in Alzheimer transgenic mice. Brain Res 2008;1214:177–87. https://doi.org/10.1016/j.brainres.2008.02.107.

Sadigh-Eteghad S, Majdi A, Mahmoudi J, Golzari SE, Talebi M. Astrocytic and microglial nicotinic acetylcholine receptors: an overlooked issue in Alzheimer's disease. J Neural Transm 2016;123:1359–67.

Sahebkar A, Serban C, Ursoniu S, Wong ND, Muntner P, Graham IM, et al. Lack of efficacy of resveratrol on C-reactive protein and selected cardiovascular risk factors—results from a systematic review and meta-analysis of randomized controlled trials. Int J Cardiol 2015;189:47–55. https://doi.org/10.1016/j.ijcard.2015.04.008.

Sahebkar A, Saboni N, Pirro M, Banach M. Curcumin: an effective adjunct in patients with statin-associated muscle symptoms? J Cachexia Sarcopenia Muscle 2017;8:19–24. https://doi.org/10.1002/jcsm.12140.

Saini N, Singh D, Sandhir R. Bacopa monnieri prevents colchicine-induced dementia by anti-inflammatory action. Metab Brain Dis 2019;34:505–18.

Santos C, Costa J, Santos J, Vaz-Carneiro A, Lunet N. Caffeine intake and dementia: systematic review and meta-analysis. J Alzheimers Dis 2010;20:S187–204. https://doi.org/10.3233/JAD-2010-091387.

Schaller S, Mauskopf J, Kriza C, Wahlster P, Kolominsky-Rabas PL. The main cost drivers in dementia: a systematic review. Int J Geriatr Psychiatry 2015;30:111–29. https://doi.org/10.1002/gps.4198.

Schmidt M, Betti G, Hensel A. Saffron in phytotherapy: pharmacology and clinical uses. Wien Med Wochenschr 2007;157:315–9.

Schneider LS. Ginkgo and AD: key negatives and lessons from GuidAge. Lancet Neurol 2012;11:836–7.

Serban C, Sahebkar A, Antal D, Ursoniu S, Banach M. Effects of supplementation with green tea catechins on plasma C-reactive protein concentrations: a systematic review and meta-analysis of randomized controlled trials. Nutrition 2015;31:1061–71. https://doi.org/10.1016/j.nut.2015.02.004.

Serban MC, Sahebkar A, Zanchetti A, Mikhailidis DP, Howard G, Antal D, et al. Effects of quercetin on blood pressure: a systematic review and meta-analysis of randomized controlled trials. J Am Heart Assoc 2016;5. https://doi.org/10.1161/JAHA.115.002713. pii: e002713.

Sharma M, Gupta YK. Chronic treatment with trans-resveratrol prevents intracerebroventricular streptozotocin induced cognitive impairment and oxidative stress in rats. Life Sci 2002;71:2489–98.

Shi J, He M, Cao J, Wang H, Ding J, Jiao Y, et al. The comparative analysis of the potential relationship between resveratrol and stilbene synthase gene family in the development stages of grapes (*Vitis quinquangularis* and *Vitis vinifera*). Plant Physiol Biochem 2014;74:24–32.

Singh B, Parsaik AK, Mielke MM, Erwin PJ, Knopman DS, Petersen RC, et al. Association of Mediterranean diet with mild cognitive impairment and Alzheimer's disease: a systematic review and meta-analysis. J Alzheimers Dis 2014;39:271–82. https://doi.org/10.3233/JAD-130830.

Singh NA, Mandal AK, Khan ZA. Potential neuroprotective properties of epigallocatechin-3-gallate (EGCG). Nutr J 2016;15:60. https://doi.org/10.1186/s12937-016-0179-4.

Sivaramakrishna C, Rao CV, Trimurtulu G, Vanisree M, Subbaraju GV. Triterpenoid glycosides from *Bacopa monnieri*. Phytochemistry 2005;66:2719–28.

Soares MB, Ramalho JB, Izaguirry AP, Pavin NF, Spiazzi CC, Schimidt HL, Mello-Carpes PB, Santos FW. Comparative effect of *Camellia sinensis* teas on object recognition test deficit and metabolic changes induced by cafeteria diet. Nutr Neurosci 2019;22(8):531–40.

Sorond FA, Hurwitz S, Salat DH, Greve DN, Fisher ND. Neurovascular coupling, cerebral white matter integrity, and response to cocoa in older people. Neurology 2013;81:904–9.

Stough C, Singh H, Zangara A. Mechanisms, efficacy, and safety of *Bacopa monnieri* (Brahmi) for cognitive and brain enhancement. Evid Based Complement Alternat Med 2015;717605. https://doi.org/10.1155/2015/717605.

Suliman NA, Mat Taib CN, Mohd Moklas MA, Adenan MI, Hidayat Baharuldin MT, Basir R. Establishing natural nootropics: recent molecular enhancement influenced by natural nootropic. Evid Based Complement Alternat Med 2016;2016:4391375. https://doi.org/10.1155/2016/4391375.

Trela BC, Waterhouse AL. Resveratrol: isomeric molar absorptivities and stability. J Agric Food Chem 1996;44:1253–7.

Tsolaki M, Karathanasi E, Lazarou I, Dovas K, Verykouki E, Karacostas A, et al. Efficacy and safety of *Crocus sativus* L. in patients with mild cognitive impairment: one year single-blind randomized, with parallel groups, clinical trial. J Alzheimers Dis 2016;54:129–33. https://doi.org/10.3233/JAD-160304.

Türközü D, Şanlier N. L-Theanin, unique amino acid of tea, and its metabolism, health effects, safety. Crit Rev Food Sci Nutr 2017;57:1681–7. https://doi.org/10.1080/10408398.2015.1016141.

Turner RS, Thomas RG, Craft S, van Dyck CH, Mintzer J, Reynolds BA, et al. Alzheimer's disease cooperative study. A randomized, double-blind, placebo-controlled trial of resveratrol for Alzheimer disease. Neurology 2015;85:1383–91.

Ullah F, Liang A, Rangel A, Gyengesi E, Niedermayer G, Münch G. High bioavailability curcumin: an anti-inflammatory and neurosupportive bioactive nutrient for neurodegenerative diseases characterized by chronic neuroinflammation. Arch Toxicol 2017;91:1623–34. https://doi.org/10.1007/s00204-017-1939-4.

Vellas B, Coley N, Ousset PJ, Berrut G, Dartigues JF, Dubois B, et al. Long-term use of standardised *Ginkgo biloba* extract for the prevention of Alzheimer's disease (GuidAge): a randomised placebo-controlled trial. Lancet Neurol 2012;11:851–9.

Venigalla M, Sonego S, Gyengesi E, Sharman MJ, Münch G. Novel promising therapeutics against chronic neuroinflammation and neurodegeneration in Alzheimer's disease. Neurochem Int 2016;95:63–74.

Von Bernhardi R, Eugenín J. Alzheimer's disease: redox dysregulation as a common denominator for diverse pathogenic mechanisms. Antioxid Redox Signal 2012;16:974–1031. https://doi.org/10.1089/ars.2011.4082.

Wahlster L, Arimon M, Nasser-Ghodsi N, Post KL, Serrano-Pozo A, Uemura K, et al. Presenilin-1 adopts pathogenic conformation in normal aging and in sporadic Alzheimer's disease. Acta Neuropathol 2013;125:187–99.

Wang Q, Xu J, Rottinghaus GE, Simonyi A, Lubahn D, Sun GY, et al. Resveratrol protects against global cerebral ischemic injury in gerbils. Brain Res 2002;958:439–47.

Wang J, Varghese M, Ono K, Yamada M, Levine S, Tzavaras N, et al. Cocoa extracts reduce oligomerization of amyloid-β: implications for cognitive improvement in Alzheimer's disease. J Alzheimers Dis 2014;41:643–50. https://doi.org/10.3233/JAD-132231.

Wierzejska R. Can coffee consumption lower the risk of Alzheimer's disease and Parkinson's disease? A literature review. Arch Med Sci 2017;13:507–14. https://doi.org/10.5114/aoms.2016.63599.

Wightman EL, Reay JL, Haskell CF, Williamson G, Dew TP, Kennedy DO. Effects of resveratrol alone or in combination with piperine on cerebral blood flow parameters and cognitive performance in human subjects: a randomised, double-blind, placebo-controlled, cross-over investigation. Br J Nutr 2014;112:203–13.

Williams RJ, Spencer JP. Flavonoids, cognition, and dementia: actions, mechanisms, and potential therapeutic utility for Alzheimer disease. Free Radic Biol Med 2012;52:35–45. https://doi.org/10.1016/j.freeradbiomed.2011.09.010.

Witte AV, Kerti L, Margulies DS, Flöel A. Effects of resveratrol on memory performance, hippocampal functional connectivity, and glucose metabolism in healthy older adults. J Neurosci 2014;34:7862–70.

Wong RH, Berry NM, Coates AM, Buckley JD, Bryan J, Kunz I, et al. Chronic resveratrol consumption improves brachial flow-mediated dilatation in healthy obese adults. J Hypertens 2013;31:1819–27.

Wong RH, Nealon RS, Scholey A, Howe PR. Low dose resveratrol improves cerebrovascular function in type 2 diabetes mellitus. Nutr Metab Cardiovasc Dis 2016;26:393–9.

Yang F, Lim GP, Begum AN, Ubeda OJ, Simmons MR, Ambegaokar SS, et al. Curcumin inhibits formation of amyloid beta oligomers and fibrils, binds plaques, and reduces amyloid in vivo. J Biol Chem 2005;280:5892–901.

Yang G, Wang Y, Sun J, Zhang K, Liu J. Ginkgo biloba for mild cognitive impairment and Alzheimer's disease: a systematic review and meta-analysis of randomized controlled trials. Curr Top Med Chem 2016;16:520–8.

Yao Y, Li J, Niu Y, Yu JQ, Yan L, Miao ZH, et al. Resveratrol inhibits oligomeric Aβ-induced microglial activation via NADPH oxidase. Mol Med Rep 2015;12:6133–9.

Yuan Q, Wang CW, Shi J, Lin ZX. Effects of *Ginkgo biloba* on dementia: an overview of systematic reviews. J Ethnopharmacol 2017;195:1–9. https://doi.org/10.1016/j.jep.2016.12.005.

Zanchetti A, Liu L, Mancia G, Parati G, Grassi G, Stramba-Badiale M, et al. Blood pressure and low-density lipoprotein-cholesterol lowering for prevention of strokes and cognitive decline: a review of available trial evidence. J Hypertens 2014;32:1741–50. https://doi.org/10.1097/HJH.0000000000000253.

Zhuang H, Kim YS, Koehler RC, Dore S. Potential mechanism by which resveratrol, a red wine constituent, protects neurons. Ann N Y Acad Sci 2003;993:276–86.

CHAPTER 14

Ways to become old: Role of lifestyle in modulation of the hallmarks of aging

Giulia Accardi and Anna Aiello
Laboratory of Immunopathology and Immunosenescence, Department of Biomedicine, Neurosciences and Advanced Diagnostics, University of Palermo, Palermo, Italy

14.1 Introduction

Among nongenetic factors, lifestyle has a primary role on health and lifespan. It may lead to a postponement of disability, resulting in a healthier aging process (Grassi et al., 2014). In particular, physical activity and diet (to whom the authors will refer to in this chapter using the term "lifestyle") impact on the molecular and functional mechanisms that have main effects on the major age-related diseases (Minuti et al., 2014; Garatachea et al., 2015). At the same time, poor and unbalanced diets as well as Western diets, characterized by a high calorie intake, full of saturated and trans fats, animal proteins, unbalanced omega-6/omega-3 ratio, refined carbohydrates, and inactive or sedentary life can accelerate aging process. These factors are chiefly responsible for unsuccessful aging and age-related diseases like obesity, diabetes, cancer, cardiovascular and Alzheimer's diseases (AD) (Büchner et al., 2013; Milte and McNaughton, 2016; Kim and Lee, 2019). Despite the widely recognized effects of lifestyle on health span, the fundamental mechanisms by whom they influence the biology of aging, age-related diseases, and longevity remain elusive (Accardi et al., 2019). However, the deep study of the so-called hallmarks of aging offers new scenarios and possible therapeutic approaches in the aging treatment. The hallmarks of aging, defined by López-Otín et al. in 2013 as the nine factors that contribute to aging and together determine the aging phenotype, are classified in three clusters: the primary hallmarks (genomic instability, telomere attrition, epigenetic alterations, and loss of proteostasis) that represent the main causes of molecular damage; the antagonistic hallmarks (deregulated nutrient sensing, mitochondrial dysfunction, and cellular senescence) that show beneficial effects at low levels but become deleterious at high levels; the integrative hallmarks (stem cell exhaustion and altered intercellular communication) that appear as a consequence of accumulating damages no longer compensable by homeostatic mechanisms (López-Otín et al., 2013). Although more research is needed, it was showed that lifestyle can influence all these hallmarks, at least partly, often involving the same molecular targets. About healthy dietary habits, the "slow-aging diets" are defined as "dietary interventions that can slow the ageing process, delaying or preventing a range

Human Aging
https://doi.org/10.1016/B978-0-12-822569-1.00009-3

of chronic age-related diseases, through the modulation of the relevant intracellular signalling pathways" (Aiello et al., 2019a). According to nutrigerontology, the diets that can slow or postpone aging process, promoting long life and healthy old age, are the dietary restriction (DR) interventions (including calorie restriction (CR), fasting, and protein or amino acids restriction), the dietary approaches to stop hypertension (DASH diet), the plant-based diets, Asiatic and Mediterranean ones. It is possible to classify them on the basis of some common characteristics. In particular, the slow-aging diets are characterized by low calorie intake, mainly vegetable proteins rather than animal, and reduction or elimination of trans and saturated fats as well as of refined carbohydrates. They are also characterized by a good balance of omega-3 and 6, vitamins, micronutrients, and antioxidants contained in fruits and vegetables, that can help to achieve a successful aging phenotype. Similarly, appropriate intake of specific foods or bioactive molecules, functional foods, and nutraceuticals, respectively, may confer health benefits, influencing the maintenance of immune homeostasis, and directly contributing to the reduction of inflammation and metabolic disorders (Aiello et al., 2019a). Although solid evidence exists about the benefits of DR interventions on almost each hallmark, it is necessary to underline that they are severe dietary regimens and not always applicable in humans. So, alternative approaches, such as a close adherence to Mediterranean or Asiatic diets, are desirable. About physical activity, it has a positive antiaging impact at cellular level, underlining its specific role in attenuating the effects of each hallmark. In particular, regular physical activity in the older population, especially aerobic and resistance training, prevents several age-related pathologies, such as sarcopenia or frailty, cardiorespiratory and metabolic diseases, and postpones cognitive decline. It is also involved in the decrease of blood pressure levels and in the improvement of metabolic function by the increase of muscle protein synthesis (Garatachea et al., 2015; Rebelo-Marques et al., 2018). See Fig. 14.1 for an overview of the main effects of dietary strategies and physical activity on each hallmark.

14.2 Primary hallmarks and lifestyle

Genomic instability, telomere attrition, epigenetic alterations, and loss of protein homeostasis, named proteostasis, belong to the first category of the hallmarks of aging. Their common characteristic is that they are all unequivocally negative.

14.2.1 Genomic instability

Nutritional genomics is the application of high-throughput functional genomic technologies in nutrition research, studying the mechanisms by which nutrients or dietary patterns and the genome interplay each other (Elliott and Ong, 2002). This interaction can

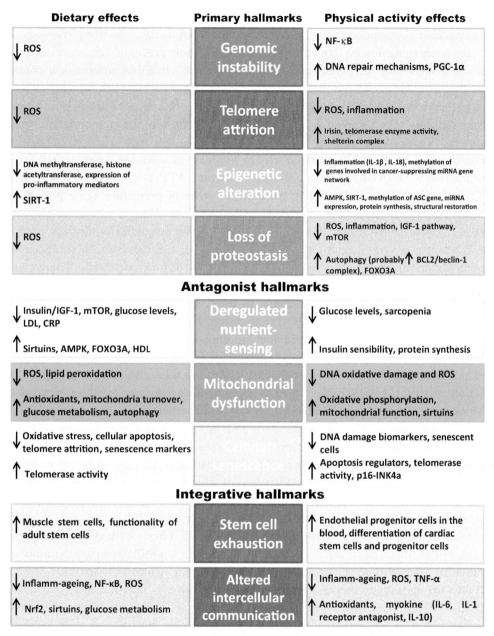

Fig. 14.1 Summary of the main effects of regular physical activity and slow-aging diets on each hallmark. (See the text for the acronyms.)

be positive or negative. The last case could be one of the causes of genomic instability, an unavoidable phenomenon that can lead to nuclear DNA damage and that physiologically characterize aging process (López-Otín et al., 2013). More precisely, "genomic instability is the tendency of the genome to undergo alterations in DNA information content through mutation" (Vijg and Montagna, 2017). Usually, repair mechanisms provide to restore the genomic damage. In some cases, these are impaired, causing a strong genomic instability. The best models to study this phenomenon are the progeroid syndromes, abnormal congenital condition, such as Werner, or Ataxia telangiectasia (AT). These are characterized by defects in genes directly or indirectly involved in genomic stability by restoration of DNA damages (e.g., Werner Syndrome ATP-Dependent Helicase, Ataxia Telangiectasia Mutated or ATM), causing the onset of premature aging phenotype, including typical age-related disorders, such as type 2 diabetes or cancer (Navarro et al., 2006). In nucleotide excision repair, defective progeroid syndromes, such as AT, reactive oxygen species (ROS) hyperproduction, and reduction in autophagy by mitochondria (named mitophagy), are present (López-Otín et al., 2016). These events lead to mitochondrial abnormalities, including ROS generation, increased transmembrane potential, and limited mitophagy. In mice, all of these can be rescued by poly (ADP-ribose) polymerase 1 (PARP1) inhibition, hyperactivated in these diseases, or external supply of the nicotinamide adenine dinucleotide (NAD^+), greatly consumed by PARP1 hyperfunctionality and key cofactor in many metabolic reactions (Fang et al., 2014; Scheibye-Knudsen et al., 2014). Among these, that one mediated by the action of sirtuin-1 (SIRT-1), a NAD^+-dependent deacetylase, regulates metabolism homeostasis and mitochondrial biogenesis that could partially explain its role in CR thus in promoting successful aging (Tang, 2016; Lee, 2019). The role of CR in lowering oxidative stress level and its interaction with nutrient-sensing pathways (NSPs) give an explanation of the indirect effect of this dietary strategy to fight genomic instability (Heydari et al., 2007). Thus the explanation of why the modulation of genomic instability can be involved in aging process, triggering or restoring different metabolic alterations by the availability of specific nutrients or by nutritional interventions. The role of physical activity in counteracting genomic instability has been studied in model organisms. For example, in rodents, aerobic exercise enhances DNA repair mechanisms and modulate peroxisome proliferator-activated receptor gamma coactivator 1-alpha (PGC-1α) signaling to suppress Nuclear Factor kappa-light-chain-enhancer of activated B cells (NF-κB) (Gomez-Cabrera et al., 2008; Leick et al., 2010). NF-κB is a well-known pathway with a key part in inflammation and aging, so favoring genomic instability (Tilstra et al., 2011). PGC-1α networks with SIRT-1 in the mitochondrial biogenesis and energy metabolism and has been identified as a potential target for mitochondrial damage, so decreasing production of ROS, involved in genomic instability (Anderson and Prolla, 1790).

14.2.2 Telomere attrition

Telomere attrition is unavoidable and strongly related to aging process. It consists in the loss of some telomere sequence during DNA replication of the first several base pairs of a linear DNA molecule (López-Otín et al., 2013). Indeed, many evidences demonstrated the association between age-related diseases and short telomere as well as studies on centenarians show the relative longer telomere compared to unsuccessful aged people (Garatachea et al., 2015; Herrmann et al., 2018; Atzmon et al., 2010). Healthy lifestyle is related to longer telomeres and lymphocytes are the most studied cells in this context because it was proven that the rate of telomere shortening is similar between them and somatic cells (Daniali et al., 2013). It seems that the constant assumption of foods rich in vitamins, minerals, polyphenols, and omega-3 fatty acids prevents or reduces the telomere shortening in lymphocytes, through the reduction of oxidative stress (D'Mello et al., 2015). Consequently, it is possible to speculate that the positive association between redox state and micronutrients or nutraceuticals (e.g., vitamins C, D, E, folate, β-carotene, zinc, magnesium, or polyphenols) has a positive impact on telomeres, reducing their attrition and protecting the chromosome, both in mice and humans (Thomas et al., 2009; Balan et al., 2018). So that, the daily consumption of antioxidant foods, such as tea, red wine, extra virgin olive oil (EVOO), or any other foods rich in polyphenols (e.g., theaflavins or resveratrol) could slow aging, as seen in different populations, likely due to the reduction of telomere attrition (Vidacek et al., 2017; Fernández del Río et al., 2016; Boccardi et al., 2013; García-Calzón et al., 2015). However, further research is needed to obtain stronger data and to establish a close causal relationship. About fats, omega-3 are well known in the prevention of cardiovascular diseases. But, more than their levels, it seems that the omega-6/omega-3 ratio may be crucial, as demonstrated by a randomized controlled trial where telomere length and this ratio in plasma were linked by an inverse proportion. In addition, dietary fibers from cereals and whole grains, belonging to the traditional Mediterranean diet, have a positive effect on telomere length in women (Vidacek et al., 2017). To adequately apply these scientific reports on daily living of humans, it is important to highlight the effects of specific compounds; however, it is even more important to evaluate the role of specific dietary patterns on telomere length. Actually, we can speculate that nutritional regimes rich in fresh and seasonal fruits, vegetables, nuts, EVOO, and fatty fish (e.g., bluefish), such as Mediterranean, DASH, Asiatic, or plant-based ones have a positive effect on telomeres length and, consequently, on aging (López-Otín et al., 2016; García-Calzón et al., 2015). Additional confirmation, in women, comes from Nurses' Health Study, which demonstrated the preservation of telomere length thanks to the adherence to Mediterranean diet (Crous-Bou et al., 2014). On the contrary, Western diets favor telomere shortening in many populations. In all cases, the explanations can be found in the high levels of inflammation and oxidative stress (Balan et al., 2018; García-Calzón et al., 2015; Lee et al., 2015). Looking at specific

DR, CR reduces telomere attrition and promotes telomerase activity in models (rodents) (Vera et al., 2013). In humans, there are few no conclusive reports due to the difficulties in carrying on this kind of nutritional patterns (Vidacek et al., 2017). For instance, Nurses' Health Study reports no association between energy intake and telomere length (Cassidy et al., 2010). An interesting datum, but in its infancy, is about the positive correlation between irisin and telomeres. Irisin is an exercise- and CR-induced thermogenic adipocytokine (Huang et al., 2017). It seems that its increased levels in plasma are in direct proportion with their length. Physical exercise, particularly aerobic exercise, has been shown to have a close association with the slowdown of telomere attrition. In particular, this relation is evident with the induction of telomerase enzyme activity and the shelterin complex. The last one is specifically associated with mammalian telomeres to distinguish the natural ends of chromosomes from sites of DNA damage as well as with the modulation of microRNAs (miRNAs) involved in the control of telomere homeostasis. The known effect of physical exercise in lowering oxidative stress and inflammation levels seems to positively affect telomere length as well, although the molecular mechanisms are not yet known (Garatachea et al., 2015). However, contrasting results exist, showing no significant associations between physical activity and telomere length. Probably, it could depend on age and sex of analyzed people, and on type and intensity of activities (Balan et al., 2018).

14.2.3 Epigenetic alterations

The role of epigenetics in aging is proven by many studies, but the molecular mechanisms involved are not so clear (Puca et al., 2018). Although there are different epigenetic modifications, the main representative regulating gene expressions are DNA methylation, covalent modifications of histones, and miRNAs. In relation to aging process, DNA methylation includes hypo or hypermethylation, and both might be either beneficial or detrimental. A study published in 2018 has strengthened the hypothesis of the modulation of epigenetic changes during intrauterine life and, likely, across generation, influencing the epigenome of newborn during the adult life and driving the aging process. Epigenetics could be the explanation of the associations between adverse environmental condition during early development and health outcomes in adulthood. The researchers analyzed DNA methylation in CpG island in whole blood of people exposed during intrauterine life to the famine of Dutch Hunger Winter compared to not exposed embryos, as control. The results demonstrated an increase in metabolic disorders in adult life, including higher body mass index, fasting glucose, serum triglycerides, and low-density lipoprotein (LDL), likely linked to epigenetic modifications (Tobi et al., 2018). Many studies, both in vivo and in vitro, about the effect of phytochemicals in foods demonstrated their role in generating epigenetic modifications with antiaging effect. Sulforaphane, contained in cruciferous vegetables and epigallocatechin-3-gallate, of which green tea is rich, act by inhibiting DNA methyltransferase and histone acetyltransferase or by modifying noncoding RNA expression (Puca et al., 2018; Li et al., 2011). EVOO and curcumin are two other important epigenetic modulators. The first one is a cornerstone of dietary habits in Mediterranean basin and the second in Southeast Asia. EVOO polyphenolic compounds can influence methylation and acetylation of genes, directly or indirectly involved in aging and metabolic diseases (Fernández del Río et al., 2016). Curcumin, of which curry is rich, inhibits the expression of pro-inflammatory mediators by affecting histone acetylation of transcription factors and methylation pattern of gene promoters associated with inflammatory response. DNA methyltransferase, histone deacetylase, histone acetyltransferase inhibitor, and miRNAs are their targets, with a consequent delay of aging process. Resveratrol and anthocyanins, mainly contained in red fruits, such as grapes and red berries, and vegetables of the same color or purple, such as cabbages, aubergines, and onions, are polyphenolic compounds with antioxidant and antiinflammatory effects but in a dose-dependent manner. Resveratrol, for example, has been identified as a potent SIRT-1 activator, mimicking CR and extending lifespan from yeast to humans. SIRT-1 leads to deacetylation of p53, NF-κB, heat shock factor 1, Forkhead box O (FOXO) 1,3,4, and PGC-1α, influencing replicative senescence, inflammation, apoptosis, metabolism, and stress resistance (Puca et al., 2018). So, overall, an "epigenetic diet" based on food contained nutraceuticals can prevent the onset of age-related diseases, thus increasing the attainment of longevity. Furthermore, CR is able to induce modulation in gene expression in different tissues through energy-sensing molecular systems, improving stress response, DNA stability, and chromatin structure preservation (Vera et al., 2013). These systems are FOXO transcription factors linked to Insulin/Insulin-like growth factor-1 (IGF-1) pathway, the sirtuins family, or AMP-activated protein kinase (AMPK) that is able to inhibit histone deacetylation (Vera et al., 2013). AMPK promotes healthy aging through physical exercise. Its activity is enhanced by training and, in muscle tissue, it is associated with nuclear exclusion of histone deacetylase and with increment of acetylation. Modulations of histone acetylation are also linked to perturbation of circadian clock, a very sensible system regulated by complex of molecules influenced by metabolic and energetic signals. Heterochromatin changes due to nutrition may be very fast and reversible, because the adaptation occurs within 3 days. However, a study in model *Caenorhabditis elegans* indicated that some metabolic states could be inherited in a transgenerational manner (Rebelo-Marques et al., 2018; Jiang et al., 2013; Rechavi et al., 2014). Moreover, in human beings, it was demonstrated that regular aerobic exercise can influence DNA methylation, playing a key role in regulation of glucose and fatty acid metabolism as well as in homeostasis (Rebelo-Marques et al., 2018). The same type of exercise induces SIRT-1 activation, leading to the regulation of the tumor suppressor p53 and NF-kB (Ntanasis-Stathopoulos et al., 2013; Kaliman et al., 2011). In addition, in lymphocytes of old people, the chronic moderate aerobic exercise increases the methylation levels of the pro-inflammatory apoptosis–associated speck-like protein caspase gene, lowering the inflammatory burden via interleukin (IL)-1β and IL-18 reduced levels (Nakajima et al., 2010). Methylation seems to be influenced by resistance exercise as well and the major involved tissues are muscle, brain, and cardiovascular system (Garatachea et al., 2015). In a meta-analysis, in a total sample of 1580 old individuals, the level of DNA methylation drastically decreased after physical exercise in genes involved in cancer-suppressing miRNA gene network (Brown, 2015). During recovery from acute exercise, a spread and transient hypomethylation in specific genes favors miRNAs expression, improving protein synthesis and structural restoration. All these results demonstrate the systemic capacity to maintain homeostasis, so to counteract the hallmarks of aging, and, possibly, could be useful to the enlargement of the homeostatic space (Garatachea et al., 2015). Thus exercise could be considered an instrument against sarcopenia and frailty, typical of aging.

14.2.4 Loss of proteostasis

Proteostasis is the protein homeostasis that can be altered in aging process. Refolding or degrading of altered proteins avoids the accumulation of aggregates and damaged fibrils with proteotoxic effect and associated with age-related diseases (e.g., neurodegenerative amyloidosis, such as AD and Parkinson's diseases). The mechanisms involved in proteostasis are autophagy, degradation by ubiquitin-proteasome, and chaperone-mediated folding, all altered during aging. Moreover, human polymorphisms in gene coding for proteasomal subunits are associated with susceptibility to metabolic disorders, such as diabetes (López-Otín et al., 2013; Rebelo-Marques et al., 2018; López-Otín et al., 2016). In human beings, the data related to the association between loss of proteostasis and nutrition or physical activity are scarce and need to be run extensively. Nonetheless, in this context, it is possible to translate the data obtained in nutritional interventions with antioxidant foods or nutraceutical compounds that lower oxidative stress levels. In fact, lower levels of ROS and of other oxidative molecules mean less damage to proteins. About physical exercise, it positively influences autophagy although few data have been published in old people. Beclin-1/B-cell lymphoma (BCL)-2 is the autophagy regulatory complex that plays a central role in triggering of the process. In human aged brain, the level of beclin-1 is lowered whereas in serum of centenarians, the best model of successful aging, it is increased, hence suggesting the association between a basal level of autophagy and longevity (Emanuele et al., 2014). In mice, aerobic exercise stimulates the beclin-1/BCL-2 complex, promoting autophagy (Colbert et al., 2004; Garber et al., 2011). Other studies in model organisms demonstrated the relation between physical activity, apoptosis, and oxidative stress with a positive outcome in lowering these pro-aging processes (Wohlgemuth et al., 2010). Moreover, aerobic exercise prevents loss of strength and muscle mass through the modulation of NSPs (e.g., insulin/IGF-1, mechanistic target of rapamycin (mTOR), and FOXO3A pathways). Consequently, a decrease in age-related diseases can be seen (e.g., cardiovascular disease and neurodegeneration) (Garatachea et al., 2015).

14.3 Antagonistic hallmarks and lifestyle

Antagonistic hallmarks represent the human body responses to molecular and cellular damage mechanisms. At low intensity, they may be beneficial and protecting the cell from injury but, at high intensity, they may induce and stimulate the aging process (López-Otín et al., 2013; Aunan et al., 2016).

14.3.1 Deregulated nutrient sensing

Alterations in NSPs have gained increasing attention in the field of aging research because they can be modulated both pharmacologically and with lifestyle interventions. The best known are insulin/IGF-1, mTOR, AMPK, and sirtuin pathways, involved in the regulation of protein synthesis, cell cycle, DNA replication, autophagy, stress response, and glucose homeostasis. The interest in NSPs arises because they are commonly deregulated in several human diseases and in aging, probably as a consequence of the chronic state of inflammation that characterized them, known as inflamm-aging (Efeyan et al., 2015; Aiello et al., 2017). However, although it remains to be determined how NSPs affect aging, several studies suggest that the limitation in nutrient intake represents the most successful intervention against the onset of aging process (Dilova et al., 2007). In fact, nutrients, such as carbohydrates or proteins, activate insulin/IGF-1 and mTOR pathways causing, in turn, an up-activation of proteins involved in aging and, in particular, in inflammatory processes. On the contrary, bioactive compounds, like nutraceuticals, upregulate sirtuin pathways that extend lifespan in model organisms, including mammals, through the activation of homeostatic genes (the so-called nutrient-sensing longevity genes) and the inhibition of pro-inflammatory ones (Fontana et al., 2010; Davinelli et al., 2012). Among healthy dietary patterns (e.g., all the slow-aging diets can moderate the deregulation of NSPs), DR patterns are the most well-known defined interventions to achieve longevity and to postpone aging in model organisms and in humans (Fontana et al., 2010). The mechanisms by which DR promotes longevity are still not completely understood. However, some studies suggest that insulin/IGF-1/mTOR/AMPK/SIRT-1 pathways are involved in (Longo et al., 2015). Epidemiologic, short-term, and observational human analyses suggest that CR could be effective in extending human lifespan (Fontana and Partridge, 2015). For example, the Pennington CALERIE Team carried out a randomized trial with healthy sedentary men and women based on six months of a medium or intensive CR, or a CR with physical activity intervention. They showed that body temperature, glucose, and insulin levels decreased with CR with or without physical activity (Heilbronn et al., 2006). The CRON study provided analyzed data about an approximately 30% CR intervention for 15 years. The authors showed lower total and LDL cholesterol as well as triglycerides, blood pressure, fasting glucose, fasting insulin levels, and higher high-density lipoprotein cholesterol levels than age-matched controls, eating in Western manner (Fontana et al., 2004). An alternative but more sustainable approach to CR is fasting. The best-characterized forms of fasting, evaluated in both models (rodents) and human studies, are intermittent fasting (IF) and prolonged fasting. Both improve health effects and protect from age-related disabilities. In humans, IF (i.e., consumption of approximately 500 kcal/day for 2 days a week) has beneficial effects on insulin, blood glucose levels, C-reactive protein, and blood pressure (Harvie et al., 2011). In models (rats) and humans, from a molecular point of view, CR and fasting result in transcriptional and posttranscriptional modifications of genes downstream of insulin/IGF-1 pathway. This implicates the activation of FOXO genes of which some polymorphisms have been associated with longer lifespan and with lower total and coronary heart diseases mortality (Willcox et al., 2016). Other studies assert that the effect of DRs on longevity is actually due to the concomitant restriction of proteins or specific amino acids (Longo et al., 2015). These dietary patterns appear sufficient to reduce IGF-1 concentrations and mTOR signaling more than simple CR (Soultoukis and Partridge, 2016). Moreover, individuals who usually followed a low-protein dietary regimen showed lower IGF-1 levels (Fontana et al., 2008). Polymorphisms in the gene encoding the IGF-1 receptor result in higher plasma concentration of IGF-1 and have been associated with extreme longevity (Suh et al., 2008). Lower plasma IGF-1 levels have also been associated with longer survival in nonagenarian women (Milman et al., 2014). On the contrary, high-protein diets have been reported to raise cardiovascular risk because protein ingestion determines elevated amino acid levels in blood and atherosclerotic plaques, stimulating mTOR pathway and exacerbating macrophage apoptosis induced by atherogenic lipids (Aiello et al., 2019a). About sirtuin pathways, the upregulation of endothelial nitric oxide synthase, which is a very important effector of CR, can also activate SIRT-1 (Xia et al., 2013). In general, sirtuins sense low energy states of cell detecting high NAD^+ levels. Consequently, they act as transcriptional effectors by controlling the acetylation state of histones, signaling nutrient scarcity and catabolism. In particular, SIRT-1 modulates responses to oxidative stress by FOXO3A deacetylation and performs a protective role in endothelial function, preventing cardiovascular diseases. Similarly, SIRT-3 is implicated in metabolism and mitochondrial function, including ROS detoxification (Imai and Guarente, 2016). Generally, activation of sirtuins modifies the acetylation/deacetylation balance that is a crucial mechanism for mediating the metabolic responses when the nutrient availability changes. So, nutritional stimuli influence mitochondrial biochemical pathways and affect the number and lifecycle of these organelles (Yang et al., 2011). Several studies have also showed the effect of exercise on glucose metabolism through increased glucose transporter type 4 production as well as through improved insulin sensibility (Mann et al., 2014). It was seen that growth hormone (GH) and IGF-1 levels are influenced by intensity, duration, and type of physical exercise and the increased muscle protein synthesis, associated with resistance exercise, is pointed out as a successful strategy to prevent age-related sarcopenia (Schwarz et al., 1996; Hartman et al., 2006).

14.3.2 Mitochondrial dysfunction

Mitochondria perform key biochemical functions essential for metabolic homeostasis and are arbiters of cell death and survival. Mutations and deletions in aged mitochondrial (mt) DNA contribute to age-related mitochondrial dysfunctions, since mtDNA is a major target for somatic mutations due to the oxidative microenvironment of the mitochondria and the lack of protective histones in the mtDNA (López-Otín et al., 2013). Oxidation of proteins and changes in the lipid composition of mitochondrial membranes are causes of their dysfunction, occurring in aging. At this regard, oxidative stress, referring to the imbalance of two opposite and antagonistic forces (e.g., production of ROS and antioxidants), has a large impact on lipid metabolism (including both cholesterol and polyunsaturated fatty acids). Lipid peroxidation leads to oxidative lipid deterioration, which alters membrane permeability and fluidity (Du et al., 2016). This results in lipid degradation (Engström et al., 2009). However, CR, via inhibition of IGF-1 and mTOR pathways and activation of AMPK and sirtuins, modulates oxidative metabolism as well as the biogenesis and turnover of mitochondria. In particular, the control of the glucose homeostasis occurs through the complex composed of PGC-1α and SIRT-1. Reduced energy increases AMP/ATP ratio and activates AMPK. It regulates PGC-1α through phosphorylation. CR and exercise also increase the level of NAD^+ in tissues thereby activate SIRT-1, which then activates PGC-1α through deacetylation (Rodgers et al., 2005). Moreover, SIRT-1 modulates mitochondrial biogenesis and improves mitophagy (Lee et al., 2008). Also, SIRT-3 targets many enzymes involved in energy metabolism, including components of the respiratory chain, tricarboxylic acid cycle, ketogenesis, and fatty acid β-oxidation. It may directly control the rate of ROS production by deacetylating manganese superoxide dismutase, a major mitochondrial antioxidant enzyme (Giralt and Villarroya, 2012). Moreover, CR lowers ROS production through enhanced mitochondrial aerobic metabolism and increases the activity of antioxidant enzymes (López-Lluch and Navas, 2016). In the same way, antioxidants, like vitamin C and vitamin E, present in different fruits and vegetables, can protect lipids from oxidation, representing an effective remedy for the treatment of obesity, insulin resistance, or other aspects of metabolic syndrome (Briganti and Picardo, 2003; Bhatti et al., 1863). Aging is associated with a reduced capacity of oxidative phosphorylation in muscle, likely due to a decline in mitochondria content and/or function and with a decreased mitochondrial enzyme activity. This is frequently accompanied by a downregulation of mRNAs, encoding mitochondrial proteins (Short et al., 2005). The regular practice of physical exercise has a positive impact on the mitochondrial function, maintaining a pool of bioenergetically useful organelles (Johannsen and Ravussin, 2009). Study carried out in an old population showed that the resistance exercise induces endogenous antioxidant defenses, decreasing DNA oxidative damage, through the PGC-1α and SIRT-1 regulators, and prevents mitochondrial alterations induced by ROS (Nilsson and Tarnopolsky, 2019).

14.3.3 Cellular senescence

Cellular senescence is strictly involved in the biology of aging and in age-related diseases (LeBrasseur et al., 2015). The locus INK4a/ARF encodes two proteins, p16INK4a and p14ARF, which acting on different targets, respectively, stopping the cell cycle and stabilizing p53. This contributes to the telomere attrition as well as to the induction of cellular senescence. The deletion of the INK4a/ARF locus has been reported in numerous types of cancer, so they are tumor suppressors. It is normally expressed at low levels in most young subject tissues but becomes hyperexpressed during the aging process. Senescent cells secrete a broad repertoire of cytokines, chemokines, matrix-remodeling proteases, and growth factors, collectively referred as senescence-associated secretory phenotype (SASP) (López-Otín et al., 2013). Multiple signals can induce a cell to enter in senescence. These include: 1) external factors like metabolic signals, such as high levels of glucose, certain fatty acids, prostanoids, hypoxia, and ROS that induce several cell replication under the control of GH or IGF-1 (Xu et al., 2015; Khosla et al., 2020); 2) internal factors, including mitochondrial dysfunction and certain genetic mutations (LeBrasseur et al., 2015). These signals trigger transcription factor cascades, including previous stated cell cycle checkpoint regulators. Few studies were carried out on the role of lifestyle in the regulation of human cellular senescence but many were conducted in ex vivo and *in vitro* models. At this regard, two of these, performed *in vitro* on endothelial cells, showed that Mediterranean diet protects the cells from oxidative stress, preventing cellular senescence, cellular apoptosis, and reducing telomere attrition (Marin et al., 2012; Capurso et al., 2019). In models, the beneficial effect of aerobic exercise on cellular senescence is associated with decreased checkpoint inhibitor vascular expression and increased telomerase activity (Werner et al., 2009). In addition, the practice of aerobic exercise reduces the expression of DNA damage biomarkers, positively correlated with p16-INK4a expression (Song et al., 2010). In another study on mice, the authors demonstrated that the exercise, initiated after a long-term fast-food diet, reduces SASP of visceral adipose tissue, suggesting that it has a beneficial role and reinforces the effect of modifiable lifestyle choices on health span (Schafer et al., 2016).

14.4 Integrative hallmarks and lifestyle

Integrative hallmarks, defined as the culprit of the phenotype, are stem cell exhaustion and altered intercellular communication (López-Otín et al., 2013).

14.4.1 Stem cell exhaustion

This hallmark is one of the final outcomes of the molecular alterations that lead to aging phenotype. Mechanistically, this feature of old organisms determines the slow or the loss of regenerative capacity of tissues, so a bad systemic adaptation to external perturbation. The reservoir of stem cell pool in aging, considering both their quality and quantity, is essential to prevent or delay the onset of age-related pathologies. Although the molecular and biochemical patterns influencing the deterioration of this pool are not clearly understood, it has been speculated that genetic, environmental, dietary, cultural, social, psychological, and economic factors can have a strong effect from in utero life (Trosko, 2008). Telomere shortening is also an important cause of stem cell decline during aging in multiple tissues (López-Otín et al., 2013). The same interventions that modulate the other hallmarks, such as dietary strategies and physical activity, could limit the exhaustion of stem cell (Valenti et al., 2020). About dietary strategies, it was demonstrated that CR increases the number of muscle stem cells and can reverse the decline of adult ones in different model organisms. The molecular and biochemical explanation of these events is not clearly elucidated, although it seems they could be linked to the modulation of NSPs, target pathways of CR (Cerletti et al., 2012; Oh et al., 2014). The effects of physical activity on stem cells have been investigated in experimental models. For instance, aerobic exercise favors cardiovascular health, increasing the number of endothelial progenitor cells in the blood and activating the differentiation of cardiac stem cells as well as progenitor cells (Sarto et al., 2007; Shen et al., 2017). So, even in this case, it could be possible to take measure to reduce this event but, although there is much evidence about the role of specific nutraceuticals, nutrients, or physical exercise, in vivo and *in vitro* as well as in model organisms, poor data are present for humans.

14.4.2 Altered intercellular communication

Altered intercellular communication provides for the change in signals between cells that can lead to diseases and disabilities of aging. It can involve all type of cellular communications, such as endocrine, neuroendocrine, neuronal, or immune-inflammatory (López-Otín et al., 2013). The prominent alteration in intercellular communication is inflamm-aging, one of the hallmarks of immunosenescence. The role of lifestyle in counteracting the other hallmarks of immunosenescence (i.e., the decreased number of peripheral blood naïve cells, with a relative increase in the frequency of memory cells) is exhaustively covered in a recent review, to whom the reader is addressed (Aiello et al., 2019b). In this review are also discussed the causes of inflamm-aging that, ultimately, stimulate NF-κB, which has a role in driving the aging process in multiple tissues (Tilstra et al., 2011). As a consequence, the use of genetic, pharmacological, or other inhibitors of the NF-κB signaling can decrease inflammation. At this regard, the activation of sirtuins has an important role in the downregulation of some genes related to inflammatory process, acting as protective factors in aging and in other inflammatory pathological conditions (Xie et al., 2013). Although a few studies have been carried out on humans, bioactive compounds, characteristic of some slow-aging and antiinflammatory diets, can upregulate sirtuin pathways also by the stimulation of the activity of the nuclear factor (erythroid-derived 2)-like 2 (Nrf2) (Rattan, 2012). This is a key transcription factor controlling many aspects of cell homoeostasis in response to oxidative and toxic insults. It mediates basal and induced transcription of antioxidant enzymes which are responsible for the clearance of ROS, providing protection against the accumulation of toxic metabolites (Kobayashi et al., 2004). Furthermore, plant polyphenols, contained in fruits, vegetables, and whole grains, act as inhibitors of NF-κB and could use as a potential antiinflammatory treatment for age-related diseases (Karunaweera et al., 2015). In mice, the inhibition of NF-κB pathway attenuates obesity and glucose intolerance due to high-fat diet. Indeed, the use of the flavonoid butein, found in some plants, inhibits NF-κB signaling. It blocks the upstream kinase inhibitor subunit-β that usually phosphorylates the inhibitor of the cytoplasmic NF-κB complex, driving the translocation of this transcription factor into the nucleus. So, in this case, the inflammatory cascade is downregulated with a reduction in adipocyte inflammation in vitro (Pandey et al., 2007; Wang et al., 2014; Benzler et al., 2015). Concerning physical activity, muscle contraction is associated with elevated myokine secretion (growth factors, cytokines, or metallopeptidases) during and after exercise, mainly aerobic one (Fiuza-Luces et al., 2014). In this case, the muscle releases IL-6 that creates healthy influence, inducing the production of antiinflammatory cytokines, IL-1 receptor antagonist, IL-10, or tumor necrosis factor (TNF) soluble receptors, while restraining pro-inflammatory cytokine such as TNF-α production (Pedersen, 2012; Ball, 2015).

14.5 Conclusion and future perspectives

In recent years, a growing body of evidence has emerged in the field of aging and its close correlation to lifestyle. In particular, nutrition or specific dietary strategies and physical activity have been well studied with the goal to delay the onset of aging phenotype and as strategies to prevent multifactorial pathologies. So, positive correlation of healthy lifestyle with slowdown of biological aging confirms the positive impact in the maintenance of the overall health status. At this regard, recent findings suggest that the total volume of sedentary behavior is a potential cancer mortality risk factor and support the public health message that adults should sit less and move more to promote longevity (Gilchrist et al., 2020). Nowadays, this assumes particular relevance due to the dramatic increase in lifespan in Western countries often not accompanied by a good health status. Since we cannot choose our genome and our future, the efforts have to be concentrated in the control of what we do to improve our lifestyle. There are two key points related to nutritional strategies and physical activity: the dose of "treatment" and the age at which it is administered. The issue of the dose should be explained considering the hormetic effect. The term "hormesis" comes from a Greek verb, with the meaning of "to stimulate." By translating the concept to human system, it is described as an adaptive response of a cell or an organism characterized by a biphasic dose-dependent response to a variety of stressors. The systemic or cellular response will be beneficial with a low to modest stimulation, detrimental in consequence to high doses (Calabrese et al., 2007). The intensity and the type of physical exercise and of dietary strategies could differentially act on aging, counteracting the hallmarks or reinforcing their harmful effects. Also, the age to start these interventions is crucial. In 1990 Barker hypothesized and explained the association between prenatal or early postnatal events and unsuccessful aging (Barker, 1990; Carpinello et al., 2018). This concept could be read in the light of more recent scientific knowledge, giving a mechanistic explanation, i.e., effect on epigenetics, as previously described. The identification of a "cure" for fight against aging seems to be a pipe dream. But, thanks to the possibility to study and to act at molecular level, the opportunity to reach a successful chronological aging in a younger biological age is a concrete reality. Even more interesting is the processing of personalized specific treatments that take account of the genetic of each individual. Unfortunately, while some molecular mechanisms have already been well elucidated in model organisms, the major hurdle is to apply and study these discoveries in humans, due to ethical issues and costs, other than to the great variability of human model. Hence, although many studies on humans are carried on (e.g., Chinese Longitudinal Healthy Longevity Survey, EPIC, Moli-sani study), the possibility to find a global strategy to become healthy centenarian is far and further researches are needed (Xu et al., 2015; DUKE, n.d.; EPIC Study, n.d.; Bonaccio et al., 2015).

References

Accardi G, Aiello A, Vasto S, Caruso C. Chance and causality in ageing and longevity. In: Caruso C, editor. Centenarians. An example of positive biology. Switzerland: Springer; 2019. p. 1–21.

Aiello A, Accardi G, Candore G, Gambino CM, Mirisola M, Taormina G, et al. Nutrient sensing pathways as therapeutic targets for healthy ageing. Expert Opin Ther Targets 2017;21:371–80. https://doi.org/10.1080/14728222.2017.1294684.

Aiello A, Caruso C, Accardi G. Slow-aging diets. In: Gu D, Dupre ME, editors. Encyclopedia of gerontology and population aging. Switzerland, AG: Springer Nature; 2019a. https://doi.org/10.1007/978-3-319-69892-2_134-121.

Aiello A, Farzaneh F, Candore G, Caruso C, Davinelli S, Gambino CM, et al. Immunosenescence and its hallmarks: how to oppose aging strategically? A review of potential options for therapeutic intervention. Front Immunol 2019b;10:2247. https://doi.org/10.3389/fimmu.2019.02247.

Anderson R, Prolla T. PGC-1alpha in aging and anti-aging interventions. Biochim Biophys Acta 1790;2009:1059–66. https://doi.org/10.1016/j.bbagen.2009.04.005.

Atzmon G, Cho M, Cawthon RM, Budagov T, Katz M, Yang X, et al. Evolution in health and medicine Sackler colloquium: genetic variation in human telomerase is associated with telomere length in Ashkenazi centenarians. Proc Natl Acad Sci U S A 2010;107(Suppl 1):1710–7. https://doi.org/10.1073/pnas.0906191106.

Aunan JR, Watson MM, Hagland HR, Søreide K. Molecular and biological hallmarks of ageing. Br J Surg 2016;103:e29–46. https://doi.org/10.1002/bjs.10053.

Balan E, Decottignies A, Deldicque L. Physical activity and nutrition: two promising strategies for telomere maintenance? Nutrients 2018;10. https://doi.org/10.3390/nu10121942. pii:E1942.

Ball D. Metabolic and endocrine response to exercise: sympathoadrenal integration with skeletal muscle. J Endocrinol 2015;224:R79–95. https://doi.org/10.1530/JOE-14-0408.

Barker DJ. The fetal and infant origins of adult disease. BMJ 1990;301:1111.

Benzler J, Ganjam GK, Pretz D, Oelkrug R, Koch CE, Legler K, et al. Central inhibition of IKKβ/NF-κB signaling attenuates high-fat diet-induced obesity and glucose intolerance. Diabetes 2015;64:2015–27. https://doi.org/10.2337/db14-0093.

Bhatti JS, Bhatti GK, Reddy PH. Mitochondrial dysfunction and oxidative stress in metabolic disorders—a step towards mitochondria based therapeutic strategies. Biochim Biophys Acta Mol Basis Dis 1863;2017:1066–77. https://doi.org/10.1016/j.bbadis.2016.11.010.

Boccardi V, Esposito A, Rizzo MR, Marfella R, Barbieri M, Paolisso G. Mediterranean diet, telomere maintenance and health status among elderly. PLoS One 2013;8:e62781. https://doi.org/10.1371/journal.pone.0062781.

Bonaccio M, Cerletti C, Iacoviello L, de Gaetano G. Mediterranean diet and low-grade subclinical inflammation: the Moli-sani study. Endocr Metab Immune Disord Drug Targets 2015;15:18–24.

Briganti S, Picardo M. Antioxidant activity, lipid peroxidation and skin diseases. What's new. J Eur Acad Dermatol Venereol 2003;17:663–9.

Brown WM. Exercise-associated DNA methylation change in skeletal muscle and the importance of imprinted genes: a bioinformatics meta-analysis. Br J Sports Med 2015;49:1567–678. https://doi.org/10.1136/bjsports-2014-094073.

Büchner N, Ale-Agha N, Jakob S, Sydlik U, Kunze K, Unfried K, et al. Unhealthy diet and ultrafine carbon black particles induce senescence and disease associated phenotypic changes. Exp Gerontol 2013;48:8–16. https://doi.org/10.1016/j.exger.2012.03.017.

Calabrese EJ, Bachmann KA, Bailer AJ, Bolger PM, Borak J, Cai L, et al. Biological stress response terminology: integrating the concepts of adaptive response and preconditioning stress within a hormetic dose-response framework. Toxicol Appl Pharmacol 2007;222:122–8.

Capurso C, Bellanti F, Lo Buglio A, Vendemiale G. The mediterranean diet slows down the progression of aging and helps to prevent the onset of frailty: a narrative review. Nutrients 2019;12. https://doi.org/10.3390/nu12010035. pii: E35.

Carpinello OJ, DeCherney AH, Hill MJ. Developmental origins of health and disease: the history of the Barker hypothesis and assisted reproductive technology. Semin Reprod Med 2018;36:177–82. https://doi.org/10.1055/s-0038-1675779.

Cassidy A, De Vivo I, Liu Y, Han J, Prescott J, Hunter DJ, et al. Associations between diet, lifestyle factors, and telomere length in women. Am J Clin Nutr 2010;91:1273–80. https://doi.org/10.3945/ajcn.2009.28947.

Cerletti M, Jang YC, Finley LW, Haigis MC, Wagers AJ. Short-term calorie restriction enhances skeletal muscle stem cell function. Cell Stem Cell 2012;10:515–9. https://doi.org/10.1016/j.stem.2012.04.002.

Colbert LH, Visser M, Simonsick EM, Tracy RP, Newman AB, Kritchevsky SB, et al. Physical activity, exercise, and inflammatory markers in older adults: findings from the health, aging and body composition study. J Am Geriatr Soc 2004;52:1098–104.

Crous-Bou M, Fung TT, Prescott J, Julin B, Du M, Sun Q, et al. Mediterranean diet and telomere length in Nurses' Health Study: population based cohort study. BMJ 2014;349. https://doi.org/10.1136/bmj.g6674. g6674.

Daniali L, Benetos A, Susser E, Kark JD, Labat C, Kimura M, et al. Telomeres shorten at equivalent rates in somatic tissues of adults. Nat Commun 2013;4:1597. https://doi.org/10.1038/ncomms2602.

Davinelli S, Willcox DC, Scapagnini G. Extending healthy ageing: nutrient sensitive pathway and centenarian population. Immun Ageing 2012;9:9. https://doi.org/10.1186/1742-4933-9-9.

Dilova I, Easlon E, Lin SJ. Calorie restriction and the nutrient sensing signaling pathways. Cell Mol Life Sci 2007;64:752–67.

D'Mello MJ, Ross SA, Briel M, Anand SS, Gerstein H, Paré G. Association between shortened leukocyte telomere length and cardiometabolic outcomes: systematic review and meta-analysis. Circ Cardiovasc Genet 2015;8:82–90. https://doi.org/10.1161/CIRCGENETICS.113.000485.

Du J, Zhu M, Bao H, Li B, Dong Y, Xiao C, et al. The role of nutrients in protecting mitochondrial function and neurotransmitter signaling: implications for the treatment of depression, PTSD, and suicidal behaviors. Crit Rev Food Sci Nutr 2016;56:2560–78.

DUKE, n.d., https://sites.duke.edu/centerforaging/programs/chinese-longitudinal-healthy-longevity-survey-clhls/ [Accessed July 23rd, 2020. 1].

Efeyan A, Comb WC, Sabatini DM. Nutrient-sensing mechanisms and pathways. Nature 2015;517:302–10. https://doi.org/10.1038/nature14190.

Elliott R, Ong TJ. Nutritional genomics. BMJ 2002;324:1438–42.

Emanuele E, Minoretti P, Sanchis-Gomar F, Pareja-Galeano H, Yilmaz Y, Garatachea N, et al. Can enhanced autophagy be associated with human longevity? Serum levels of the autophagy biomarker beclin-1 are increased in healthy centenarians. Rejuvenation Res 2014;17:518–24. https://doi.org/10.1089/rej.2014.1607.

Engström K, Saldeen AS, Yang B, Mehta JL, Saldeen T. Effect of fish oils containing different amounts of EPA, DHA, and antioxidants on plasma and brain fatty acids and brain nitric oxide synthase activity in rats. Ups J Med Sci 2009;114:206–13. https://doi.org/10.3109/03009730903268958.

EPIC study, n.d., https://epic.iarc.fr [Accessed July 23rd, 2020].

Fang EF, Scheibye-Knudsen M, Brace LE, Kassahun H, SenGupta T, Nilsen H, et al. Defective mitophagy in XPA via PARP-1 hyperactivation and NAD(+)/SIRT1 reduction. Cell 2014;157:882–96. https://doi.org/10.1016/j.cell.2014.03.026.

Fernández del Río L, Gutiérrez-Casado E, Varela-López A, Villalba JM. Olive oil and the hallmarks of aging. Molecules 2016;21:163. https://doi.org/10.3390/molecules21020163.

Fiuza-Luces C, Delmiro A, Soares-Miranda L, González-Murillo Á, Martínez-Palacios J, Ramírez M, et al. Exercise training can induce cardiac autophagy at end-stage chronic conditions: insights from a graft-versus-host-disease mouse model. Brain Behav Immun 2014;39:56–60. https://doi.org/10.1016/j.bbi.2013.11.007.

Fontana L, Partridge L. Promoting health and longevity through diet: from model organisms to humans. Cell 2015;161:106–18. https://doi.org/10.1016/j.cell.2015.02.020.

Fontana L, Meyer TE, Klein S, Holloszy JO. Long-term calorie restriction is highly effective in reducing the risk for atherosclerosis in humans. Proc Natl Acad Sci U S A 2004;101:6659–63.

Fontana L, Weiss EP, Villareal DT, Klein S, Holloszy JO. Long-term effects of calorie or protein restriction on serum IGF-1 and IGFBP-3 concentration in humans. Aging Cell 2008;7:681–7.

Fontana L, Partridge L, Longo VD. Extending healthy life span- -from yeast to humans. Science 2010;328:321–6. https://doi.org/10.1126/science.1172539.

Garatachea N, Pareja-Galeano H, Sanchis-Gomar F, Santos-Lozano A, Fiuza-Luces C, Morán M, et al. Exercise attenuates the major hallmarks of aging. Rejuvenation Res 2015;18:57–89. https://doi.org/10.1089/rej.2014.1623.

Garber CE, Blissmer B, Deschenes MR, Franklin BA, Lamonte MJ, Lee IM, et al. Quantity and quality of exercise for developing and maintaining cardiorespiratory, musculoskeletal, and neuromotor fitness in apparently healthy adults: guidance for prescribing exercise. Med Sci Sports Exerc 2011;43:1334–59. https://doi.org/10.1249/MSS.0b013e318213fefb.

García-Calzón S, Martínez-González MA, Razquin C, Corella D, Salas-Salvadó J, Martínez JA, et al. Pro12Ala polymorphism of the PPARγ2 gene interacts with a mediterranean diet to prevent telomere shortening in the PREDIMED-NAVARRA randomized trial. Circ Cardiovasc Genet 2015;8:91–9. https://doi.org/10.1161/CIRCGENETICS.114.000635.

Gilchrist SC, Howard VJ, Akinyemiju T, et al. Association of sedentary behavior with cancer mortality in middle-aged and older US adults [published online ahead of print, 2020 Jun 18]. JAMA Oncol 2020; e202045. https://doi.org/10.1001/jamaoncol.2020.2045.

Giralt A, Villarroya F. SIRT3, a pivotal actor in mitochondrial functions: metabolism, cell death and aging. Biochem J 2012;444:1–10. https://doi.org/10.1042/BJ20120030.

Gomez-Cabrera M-C, Domenech E, Viña J. Moderate exercise is an anti-oxidant: upregulation of antioxidant genes by training. Free Radic Biol Med 2008;44:126–31. https://doi.org/10.1016/j.freeradbiomed.2007.02.00134.

Grassi C, Landi F, Delogu G. Lifestyles and ageing: targeting key mechanisms to shift the balance from unhealthy to healthy ageing. Stud Health Technol Inform 2014;203:99–111.

Hartman JW, Moore DR, Phillips SM. Resistance training reduces whole-body protein turnover and improves net protein retention in untrained young males. Appl Physiol Nutr Metab 2006;31:557–64.

Harvie MN, Pegington M, Mattson MP, Frystyk J, Dillon B, Evans G, et al. The effects of intermittent or continuous energy restriction on weight loss and metabolic disease risk markers: a randomized trial in young overweight women. Int J Obes (Lond) 2011;35:714–27. https://doi.org/10.1038/ijo.2010.171.

Heilbronn LK, de Jonge L, Frisard MI, DeLany JP, Larson-Meyer DE, Rood J, et al. Effect of 6-month calorie restriction on biomarkers of longevity, metabolic adaptation, and oxidative stress in overweight individuals: a randomized controlled trial. JAMA 2006;295:1539–48. Erratum in: JAMA. 2006;295:2482.

Herrmann M, Pusceddu I, März W, Herrmann W. Telomere biology and age-relate diseases. Clin Chem Lab Med 2018;56:1210–22. https://doi.org/10.1515/cclm-2017-0870.

Heydari AR, Unnikrishnan A, Lucente LV, Richardson A. Caloric restriction and genomic stability. Nucleic Acids Res 2007;35:7485–96.

Huang J, Wang S, Xu F, Wang D, Yin H, Lai Q, et al. Exercise training with dietary restriction enhances circulating irisin level associated with increasing endothelial progenitor cell number in obese adults: an intervention study. Peer J 2017;5:e3669. https://doi.org/10.7717/peerj.3669.

Imai SI, Guarente L. It takes two to tango: NAD(+) and sirtuins in aging/longevity control. NPJ Aging Mech Dis 2016;2:16017. https://doi.org/10.1038/npjamd.2016.17.

Jiang N, Du G, Tobias E, Wood JG, Whitaker R, Neretti N, et al. Dietary and genetic effects on age-related loss of gene silencing reveal epigenetic plasticity of chromatin repression during aging. Aging (Albany NY) 2013;5:813–24.

Johannsen DL, Ravussin E. The role of mitochondria in health and disease. Curr Opin Pharmacol 2009;9:780–6. https://doi.org/10.1016/j.coph.2009.09.002.

Kaliman P, Párrizas M, Lalanza JF, Camins A, Escorihuela RM, Pallàs M. Neurophysiological and epigenetic effects of physical exercise on the aging process. Ageing Res Rev 2011;10:475–86. https://doi.org/10.1016/j.arr.2011.05.002.

Karunaweera N, Raju R, Gyengesi E, Münch G. Plant polyphenols as inhibitors of NF-κB induced cytokine production-a potential anti-inflammatory treatment for Alzheimer's disease? Front Mol Neurosci 2015;8:24. https://doi.org/10.3389/fnmol.2015.00024.

Khosla S, Farr JN, Tchkonia T, Kirkland JL. The role of cellular senescence in ageing and endocrine disease. Nat Rev Endocrinol 2020. https://doi.org/10.1038/s41574-020-0335-y. [Epub ahead of print].

Kim Y, Lee E. The association between elderly people's sedentary behaviors and their health-related quality of life: focusing on comparing the young-old and the old-old. Health Qual Life Outcomes 2019;17:131. https://doi.org/10.1186/s12955-019-1191-0.

Kobayashi A, Ohta T, Yamamoto M. Unique function of the Nrf2-Keap1 pathway in the inducible expression of antioxidant and detoxifying enzymes. Methods Enzymol 2004;378:273–86.

LeBrasseur NK, Tchkonia T, Kirkland JL. Cellular senescence and the biology of aging, disease, and frailty. Nestle Nutr Inst Workshop Ser 2015;83:11–8. https://doi.org/10.1159/000382054.

Lee IH. Mechanisms and disease implications of sirtuin-mediated autophagic regulation. Exp Mol Med 2019;51:1–11. https://doi.org/10.1038/s12276-019-0302-7.

Lee IH, Cao L, Mostoslavsky R, Lombard DB, Liu J, Bruns NE, et al. A role for the NAD-dependent deacetylase Sirt1 in the regulation of autophagy. Proc Natl Acad Sci U S A 2008;105:3374–9. https://doi.org/10.1073/pnas.0712145105.

Lee JY, Jun NR, Yoon D, Shin C, Baik I. Association between dietary patterns in the remote past and telomere length. Eur J Clin Nutr 2015;69:1048–52.

Leick L, Lyngby SS, Wojtasewski JF, Pilegaard H. PGC-1α is required for training-induced prevention of age-associated decline in mitochondrial enzymes in mouse skeletal muscle. Exp Gerontol 2010;45:336–42. https://doi.org/10.1016/j.exger.2010.01.01135.

Li Y, Daniel M, Tollefsbol TO. Epigenetic regulation of caloric restriction in aging. BMC Med 2011;9:98. https://doi.org/10.1186/1741-7015-9-98.

Longo VD, Antebi A, Bartke A, Barzilai N, Brown-Borg HM, Caruso C, et al. Interventions to slow aging in humans: are we ready? Aging Cell 2015;14:497–510. https://doi.org/10.1111/acel.12338.

López-Lluch G, Navas P. Calorie restriction as an intervention in ageing. J Physiol 2016;594:2043–60. https://doi.org/10.1113/JP270543.

López-Otín C, Blasco MA, Partridge L, Serrano M, Kroemer G. The hallmarks of aging. Cell 2013;153:1194–217. https://doi.org/10.1016/j.cell.2013.05.039.

López-Otín C, Galluzzi L, Freije JMP, Madeo F, Kroemer G. Metabolic control of longevity. Cell 2016;166:802–21. https://doi.org/10.1016/j.cell.2016.07.031.

Mann S, Beedie C, Balducci S, Zanuso S, Allgrove J, Bertiato F, et al. Changes in insulin sensitivity in response to different modalities of exercise: a review of the evidence. Diabetes Metab Res Rev 2014;30:257–68. https://doi.org/10.1002/dmrr.2488.

Marin C, Delgado-Lista J, Ramirez R, Carracedo J, Caballero J, Perez-Martinez P, et al. Mediterranean diet reduces senescence-associated stress in endothelial cells. Age (Dordr) 2012;34:1309–16. https://doi.org/10.1007/s11357-011-9305-6.

Milman S, Atzmon G, Huffman DM, Wan J, Crandall JP, Cohen P, et al. Low insulin-like growth factor-1 level predicts survival in humans with exceptional longevity. Aging Cell 2014;13:769–71. https://doi.org/10.1111/acel.12213.

Milte CM, McNaughton SA. Dietary patterns and successful ageing: a systematic review. Eur J Nutr 2016;55:423–50. https://doi.org/10.1007/s00394-015-1123-7.

Minuti A, Patrone V, Giuberti G, Spigno G, Pietri A, Battilani P, et al. Nutrition and ageing. Stud Health Technol Inform 2014;203:112–21.

Nakajima K, Takeoka M, Mori M, Hashimoto S, Sakurai A, Nose H, et al. Exercise effects on methylation of ASC gene. Int J Sports Med 2010;31:671–5. https://doi.org/10.1055/s-0029-1246140.

Navarro CL, Cau P, Lévy N. Molecular bases of progeroid syndromes. Hum Mol Genet 2006;15:R151–61.

Nilsson MI, Tarnopolsky MA. Mitochondria and aging-the role of exercise as a countermeasure. Biology (Basel) 2019;8. https://doi.org/10.3390/biology8020040. pii: E40.

Ntanasis-Stathopoulos J, Tzanninis JG, Philippou A, Koutsilieris M. Epigenetic regulation on gene expression induced by physical exercise. J Musculoskelet Neuronal Interact 2013;13:133–46.

Oh J, Lee YD, Wagers AJ. Stem cell aging: mechanisms, regulators and therapeutic opportunities. Nat Med 2014;20:870–80. https://doi.org/10.1038/nm.3651.

Pandey MK, Sandur SK, Sung B, Sethi G, Kunnumakkara AB, Aggarwal BB. Butein, a tetrahydroxychalcone, inhibits nuclear factor (NF)-kappaB and NF-kappaB-regulated gene expression through direct inhibition of IkappaBalpha kinase beta on cysteine 179 residue. J Biol Chem 2007;282:17340–50.

Pedersen BK. Muscular interleukin-6 and its role as an energy sensor. Med Sci Sports Exerc 2012;44:392–6. https://doi.org/10.1249/MSS.0b013e31822f94ac.

Puca AA, Spinelli C, Accardi G, Villa F, Caruso C. Centenarians as a model to discover genetic and epigenetic signatures of healthy ageing. Mech Ageing Dev 2018;174:95–102. https://doi.org/10.1016/j.mad.2017.10.004.

Rattan SI. Rationale and methods of discovering hormetins as drugs for healthy ageing. Expert Opin Drug Discovery 2012;7:439–48. https://doi.org/10.1517/17460441.2012.677430.

Rebelo-Marques A, De Sousa LA, Andrade R, Ribeiro CF, Mota-Pinto A, Carrilho F, et al. Aging hallmarks: the benefits of physical exercise. Front Endocrinol (Lausanne) 2018;9:258. https://doi.org/10.3389/fendo.2018.00258.

Rechavi O, Houri-Ze'evi L, Anava S, Goh WSS, Kerk SY, Hannon GJ, et al. Starvation-induced transgenerational inheritance of small RNAs in C elegans. Cell 2014;158:277–87. https://doi.org/10.1016/j.cell.2014.06.020.

Rodgers JT, Lerin C, Haas W, Gygi SP, Spiegelman BM, Puigserver P. Nutrient control of glucose homeostasis through a complex of PGC-1alpha and SIRT1. Nature 2005;434:113–8.15744310.

Sarto P, Balducci E, Balconi G, Fiordaliso F, Merlo L, Tuzzato G, et al. Effects of exercise training on endothelial progenitor cells in patients with chronic heart failure. J Card Fail 2007;13:701–8.

Schafer MJ, White TA, Evans G, Tonne JM, Verzosa GC, Stout MB, et al. Exercise prevents diet-induced cellular senescence in adipose tissue. Diabetes 2016;65:1606–15. https://doi.org/10.2337/db15-0291.

Scheibye-Knudsen M, Mitchell SJ, Fang EF, Iyama T, Ward T, Wang J, et al. A high-fat diet and NAD(+) activate Sirt1 to rescue premature aging in cockayne syndrome. Cell Metab 2014;20:840–55. https://doi.org/10.1016/j.cmet.2014.10.005.

Schwarz AJ, Brasel JA, Hintz RL, Mohan S, Cooper DM. Acute effect of brief low- and high-intensity exercise on circulating insulin-like growth factor (IGF) I, II, and IGF-binding protein-3 and its proteolysis in young healthy men. J Clin Endocrinol Metab 1996;81:3492–7.

Shen L, Wang H, Bei Y, Cretoiu D, Cretoiu SM, Xiao J. Formation of new cardiomyocytes in exercise. Adv Exp Med Biol 2017;999:91–102. https://doi.org/10.1007/978-981-10-4307-9_6.

Short KR, Bigelow ML, Kahl J, Singh R, Coenen-Schimke J, Raghavakaimal S, et al. Decline in skeletal muscle mitochondrial function with aging in humans. Proc Natl Acad Sci U S A 2005;102:5618–23.

Song Z, von Figura G, Liu Y, Kraus JM, Torrice C, Dillon P, Rudolph-Watabe M, Ju Z, Kestler HA, Sanoff H, Lenhard RK. Lifestyle impacts on the aging-associated expression of biomarkers of DNA damage and telomere dysfunction in human blood. Aging Cell 2010;9:607–15. https://doi.org/10.1111/j.1474-9726.2010.00583.x.

Soultoukis GA, Partridge L. Dietary protein, metabolism, and aging. Annu Rev Biochem 2016;85:5–34. https://doi.org/10.1146/annurev-biochem-060815-014422.

Suh Y, Atzmon G, Cho MO, Hwang D, Liu B, Leahy DJ, et al. Functionally significant insulin-like growth factor I receptor mutations in centenarians. Proc Natl Acad Sci U S A 2008;105:3438–42. https://doi.org/10.1073/pnas.0705467105.

Tang BL. Sirt1 and the mitochondria. Mol Cells 2016;39:87–95. https://doi.org/10.14348/molcells.2016.2318.

Thomas P, Wang YJ, Zhong JH, Kosaraju S, O'Callaghan NJ, Zhou XF, et al. Grape seed polyphenols and curcumin reduce genomic instability events in a transgenic mouse model for Alzheimer's disease. Mutat Res 2009;661:25–34.

Tilstra JS, Clauson CL, Niedernhofer LJ, Robbins PD. NF-κB in aging and disease. Aging Dis 2011;2:449–65.

Tobi EW, Slieker RC, Luijk R, Dekkers KF, Stein AD, Xu KM, et al. DNA methylation as a mediator of the association between prenatal adversity and risk factors for metabolic disease in adulthood. Sci Adv 2018;4:eaao4364. https://doi.org/10.1126/sciadv.aao4364.

Trosko JE. Role of diet and nutrition on the alteration of the quality and quantity of stem cells in human aging and the diseases of aging. Curr Pharm Des 2008;14:2707–18.

Valenti MT, Dalle Carbonare L, Dorelli G, Mottes M. Effects of physical exercise on the prevention of stem cells senescence. Stem Cell Rev Rep 2020;16:33–40. https://doi.org/10.1007/s12015-019-09928-w.

Vera E, de Bernardes Jesus B, Foronda M, Flores JM, Blasco MA. Telomerase reverse transcriptase synergizes with calorie restriction to increase health span and extend mouse longevity. PLoS One 2013;8:e53760. https://doi.org/10.1371/journal.pone.0053760.

Vidacek NŠ, Nanic L, Ravlic S, Sopta M, Geric M, Gajski G, et al. Telomeres, nutrition, and longevity: can we really navigate our aging? J Gerontol A Biol Sci Med Sci 2017;73:39–47. https://doi.org/10.1093/gerona/glx082.

Vijg J, Montagna C. Genome instability and aging: cause or effect? Transl Med Aging 2017;1:5–11. https://doi.org/10.1016/j.tma.2017.09.003.

Wang Z, Lee Y, Eun JS, Bae EJ. Inhibition of adipocyte inflammation and macrophage chemotaxis by butein. Eur J Pharmacol 2014;738:40–8. https://doi.org/10.1016/j.ejphar.2014.05.031.

Werner C, Fürster T, Widmann T, Pöss J, Roggia C, Hanhoun M, et al. Physical exercise prevents cellular senescence in circulating leukocytes and in the vessel wall. Circulation 2009;120:2438–47. https://doi.org/10.1161/CIRCULATIONAHA.109.861005.

Willcox BJ, Tranah GJ, Chen R, Morris BJ, Masaki KH, He Q, et al. The FoxO3 gene and cause-specific mortality. Aging Cell 2016;15:617–24. https://doi.org/10.1111/acel.12452.

Wohlgemuth SE, Seo AY, Marzetti E, Lees HA, Leeuwenburgh C. Skeletal muscle autophagy and apoptosis during aging: effects of calorie restriction and life-long exercise. Exp Gerontol 2010;45:138–48. https://doi.org/10.1016/j.exger.2009.11.002.

Xia N, Strand S, Schlufter F, Siuda D, Reifenberg G, Kleinert H, et al. Role of SIRT1 and FOXO factors in eNOS transcriptional activation by resveratrol. Nitric Oxide 2013;32:29–35. https://doi.org/10.1016/j.niox.2013.04.001.

Xie J, Zhang X, Zhang L. Negative regulation of inflammation by SIRT1. Pharmacol Res 2013;67:60–7. https://doi.org/10.1016/j.phrs.2012.10.010.

Xu M, Palmer AK, Ding H, Weivoda MM, Pirtskhalava T, White TA, et al. Targeting senescent cells enhances adipogenesis and metabolic function in old age. Elife 2015;4:e12997. https://doi.org/10.7554/eLife.12997.

Yang L, Vaitheesvaran B, Hartil K, Robinson AJ, Hoopmann MR, Eng JK, et al. The fasted/fed mouse metabolic acetylome: N6-acetylation differences suggest acetylation coordinates organ-specific fuel switching. J Proteome Res 2011;10:4134–49. https://doi.org/10.1021/pr200313x.

CHAPTER 15

Nutritional biomarkers in aging research

Sergio Davinelli and Giovanni Scapagnini

Department of Medicine and Health Sciences "V. Tiberio", University of Molise, Campobasso, Italy

15.1 Introduction

Aging is now considered as a gradual and progressive deterioration of integrity across multiple organ systems. As the aged population expands, the economic burden of care and treatment of age-related diseases increases exponentially (Kaeberlein, 2013). Nutritional status is considered one of the key determinants of health and several age-related diseases of our time have a clear link with lifestyle factors, including the diet (Fig. 15.1). During the last decades the potential of nutrition for the prevention or treatment of diseases or risk factors for disease was rediscovered. These developments contributed to the shift of nutrition science into a highly multidisciplinary area of research and emerging data indicate that optimal nutrition may serve as a potential avenue to slowing the progression of aging and reducing the incidence of debilitating diseases (Aiello et al., 2016). Diet and the many bioactive substances present in food represent a novel target for interventions that may promote healthy aging and prevent age-related diseases. However, the measurement of nutritional biomarkers related to aging process is not straightforward. There is no consensus regarding which biomarkers best reflect the link between nutrition and healthy aging. Therefore the major challenge is the characterization of suitable biomarkers to assess this connection. Although we are not been able to define an ideal biomarker yet, what a biomarker for aging should be or predict is quite broadly defined. A biomarker should not only (i) reflect some basic property of aging, but also (ii) be reproducible in cross-species comparison, (iii) change independently of the passage of chronological time (so that the biomarker indicates biological rather than chronological age), (iv) be obtainable by noninvasive means, and (v) be measurable during a short interval of lifespan. A biomarker should reflect the underlying aging process rather than disease (Warner, 2004). A nutritional biomarker is a parameter that reflects biological consequence of dietary intake or adherence to dietary patterns and should indicate the nutritional status with respect to intake or metabolism of dietary constituents (Potischman, 2003). It should be also highlighted that biochemical markers of nutrient status have less error than dietary assessment of nutrient status (i.e., errors in completion of the instrument and food composition tables) (Kaaks et al., 1997). Moreover, for some nutrients, dietary data are inadequate because

Human Aging
https://doi.org/10.1016/B978-0-12-822569-1.00016-0

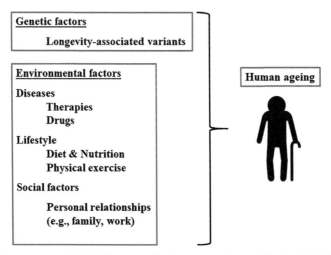

Fig. 15.1 Aging is a complex biological process that has not yet been fully elucidated. Human aging is affected not only by genetic factors, but also by various environmental factors such as diseases, lifestyle, and social factors.

of limitations in food composition data in dedicated databases (e.g., USDA Food Composition Database), whereas biomarkers of nutritional status related to these nutrients are available. More importantly, nutritional biomarkers provide a more proximal measure of nutrient status than dietary intake data for disease outcomes (Potischman, 2003). A nutritional biomarker can be objectively measured in different biological samples and can be used as an indicator of nutritional status with respect to the intake or metabolism of dietary constituents. Ideally, a nutritional biomarker of aging should objectively reflect the influence of dietary patterns and/or individual food components on the aging process and establish a causal relationship with functional performance. "Omics" techniques have opened new research avenues in nutrition and aging (Fig. 15.2). Advances in DNA sequencing techniques, microarray technologies, mass spectrometry, and nuclear magnetic resonance have facilitated the analysis of multiple parameters.

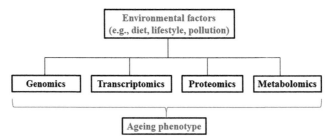

Fig. 15.2 Genomics, transcriptomics, and proteomics platforms can be used to identify and discover biomarkers associated with aging phenotype.

"Omics" platforms have provided unprecedented opportunities to explore the complex relationships between nutrition and aging, particularly to investigate the role of dietary components in health maintenance or in disease development (Kussmann et al., 2006). The purpose of this chapter is to highlight, both from a conceptual and a clinical perspective, a number of new developments in biomarker research that are taking place at the interface of aging and nutrition.

Biochemical measures can provide a useful tool for estimating the exposure to a particular nutrient of interest and assessing health risks. Randomized controlled trials and observational studies indicate that particular nutrients have beneficial effects on aging, including antioxidant phytochemicals, omega-3 polyunsaturated fatty acids (ω 3-PUFA), zinc, and vitamins A, B12, C, D, and E (Davinelli et al., 2016; Aiello et al., 2019; Sheats et al., 2015). Biochemical markers of dietary intake have been developed to avoid substantial measurement errors of dietary assessment techniques, such as food frequency questionnaires, 24-h recall, and weighed food record (Combs et al., 2013). Biochemical markers of dietary exposure can provide measures of nutritional status and exposure to bioactive molecules in foods, and therefore can be used as indicators of food intake. Moreover, measurement of biomarkers allows the identification of nutrient deficiencies (Potischman and Freudenheim, 2003). These biomarkers can be measured from blood, urine, or tissue and the concentration of a given marker reflects intake of a particular dietary component. The development of a metabolomic methodology makes urine samples and blood fractions the two most relevant biological fluids for measuring nutritional biomarkers but urine is probably the most used biological source in nutritional studies and long-term monitoring (O'Gorman et al., 2013). Nutrient biomarkers can be used to estimate intake of a wide range of dietary components, including overall fruit and vegetable intake (Baldrick et al., 2011), cruciferous vegetables (Andersen et al., 2014), whole grain cereals (Ross et al., 2012), tea and wine (Mennen et al., 2006), or coffee (Rothwell et al., 2014). Therefore biochemical analyses of nutrient biomarkers can provide an objective and sensitive assessment of a wide range of dietary components. In the following sections, we discuss the most promising food components in aging research and their causal relationship with age-related diseases.

15.2 Minerals (zinc and selenium)

Minerals, such as zinc and selenium, may play an important role in lowering oxidative stress in cells of older subjects, and supplementing the older people diet with these compounds can improve the immune response and reduce oxidative stress markers (Richard and Roussel, 1999). Although these compounds are needed in trace quantities, because the hemodynamic range is small, the maintenance of a correct amount and balance is very rare in older people.

15.2.1 Zinc

One of the main minerals related to physiologic processes associated with aging is zinc. Currently, there is not a sensitive and specific biomarker to detect zinc deficiency in humans. Low plasma or serum zinc concentrations are typically used as indicators of zinc status in populations and in intervention studies, but they have a number of limitations, including lack of sensitivity to detect marginal zinc deficiency (Gibson et al., 2008). A meta-analysis examining the efficacy of potential biomarkers of zinc status was undertaken by Lowe et al. This study presented an analysis of data from more than 32 potential biomarkers; however, for many biomarkers, there was insufficient evidence to assess their reliability. Lowe et al. concluded that plasma zinc concentration responds to an increase in intake over short periods, but that the homeostatic mechanisms that act to maintain plasma zinc concentration within the physiological range may prevent high plasma concentrations from being sustained over a prolonged period (Lowe et al., 2009). It was proposed that there may be biomarkers based on zinc transporters, however, this has not been confirmed. Candidates based on proteomic and metabolomic techniques are of current research interest (Ryu et al., 2012). Despite this, decreasing plasma zinc levels occur with age in both genders. The reason is due to intestinal malabsorption and inadequate diet (Mocchegiani, 2007). The recommended intake for adults amounts to 10 mg/day for men and 7 mg/day for women ("Scientific Opinion on Dietary Reference Values for Zinc," 2014). Zinc deficiency can increase incidence of age-related degenerative diseases and impair many molecular processes, such as signal transduction, apoptosis, proliferation, and differentiation of cellular components of the immune system (Prasad, 2008). Superoxide dismutases (SOD) are enzymes which play a critical part in cell defense against the damaging effects of free oxygen radicals. SOD enzymes are divided into different groups according to their capabilities for binding specific metal cofactors, including zinc. A lower SOD activity was detected in aged subjects and a negative correlation of plasma zinc level and antioxidant enzyme activity to age was observed (Sfar et al., 2009).

15.2.2 Selenium

Selenium is an essential micronutrient for humans and has its biological function mainly as a part of selenoproteins, such as glutathione peroxidases (GPx). In older adults, a lower level of selenium may contribute to inflammation and a higher mortality rate (Walston et al., 2006). On the basis of selenium serum/plasma levels of above 100 µg/L, an inverse relationship to a decreased risk of certain types of cancer (e.g., prostate cancer) and total mortality rate is suggested (Clark et al., 1998). The recommended selenium dietary intake is from 30 to 70 µg/day for adults (Combs et al., 2011). Methods to measure selenium intake are imprecise, mainly because of large variability in the selenium content of foods. However, various markers of intake or status are used, including concentrations of selenium in blood cells or body fluids, and concentration of selenoproteins or activity of selenoenzymes. Plasma, blood cells, and whole blood selenium concentration, plasma selenoprotein P (SEPP1) concentration, and GPx activity (assessed in plasma, red blood cells, or whole blood) have been considered useful biomarkers of selenium intake/status (Hurst et al., 2013). GPx enzymes are part of the human antioxidant network and contribute to protecting the organism from oxidative damage. Low activity levels of the GPx enzymes have been reported in oxidative stress-related diseases, such as cardiovascular disease (Kok et al., 1989). The maintenance of an optimal selenium status has the potential to protect against oxidative processes, including lipid peroxidation, and could eventually prevent chronic inflammation and cardiovascular disorders (Eaton et al., 2010). Additionally, SEPP1 plays a central role for selenium transport and homeostasis. This selenoprotein appears the most informative biomarker of selenium function on the basis of its role in selenium transport and metabolism and its response to different forms of ingested selenium (Hurst et al., 2010). SEPP1 also functions as an antioxidant that protects cells from oxidative damage and it has been also implicated in the regulation of glucose metabolism and insulin sensitivity (Misu et al., 2010).

15.3 Vitamins

Vitamins are a group of organic compounds that are vital to metabolism of human beings. Four vitamins (vitamins A, D, E, and K) are fat soluble and stay in the body for a long period of time, while nine of them (vitamins C, B1, B2, B3, B5, B6, B9, B12, and biotin) are water soluble and stay in the body much shorter (Lieberman and Bruning, 2007). Most of the vitamins are obtained from food or supplement, except vitamin D which is produced within the body (Glossmann, 2010). Aging is associated with increased risk for low vitamin consumption, and deficiency of vitamins can cause serious medical conditions. Although there are multiple biomarkers for each vitamin deficiency, the best biomarker, in many cases, is the direct measurement of the vitamin. When a vitamin is evaluated as a biomarker for its status and deficiency, the analytical considerations are as follows: (1) which form(s) of the vitamin within its production and metabolism pathways are most representative of its status, (2) whether these form(s) are suitable for a particular test, and (3) where the vitamin should be tested (bodily fluids such as blood and urine, or tissues) (Zhang and Crowther, 2017). In the following sections, we discuss vitamins that are deemed essential to nutrition and aging.

15.3.1 Antioxidant vitamins

Vitamin C, vitamin E, and vitamin A, often referred to as "antioxidant vitamins," have been suggested to limit oxidative damage in humans and lower the risk of certain chronic diseases. In observational studies, people with a high intake of antioxidant vitamins by regular diet or as food supplements generally have a lower risk of major chronic disease, such as myocardial infarction or stroke, than people who are low consumers of antioxidant vitamins (Thomas, 2006).

The primary source of vitamin C is fruits and vegetables, and the recommended intake of vitamin C is established to be 100 mg/day for adults (Levine et al., 1996). Older people as well as people suffering from diabetes or hypertension have lowered plasma vitamin C levels (Simon, 1992). A strong inverse trend for all-cause mortality and blood vitamin C is known (Fletcher et al., 2003). Moreover, a vitamin C rich diet has protective effects against the risk of some types of cancer (Kaaks et al., 1997), and an adequate intake of vitamin C seems to support cognition and may prevent from Alzheimer's disease (Harrison et al., 2014). Vitamin C level in the blood has been considered a reasonable surrogate for dietary vitamin C intake and an index of the circulating vitamin C available to tissues (Dehghan et al., 2007). Total plasma vitamin C can readily be determined by automated colorimetric assays. However, high-performance liquid chromatography (HPLC) methods with electrochemical detection, which provide necessary sensitivity and specificity, are generally used to quantify serum vitamin C concentrations (McCoy et al., 2005). Liquid chromatography-tandem mass spectrometry (LC-MS/MS)-based methodology has also been explored.

Vitamin E occurs in eight natural forms as tocopherols (α, β, γ, δ) and tocotrienols (α, β, γ, δ), all of which possess potent antioxidant properties. Lipid-rich plant products, such as nuts, seeds, and grain are a main dietary source of vitamin E. Estimated values for an adequate tocopherol supply amount to 15–12 mg tocopherol equivalents per day for men and 12–11 mg tocopherol equivalents per day for women. At a molecular level, vitamin E and some of its metabolites have shown capacity of regulating cell signaling and modulating gene transcription involved in oxidative processes and inflammation. If many epidemiological studies have given positive results, showing prevention by high vitamin E containing diets of cardiovascular events, neurodegenerative disease, macular degeneration, and cancer, the clinical confirmatory intervention studies were mostly negative (Azzi, 2017). A meta-analysis demonstrated that supplementation with vitamin E appears to have no effect on all-cause mortality and therefore should not be recommended as a means of improving longevity (Abner et al., 2011). However, healthy centenarians exhibit higher plasma vitamin E levels and an improved activity of antioxidant endogenous enzymes, compared to noncentenarians (Mecocci et al., 2000). To date, HPLC and gas chromatography (GC)-MS methodology has high precision and accuracy in the detection of vitamin E analogues in plasma (Dragsted, 2008).

The most common forms of preformed vitamin A in the human diet are retinol and retinyl esters. Vitamin A is essential for various human functions, among which the best known is vision health. The earliest symptom of vitamin A deficiency is impaired dark adaptation known as night blindness or nyctalopia. Prolonged vitamin A deficiency is characterized by changes in the cells of the cornea that result in corneal ulcers, scarring, and blindness (Wiseman et al., 2017). Direct fluorescence methods for assessing the retinol level in plasma or in dried blood are feasible because of the high intensity of retinol fluorescence. However, with the advent of HPLC, today retinol can be determined in serum routinely by direct- or reversed-phase liquid chromatography (Dragsted, 2008).

15.3.2 Vitamin D

Vitamin D is maybe the most discussed vitamin in recent years because its deficiency is a major public health problem worldwide in all age groups. In particular, almost all dark-skinned individuals residing in northern latitudes are particularly deficient (Palacios and Gonzalez, 2014). Vitamin D promotes calcium absorption and helps in maintaining adequate serum calcium level to enable normal mineralization of bone. It is also needed for bone growth and bone remodeling by osteoblasts and osteoclasts (Cranney et al., 2007). Vitamin D was long considered responsible only for protecting against rickets, but it has now been shown to be involved in a myriad of functions. Extensive evidence shows that vitamin D deficiency causes—or has been associated with—a large number of diseases that affect healthy aging, such as all-cause mortality, cancer, cardiovascular disease, diabetes, and neurodegeneration. There have been clinical trials, and meta-analyses providing high level of evidence of the efficacy of vitamin D for preventing and/or managing significant health problems in older adults (Johnson and Kimlin, 2006). Vitamin D exists in two forms in humans: vitamin D2 and vitamin D3. Vitamin D3, the major form of the two, may either be synthesized in skin or obtained from diet. The human body does not produce vitamin D2; therefore, the only source is dietary intake. Normally vitamin D2 levels in the blood are very low and hardly detectable. Through enzymatic reactions, vitamin D (including both D2 and D3) is first converted in the liver into 25-hydroxyvitamin D (25(OH)D), the major circulating form of vitamin D, and then in the kidney into 1,25-dihy-droxyvitamin D (1,25(OH)2D), the active form of vitamin D. The only way to determine whether a person is vitamin D deficient or sufficient is to measure their circulating level of 25(OH)D. There are a variety of assays used to measure 25(OH)D. The radioimmunoassays and competitive protein binding assays for 25(OH)D are useful in detecting vitamin D deficiency and sufficiency. However, these assays have several technical difficulties especially if they are not run routinely. Today, laboratories have now switched to LC-MS which measures both $25(OH)D_2$ and $25(OH)D_3$ quantitatively (Holick, 2009).

15.4 Polyunsaturated fatty acids

The n–3 and n–6 polyunsaturated fatty acids (PUFA) are among the most studied nutrients in human metabolism. Dietary sources that are rich in PUFA include many vegetable oils, nuts, seeds, and certain types of fish. The importance of the PUFA to human health has been linked to their involvement in multiple biochemical functions, including synthesis of inflammatory mediators, cell membrane fluidity, intracellular signaling, and gene expression. Long-chain PUFA appear to play a crucial role in specialized cells and tissues such as brain, retina, heart, and liver. The dietary intake of these essential fatty acids promotes the synthesis of arachidonic acid (AA; 20:4n-6), eicosapentaenoic acid (EPA; 20:5n-3), and docosahexaenoic acid (22:6n-3; DHA) that regulate diverse homeostatic processes by acting on the synthesis of bioactive signaling lipids called eicosanoids. However, n-3 and n-6 PUFA have opposing effects on metabolic functions in the body. In general, AA synthesizes the pro-inflammatory eicosanoids, while EPA and DHA induce the synthesis of antiinflammatory eicosanoids (Schmitz and Ecker, 2008). With the evolution in lipidomic techniques, it has been reported that PUFA play an important role in several conditions, such as cardiovascular diseases, cancer, depression, insulin resistance, nonalcoholic fatty liver disease (NAFLD), and neurodegeneration. The measurement of PUFA levels is not straightforward, and a wide variety of indices have been used in clinical studies. However, the n-3 index and AA/EPA ratio are the most promising biomarkers associated with PUFA metabolism (Davinelli et al., 2020b).

15.4.1 The n-3 index

The n–3 index was defined as the amount of EPA plus DHA in red blood cell (RBC) membranes, expressed as the percentage of total RBC membrane fatty acids. The n–3 index has been shown to be a stable biomarker of dietary intake and a valid surrogate of tissue long-chain n–3 PUFA. This index was proposed as a potential risk factor for cardiometabolic disease, and cross-sectional and prospective studies have supported its clinical utility (Jackson and Harris, 2018). Clinical studies show that a protective target level for the n–3 index appears to be about 8%, and the level associated with the increased disease risk is <4% (Fig. 15.3) (Harris et al., 2017). It was found an association between n–3 PUFA supplementation and a lower risk of cognitive decline in Alzheimer's disease

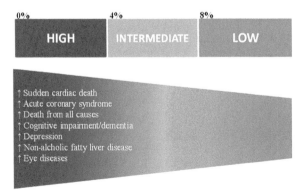

Fig. 15.3 The protective target level for the n-3 index appeared to be about 8%, and the level associated with increased risk of disease is <4%. However, these values have yet to be prospectively tested in large human studies.

patients. This is supported by a study where the n–3 index was correlated with cognitive function (Pottala et al., 2014). Only few clinical studies used the n–3 index to monitor PUFA metabolism in cancer. However, DHA supplementation was associated with increased levels of n–3 index in breast cancer patients (Molfino et al., 2017). NAFLD is common in older people and a cross-sectional study demonstrated an inverse association between n–3 index and NAFLD in older adults, supporting a relationship between n–3 index and NAFLD (Rose et al., 2016).

15.4.2 The AA/EPA ratio

Although the interaction between AA and EPA is complex and still not properly understood, several findings support the hypothesis that the balance between AA and EPA is important to regulate the synthesis of inflammatory mediators. Chronic low-grade inflammation creates a microenvironment suitable to the development of age-related diseases. Since AA and EPA are the precursors of important inflammatory mediators (i.e. eicosanoids), AA/EPA ratio provides an indication of the levels of inflammation in the body (Tutino et al., 2019). Moreover, the AA/EPA ratio is related to food intake and it is one of the most reliable indicators of PUFA nutritional status. The AA/EPA ratio has been found to be more closely associated with the pathophysiology of several diseases. Epidemiological and clinical studies have shown that a lower AA/EPA ratio is associated with decreased risk of coronary artery disease, acute coronary syndrome, myocardial infarction, stroke, chronic heart failure, and peripheral artery disease (Davinelli et al., 2020b). It was also demonstrated that high levels of AA/EPA ratio in metastatic patients induce an inflammatory microenvironment more susceptible to tumor progression (Tutino et al., 2019). Depression is a common and disabling psychiatric disorder in later life. Higher levels of AA/EPA ratio were positively associated with depression severity, suggesting a potential role for the balance of AA/EPA in mood disorders. A cross-sectional study involving 2529 subjects demonstrated that a higher AA/EPA ratio was associated with an increased risk of the presence of depressive symptoms in individuals with higher C-reactive protein (i.e., inflammatory biomarker) levels (Shibata et al., 2018). A limitation of the AA/EPA ratio is that a clinically useful threshold for identification of at-risk patients has not been clearly defined in large and prospective studies.

15.5 Carotenoids

Carotenoids are an important subgroup of phytonutrients in plant-derived food. These compounds are lipophilic antioxidants. Among the more than 600 carotenoids identified in nature, nutrition research has focused primarily on α-carotene, β-carotene, β-cryptoxanthin, lutein, zeaxanthin, and lycopene because of their prevalence in both food and the body. Lycopene, α- and β-carotene appear predominantly in red, orange, and yellow fruit and vegetables, whereas lutein and zeaxanthin occur mainly in green-leaf vegetables. Optimal levels of intake for three carotenoids were as follows: β-carotene more than 4.0 mg/day for men and more than 4.4 mg/day for women, α-carotene more than 0.6 mg/day for men and more than 0.7 mg/day for women, and lutein more than 3.3 mg/day for both males and females (Panel on Dietary Antioxidants and Related Compounds, 2000). As for other nutrients and phytochemicals, blood concentrations of carotenoids provide direct relationships between nutrient input, status, and availability. Tissues also reflect exposure to carotenoids. Although requires invasive approaches (i.e., biopsies), adipose tissue is thought to be a stable depot of carotenoids. Carotenoids accumulate in human skin and supplementation leads to increases in carotenoids content. Therefore skin carotenoid status (measured by reflectance methods or Raman spectroscopy) may be used as an objective biomarker of dietary intake. Noninvasive tests (i.e., heteroflicker photometry) have been used to measure macular pigment optical density, which provides information on long-term lutein (and zeaxanthin) exposure. In clinical practice, carotenoids analysis (mostly β-carotene) is not routinely but exceptionally performed and the best practice includes the use of fasting serum. C18 columns and UV/Vis detection are recommended for the analysis of β-carotene, although mass spectrometry is recommended as an alternative detection method (Granado-Lorencio et al., 2017). The most prevalent carotenoids in human serum are the same as those most commonly found in the diet: β-carotene, lycopene, and lutein (Khachik et al., 1997). The concentrations of various carotenoids in human serum and tissues are highly variable and likely depend on a number of factors such as food sources, efficiency of absorption, and amount of fat in the diet (Nebeling et al., 1997). However, increased levels of plasma/serum carotenoids have many different positive effects on age-related pathological conditions, such as eye disorders, neurodegenerative diseases, cardiovascular disorders, osteoporosis, and cancer (Tan and Norhaizan, 2019).

15.5.1 Lycopene, α- and β-carotene

Lycopene was shown to upregulate the expression of antioxidant and detoxifying enzymes via the activation of the nuclear factor E2-related factor 2 (Nrf2)-dependent pathway (Yang et al., 2012). Prospective cohort studies have suggested that lycopene–rich diets were associated with significant reductions in the risk of prostate cancer, particularly more aggressive forms (Giovannucci, 2002). A meta-analysis has demonstrated an inverse relationship between lycopene intake and cardiovascular disease risk (Cheng et al., 2019). α-Carotene and β-carotene are provitamin A carotenoids, meaning they can be converted in the body to vitamin A. Several studies found that higher serum/plasma concentrations of β-carotene were inversely associated with several cardiovascular risk factors, including C-reactive protein, total homocysteine, insulin resistance, hyperuricemia, and metabolic syndrome (Wang et al., 2008; Wang et al., 2014). Moreover, Dai et al. observed that dietary intakes of α- and β-carotene were inversely associated with hip fracture risk in men

(Dai et al., 2014). Some clinical trials have also found that β-carotene supplementation improves biomarkers of immune function in older man (Santos et al., 1998), and slows the progression of age-related cataracts (Group et al., 2002).

15.5.2 Lutein and zeaxanthin

Numerous findings suggest that high dietary consumption of lutein and zeaxanthin is likely to protect against age-related macular degeneration. Indeed, these carotenoids are found in the macula, both of dietary origin, and are involved in vision and protection against light-induced oxidative damage (Krinsky et al., 2003). Increasing dietary intake of lutein and zeaxanthin was shown to enhance their serum concentration and macular pigment density, lowering the risk of age-related macular degeneration (Marse-Perlman et al., 2001). Cross-sectional and retrospective case-control studies found that higher levels of lutein and zeaxanthin in the blood and retina were associated with a lower incidence of this pathological condition (Bone et al., 2001; Gale et al., 2003). A large-scale clinical trial randomized 3306 subjects to various combinations of carotenoids. Supplementation with lutein and zeaxanthin significantly reduced the risk of progression to late age-related macular degeneration, compared to supplementation with β-carotene (Chew et al., 2014).

15.6 Polyphenols

Polyphenols are a large family of phytochemicals with over 8000 structural variants. These molecules are secondary metabolites of plants. The main polyphenol subgroup are flavonoids, while stilbenes and lignans are less common (Fig. 15.4). Polyphenols are an unavoidable component in the human diet and evidence suggests that their consumption is associated with the beneficial modulation of a number of health-related variables (Davinelli et al., 2020a). However, the health benefits of polyphenols and their role in modulating the risk of high-prevalence diseases are difficult to demonstrate due to the wide variability of chemical structures, biological actions, and complexity of estimating the polyphenol content in foods and biological fluids. Native polyphenols are subjected to extensive metabolism following ingestion, and the identification of circulating metabolites, which serve as potential biomarkers of intake, has been undertaken for a small number of polyphenols. In the following sections, we briefly discuss the main beneficial effects of flavonoids and factors affecting choice of potential biomarkers of polyphenols intake.

15.6.1 Flavonoids

Flavonoids are classified into 12 major subclasses based on chemical structures, six of which, namely anthocyanidins, flavan-3-ols, flavonols, flavones, flavanones, and

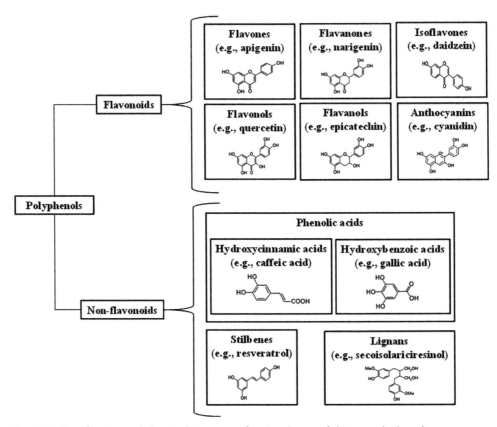

Fig. 15.4 Classification and chemical structure of major classes of dietary polyphenols.

isoflavones are of dietary significance. These compounds can be rapidly transformed by phase II detoxification enzymes, increasing their excretion in the urine (Zhang et al., 2014). A high intake of flavonoids was associated with lower concentrations of pro-inflammatory cytokines and markers of oxidative stress measured in urine and plasma (Cassidy et al., 2015; Davinelli et al., 2018b). An increasing number of intervention trials with flavonoids reported reduced levels of markers of cardiovascular disease such as blood pressure, low density lipoprotein (LDL)-cholesterol, and oxidized LDL (Qin et al., 2009; Davinelli et al., 2018a). Other benefits include reductions in arterial stiffness and serum concentrations of markers of vasoconstriction (Grassi et al., 2012). A meta-analysis of prospective cohort studies found a reduced risk of breast cancer in women with high iso-flavone intakes (Dong and Qin, 2011). Equol, one of the major metabolites produced by isoflavones, improves menopause-related quality of life in healthy women (Davinelli et al., 2017). Dietary flavonoids and/or their metabolites have been shown to cross the blood-brain barrier and exert preventive effects against cognitive impairments associated with aging (Vauzour et al., 2008). Randomized controlled trials reported that flavonoids might enhance some aspects of cognitive function, increasing cerebral blood flow velocity or improving measures of cognitive process speed, flexibility, and verbal fluency (Sorond et al., 2008; Mastroiacovo et al., 2015). The dentate gyrus is a region in the hippocampal formation whose function declines in association with human aging and is considered to be a possible source of age-related memory decline. By using functional magnetic resonance imaging, a high-flavanol cocoa intervention was found to enhance dentate gyrus function in 50–69-year-old subjects (Brickman et al., 2014).

15.6.2 Biochemical assessment of polyphenols intake

The choice of biomarker for the estimation of polyphenol intake is complex due to the wide variability of chemical structures. The circulating concentrations of native forms of polyphenols and their metabolites are in the nanomolar to low micromolar range and represent only a very small percentage of the amount consumed. If metabolites are used as biomarkers in polyphenol intake, it is crucial that analytical techniques for their measurement in biological samples are both sensitive and reliable. The most common quantification approach is to use HPLC coupled with UV spectra, fluorescence, or electrochemical detection. However, recent application of ultra HPLC-MS/MS has provided a useful tool for the identification and quantification of polyphenols and their metabolites in biological fluids such as plasma and urine (Hakeem Said et al., 2020). To measure a biomarker of polyphenol intake, an important factor to consider is the kinetic profile of its metabolism and appearance in the biological fluid. Urine sampling is particularly useful for polyphenols with short half-lives, where plasma measurements may fail to monitor even acute intake. Therefore, for the majority of flavonoids, which possess short plasma half-lives, quantification of their output in urine maybe a better approach (Rechner et al., 2002). To date, the majority of biomarkers are those deriving from phase I and II metabolism of flavonoids and other polyphenols in the small intestine and liver. Metabolites such as glucuronide and sulfate conjugates (and related O-methylated forms) are commonly used to determine polyphenol absorption in human subjects and represent excellent biomarkers. There is also great potential for glutathionyl and cysteinyl conjugates, as they are present in significant amounts in both blood and urine (Spencer et al., 2008).

15.7 Molecular biomarkers of aging and nutrition

Human aging is associated with a complex network of interactions between pathways, processes, and molecules, implying interactive mechanisms across different cells, tissues, and organs. Aging biomarkers are the cornerstone of nutritional research to investigate the functional effects of nutrition on aging process and age-related diseases. Systems approaches and high-throughput technologies allow to obtain a comprehensive and in-depth view of the physiology/pathology of an individual and open the possibility to explore the complex relationships between nutrition and aging, particularly to investigate the role of dietary components in health maintenance or in disease development (Corella et al., 2018).

15.7.1 DNA and chromosomes

Telomeres are ribonucleoprotein complexes at the end of chromosomes that prevent the loss of genomic DNA, protecting its physical integrity. Telomere length is considered to be a biomarker of aging; shorter telomeres are associated with a decreased lifespan and increased rates of developing age-related chronic diseases (Calado and Young, 2009). Given that there is an existing link between "inflammatory/oxidative status" and telomere attrition, it is plausible that consuming antioxidant-rich foods may have important health benefits by helping to counteract telomere attrition (Davinelli et al., 2019). Even though evidence from trials is very limited, several observational studies have observed that adherence to plant-rich dietary patterns and consumption of plant-based foods may play an important role for telomere maintenance (Crous-Bou et al., 2019). The relationship between aging and epigenetic modifications is well established. Age-related changes in DNA methylation patterns are among the best-studied aging biomarkers. Bioactive food components such as resveratrol, curcumin, sulforaphane, and tea polyphenols can modulate epigenetic patterns by altering the levels of enzymes that catalyze DNA methylation (Davinelli et al., 2014).

15.7.2 RNA and transcriptome

With rapid progress in RNA sequencing technology, it has begun to be applied to the study of biomarkers of aging. A recent study used whole-blood gene expression profiles from 14,983 individuals to identify 1497 genes with age-dependent differential expression, suggesting that transcriptome signatures can be used to measure aging (Peters et al., 2015). Interestingly, changes in the transcriptome have been observed after consumption of diets rich in n-3 PUFA, polyphenols, or other dietary interventions (Afman et al., 2014). Although promising, these findings still need to be validated in other studies to be applied in aging research. MicroRNA (miRNA) are a class of small (21- to 23-nucleotide) noncoding RNA. miRNAs have emerged as crucial regulators of many processes related to nutrition, including nutritional regulation of aging and disease-related pathways (Martínez-Jiménez et al., 2018). Dietary modulation of miRNA expression has been shown to influence various diseases, such as type 2 diabetes, obesity, or hepatic steatosis. Food components have been shown to modulate the expression of miRNAs (Ross and Davis, 2014). For example, diets rich in PUFA have been shown to modulate specific miRNA, and some of them were associated with markers of inflammation and metabolic health (Ortega et al., 2015).

15.7.3 Metabolism

Many of the genes that act as key regulators of lifespan also have known functions in nutrient sensing, and thus are called "nutrient-sensing longevity genes." Some examples of nutrient-sensing pathways (NSP) involved in the longevity response are the mechanistic target of rapamycin (mTOR), AMP kinase, sirtuins, and insulin/insulin-like growth factor 1 (IGF-1) signaling (IIS) (Davinelli et al., 2012). Several studies have suggested that both in animal models and humans, dietary interventions can prevent or decrease various age-related diseases, by positively regulating aging process through the modulation of NSP. These NSP are modulated by caloric restriction, carbohydrates, proteins, or polyphenols that trigger a downstream activation of genes involved in aging process. For example, through the downregulation of IGF-1 and mTOR cascade or the upregulation of sirtuins, one can extend lifespan in various model organisms, including mammals (Aiello et al., 2017). Advanced glycation end products (AGE) are a heterogeneous group of bioactive molecules that are formed by nonenzymatic glycation of proteins, lipids, and nucleic acids. Accumulation of AGE in aging tissues leads to inflammation, apoptosis, diabetes, and other age-related disorders (Brownlee, 1995). The use of proteomics (e.g., mass spectrometry) in nutritional research has been useful to identify specific AGE as potential biomarkers for changes in glucose metabolism related to diabetes and/or aging (Chiu et al., 2018).

15.7.4 Mitochondria

Aging is generally associated with a decline in mitochondrial function and an accumulation of damaged mitochondria. Although free radicals, the source of oxidative stress, are mainly produced in mitochondria, dysfunctional mitochondria can contribute to aging independently of reactive oxygen species. Indeed, the tight coordination between mitochondrial biogenesis and mitophagy (i.e., degradation of damaged mitochondria by autophagy) can be dysregulated during aging, critically influencing whole-body metabolism, health, and lifespan (Moehle et al., 2019). Several nutritional approaches that promote cell survival, such as caloric restriction and the dietary intake of polyphenols, ameliorate mitochondrial dysfunction, thus increasing mitochondrial biogenesis and mitophagy. For example, a few prominent examples of dietary polyphenols that improve mitochondrial biogenesis include quercetin, curcumin, myricitrin, ellagitannin, and epigallocatechin-3-gallate (EGCG) (Davinelli et al., 2020a). A number of studies have identified pomegranate and its bioactive compounds as mitophagy modulators. Urolithin A is the most prevalent ellagitannin-derived metabolite in humans and promotes mitophagy in nematodes and mammals, leading to a general improvement of lifespan and mitochondrial dysfunction. In particular, urolithin A maintains mitochondrial respiratory capacity with age and induces mitophagy by increasing the expression of autophagy genes (Ryu et al., 2016).

15.7.5 Cell senescence

The gradual accumulation of senescent cells in mitotic tissues is thought be one of the causal factors of aging. Therefore the biomarkers of cell senescence can also be used as markers. The most widely used marker is senescence-associated β-galactosidase (SAβ-gal); however, other senescent cell markers include activated and persistent DNA-damage response, telomere shortening and dysfunction, p16^{INK4A}, and senescence-associated secretory phenotype (SASP) (i.e., inflammation and altered intercellular communication) (Bernardes de Jesus and Blasco, 2012). A number of bioactive food compounds have been investigated for their antisenescence and antiaging potential in cellular and animal models as well as in humans (Gurău et al., 2018). Many dietary phytochemicals have been screened for their ability to inhibit senescence. Fisetin, a naturally occurring flavonoid, was the most potent senolytic compound in cultured senescent murine and human fibroblasts. Moreover, administration of fisetin to wild-type mice late in life restored tissue homeostasis, reduced age-related pathology, and extended median and maximum lifespan (Yousefzadeh et al., 2018). Several other flavonoids such as quercetin, naringenin, apigenin, and kaempferol have been reported to exert anti-SASP effects on human cells (Lim et al., 2015). Moreover, drugs able to selectively kill senescent cells have been identified including a senolytic cocktail of quercetin and dasatinib, which improves many aspects of aging in mouse models of accelerated and natural aging (Roos et al., 2016).

15.8 Conclusions and future perspectives

A new scientific discipline, nutrigerontology, examines the impact of nutrients, foods, and dietary habits on the ability to achieve successful aging and longevity. Aging is a multidimensional process and nutritional biomarkers could be used to prevent, monitor, and identify the development of age-related diseases. Here, we summarized the current knowledge of how various nutritional parameters are directly affected and altered during aging and to what extent these indicators are useful to evaluate the aging process. Due to the large number of factors involved and the complexity of the aging process, there is not yet a "pure" biomarker to investigate the relationship between aging and nutrition. There is a critical need to discover novel biomarkers and validate existing biomarkers in large-scale randomized clinical trials to translate them into primary prevention against age-related diseases and premature aging. To date, in the absence of the ideal biomarker, the best approach might be a combination of various biomarkers rather than a single biomarker. Holistic approaches using omics platforms, capable of gathering high amounts of data, appear to be very useful to identify and develop new biomarkers and enhance our understanding of the role of food in health and disease.

References

Azzi A. Many tocopherols, one vitamin E. Mol Aspects Med 2017;61:92–103. https://doi.org/10.1016/j.mam.2017.06.004.

Abner EL, Schmitt FA, Mendiondo MS, Marcum JL, Kryscio RJ. Vitamin E and all-cause mortality: a meta-analysis. Curr Aging Sci 2011;4:158–70.

Afman L, Milenkovic D, Roche HM. Nutritional aspects of metabolic inflammation in relation to health—insights from transcriptomic biomarkers in PBMC of fatty acids and polyphenols. Mol Nutr Food Res 2014;58:1708–20. https://doi.org/10.1002/mnfr.201300559.

Aiello A, Accardi G, Candore G, Carruba G, Davinelli S, Passarino G, Scapagnini G, Vasto S, Caruso C. Nutrigerontology: a key for achieving successful ageing and longevity. Immun Ageing 2016;13:1–5. https://doi.org/10.1186/s12979-016-0071-2.

Aiello A, Accardi G, Candore G, Gambino CM, Mirisola M, Taormina G, Virruso C, Caruso C. Nutrient sensing pathways as therapeutic targets for healthy ageing. Expert Opin Ther Targets 2017;21:371–80. https://doi.org/10.1080/14728222.2017.1294684.

Aiello A, Farzaneh F, Candore G, Caruso C, Davinelli S, Gambino CM, Ligotti ME, Zareian N, Accardi G. Immunosenescence and its hallmarks: how to oppose aging strategically? A review of potential options for therapeutic intervention. Front Immunol 2019;10. https://doi.org/10.3389/fimmu.2019.02247.

Andersen M-BS, Kristensen M, Manach C, Pujos-Guillot E, Poulsen SK, Larsen TM, Astrup A, Dragsted L. Discovery and validation of urinary exposure markers for different plant foods by untargeted metabolomics. Anal Bioanal Chem 2014;406:1829–44. https://doi.org/10.1007/s00216-013-7498-5.

Baldrick FR, Woodside JV, Elborn JS, Young IS, McKinley MC. Biomarkers of fruit and vegetable intake in human intervention studies: a systematic review. Crit Rev Food Sci Nutr 2011;51:795–815. https://doi.org/10.1080/10408398.2010.482217.

Bernardes de Jesus B, Blasco MA. Assessing cell and organ senescence biomarkers. Circ Res 2012;111:97–109. https://doi.org/10.1161/CIRCRESAHA.111.247866.

Bone RA, Landrum JT, Mayne ST, Gomez CM, Tibor SE, Twaroska EE. Macular pigment in donor eyes with and without AMD: a case-control study. Invest Ophthalmol Vis Sci 2001;42:235–40.

Brickman AM, Khan UA, Provenzano FA, Yeung L-K, Suzuki W, Schroeter H, Wall M, Sloan RP, Small SA. Enhancing dentate gyrus function with dietary flavanols improves cognition in older adults. Nat Neurosci 2014;17:1798–803. https://doi.org/10.1038/nn.3850.

Brownlee M. Advanced protein glycosylation in diabetes and aging. Annu Rev Med 1995;46:223–34. https://doi.org/10.1146/annurev.med.46.1.223.

Calado RT, Young NS. Telomere diseases. N Engl J Med 2009;361:2353–65. https://doi.org/10.1056/NEJMra0903373.

Cassidy A, Rogers G, Peterson JJ, Dwyer JT, Lin H, Jacques PF. Higher dietary anthocyanin and flavonol intakes are associated with anti-inflammatory effects in a population of US adults. Am J Clin Nutr 2015;102:172–81. https://doi.org/10.3945/ajcn.115.108555.

Cheng HM, Koutsidis G, Lodge JK, Ashor AW, Siervo M, Lara J. Lycopene and tomato and risk of cardiovascular diseases: a systematic review and meta-analysis of epidemiological evidence. Crit Rev Food Sci Nutr 2019;59:141–58. https://doi.org/10.1080/10408398.2017.1362630.

Chew EY, Clemons TE, SanGiovanni JP, Danis RP, Ferris FL, Elman MJ, Antoszyk AN, Ruby AJ, Orth D, Bressler SB, Fish GE, Hubbard GB, Klein ML, Chandra SR, Blodi BA, Domalpally A, Friberg T, Wong WT, Rosenfeld PJ, Agrón E, Toth CA, Bernstein PS, Sperduto RD. Secondary analyses of the effects of lutein/zeaxanthin on age-related macular degeneration progression: AREDS2 report no. 3. JAMA Ophthalmol 2014;132:142–9. https://doi.org/10.1001/jamaophthalmol.2013.7376.

Chiu C-J, Rabbani N, Rowan S, Chang M-L, Sawyer S, Hu FB, Willett W, Thornalley PJ, Anwar A, Bar L, Kang JH, Taylor A. Studies of advanced glycation end products and oxidation biomarkers for type 2 diabetes. Biofactors 2018;44:281–8. https://doi.org/10.1002/biof.1423.

Clark D, Krongrad Jr. C, Turnbull S, Witherington H, Janosko C, Borosso F, Rounder. Decreased incidence of prostate cancer with selenium supplementation: results of a double-blind cancer prevention trial. BJU Int 1998;81:730–4. https://doi.org/10.1046/j.1464-410x.1998.00630.x.

Combs GF, Trumbo PR, McKinley MC, Milner J, Studenski S, Kimura T, Watkins SM, Raiten DJ. Bio-markers in nutrition: new frontiers in research and application. Ann N Y Acad Sci 2013;1278:1–10. https://doi.org/10.1111/nyas.12069.

Combs GF, Watts JC, Jackson MI, Johnson LK, Zeng H, Scheett AJ, Uthus EO, Schomburg L, Hoeg A, Hoefig CS, Davis CD, Milner JA. Determinants of selenium status in healthy adults. Nutr J 2011;10:75. https://doi.org/10.1186/1475-2891-10-75.

Corella D, Coltell O, Macian F, Ordovás JM. Advances in understanding the molecular basis of the Mediterranean diet effect. Annu Rev Food Sci Technol 2018;9:227–49. https://doi.org/10.1146/annurev-food-032217-020802.

Cranney A, Horsley T, O'Donnell S, Weiler H, Puil L, Ooi D, Atkinson S, Ward L, Moher D, Hanley D, Fang M, Yazdi F, Garritty C, Sampson M, Barrowman N, Tsertsvadze A, Mamaladze V. Effectiveness and safety of vitamin D in relation to bone health. Evid Rep Technol Assess (Full Rep) 2007;1–235.

Crous-Bou M, Molinuevo J-L, Sala-Vila A. Plant-rich dietary patterns, plant foods and nutrients, and telomere length. Adv Nutr 2019;10:S296–303. https://doi.org/10.1093/advances/nmz026.

Dai Z, Wang R, Ang L-W, Low Y-L, Yuan J-M, Koh W-P. Protective effects of dietary carotenoids on risk of hip fracture in men: the Singapore Chinese health study. J Bone Miner Res 2014;29:408–17. https://doi.org/10.1002/jbmr.2041.

Davinelli S, Calabrese V, Zella D, Scapagnini G. Epigenetic nutraceutical diets in Alzheimer's disease. J Nutr Health Aging 2014;18:800–5. https://doi.org/10.1007/s12603-014-0552-y.

Davinelli S, Corbi G, Righetti S, Sears B, Olarte HH, Grassi D, Scapagnini G. Cardioprotection by cocoa polyphenols and ω-3 fatty acids: a disease-prevention perspective on aging-associated cardiovascular risk. J Med Food 2018a;21:1060–9. https://doi.org/10.1089/jmf.2018.0002.

Davinelli S, Corbi G, Zarrelli A, Arisi M, Calzavara-Pinton P, Grassi D, De Vivo I, Scapagnini G. Short-term supplementation with flavanol-rich cocoa improves lipid profile, antioxidant status and positively influences the AA/EPA ratio in healthy subjects. J Nutr Biochem 2018b;61:33–9. https://doi.org/10.1016/j.jnutbio.2018.07.011.

Davinelli S, De Stefani D, De Vivo I, Scapagnini G. Polyphenols as caloric restriction mimetics regulating mitochondrial biogenesis and mitophagy. Trends Endocrinol Metab 2020a;31:536–50. https://doi.org/10.1016/j.tem.2020.02.011.

Davinelli S, Intrieri M, Corbi G, Scapagnini G. Metabolic indices of polyunsaturated fatty acids: current evidence, research controversies, and clinical utility. Crit Rev Food Sci Nutr 2020b;0:1–16. https://doi.org/10.1080/10408398.2020.1724871.

Davinelli S, Maes M, Corbi G, Zarrelli A, Willcox DC, Scapagnini G. Dietary phytochemicals and neuro-inflammaging: from mechanistic insights to translational challenges. Immun Ageing 2016;13:16. https://doi.org/10.1186/s12979-016-0070-3.

Davinelli S, Scapagnini G, Marzatico F, Nobile V, Ferrara N, Corbi G. Influence of equol and resveratrol supplementation on health-related quality of life in menopausal women: a randomized, placebo-controlled study. Maturitas 2017;96:77–83. https://doi.org/10.1016/j.maturitas.2016.11.016.

Davinelli S, Trichopoulou A, Corbi G, De Vivo I, Scapagnini G. The potential nutrigeroprotective role of Mediterranean diet and its functional components on telomere length dynamics. Ageing Res Rev 2019;49:1–10. https://doi.org/10.1016/j.arr.2018.11.001.

Davinelli S, Willcox DC, Scapagnini G. Extending healthy ageing: nutrient sensitive pathway and centenarian population. Immun Ageing 2012;9:9. https://doi.org/10.1186/1742-4933-9-9.

Dehghan M, Akhtar-Danesh N, McMillan CR, Thabane L. Is plasma vitamin C an appropriate biomarker of vitamin C intake? A systematic review and meta-analysis. Nutr J 2007;6:41. https://doi.org/10.1186/1475-2891-6-41.

Dong J-Y, Qin L-Q. Soy isoflavones consumption and risk of breast cancer incidence or recurrence: a meta-analysis of prospective studies. Breast Cancer Res Treat 2011;125:315–23. https://doi.org/10.1007/s10549-010-1270-8.

Dragsted LO. Biomarkers of exposure to vitamins a, C, and E and their relation to lipid and protein oxidation markers. Eur J Nutr 2008;47:3–18. https://doi.org/10.1007/s00394-008-2003-1.

Eaton CB, Baki ARA, Waring ME, Roberts MB, Lu B. The association of low selenium and renal insufficiency with coronary heart disease and all-cause mortality: NHANES III follow-up study. Atherosclerosis 2010;212:689–94. https://doi.org/10.1016/j.atherosclerosis.2010.07.008.

Fletcher AE, Breeze E, Shetty PS. Antioxidant vitamins and mortality in older persons: findings from the nutrition add-on study to the Medical Research Council Trial of Assessment and Management of Older People in the community. Am J Clin Nutr 2003;78:999–1010. https://doi.org/10.1093/ajcn/78.5.999.

Gale CR, Hall NF, Phillips DIW, Martyn CN. Lutein and zeaxanthin status and risk of age-related macular degeneration. Invest Ophthalmol Vis Sci 2003;44:2461–5. https://doi.org/10.1167/iovs.02-0929.

Gibson RS, Hess SY, Hotz C, Brown KH. Indicators of zinc status at the population level: a review of the evidence. Br J Nutr 2008;99:S14–23. https://doi.org/10.1017/S0007114508006818.

Giovannucci E. A review of epidemiologic studies of tomatoes, lycopene, and prostate cancer. Exp Biol Med (Maywood) 2002;227:852–9. https://doi.org/10.1177/153537020222701003.

Glossmann HH. Origin of 7-dehydrocholesterol (provitamin D) in the skin. J Invest Dermatol 2010;130:2139–41. https://doi.org/10.1038/jid.2010.118.

Granado-Lorencio F, Blanco-Navarro I, Pérez-Sacristán B, Hernández-Álvarez E. Biomarkers of carotenoid bioavailability. Food Res Int Carotenoids Food health 2017;99:902–16. https://doi.org/10.1016/j.foodres.2017.03.036.

Grassi D, Desideri G, Necozione S, Ruggieri F, Blumberg J, Stornello M, Ferri C. Protective effects of flavanol-rich dark chocolate on endothelial function and wave reflection during acute hyperglycemia. Hypertension 2012;60:827–32. https://doi.org/10.1161/HYPERTENSIONAHA.112.193995.

Group TR, Chylack LT, Brown NP, Bron A, Hurst M, Köpcke W, Thien U, Schalch W. The Roche European American Cataract Trial (REACT): a randomized clinical trial to investigate the efficacy of an oral antioxidant micronutrient mixture to slow progression of age-related cataract. Ophthalmic Epidemiol 2002;9:49–80. https://doi.org/10.1076/opep.9.1.49.1717.

Gurău F, Baldoni S, Prattichizzo F, Espinosa E, Amenta F, Procopio AD, Albertini MC, Bonafè M, Olivieri F. Anti-senescence compounds: a potential nutraceutical approach to healthy aging. Ageing Res Rev 2018;46:14–31. https://doi.org/10.1016/j.arr.2018.05.001.

Hakeem Said I, Truex JD, Heidorn C, Retta MB, Petrov DD, Haka S, Kuhnert N. LC-MS/MS based molecular networking approach for the identification of cocoa phenolic metabolites in human urine. Food Res Int 2020;132:109119. https://doi.org/10.1016/j.foodres.2020.109119.

Harris WS, Gobbo LD, Tintle NL. The Omega-3 index and relative risk for coronary heart disease mortality: estimation from 10 cohort studies. Atherosclerosis 2017;262:51–4. https://doi.org/10.1016/j.atherosclerosis.2017.05.007.

Harrison FE, Bowman GL, Polidori MC. Ascorbic acid and the brain: rationale for the use against cognitive decline. Nutrients 2014;6:1752–81. https://doi.org/10.3390/nu6041752.

Holick MF. Vitamin d status: measurement, interpretation and clinical application. Ann Epidemiol 2009;19:73–8. https://doi.org/10.1016/j.annepidem.2007.12.001.

Hurst R, Armah CN, Dainty JR, Hart DJ, Teucher B, Goldson AJ, Broadley MR, Motley AK, Fairweather-Tait SJ. Establishing optimal selenium status: results of a randomized, double-blind, placebo-controlled trial. Am J Clin Nutr 2010;91:923–31. https://doi.org/10.3945/ajcn.2009.28169.

Hurst R, Collings R, Harvey LJ, King M, Hooper L, Bouwman J, Gurinovic M, Fairweather-Tait SJ. EURRECA—estimating selenium requirements for deriving dietary reference values. Crit Rev Food Sci Nutr 2013;53:1077–96. https://doi.org/10.1080/10408398.2012.742861.

Jackson KH, Harris WS. Blood fatty acid profiles: new biomarkers for cardiometabolic disease risk. Curr Atheroscler Rep 2018;20:22. https://doi.org/10.1007/s11883-018-0722-1.

Johnson MA, Kimlin MG. Vitamin D, aging, and the 2005 dietary guidelines for Americans. Nutr Rev 2006;64:410–21. https://doi.org/10.1111/j.1753-4887.2006.tb00226.x.

Kaaks R, Riboli E, Sinha R. Biochemical markers of dietary intake. IARC Sci Publ 1997;103–26.

Kaeberlein M. Longevity and aging. F1000Prime Rep 2013;5. https://doi.org/10.12703/P5-5.

Khachik F, Spangler CJ, Smith JC, Canfield LM, Steck A, Pfander H. Identification, quantification, and relative concentrations of carotenoids and their metabolites in human milk and serum. Anal Chem 1997;69:1873–81. https://doi.org/10.1021/ac961085i.

Kok FJ, Hofman A, Witteman JCM, de Bruijn AM, Kruyssen DHCM, de Bruin M, Valkenburg HA. Decreased selenium levels in acute myocardial infarction. JAMA 1989;261:1161–4. https://doi.org/10.1001/jama.1989.03420080081035.

Krinsky NI, Landrum JT, Bone RA. Biologic mechanisms of the protective role of lutein and zeaxanthin in the eye. Annu Rev Nutr 2003;23:171–201. https://doi.org/10.1146/annurev.nutr.23.011702.073307.

Kussmann M, Raymond F, Affolter M. OMICS-driven biomarker discovery in nutrition and health. J Biotechnol 2006;124:758–87. https://doi.org/10.1016/j.jbiotec.2006.02.014. Highlights from ECB12.

Levine M, Conry-Cantilena C, Wang Y, Welch RW, Washko PW, Dhariwal KR, Park JB, Lazarev A, Graumlich JF, King J, Cantilena LR. Vitamin C pharmacokinetics in healthy volunteers: evidence for a recommended dietary allowance. PNAS 1996;93:3704–9. https://doi.org/10.1073/pnas.93.8.3704.

Lieberman S, Bruning N. The Real vitamin and mineral book: a definitive guide to designing your personal supplement program. Penguin.

Lim H, Park H, Kim HP. Effects of flavonoids on senescence-associated secretory phenotype formation from bleomycin-induced senescence in BJ fibroblasts. Biochem Pharmacol 2015;96:337–48. https://doi.org/10.1016/j.bcp.2015.06.013.

Lowe NM, Fekete K, Decsi T. Methods of assessment of zinc status in humans: a systematic review. Am J Clin Nutr 2009;89:2040S–2051S. https://doi.org/10.3945/ajcn.2009.27230G.

Marse-Perlman JA, Fisher AI, Klein R, Palta M, Block G, Millen AE, Wright JD. Lutein and zeaxanthin in the diet and serum and their relation to age-related maculopathy in the Third National Health and Nutrition Examination Survey. Am J Epidemiol 2001;153:424–32. https://doi.org/10.1093/aje/153.5.424.

Martínez-Jiménez VdC, Méndez-Mancilla A, Portales-Pérez DP. miRNAs in nutrition, obesity, and cancer: the biology of miRNAs in metabolic disorders and its relationship with cancer development. Mol Nutr Food Res 2018;62:1600994. https://doi.org/10.1002/mnfr.201600994.

Mastroiacovo D, Kwik-Uribe C, Grassi D, Necozione S, Raffaele A, Pistacchio L, Righetti R, Bocale R, Lechiara MC, Marini C, Ferri C, Desideri G. Cocoa flavanol consumption improves cognitive function, blood pressure control, and metabolic profile in elderly subjects: the Cocoa, Cognition, and Aging (CoCoA) Study—a randomized controlled trial. Am J Clin Nutr 2015;101:538–48. https://doi.org/10.3945/ajcn.114.092189.

McCoy LF, Bowen MB, Xu M, Chen H, Schleicher RL. Improved HPLC assay for measuring serum vitamin C with 1-methyluric acid used as an electrochemically active internal standard. Clin Chem 2005;51:1062–4. https://doi.org/10.1373/clinchem.2004.046904.

Mecocci P, Polidori MC, Troiano L, Cherubini A, Cecchetti R, Pini G, Straatman M, Monti D, Stahl W, Sies H, Franceschi C, Senin U. Plasma antioxidants and longevity: a study on healthy centenarians. Free Radic Biol Med 2000;28:1243–8. https://doi.org/10.1016/S0891-5849(00)00246-X.

Mennen LI, Sapinho D, Ito H, Bertrais S, Galan P, Hercberg S, Scalbert A. Urinary flavonoids and phenolic acids as biomarkers of intake for polyphenol-rich foods. Br J Nutr 2006;96:191–8. https://doi.org/10.1079/BJN20061808.

Misu H, Takamura T, Takayama H, Hayashi H, Matsuzawa-Nagata N, Kurita S, Ishikura K, Ando H, Takeshita Y, Ota T, Sakurai M, Yamashita T, Mizukoshi E, Yamashita T, Honda M, Miyamoto K, Kubota T, Kubota N, Kadowaki T, Kim H-J, Lee I, Minokoshi Y, Saito Y, Takahashi K, Yamada Y, Takakura N, Kaneko S. A liver-derived secretory protein, selenoprotein P, causes insulin resistance. Cell Metab 2010;12:483–95. https://doi.org/10.1016/j.cmet.2010.09.015.

Mocchegiani E. Zinc and ageing: third Zincage conference. Immun Ageing 2007;4:5. https://doi.org/10.1186/1742-4933-4-5.

Moehle EA, Shen K, Dillin A. Mitochondrial proteostasis in the context of cellular and organismal health and aging. J Biol Chem 2019;294:5396–407. https://doi.org/10.1074/jbc.TM117.000893.

Molfino A, Amabile MI, Mazzucco S, Biolo G, Farcomeni A, Ramaccini C, Antonaroli S, Monti M, Muscaritoli M. Effect of oral docosahexaenoic acid (DHA) supplementation on DHA levels and omega-3 index in red blood cell membranes of breast cancer patients. Front Physiol 2017;8. https://doi.org/10.3389/fphys.2017.00549.

Nebeling LC, Forman MR, Graubard BI, Snyder RA. Changes in carotenoid intake in the United States: the 1987 and 1992 National Health Interview Surveys. J Am Diet Assoc 1997;97:991–6. https://doi.org/10.1016/S0002-8223(97)00239-3.

O'Gorman A, Gibbons H, Brennan L. Metabolomics in the identification of biomarkers of dietary intake. Comput Struct Biotechnol J 2013;4. https://doi.org/10.5936/csbj.201301004.

Ortega FJ, Cardona-Alvarado MI, Mercader JM, Moreno-Navarrete JM, Moreno M, Sabater M, Fuentes-Batllevell N, Ramírez-Chávez E, Ricart W, Molina-Torres J, Pérez-Luque EL, Fernández-Real JM. Circulating profiling reveals the effect of a polyunsaturated fatty acid-enriched diet on common microRNAs. J Nutr Biochem 2015;26:1095–101. https://doi.org/10.1016/j.jnutbio.2015.05.001.

Palacios C, Gonzalez L. Is vitamin D deficiency a major global public health problem? J Steroid Biochem Mol Biol 2014;144PA:138–45. https://doi.org/10.1016/j.jsbmb.2013.11.003.

Panel on Dietary Antioxidants and Related Compounds, Subcommittee on Upper Reference Levels of Nutrients, Subcommittee on Interpretation and Uses of Dietary Reference Intakes, Standing Committee on the Scientific Evaluation of Dietary Reference Intakes, Food and Nutrition Board, Institute of Medicine. Dietary reference intakes for vitamin C, vitamin E, selenium, and carotenoids. Washington, DC: National Academies Press; 2000. https://doi.org/10.17226/9810.

Peters MJ, Joehanes R, Pilling LC, Schurmann C, Conneely KN, Powell J, Reinmaa E, Sutphin GL, Zhernakova A, Schramm K, Wilson YA, Kobes S, Tukiainen T, Ramos YF, Göring HHH, Fornage M, Liu Y, Gharib SA, Stranger BE, De Jager PL, Aviv A, Levy D, Murabito JM, Munson PJ, Huan T, Hofman A, Uitterlinden AG, Rivadeneira F, van Rooij J, Stolk L, Broer L, Verbiest MMPJ, Jhamai M, Arp P, Metspalu A, Tserel L, Milani L, Samani NJ, Peterson P, Kasela S, Codd V, Peters A, Ward-Caviness CK, Herder C, Waldenberger M, Roden M, Singmann P, Zeilinger S, Illig T, Homuth G, Grabe H-J, Völzke H, Steil L, Kocher T, Murray A, Melzer D, Yaghootkar H, Bandinelli S, Moses EK, Kent JW, Curran JE, Johnson MP, Williams-Blangero S, Westra H-J, McRae AF, Smith JA, Kardia SLR, Hovatta I, Perola M, Ripatti S, Salomaa V, Henders AK, Martin NG, Smith AK, Mehta D, Binder EB, Nylocks KM, Kennedy EM, Klengel T, Ding J, Suchy-Dicey AM, Enquobahrie DA, Brody J, Rotter JI, Chen Y-DI, Houwing-Duistermaat J, Kloppenburg M, Slagboom PE, Helmer Q, den Hollander W, Bean S, Raj T, Bakhshi N, Wang QP, Oyston LJ, Psaty BM, Tracy RP, Montgomery GW, Turner ST, Blangero J, Meulenbelt I, Ressler KJ, Yang J, Franke L, Kettunen J, Visscher PM, Neely GG, Korstanje R, Hanson RL, Prokisch H, Ferrucci L, Esko T, Teumer A, van Meurs JBJ, Johnson AD. The transcriptional landscape of age in human peripheral blood. Nat Commun 2015;6. https://doi.org/10.1038/ncomms9570.

Potischman N. Biologic and methodologic issues for nutritional biomarkers. J Nutr 2003;133:875S–880S. https://doi.org/10.1093/jn/133.3.875S.

Potischman N, Freudenheim JL. Biomarkers of nutritional exposure and nutritional status: an overview. J Nutr 2003;133:873S–874S. https://doi.org/10.1093/jn/133.3.873S.

Pottala JV, Yaffe K, Robinson JG, Espeland MA, Wallace R, Harris WS. Higher RBC EPA + DHA corresponds with larger total brain and hippocampal volumes: WHIMS-MRI study. Neurology 2014;82:435–42. https://doi.org/10.1212/WNL.0000000000000080.

Prasad AS. Clinical, immunological, anti-inflammatory and antioxidant roles of zinc. Experimental gerontology. Zinc Ageing (ZINCAGE Project) 2008;43:370–7. https://doi.org/10.1016/j.exger.2007.10.013.

Qin Y, Xia M, Ma J, Hao Y, Liu J, Mou H, Cao L, Ling W. Anthocyanin supplementation improves serum LDL- and HDL-cholesterol concentrations associated with the inhibition of cholesteryl ester transfer protein in dyslipidemic subjects. Am J Clin Nutr 2009;90:485–92. https://doi.org/10.3945/ajcn.2009.27814.

Rechner AR, Kuhnle G, Bremner P, Hubbard GP, Moore KP, Rice-Evans CA. The metabolic fate of dietary polyphenols in humans. Free Radic Biol Med 2002;33:220–35. https://doi.org/10.1016/S0891-5849(02)00877-8.

Richard M-J, Roussel A-M. Micronutrients and ageing: intakes and requirements. Proc Nutr Soc 1999;58:573–8. https://doi.org/10.1017/S0029665199000750.

Roos CM, Zhang B, Palmer AK, Ogrodnik MB, Pirtskhalava T, Thalji NM, Hagler M, Jurk D, Smith LA, Casaclang-Verzosa G, Zhu Y, Schafer MJ, Tchkonia T, Kirkland JL, Miller JD. Chronic senolytic treatment alleviates established vasomotor dysfunction in aged or atherosclerotic mice. Aging Cell 2016;15:973–7. https://doi.org/10.1111/acel.12458.

Rose M, Veysey M, Lucock M, Niblett S, King K, Baines S, Garg ML. Association between erythrocyte omega-3 polyunsaturated fatty acid levels and fatty liver index in older people is sex dependent. J Nutr Intermed Metab 2016;5:78–85. https://doi.org/10.1016/j.jnim.2016.04.007.

Ross AB, Bourgeois A, Macharia HN, Kochhar S, Jebb SA, Brownlee IA, Seal CJ. Plasma alkylresorcinols as a biomarker of whole-grain food consumption in a large population: results from the WHOLEheart Intervention Study. Am J Clin Nutr 2012;95:204–11. https://doi.org/10.3945/ajcn.110.008508.

Ross SA, Davis CD. The emerging role of microRNAs and nutrition in modulating health and disease. Annu Rev Nutr 2014;34:305–36. https://doi.org/10.1146/annurev-nutr-071813-105729.

Rothwell JA, Fillâtre Y, Martin J-F, Lyan B, Pujos-Guillot E, Fezeu L, Hercberg S, Comte B, Galan P, Touvier M, Manach C. New biomarkers of coffee consumption identified by the non-targeted metabolomic profiling of cohort study subjects. PLoS ONE 2014;9:e93474. https://doi.org/10.1371/journal.pone.0093474.

Ryu D, Mouchiroud L, Andreux PA, Katsyuba E, Moullan N, Nicolet-dit-Félix AA, Williams EG, Jha P, Lo Sasso G, Huzard D, Aebischer P, Sandi C, Rinsch C, Auwerx J. Urolithin A induces mitophagy and prolongs lifespan in *C. elegans* and increases muscle function in rodents. Nat Med 2016;22:879–88. https://doi.org/10.1038/nm.4132.

Ryu M-S, Guthrie GJ, Maki AB, Aydemir TB, Cousins RJ. Proteomic analysis shows the upregulation of erythrocyte dematin in zinc-restricted human subjects. Am J Clin Nutr 2012;95:1096–102. https://doi.org/10.3945/ajcn.111.032862.

Santos MS, Gaziano JM, Leka LS, Beharka AA, Hennekens CH, Meydani SN. Beta-carotene-induced enhancement of natural killer cell activity in elderly men: an investigation of the role of cytokines. Am J Clin Nutr 1998;68:164–70. https://doi.org/10.1093/ajcn/68.1.164.

Schmitz G, Ecker J. The opposing effects of n-3 and n-6 fatty acids. Prog Lipid Res 2008;47:147–55. https://doi.org/10.1016/j.plipres.2007.12.004.

Scientific Opinion on Dietary Reference Values for Zinc. EFSA J 2014;12:3844. https://doi.org/10.2903/j.efsa.2014.3844.

Sfar S, Jawed A, Braham H, Amor S, Laporte F, Kerkeni A. Zinc, copper and antioxidant enzyme activities in healthy elderly Tunisian subjects. Exp Gerontol 2009;44:812–7. https://doi.org/10.1016/j.exger.2009.10.008.

Sheats JL, Winter SJ, King AC. Nutrition interventions for aging populations. In: Bales CW, Locher JL, Saltzman E, editors. Handbook of clinical nutrition and aging, nutrition and health. New York, NY: Springer; 2015. p. 3–19. https://doi.org/10.1007/978-1-4939-1929-1_1.

Shibata M, Ohara T, Yoshida D, Hata J, Mukai N, Kawano H, Kanba S, Kitazono T, Ninomiya T. Association between the ratio of serum arachidonic acid to eicosapentaenoic acid and the presence of depressive symptoms in a general Japanese population: the Hisayama study. J Affect Disord 2018;237:73–9. https://doi.org/10.1016/j.jad.2018.05.004.

Simon JA. Vitamin C and cardiovascular disease: a review. J Am Coll Nutr 1992;11:107–25.

Sorond FA, Lipsitz LA, Hollenberg NK, Fisher ND. Cerebral blood flow response to flavanol-rich cocoa in healthy elderly humans. Neuropsychiatr Dis Treat 2008;4:433–40.

Spencer JPE, Abd El Mohsen MM, Minihane A-M, Mathers JC. Biomarkers of the intake of dietary polyphenols: strengths, limitations and application in nutrition research. Br J Nutr 2008;99:12–22. https://doi.org/10.1017/S0007114507798938.

Tan BL, Norhaizan ME. Carotenoids: how effective are they to prevent age-related diseases? Molecules 2019;24. https://doi.org/10.3390/molecules24091801.

Thomas DR. Vitamins in aging, health, and longevity. Clin Interv Aging 2006;1:81–91. https://doi.org/10.2147/ciia.2006.1.1.81.

Tutino V, De Nunzio V, Caruso MG, Veronese N, Lorusso D, Di Masi M, Benedetto ML, Notarnicola M. Elevated AA/EPA ratio represents an inflammatory biomarker in tumor tissue of metastatic colorectal cancer patients. Int J Mol Sci 2019;20. https://doi.org/10.3390/ijms20082050.

Vauzour D, Vafeiadou K, Rodriguez-Mateos A, Rendeiro C, Spencer JPE. The neuroprotective potential of flavonoids: a multiplicity of effects. Genes Nutr 2008;3:115–26. https://doi.org/10.1007/s12263-008-0091-4.

Walston J, Xue Q, Semba RD, Ferrucci L, Cappola AR, Ricks M, Guralnik J, Fried LP. Serum antioxidants, inflammation, and total mortality in older women. Am J Epidemiol 2006;163:18–26. https://doi.org/10.1093/aje/kwj007.

Wang L, Gaziano JM, Norkus EP, Buring JE, Sesso HD. Associations of plasma carotenoids with risk factors and biomarkers related to cardiovascular disease in middle-aged and older women. Am J Clin Nutr 2008;88:747–54. https://doi.org/10.1093/ajcn/88.3.747.

Wang Y, Chung S-J, McCullough ML, Song WO, Fernandez ML, Koo SI, Chun OK. Dietary carotenoids are associated with cardiovascular disease risk biomarkers mediated by serum carotenoid concentrations. J Nutr 2014;144:1067–74. https://doi.org/10.3945/jn.113.184317.

Warner HR. The future of aging interventions: current status of efforts to measure and modulate the biological rate of aging. J Gerontol A Biol Sci Med Sci 2004;59:B692–6. https://doi.org/10.1093/gerona/59.7.B692.

Wiseman EM, Dadon SB-E, Reifen R. The vicious cycle of vitamin a deficiency: a review. Crit Rev Food Sci Nutr 2017;57:3703–14. https://doi.org/10.1080/10408398.2016.1160362.

Yang C-M, Huang S-M, Liu C-L, Hu M-L. Apo-8′-lycopenal induces expression of HO-1 and NQO-1 via the ERK/p38-Nrf2-ARE pathway in human HepG2 cells. J Agric Food Chem 2012;60:1576–85. https://doi.org/10.1021/jf204451n.

Yousefzadeh MJ, Zhu Y, McGowan SJ, Angelini L, Fuhrmann-Stroissnigg H, Xu M, Ling YY, Melos KI, Pirtskhalava T, Inman CL, McGuckian C, Wade EA, Kato JI, Grassi D, Wentworth M, Burd CE, Arriaga EA, Ladiges WL, Tchkonia T, Kirkland JL, Robbins PD, Niedernhofer LJ. Fisetin is a senotherapeutic that extends health and lifespan. EBioMedicine 2018;36:18–28. https://doi.org/10.1016/j.ebiom.2018.09.015.

Zhang H, Yu D, Sun J, Liu X, Jiang L, Guo H, Ren F. Interaction of plant phenols with food macronutrients: characterisation and nutritional–physiological consequences. Nutr Res Rev 2014;27:1–15. https://doi.org/10.1017/S095442241300019X.

Zhang S(W), Crowther J. Biomarkers for vitamin status and deficiency. In: Targeted biomarker quantitation by LC–MS. John Wiley & Sons, Ltd.; 2017. p. 331–45. https://doi.org/10.1002/9781119413073.ch21

CHAPTER 16

The role of cytomegalovirus in organismal and immune aging

Christopher P. Coplen, Mladen Jergović, and Janko Nikolich-Žugich ☆
Department of Immunobiology, University of Arizona Center on Aging, University of Arizona College of Medicine-Tucson, Tucson, AZ, United States

16.1 Introduction

Cytomegalovirus (CMV) is a ubiquitous beta herpes virus that infects the majority (60%–90%, depending on the population) of humans globally, and that, like the other members of its genus, then persists for life. Significant differences in seroepidemiology exist both between and within populations: incidence of infection is known to range from ∼100% of adults and children in undeveloped countries to <30% in younger populations in other areas globally (Mocarski et al., 2013). The rate of seroprevalence is known to correlate with many factors, including sex, socioeconomic factors, race, and age in both developed and less developed countries. In immunocompetent hosts, both primary CMV infection and reactivation are largely asymptomatic or induce only very limited disease. In sharp contrast, much more severe disease can emerge in hosts with immune deficiencies and in congenital infection. As such, CMV has become widely appreciated as a pathogen of global health concern. For example, in the developing world the incidence of congenital CMV infection is ∼1%–5%. Whether primary or established persistent infection is the major contributor to congenital infection remains poorly understood (Adler, 1989; McVoy and Adler, 1989).

As a likely result of millennia of coevolution with its host, CMV has developed an exceptional capacity for immune evasion (discussed later), and as such has no problem persisting for the lifetime of the host, or even reinfecting a host that already carries it and also has a whole slew of immune mechanisms directed against the virus. Following brief viremia during primary infection, the virus becomes latent in a small subset of broadly distributed cell types (CD34+ hematopoietic progenitors, monocyte-macrophage lineage including the dendritic cells, endothelial cells, and perhaps others), and reaches a homeostatic balance with the host immune system. It is debatable whether the virus in these cells actually establishes true dormancy, and its status has been more

☆ Supported in part by USPHS awards AG052359, AG020719 and the Bowman Professorship in Medical Sciences to J.N-Ž.

recently described as "smoldering infection" (Dupont and Reeves, 2016). Indeed, there is both virological and immunological evidence that the virus remains in a quasi-dormant state with frequent periodical reactivation, perhaps in different parts of the body, whereby CMV will resume transcription, translation, and, if allowed, even replication of infectious virions to facilitate dissemination into additional tissues and/or hosts. A multitude of direct triggers of reactivation have been identified, including bacterial and viral infections, inflammatory cytokines (e.g. tumor necrosis factor-α and interferon(IFN)-γ), ionizing radiation, and other stressors (Sinclair and Sissons, 2006). While these reactivation events are largely subclinical in immunocompetent hosts, exertion on the part of the immune system to restrain the virus is evident—in healthy monozygotic twins discordant for the CMV infection, CMV alone accounts for over 50% of phenotypic, transcriptional, and cytokine variability (Brodin et al., 2015).

16.2 Host immune response to CMV

Traditionally, the host immune response to CMV has been delineated into two chronological phases, corresponding to the two major phases of CMV infection. These phases of infection include the primary acute infection, persistent tissue-localized replication, and multisite latency characterized by percolating reactivation. Distinct arms of the host immune defense are utilized to control the infection during each stage (rev. in La Rosa and Diamond, 2012). Current understanding of the primary acute infection points to localized infection and replication in the exposed mucosal site, leading to dissemination and systemic viral replication in many peripheral tissues. Control during this stage of infection is mediated initially by several subsets of lymphocytes, including the innate immune natural killer (NK) cells, dendritic cells (DCs), and monocytes. In this phase, NK cells and type-I IFN (IFN-I) are particularly important in restricting CMV replication during early infection. In addition to producing IFN-I, DCs also secrete high levels of interleukin(IL)-12 and -18, which have been demonstrated to contribute to the subsequent priming of anti-CMV CD8 and T helper-1 CD4 T cell responses. During the first weeks of infection, the primary effector (particularly CD8) T cell response helps control of systemic CMV replication and is characterized by the canonical expansion and contraction of certain antigen-specific CD4 and CD8 T cell populations specific for the acute-phase CMV antigens. Antibodies against CMV are also produced, although it remains debatable whether they are necessary for the resolution of primary infection and/or control of viral spread in the cases of reactivation.

During the second phase of CMV infection, when the virus establishes latency and then continues to reactivate from the latent niches, T cell responses begin to significantly deviate from those seen in other acute and even persistent infections. Experiments in mice have demonstrated that some CD8 T cell populations contract and form a stable memory pool following the initial weeks of infection, as is the case in typical acute infections. By contrast, other populations do not significantly contract, and instead begin to expand gradually and appreciably in absolute numbers in an antigen-specific manner during this phase of infection. The simplest model for this phenomenon has been postulated to rely upon antigen reencounter via viral reactivation, which elicits proliferation and phenotypic remodeling of the standing pool of CMV-specific central memory CD8 T cells. In this model, reactivation events first drive the expression of immediate early and early viral proteins, and consequently the reexposure of antigen-specific T cells to these viral epitopes, that have been named "inflationary" epitopes or antigens. Their corresponding "inflationary" memory T cells acquire a distinct functional and phenotypic profile, as compared to the CD8 T cell memory compartment seen in other infections (Klenerman, 2018; Newell et al., 2012). They primarily possess an effector-memory phenotype when assessed by flow cytometry, they often downregulate costimulatory molecules like CD27 (in mice and humans) and CD28 (in humans, and occasionally in mice) and, in contrast to the functional exhaustion phenotype that arises in other persistent infections, retain their effector functionality as measured by cytokine release upon stimulation against their cognate epitope, even in old individuals (Doering et al., 2012). "Inflationary" T cells can comprise a significant fraction of the circulating antigen-experienced T cells in a CMV positive person—upwards of 20% of the effector memory T cell compartment can be observed to be directed against a single CMV epitope, and >50% of total memory compartment can be specific for different CMV epitopes in an individual. By contrast, acute "noninflationary" epitopes are not produced unless there is a rare occurrence of full reactivation with viremia (i.e. production of infectious virions), presumably because these "noninflationary" epitopes originate from proteins synthetized late in the viral cycle (Klenerman, 2018).

These dynamics present a challenge to the host, because every cycle of reactivation stimulates an already expanded population of T cells, and, if left unchecked, would lead to uncontrolled accumulation of CMV-specific T cells. While indeed the effector memory compartment increases in absolute numbers in subjects with CMV, and may increase a bit more with aging (although that has not been precisely determined so far), at least two mechanisms combine to curtail this problem and to limit the size of the CMV-specific T cell pool. First, many of the stimulated cells become highly differentiated and consequently lose the ability to proliferate, which in humans is typically marked by two phenotypes: the $CD28^- CD95^{hi} CCR7^- CD45^-$ T effector memory (Tem) and, even more so, the $CD28^- CD95^{hi} CCR7^- CD45^+$ T effector memory cells reexpressing CD45RA (Temra). Both subsets are functional as they produce cytokines and lytic molecules involved in CMV control, and they represent an evolutionary adaptation to virus persistence and reactivation. Moreover, these cells exhibit T cell receptor with lower affinity for antigen in both humans (Griffiths et al., 2013) and mice (Schober et al., 2020), where it has been shown that the clones with lower avidity replace the high-avidity clones over the course of infection. The authors suggested that this may occur via cellular senescence.

This is precisely the opposite to what happens with acute immune responses, where the best affinity/avidity clones tend to dominate. Thereby, the immune system makes sure that it will not be overly stimulated and thereby put out of homeostasis by CMV.

Recent studies have demonstrated that the immunomodulatory functions of CMV may persist during latency both within and outside of the context of reactivation. A substantial fraction of the CMV genome appears to be dedicated to immune evasion that targets host immunity through many mechanisms, including influencing the regulation of CD8 and CD4 T cell function, impairing NK cell killing of infected cells, and altering cytokine-mediated signaling (Ahn et al., 1996; Basta and Bennink, 2003; Benedict, 1999; Jackson et al., 2017). Coupled with the lifelong persistent nature of CMV infection and memory inflation mentioned before, CMV's capacity to evade immune recognition and elimination has led to a number of hypotheses on the impact of CMV on immune aging, immune function over the lifespan, and to human chronic disease in otherwise healthy individuals. These issues are under intense investigation, and the summary of our current knowledge is presented later.

16.3 CMV, longevity, and chronic diseases

The most germane question related to CMV is a simple one—is CMV good or bad for you? Does CMV carriage adversely impact lifespan? If so, is CMV adversely affecting certain chronic disease onset and/or severity? These possibilities were suggested based on several studies that reported associations between CMV and all-cause mortality, cardiovascular diseases, diabetes, frailty, and other diseases of aging in some cohorts; however, other studies found little or no association between CMV and either lifespan or any of the earlier mentioned clinical manifestations (Adler et al., 2011; Blankenberg et al., 2001; Chen et al., 2012; Fagerberg et al., 1999; Gkrania-Klotsas et al., 2012, 2013; Gow et al., 2013; Haeseker et al., 2013; Li et al., 2011; Matheï et al., 2015; Moro-García et al., 2012; Savva et al., 2013; Schmaltz et al., 2005; Simanek et al., 2011; Spyridopoulos et al., 2016; Wang et al., 2010; Zhu et al., 2000).

How could CMV adversely impact health and longevity? Obviously, CMV could reactivate and kill its host, but we know that this simply does not happen in immunocompetent individuals. Indeed, immunity to CMV has been found to be exceptionally robust even in the old age (Čičin-Šain et al., 2011; Lelic et al., 2012), and it is clear that, unlike the immunity to chronic persistent infection, CMV-specific responses show no exhaustion of their functional potential. Another mechanism by which CMV could impact many diseases of aging would be via induction of inflammatory responses upon each occurrence of reactivation. The accumulation of these lifelong bouts of inflammation would then have the potential to produce adverse effects, because elevated inflammation is known to drive or potentiate many of the diseases of aging (Ferrucci et al., 2005). However, evidence linking CMV to an increase in systemic proinflammatory markers remains scarce and controversial.

Recently, Chen, Maier, and colleagues revisited the issue of CMV association between all-cause and cardiovascular mortality. They performed a powerful and conclusive study using five longitudinal cohorts, totaling over 10,000 older community-dwelling adults over a broad range of ages. A large number of covariates were used to adjust for a wide array of confounding factors; CMV status was assessed both as binary seropositive/seronegative (antibodies present or not) value, and also in a quantitative manner, using quartiles of antibody titers, in which case the authors used both the seronegative group, as well as the lowest quartile group, as reference groups against which the highest quartile group was compared. Longitudinal studies included the Leiden Longevity Study F2 (LLS F2, familial longevity study of 2429 >60-year-old offspring of nonagenarians and their partners); the PROSPER study (PROspective Study of Pravastatin in the Elderly at Risk), with 5639 subjects at-risk of cardiovascular disease; the longitudinal study of Danish twins >70 years of age (604 twins); the Leiden 85-Plus study (549 subjects aged 85 and over at enrollment); and the Leiden Longitudinal Study F1—the 901 nonagenarian parents of the LLS F2 cohort described before. Just over a third of all subjects died within the observation period, of which 579 died of cardiovascular causes. CMV seropositivity by itself was not linked to either all-cause or cardiovascular mortality. Initially the highest anti-CMV IgG quartile group showed increased all-cause mortality compared to seronegatives; however, this effect disappeared when the analysis was adjusted for confounders. Of importance, having high anti-CMV IgG (even in the highest quartile) did not correlate to cardiovascular mortality.

Because the abovementioned study dominantly enrolled Caucasian-origin subjects, its results may or may not extend to other ethnic and racial groups. Within that limitation and within the limits of the accuracy of death certificates, this study conclusively demonstrates that CMV is not associated with all-cause or cardiovascular mortality in community-dwelling older adults of dominantly Caucasian origin. Therefore it may be that the associations between CMV and various chronic diseases actually mean that CMV seropositivity or titers are a consequence, and not a cause, of these diseases. Consistent with this, Maier's group had addressed the earlier mentioned hypothesis on CMV-induced inflammation as the potential driver of disease, and found no association between the levels of C-reactive protein as one measure of inflammation, CMV, and cardiovascular and general mortality (Muhlestein et al., 2000). It is always important to remember that CMV is a master in evading the immune response (rev. Jackson et al., 2017) and that this virus may establish latency in only very few of the susceptible cells that it infects (rev. in Leng et al., 2017). All of this is consistent with the behavior of a virus that has adapted exceptionally well to its host over millions of years and that consequently has evolved not to harm it within the confines of a fine dance with its immune system.

16.4 CMV and immune aging

This then brings us to the nuanced and manifold interactions between CMV and the immune system itself. In 1999 Looney, Abraham, and coworkers reported that CMV-seropositive individuals accumulate significant amounts of CD28⁻ memory T cells (Looney et al., 1999). This gave rise to the concept whereby CMV may be not only associated with, but also responsible for, substantial phenomenology associated with aging of the immune system, and in particular, of its T cell lineage, a view originally proposed by Pawelec et al. (2004, 2005, 2010). Intense research followed, and many studies demonstrated beyond any doubt that CMV powerfully modulates many parameters of the immune system (rev in. Nikolich-Žugich et al., 2020). One particularly incisive study on 105 pairs of monozygotic twins (briefly mentioned before) used high-resolution immunophenotyping coupled with cytokine/chemokine analysis (Brodin et al., 2015). Authors showed that monozygotic twins discordant for CMV showed significant differences in 58% of the measured immune parameters, strongly implicating CMV as the major modulator of the human immune system.

The past two decades also brought about studies that also very clearly demonstrated that CMV may compromise various aspects of immune function in older mice and humans (Cicin-Sain et al., 2012; Khan et al., 2002, 2004; Beswick et al., 2012; Mekker et al., 2012; Smithey et al., 2012). However, a careful analysis of immune protection revealed that such changes were rather subtle, and that they failed to lead to true adverse effects, namely to an actual increase in mortality from infections in CMV-infected individuals over mock-infected or uninfected controls (Marandu et al., 2015). Perhaps even more importantly, over the same period of time, evidence had accumulated that CMV may unexpectedly provide benefits to the adult (Barton et al., 2007; Furman et al., 2015), and even aging (Smithey et al., 2018) host, by augmenting immune defense against other infections. All this has led to an increasingly embraced view that CMV interactions with its mammalian host may not have a clear negative impact upon the immune system (rev. in Nikolich-Žugich et al., 2020), and that the relationship between CMV and its immunocompetent hosts abounds in nuances, with some costs and some benefits that on balance appear compatible with long and healthy coexistence.

16.5 Conclusion and future perspectives

It is clear that CMV is an exceptionally well adapted virus to its mammalian host, and evidence so far suggests that it is not pathogenic in the vast majority of its hosts. The virus appears compatible with long life and preserved immune function, and there is evidence that it may help the host combat other infections. However, given the complexities and technical difficulties in studying quantitative aspects of CMV: host interactions, it may be premature to drop our guard on this virus. It has been able to surprise virologists and immunologists over and over, and we therefore eagerly await the next round of studies that will undoubtedly shed further light on CMV's role in aging, health, and diseases. After all, it is worth remembering that the virus has millions of years of evolutionary advantage over any given scientist trying to study it.

References

Adler SP. Cytomegalovirus and child day care. N Engl J Med 1989;321(19):1290–6. https://doi.org/10.1056/NEJM198911093211903.

Adler SP, Best AM, Marshall B, Vetrovec GW. Infection with cytomegalovirus is not associated with premature mortality. Infect Dis Rep 2011;3(2). https://doi.org/10.4081/idr.2011.e17.

Ahn K, Angulo A, Ghazal P, Peterson PA, Yang Y, Früh K. Human cytomegalovirus inhibits antigen presentation by a sequential multistep process. Proc Natl Acad Sci U S A 1996;93(20):10990–5. https://doi.org/10.1073/pnas.93.20.10990.

Barton ES, White DW, Cathelyn JS, Brett-McClellan KA, Engle M, Diamond MS, Miller VL, Virgin IV HW. Herpesvirus latency confers symbiotic protection from bacterial infection. Nature 2007;447(7142):326–9. https://doi.org/10.1038/nature05762.

Basta S, Bennink JR. A survival game of hide and seek: cytomegaloviruses and MHC class I antigen presentation pathways. Viral Immunol 2003;16(3):231–42. https://doi.org/10.1089/088282403322396064.

Benedict CA. Cutting edge: a novel viral TNF receptor superfamily member in virulent strains of human cytomegalovirus. J Immunol 1999;62(12):6967–70.

Beswick M, Pachnio A, Lauder SN, Sweet C, Moss PA. Antiviral therapy can reverse the development of immune senescence in elderly mice with latent cytomegalovirus infection. J Virol 2012;779–89. https://doi.org/10.1128/jvi.02427-12.

Blankenberg S, Rupprecht HJ, Bickel C, Espinola-Klein C, Rippin G, Hafner G, Ossendorf M, Steinhagen K, Meyer J. Cytomegalovirus infection with interleukin-6 response predicts cardiac mortality in patients with coronary artery disease. Circulation 2001;103(24):2915–21. https://doi.org/10.1161/01.CIR.103.24.2915.

Brodin P, Jojic V, Gao T, Bhattacharya S, Angel CJL, Furman D, Shen-Orr S, Dekker CL, Swan GE, Butte AJ, Maecker HT, Davis MM. Variation in the human immune system is largely driven by non-heritable influences. Cell 2015;160(1–2):37–47. https://doi.org/10.1016/j.cell.2014.12.020.

Chen S, De Craen AJM, Raz Y, Derhovanessian E, Vossen ACTM, Westendorp RGJ, Pawelec G, Maier AB. Cytomegalovirus seropositivity is associated with glucose regulation in the oldest old. Results from the Leiden 85-plus study. Immun Ageing 2012;9. https://doi.org/10.1186/1742-4933-9-18.

Čičin-Šain L, Sylwester AW, Hagen SI, Siess DC, Currier N, Legasse AW, Fischer MB, Koudelka CW, Axthelm MK, Nikolich-Žugich J, Picker LJ. Cytomegalovirus-specific T cell immunity is maintained in immunosenescent rhesus macaques. J Immunol 2011;187(4):1722–32. https://doi.org/10.4049/jimmunol.1100560.

Cicin-Sain L, Brien JD, Uhrlaub JL, Drabig A, Marandu TF, Nikolich-Zugich J. Cytomegalovirus infection impairs immune responses and accentuates T-cell pool changes observed in mice with aging. PLoS Pathog 2012;8(8). https://doi.org/10.1371/journal.ppat.1002849.

Doering TA, Crawford A, Angelosanto JM, Paley MA, Ziegler CG, Wherry EJ. Network analysis reveals centrally connected genes and pathways involved in CD8 + T cell exhaustion versus memory. Immunity 2012;37(6):1130–44. https://doi.org/10.1016/j.immuni.2012.08.021.

Dupont L, Reeves MB. Cytomegalovirus latency and reactivation: recent insights into an age old problem. Rev Med Virol 2016;75–89. https://doi.org/10.1002/rmv.1862.

Fagerberg B, Gnarpe J, Gnarpe H, Agewall S, Wikstrand J. Chlamydia pneumoniae but not cytomegalovirus antibodies are associated with future risk of stroke and cardiovascular disease: a prospective study in middle-aged to elderly men with treated hypertension. Stroke 1999;30(2):299–305. https://doi.org/10.1161/01.STR.30.2.299.

Ferrucci L, Corsi A, Lauretani F, Bandinelli S, Bartali B, Taub DD, Guralnik JM, Longo DL. The origins of age-related proinflammatory state. Blood 2005;105(6):2294–9. https://doi.org/10.1182/blood-2004-07-2599.

Furman D, Jojic V, Sharma S, Shen-Orr SS, Angel CJL, Onengut-Gumuscu S, Kidd BA, Maecker HT, Concannon P, Dekker CL, Thomas PG, Davis MM. Cytomegalovirus infection enhances the immune response to influenza. Sci Transl Med 2015;7(281). https://doi.org/10.1126/scitranslmed.aaa2293.

Gkrania-Klotsas E, Langenberg C, Sharp SJ, Luben R, Khaw KT, Wareham NJ. Higher immunoglobulin G antibody levels against cytomegalovirus are associated with incident ischemic heart disease in the population-based EPIC-norfolk cohort. J Infect Dis 2012;206(12):1897–903. https://doi.org/10.1093/infdis/jis620.

Gkrania-Klotsas E, Langenberg C, Sharp SJ, Luben R, Khaw KT, Wareham NJ. Seropositivity and higher immunoglobulin g antibody levels against cytomegalovirus are associated with mortality in the population-based European prospective investigation of cancer-norfolk cohort. Clin Infect Dis 2013;56(10):1421–7. https://doi.org/10.1093/cid/cit083.

Gow AJ, Firth CM, Harrison R, Starr JM, Moss P, Deary IJ. Cytomegalovirus infection and cognitive abilities in old age. Neurobiol Aging 2013;34(7):1846–52. https://doi.org/10.1016/j.neurobiolaging.2013.01.011.

Griffiths SJ, Riddell NE, Masters J, Libri V, Henson SM, Wertheimer A, Wallace D, Sims S, Rivino L, Larbi A, Kemeny DM, Nikolich-Zugich J, Kern F, Klenerman P, Emery VC, Akbar AN. Age-associated increase of low-avidity cytomegalovirus-specific CD8 + T cells that re-express CD45RA. J Immunol 2013;190(11):5363–72. https://doi.org/10.4049/jimmunol.1203267.

Haeseker MB, Pijpers E, Dukers-Muijrers NHTM, Nelemans P, Hoebe CJPA, Bruggeman CA, Verbon A, Goossens VJ. Association of cytomegalovirus and other pathogens with frailty and diabetes mellitus, but not with cardiovascular disease and mortality in psycho-geriatric patients; a prospective cohort study. Immun Ageing 2013;10(1). https://doi.org/10.1186/1742-4933-10-30.

Jackson SE, Redeker A, Arens R, van Baarle D, van den Berg SPH, Benedict CA, Čičin-Šain L, Hill AB, Wills MR. CMV immune evasion and manipulation of the immune system with aging. GeroScience 2017;39(3):273–91. https://doi.org/10.1007/s11357-017-9986-6.

Khan N, Shariff N, Cobbold M, Bruton R, Ainsworth JA, Sinclair AJ, Nayak L, Moss PAH. Cytomegalovirus seropositivity drives the CD8 T cell repertoire toward greater clonality in healthy elderly individuals. J Immunol 2002;169(4):1984–92. https://doi.org/10.4049/jimmunol.169.4.1984.

Khan N, Hislop A, Gudgeon N, Cobbold M, Khanna R, Nayak L, Rickinson AB, Moss PAH. Herpesvirus-specific CD8 T cell immunity in old age: cytomegalovirus impairs the response to a coresident EBV infection. J Immunol 2004;173(12):7481–9. https://doi.org/10.4049/jimmunol.173.12.7481.

Klenerman P. The (gradual) rise of memory inflation. Immunol Rev 2018;283(1):99–112. https://doi.org/10.1111/imr.12653.

La Rosa C, Diamond DJ. The immune response to human CMV. Futur Virol 2012;7(3):279–93. https://doi.org/10.2217/fvl.12.8.

Lelic A, Verschoor CP, Ventresca M, Parsons R, Evelegh C, Bowdish D, Betts MR, Loeb MB, Bramson JL. The polyfunctionality of human memory CD8 + T cells elicited by acute and chronic virus infections is not influenced by age. PLoS Pathog 2012;8(12). https://doi.org/10.1371/journal.ppat.1003076.

Leng SX, Kamil J, Purdy JG, Lemmermann NA, Reddehase MJ, Goodrum FD. Recent advances in CMV tropism, latency, and diagnosis during aging. GeroScience 2017;39(3):251–9. https://doi.org/10.1007/s11357-017-9985-7.

Li S, Zhu J, Zhang W, Chen Y, Zhang K, Popescu LM, Ma X, Bond Lau W, Rong R, Yu X, Wang B, Li Y, Xiao C, Zhang M, Wang S, Yu L, Chen AF, Yang X, Cai J. Signature microRNA expression profile of essential hypertension and its novel link to human cytomegalovirus infection. Circulation 2011;124(2):175–84. https://doi.org/10.1161/CIRCULATIONAHA.110.012237.

Looney RJ, Falsey A, Campbell D, Torres A, Kolassa J, Brower C, McCann R, Menegus M, McCormick K, Frampton M, Hall W, Abraham GN. Role of cytomegalovirus in the T cell changes seen in elderly individuals. Clin Immunol 1999;90(2):213–9. https://doi.org/10.1006/clim.1998.4638.

Marandu TF, Oduro JD, Borkner L, Dekhtiarenko I, Uhrlaub JL, Drabig A, Kröger A, Nikolich-Zugich J, Cicin-Sain L. Immune protection against virus challenge in aging mice is not affected by latent herpesviral infections. J Virol 2015;89(22):11715–7. https://doi.org/10.1128/JVI.01989-15.

Matheï C, Adriaensen W, Vaes B, Van Pottelbergh G, Wallemacq P, Degryse J. No relation between CMV infection and mortality in the oldest old: results from the Belfrail study. Age Ageing 2015;44(1):130–5. https://doi.org/10.1093/ageing/afu094.

McVoy MA, Adler SP. Immunologic evidence for frequent age-related cytomegalovirus reactivation in seropositive immunocompetent individuals. J Infect Dis 1989;160(1):1–10. https://doi.org/10.1093/infdis/160.1.1.

Mekker A, Tchang VS, Haeberli L, Oxenius A, Trkola A, Karrer U. Immune senescence: relative contributions of age and cytomegalovirus infection. PLoS Pathog 2012;8(8). https://doi.org/10.1371/journal.ppat.1002850.

Mocarski ES, Shenk T, Griffiths PD, Pass RF, Cytomegaloviruses. Fields virology. 6th ed. Wolters Kluwer; 2013.

Moro-García MA, Alonso-Arias R, López-Vázquez A, Suárez-García FM, Solano-Jaurrieta JJ, Baltar J, López-Larrea C. Relationship between functional ability in older people, immune system status, and intensity of response to CMV. Age 2012;34(2):479–95. https://doi.org/10.1007/s11357-011-9240-6.

Muhlestein JB, Horne BD, Carlquist JF, Madsen TE, Bair TL, Pearson RR, Anderson JL. Cytomegalovirus seropositivity and C-reactive protein have independent and combined predictive value for mortality in patients with angiographically demonstrated coronary artery disease. Circulation 2000;102(16):1917–23. https://doi.org/10.1161/01.CIR.102.16.1917.

Newell EW, Sigal N, Bendall SC, Nolan GP, Davis MM. Cytometry by time-of-flight shows combinatorial cytokine expression and virus-specific cell niches within a continuum of CD8 + T cell phenotypes. Immunity 2012;36(1):142–52. https://doi.org/10.1016/j.immuni.2012.01.002.

Nikolich-Žugich J, Čicin-Šain L, Collins-McMillen D, Jackson S, Oxenius A, Sinclair J, Snyder C, Wills M, Lemmermann N. Advances in cytomegalovirus (CMV) biology and its relationship to health, diseases, and aging. GeroScience 2020;42(2):495–504. https://doi.org/10.1007/s11357-020-00170-8.

Pawelec G, Akbar A, Caruso C, Effros R, Grubeck-Loebenstein B, Wikby A. Is immunosenescence infectious? Trends Immunol 2004;25(8):406–10. https://doi.org/10.1016/j.it.2004.05.006.

Pawelec G, Akbar A, Caruso C, Grubeck-Loebenstein B, Solana R, Wikby A. Human immunosenescence: is it infectious? Immunol Rev 2005;205:257–68. https://doi.org/10.1111/j.0105-2896.2005.00271.x.

Pawelec G, Akbar A, Beverley P, Caruso C, Derhovanessian E, Fülöp T, Griffiths P, Grubeck-Loebenstein B, Hamprecht K, Jahn G, Kern F, Koch SD, Larbi A, Maier AB, Macallan D, Moss P, Samson S, Strindhall J, Trannoy E, Wills M. Immunosenescence and cytomegalovirus: where do we stand after a decade? Immun Ageing 2010;7. https://doi.org/10.1186/1742-4933-7-13.

Savva GM, Pachnio A, Kaul B, Morgan K, Huppert FA, Brayne C, Moss PAH. Cytomegalovirus infection is associated with increased mortality in the older population. Aging Cell 2013;12(3):381–7. https://doi.org/10.1111/acel.12059.

Schmaltz HN, Fried LP, Xue QL, Walston J, Leng SX, Semba RD. Chronic cytomegalovirus infection and inflammation are associated with prevalent frailty in community-dwelling older women. J Am Geriatr Soc 2005;53(5):747–54. https://doi.org/10.1111/j.1532-5415.2005.53250.x.

Schober K, Voit F, Grassmann S, Müller TR, Eggert J, Jarosch S, Weißbrich B, Hoffmann P, Borkner L, Nio E, Fanchi L, Clouser CR, Radhakrishnan A, Mihatsch L, Lückemeier P, Leube J, Dössinger G, Klein L, Neuenhahn M, … Busch DH. Reverse TCR repertoire evolution toward dominant low-affinity clones during chronic CMV infection. Nat Immunol 2020;21(4):434–41. https://doi.org/10.1038/s41590-020-0628-2.

Simanek AM, Dowd JB, Pawelec G, Melzer D, Dutta A, Aiello AE. Seropositivity to cytomegalovirus, inflammation, all-cause and cardiovascular disease-related mortality in the United States. PLoS ONE 2011;6(2). https://doi.org/10.1371/journal.pone.0016103.

Sinclair J, Sissons P. Latency and reactivation of human cytomegalovirus. J Gen Virol 2006;87(7):1763–79. https://doi.org/10.1099/vir.0.81891-0.

Smithey MJ, Li G, Venturi V, Davenport MP, Nikolich-Žugich J. Lifelong persistent viral infection alters the naive T cell pool, impairing CD8 T cell immunity in late life. J Immunol 2012;189(11):5356–66. https://doi.org/10.4049/jimmunol.1201867.

Smithey MJ, Venturi V, Davenport MP, Buntzman AS, Vincent BG, Frelinger JA, Nikolich-Žugich J. Lifelong CMV infection improves immune defense in old mice by broadening the mobilized TCR repertoire against third-party infection. Proc Natl Acad Sci U S A 2018;115(29):E6817–25. https://doi.org/10.1073/pnas.1719451115.

Spyridopoulos I, Martin-Ruiz C, Hilkens C, Yadegarfar ME, Isaacs J, Jagger C, Kirkwood T, von Zglinicki T. CMV seropositivity and T-cell senescence predict increased cardiovascular mortality in octogenarians: results from the Newcastle 85 + study. Aging Cell 2016;15(2):389–92. https://doi.org/10.1111/acel.12430.

Wang GC, Kao WHL, Murakami P, Xue QL, Chiou RB, Detrick B, McDyer JF, Semba RD, Casolaro V, Walston JD, Fried LP. Cytomegalovirus infection and the risk of mortality and frailty in older women: a prospective observational cohort study. Am J Epidemiol 2010;171(10):1144–52. https://doi.org/10.1093/aje/kwq062.

Zhu J, Quyyumi AA, Norman JE, Csako G, Waclawiw MA, Shearer GM, Epstein SE. Effects of total pathogen burden on coronary artery disease risk and C-reactive protein levels. Am J Cardiol 2000;85(2):140–6. https://doi.org/10.1016/S0002-9149(99)00653-0.

CHAPTER 17

Ethics of aging

Lucia Craxì
Department of Biomedicine, Neuroscience and Advanced Diagnostics (Bi.N.D.), Section of Pathology, University of Palermo, Palermo, Italy

17.1 Population aging: The challenges

With a steady increase in the number of old people worldwide and a much higher percentage of them as part of the overall population, we are on the verge of becoming a mass geriatric society.

In 2018, for the first time in history, persons aged 65 or above outnumbered children under 5 years of age globally. According to the *World Population Prospects 2019* of the United Nations (United Nations and Department of Economic and Social Affairs, 2019), by 2050 one out six people in the world will be over age 65 (16%), up from one out 11 in 2019 (9%), and the number of persons aged 80 years or over is projected to triple, from 143 million to 426 million. By 2050, one out four persons living in Europe and Northern America could be aged 65 or over (World Health Organization, 2015; Bloom, 2011; Christensen et al., 2009).

These gains in extra years are not only a prolongation of individual lives but also result in demographic transitions that will challenge economic, social, and geopolitical systems (housing, nutrition, transportation, infrastructures, etc.) and chiefly our healthcare systems (clinical care, nutrition, care dependence, long-term care and social support, advanced directives, health promotion, cost–effectiveness/economic evaluation) in many countries.

Demographic factors such as population aging are crucial in shaping social, economic, and political trends, but population dynamics alone do not determine the course of human affairs. Population aging calls for the reassessment of existing policies and programs and their realignment to fit both the new demographic reality and the ethical principles and social values that define intergenerational relationships.

Western societies will have to reconsider their former policies and deal with the implications of the growing number of older people, paying attention to the major moral issue of justice between generations and how scarce resources have to be distributed among different age groups. At the same time national governments need to develop new strategies and policies to support people in living not only longer but also healthier lives (World Health Organization, 2017a), considering that older members of society are

Human Aging
https://doi.org/10.1016/B978-0-12-822569-1.00018-4

negatively affected not only by their inevitable physical decline, but also by concomitant social factors such as inability to work, poverty, loss of friends, paternalism, and the pathologizing of aging.

Population changes require equally profound modifications in the way health policies for aging populations are formulated and services are provided. But new policies require first of all to rethink the entire approach to aging and to focus on related ethical issues, in order to make consistent and unvarying choices.

17.2 Moral and social attitudes to old age

17.2.1 The cultural background

Since the attitudes of the members of a society shape the policies that govern it, bias, prejudice, and stereotypes can interfere with effective policy formulation (Butler, 1980) and limit the way problems are conceptualized.

Therefore understanding public perceptions, attitudes, and priorities concerning older adults as members of communities and of a society, and concerning personal, familial, intergenerational, and public responsibilities, is part of the task of an ethical assessment.

Some cultural trends which characterize modern western culture (individualism, cult of youth, distance from death) strongly affect the image of old age that we have as a society. We live in an aging society that doesn't accept aging.

The strong current emphasis on individualism and autonomy as 'independence' has had a negative impact on older people who find themselves increasingly in need of help (Neubauer et al., 2017; Holstein et al., 2010).

Moreover, in a society characterized by the cult of youth and by attempt to get complete control of our body, elders embody our deepest fears: losing control of our body and stopping to be masters of our mind (Kass, 2003).

Last but not least, modern western societies are characterized by an estrangement from death, which started at the beginning of the 20th century, when death stopped being part of everyday's life experience and started happening in hospitals. Together with the physical distance of the places of dying—from the house to the hospital—comes a distance from those who evoke feelings of death, the older persons (Aries, 1974).

All these factors together contribute to fuelling attitudes of real discrimination against older people, as they evoke a multitude of feelings about dependence and mortality (Pope, 1999). This process of discrimination against older people is called ageism, and it is so pervasive and disruptive that dismantling it is one of the goals of the United Nations Sustainable Development Goals for 2030 (United Nations, 2015).

17.2.2 Ageism

First described by Butler (1975) in an analogy to sexism and racism, ageism can be seen as a process of systematic stereotyping and discrimination against people because they are old, just as racism and sexism accomplish this for skin color and gender.

Ageism can take many forms, including prejudicial attitudes, discriminatory practices, or institutional policies and practices that perpetuate stereotypical beliefs. Discrimination against older people is often based on pervasive negative stereotypes and on implicitly (subconscious) or explicitly (conscious) held views on cognitive, affective, and behavioral stereotypes. Old people are categorized as senile, rigid in thought and manner, old-fashioned in morality and skills.

Ageism can be externalized (e.g., by younger people toward older people) or internalized (by older people toward themselves). In the first case it allows the younger generations to see older people as different than themselves; thus they subtly cease to identify with their elders as human beings (Gendron et al., 2018). In the latter case, older people may feel that they are a burden and perceive their lives as less valuable because of their age. It has been shown that ageism can cause lowered levels of self-efficacy, decreased productivity, higher risk for depression and social isolation, and cardiovascular stress (Levy et al., 1999–2000). Moreover, these stereotypes can become a self-fulfilling prophecy, reinforcing the inaction and deficits that result from their internalization.

These negative stereotypes are so pervasive that even those who outwardly express the best of intentions may have difficulty avoiding engaging in negative actions and expressions. People may, indeed, express idealized positive attitudes to old age but in everyday life act according to hidden negative attitudes toward older people.

Furthermore, negative ageist attitudes are often seen as humorous and based in some degree of fact; thus the humor is often mistakenly assumed to counteract any negative effects on the older person.

17.2.3 Vulnerability

The notion of vulnerability is usually described as an increased susceptibility to harm. More restrictively it can be considered a specific context-dependent susceptibility to harm and exploitation, and a limited capacity of autonomy (Goodin, 1985; Macklin, 2003).

The traditional concept of vulnerability as a necessary marker of old age has been questioned: old age as such is not a sufficient criterion for being considered vulnerable (Bozzaro et al., 2018). Applying this concept of vulnerability to all older people simply due to their advanced chronological age is problematic for several reasons: first, because it stems from and consolidates negative stereotypes of aging; second, because vulnerability is often understood only in terms of failing to attain or retain autonomous agency, and third, because aging is not a constant state but a process that can develop in a variety of different ways depending on an individual's resources as well as on societal, political, financial, and cultural framework conditions (Levine et al., 2004).

Therefore old age alone should not be considered a general marker of vulnerability. Rather, it should encourage special attention in order to determine whether older people are more susceptible to experiencing manifestations of vulnerability for some specific reasons, which could include their physical or cognitive constitution, or perhaps their social situation. As a matter of fact, such a differentiated perspective actually seems to be taking shape in the current bioethical discourse (Rogers et al., 2012; Hurst, 2008; Martin et al., 2014).

17.3 Ethics of aging

17.3.1 A new subfield of bioethics

In the last decades bioethics has been focused mainly on the issues of the beginning and of the very end of life, but the 20 or 30 years between retirement and death are still very incompletely theorized (Holm, 2013). On the contrary, such an important area of human life requires the creation of a specific ethical approach, to deal with contemporary issues at the cutting edge of aging, responsibility, and the good life (World Health Organization, 2017b). While well-known ethical frameworks in clinical practice, research, and in public health exist and could be used as a starting point for developing a framework for healthy aging, the important differences and nuances relevant to aging must first be identified and explored.

Recently the "Ethics of ageing" is starting to emerge as a specific subfield of applied ethics and a few boundary works attempted to set the scope for this field of investigation (Wareham, 2018; Fenech, 2003; Jecker, 1991), trying to sketch the range of issues that could be covered and to clarify its conceptual boundaries. Nonetheless, there is still a long way to go to reach a single definition of what the scope of this subfield is and a preliminary investigation into the very concepts of age and aging in the ethical field is still needed.

17.3.2 Age and aging

Before discussing the problems of older people, a definition of the "older person" is required. Various thresholds have been established by the scientific community and different groups have been identified (e.g., "younger old" and "older old"), but a comprehensive definition of old age in bioethics is still needed, since chronological age is too simplistic a way of defining groups of older people.

In 1977 George Engel published on Science a classical article: *The need for a new medical model: a challenge for biomedicine*. In his article he highlighted that "The dominant model of disease today is biomedical, and it leaves no room within its framework for the social, psychological, and behavioral dimensions of illness" (Engel, 1977). That's particularly true when we talk about aging, a physiological phenomenon with pathological implications.

As a proof of the inadequacy of biomedical model alone in the ethics of aging we have to work with many different definitions of age (chronological age, biological age, social age, functional age, and subjective age) and with different definitions of aging, going beyond a definition of aging as a purely biological phenomenon (López-Otín et al., 2013) and considering it as a complex, multidirectional process that encompasses maintenance, growth, and decline as well as cultural factors that influence development (Baltes, 1987).

17.3.3 Field of investigation

When we think about ethics of aging it's not surprising that many of us think about well-known ethical issues associated with end-of-life care and dying. Of course, an ethics of aging will explore questions related to end-of-life decision-making and the challenges that can arise, particularly in a healthcare setting, when addressing conditions commonly associated with older age. But aging is a part of life, therefore the focus of the ethics of aging cannot be so narrow.

As aging is a socially and biologically fundamental area, the range of ethical issues covered by this subfield could be very wide and can be roughly grouped into issues concerning justice and intergenerational relationships and issues related to autonomy and dignity. Some areas immediately stand out, such as dementia and decision-making capacity, assistive medical technology, use of digital media in the care of older people, intergenerational ethics (with special attention to healthcare rationing), and healthy environments; other problems may need to be identified, and still require conceptual and ethical exploration.

The entire range of issues still needs to be explored, in order to increase awareness on ethical issues related to aging, help to reshape moral and social attitudes to old age, and build a proper ethical framework that could be a decision tool for stakeholders, including policy-makers and care providers.

17.4 Fair allocation of medical resources

Population aging, along with the increasing demands in health resources and their growing cost, has brought to the fore the issue of justice between generations and how scarce resources should be distributed among different age groups.

The issue of resource allocation is one of the widest, most debated, and complex in the ethics of aging; it extends far beyond the scope of access to health resources—including housing, pensions, infrastructures, etc.—and it covers both the issue of rights and duties between generations, and the right to good aging.

Good aging requires, indeed, a focus on the circumstances and technologies that may impact on the length of life and well-being of the aging person; therefore a fair allocation of resources impacts deeply on the process of good aging itself, bringing into play the issue of the social determinants of health.

Experiences of older adults differ vastly due to social determinants of health that include the role of wealth throughout life in shaping access to services and environments in late life. While bioethics has tended to focus on access to healthcare and on health financing, policy research suggests that reducing aging-related inequalities calls for attention to housing, the built environment, social spending, and the role of the private sector and of community members in planning and designing decisions.

17.4.1 Different approaches to the allocation of medical resources

Resource allocation is an issue of distributive justice. Justice is concerned with the equitable distribution of benefits and burdens to individuals in social institutions, and how the rights of various individuals are realized (Spagnolo et al., 2004; Forni, 2016).

The approach to the issue of fair distribution of healthcare resources differs considerably according to the country, depending on the cultural context, the political system, and their underlying values and principles.

For historical and cultural reasons, countries like the United States have a healthcare system based on individualistic liberalism (conservative liberalism). According to this approach, as freedom is the main value, the government's role is protecting individual liberties and therefore giving space to free market: the healthcare system is private and access to care depends on the ability to pay (through private insurances) and on individual choice (Bodenheimer, 2005). In this context, the risk of harm to potentially vulnerable populations such as older people is unquestionably higher. In order to protect people's right to equally receive essential services such as healthcare, communitarian liberals conceived in the 60's Medicare (Falk, 1973)—an insurance plan offered by the federal government—and later expanded it with the Affordable Care Act, a comprehensive healthcare reform law enacted in 2010. Even though Medicare represents a step forward in the protection of older people, it still has many critical points. Inspired by the idea of coresponsibility, Medicare provides for a form of copayment for a large part of the interventions covered, which means that many destitute older persons, unable to take out supplementary private policies, remain uninsured for important health services. Moreover, Medicare doesn't cover the expenses of many interventions that could be important for the well-being of older people (e.g., dental care) and others that could be essential, such as long-term care.

Different cultural backgrounds and historical paths pushed many European countries to a diverse approach to the issue of justice and fair allocation of resources. After the Second World War they developed a political system mainly based on an egalitarian approach that led to the birth of the Welfare State. Accordingly, countries such as Italy, Sweden, and Great Britain later developed public healthcare systems based on universal health coverage (Sidel, 1978).

According to an egalitarian approach, equality (all humans are equal in fundamental worth regardless of sex, race, age, etc.) is both the main value and the most important goal to achieve, when dealing with the essential needs of every human being. Every individual, consequently, has the right to the highest attainable standard of health (Comitato Nazionale per la Bioetica, 2001) and the government should, therefore, take action to guarantee access to care and the same opportunities for everyone. According to this approach, equity is both an ethical norm and a requirement for protection against discrimination if there is status denigration because of age, sex, race, ethnicity, or social economic status.

When healthcare resources are sufficient for the needs of all relevant patients, egalitarianism adopts a "first come, first served" criterion, which allocates resources preferentially to those arriving first for medical attention.

But what happens if resources are not sufficient and their availability can't be increased? According to the principle of temporal neutrality (Savulescu and Goold, 2008), when a harm occurs it should not make a moral difference, and therefore adopting the "first come, first served" criterion would be unfair. Therefore the selection of patients for access or prioritization to a medical intervention becomes needs based, which means that it gives priority to the medical need.

On this basis, older patients are to be considered as any other individual and an age-based selection would be unfair and discriminatory (Fenech, 2003). In order to respect the principle of equity, health providers and health systems should therefore be able to identify and address the distinct care needs of aged adults (McNally and Lahey, 2015).

A needs-based approach, though fair, has got—just like any other approach—its shortcomings. The choice that privileges the patients who are most in need, indeed, is not always the best choice for a specific individual.

With a growing number of older patients, an increasing demand in health resources, and a steady increase in drug prices, many countries are facing a huge problem of distributive justice. On the one hand, they need to be able to guarantee individual rights, among which the right to health in its broader sense; on the other hand, they need to provide equal access to the resources and keep the distribution system sustainable.

17.4.2 Strategies to reduce rising healthcare costs for older people

In order to reduce rising healthcare costs for older people, many different strategies have been proposed (World Health Organization, 2017b). As the highest costs in any person's life are usually generated during the last year of life and many people prefer not to be treated aggressively at any cost, limiting life-prolonging treatment according to individual preferences might be an acceptable strategy. Another strategy would be to focus on healthy aging and disease prevention throughout the life course in order to prevent chronic diseases in late life.

Nowadays by using better diagnostic instruments, we can detect more abnormalities or detect them at an earlier stage; however, there is no evidence that all "abnormalities" must be treated and some of them may be irrelevant for patients, especially for older people. Moreover, diagnostic instruments or algorithms may produce misleading empirical data because the normal physiological values of older people vary widely. Thus, in high-income countries, older people undergo more diagnostic tests and, consequently, appear to need more healthcare, even though their physiological and functional capacities are within the normal range. In contrast, some conditions, such as frailty, are not diagnosed as diseases but should be considered just as relevant for medical care and prevention. The individual body parts may not be diseased, but lack of reserves makes the body vulnerable. Older people must therefore receive comprehensive geriatric assessments, focusing on function rather than on disease (Cesari et al., 2016). Thus a healthcare system for older people could be inadequate if it is based on a single disease or chronological age. We should shift to novel, integrated models of care that focus on meaningful outcomes for patients, such as function instead of disease.

17.4.3 Conditions of dramatically scarce medical resources

In conditions of dramatically scarce medical resources, additional criteria—inspired to a "soft" utilitarian approach—could be introduced to govern access and prioritization, in order to maximize the achievable benefit in terms of lives saved, thus optimizing the use of available resources (Craxì et al., 2020).

This is a basic principle endorsed by triage in settings of overwhelming medical need (e.g. disasters, battlefields, or pandemics) and also applied in other fields of dramatically scarce resources, such as organ transplants (Persad et al., 2009). It is supported by both popular intuition and multiple ethical theories (Arora et al., 2016; Emanuel et al., 2020).

According to utilitarianism, resources should be distributed to bring about the most good: the greatest good to the greatest number. But nonutilitarian theories can also recognize the importance of this principle. According to the egalitarianist approach proposed by Rawls (1971), the right distribution is the one we would choose from behind a "veil of ignorance," that is if we did not know who we would be in society. From behind the veil, rational self-interest requires that you choose the policy that gives you the greatest chance of surviving. We should save more lives rather than fewer, other things being equal. We can call this the moral requirement to save the greatest number (Savulescu et al., 2020).

The utilitarian criterion to maximize the lives saved is different from the "Fair Innings" approach. This latter approach reflects the idea that everyone is entitled to a span of life years that we consider a reasonable life for a person to have had, a fair innings (Harris, 1970; Callahan, 1987). According to this argument, younger persons have stronger claims to life-saving interventions than older persons, because they have had fewer opportunities to experience life. The implication is that saving 1 year of life for a young person is valued more than saving 1 year of life for an older person.

The utilitarian approach is completely different from the fair innings approach because, given that the survival benefit for an intervention is the same for young and older patients, it grants the opportunity to have access to older patients independently from their age. Using patient characteristics (such as age or frailty) to estimate prognosis, in order to maximize the number of lives saved, is not discriminatory, unless a characteristic is used to systematically disadvantage a group. For example, age per se (without consideration of prognosis) would be ageist and arguably unlawful discrimination. However, using probability of survival is an ethically defensible criterion.

If we accept the criterion of maximizing the lives saved according to the probability of survival in settings of overwhelming medical need, we need to be aware that the hardest part is finding out reliable markers to predict the probability of survival. As biological aging is a result of an accumulation of a wide variety of molecular and cellular damage over time, biological age might be informative about the health of individuals; however, there is no reliable biomarker of biological age. Other markers such as chronological age or frailty could instead not be reliable enough. Moreover, consideration of ethical and legal themes relating to frailty must engage with the concern that frailty may be a pejorative concept that validates and reinforces the disadvantage and vulnerability of aged adults.

Only a combination of different markers and a proper scoring system could be therefore complete and could prevent unjustified discriminations.

17.5 Conclusion and future perspectives

In the coming decades the world will face an unprecedented change in the composition of its population that will reshape and challenge social and political systems. Unique in human history, this change, unlike most of the changes that societies will experience during the next 50 years, is largely predictable.

Population aging, with advances in medical technology, ensures that the allocation problem will continue to affect societies as far as one can see into the future. Currently this issue has mostly been formulated in terms of challenges created by increasing costs, and the focus has been squarely on life-prolonging treatments. In doing so, we're missing many other important issues which should be examined and need a proper ethical analysis.

Many considerations still need further investigation such as changing expectations of medical care in older age, ensuring real possibilities for the participation of older people in social life, promoting an age-friendly environment that supports older people, ensuring the absence of discrimination and abuse in both personal relations and social structures, having a deep understanding of the life course, especially the meaning of older age, and respect for its special existential dimensions.

As a society we will be responsible for ensuring a fair distribution of rights, burdens, and responsibilities between generations and giving the opportunity to older people to live fully a phase of life that should not only be seen as a long wait for death.

Policies should aim to ensure that the fundamental rights of each individual are safeguarded and, in this respect, a key element will be effective planning. COVID-19 pandemic, for instance, made it dramatically clear that we are unable to effectively protect our most vulnerable older persons without effective planning and appropriate investment in long-term care facilities, where a real massacre of innocents took place (Werner et al., 2020; Burki, 2020).

Effective planning and fair policies should be based on a deep awareness of the priorities to be assigned, of the vulnerabilities to be safeguarded, but also of the prejudices that could affect our choices. Hence, an extensive ethical assessment on the issues related to aging will be essential to clarify what values we want to defend as a society and what principles we want to respect, in order to implement transparent and consistent policies.

References

Aries P. Western attitudes towards death. Baltimore: The Johns Hopkins University Press; 1974.

Arora C, Savulescu J, Maslen H, Selgelid M, Wilkinson D. The intensive care lifeboat: a survey of lay attitudes to rationing dilemmas in neonatal intensive care. BMC Med Ethics 2016;17:69–77.

Baltes PB. Theoretical propositions of lifespan developmental psychology: on the dynamics between growth and decline. Dev Psychol 1987;23:611–26.

Bloom DE. 7 billion and counting. Science 2011;333(6042):562–9. https://doi.org/10.1126/science.1209290.

Bodenheimer T. The political divide in health care: a liberal perspective. Health Aff 2005;24(6):1426–35.

Bozzaro C, Boldt J, Schweda M. Are older people a vulnerable group? Philosophical and bioethical perspectives on ageing and vulnerability. Bioethics 2018;32(4):233–9.

Burki T. England and Wales see 20000 excess deaths in care homes. Lancet 2020;395(10237):1602.

Butler RN. Why survive? Being old in America. New York: Harper & Row; 1975.

Butler RN. Ageism: a foreword. J Soc Issues 1980;36(2):8–11.

Callahan D. Setting limits. Medical goals in an aging society. New York: Simon and Schuster; 1987.

Cesari M, Marzetti E, Thiem U, Pérez-Zepeda MU, Abellan Van Kan G, Landi F, et al. The geriatric management of frailty as paradigm of "The end of the disease era" Eur J Intern Med 2016;31:11–4.

Christensen K, Doblhammer G, Rau R, Vaupel JW. Ageing populations: the challenges ahead. Lancet 2009;374(9696):1196–208. https://doi.org/10.1016/S0140-6736(09)61460-4.

Comitato Nazionale per la Bioetica. Orientamenti Bioetici per l'Equità nella Salute, http://bioetica.governo.it/media/3585/p49_2001_equita-nella-salute_it.pdf; 2001. [Accessed 19 July 2020].

Craxì L, Vergano M, Savulescu J, et al. Rationing in a pandemic: lessons from Italy. Asian Bioeth Rev 2020; https://doi.org/10.1007/s41649-020-00127-1. Epub ahead of print.

Emanuel EJ, Persad G, Upshur R, Thome B, Parker M, Glickman A, et al. Fair allocation of scarce medical resources in the time of Covid-19. N Engl J Med 2020;382:2049–55.

Engel GL. The need for a new medical model: a challenge for biomedicine. Science 1977;196:129–36.

Falk IS. Medical care in the U.S.A.: 1932–1972. Milbank Q 1973;51(1):1–32.

Fenech FF. Ethical issues in ageing. Clin Med 2003;3:232–4.

Forni L. La sfida della giustizia in sanità. Salute, equità, risorse. Torino: G. Giappichelli editore; 2016.

Gendron TL, Inker J, Welleford EA. A theory of relational ageism: a discourse analysis of the 2015 white house conference on aging. Gerontologist 2018;58(2):242–50.

Goodin R. Protecting the vulnerable. Chicago: University of Chicago Press; 1985.

Harris J. The value of life, an introduction to medical ethics. London: Routledge & Keegan Paul; 1970.

Holm S. The implicit anthropology of bioethics and the problem of the aging person. In: Schermer M, Pinxten W, editors. Ethics, health policy and (anti-) aging: mixed blessings. Dordrecht: Springer; 2013. p. 59–71.

Holstein M, Parks J, Waymack M. Ethics, ageing and society: the critical turn. New York: Springer; 2010.

Hurst SA. Vulnerability in research and health care; describing the elephant in the room? Bioethics 2008;22 (4):191–202.

Jecker NS, editor. Aging and ethics: philosophical problems in gerontology. New York: Springer; 1991.

Kass LR. Ageless bodies, happy souls: biotechnology and the pursuit of perfection. New Atlantis 2003;352:9–28.

Levine C, Faden R, Grady C, Hammerschmidt D, Eckenwiler L, Sugarman J, et al. The limitations of 'vulnerability' as a protection for human research participants. Am J Bioeth 2004;4(3):44–9.

Levy B, Ashman O, Dror I. To be or not to be: the effects of aging stereotypes on the will to live. Omega (Westport) 1999–2000;40(3):409–20.

López-Otín C, Blasco MA, Partridge L, Serrano M, Kroemer G. The hallmarks of aging. Cell 2013;153:1194–217.

Macklin R. Bioethics, vulnerability and protection. Bioethics 2003;17(5–6):472–86.

Martin A, Tavaglione N, Hurst S. Resolving the conflict: some clarifications of vulnerability in health care. Kennedy Inst Ethics J 2014;24(1):51–72.

McNally M, Lahey W. Frailty's place in ethics and law: some thoughts on equality and autonomy and on limits and possibilities for aging citizens. Interdiscip Top Gerontol Geriatr 2015;41:174–85.

Neubauer AB, Schilling OK, Wahl HW. What do we need at the end of life? Competence, but not autonomy, predicts intraindividual fluctuations in subjective well-being in very old age. J Gerontol B Psychol Sci Soc Sci 2017;72(3):425–35.

Persad G, Wertheimer A, Emanuel EJ. Principles for allocation of scarce medical interventions. Lancet 2009 Jan 31;373(9661):423–31.

Pope A. The elderly in modern society: a cultural psychological reading. Janus Head 1999;1:223–32.

Rawls J. A theory of justice. Cambridge: Harvard University Press; 1971.

Rogers W, Mackenzie C, Dodds S. Why bioethics needs a concept of vulnerability. Int J Fem Approaches Bioeth 2012;5(2):11–38.

Savulescu J, Goold I. Freezing eggs for lifestyle reasons. Am J Bioeth 2008;8:32–5.

Savulescu J, Vergano M, Craxì L, Wilkinson D. An ethical algorithm for rationing life sustaining treatment during the COVID-19 pandemic. Br J Anaest 2020;1. https://doi.org/10.1016/j.bja.2020.05.028. Epub ahead of print.

Sidel VW. The right to health care: an international perspective. In: Bandman EL, Bandman B, editors. Bioethics and human rights. Boston: Little, Brown and Company; 1978. p. 341–50.

Spagnolo AG, Sacchini D, Pessina A, Lenoci M. Etica e giustizia in sanità. Questioni generali, aspetti metodologici e organizzativi. Mac Graw Hill: Milano; 2004.

United Nations. Sustainable development goals for 2030, https://sustainabledevelopment.un.org/sdgsproposal.html; 2015. [Accessed 16 July 2020].

United Nations, Department of Economic and Social Affairs. World population prospects 2019, https://population.un.org/wpp/Publications/Files/WPP2019_Highlights.pdf; 2019. [Accessed 16 July 2020].

Wareham CS. What is the ethics of ageing? J Med Ethics 2018;44(2):129.

Werner RM, Hoffman AK, Coe NB. Long-term care policy after covid-19—solving the nursing home crisis. N Engl J Med 2020. https://doi.org/10.1056/NEJMp2014811. Epub ahead of print.

World Health Organization. World report on ageing and health, https://apps.who.int/iris/bitstream/handle/10665/186463/9789240694811_eng.pdf;
jsessionid=5C8AE39281641A3FFB4284F52A89834A?sequence=1; 2015. [Accessed 16 July 2020].

World Health Organization. Global strategy and action plan on health and ageing, https://www.who.int/ageing/WHO-GSAP-2017.pdf?ua=1; 2017. [Accessed 16 July 2020].

World Health Organization. Developing an ethical framework for healthy ageing, https://apps.who.int/iris/bitstream/handle/10665/259932/WHO-HIS-IER-REK-GHE-2017.4-eng.pdf?sequence=1; 2017. [Accessed 16 July 2020].

CHAPTER 18

Conclusions. Slowing aging and fighting age-related diseases, from bench to bedside?

Calogero Caruso and Giuseppina Candore
Laboratory of Immunopathology and Immunosenescence, Department of Biomedicine, Neurosciences and Advanced Diagnostics, University of Palermo, Palermo, Italy

18.1 Introduction

Nowadays, individuals in developed countries live much longer today than in the past. However, they are not exempt from the aging process. Many countries show an increase in the older population with a concomitant increase in the prevalence of age-related diseases and health expenditure. Over the course of the 20th century, life expectancy at birth increased by 30 years in developed countries, initially due to the reduction in neonatal and infant mortality, and, then, due to the decline in mortality in middle and older age. In 1900 about 40% of children born in countries for which reliable data exist had a life expectancy over 65 years. Today, in these same countries, more than 88% of all newborns will live longer than 65 years and at least 44% will live over 85 years. This increased life expectancy is closely associated with socioeconomic development which has enabled greater availability of food and clean water, better housing and living conditions, reduced exposure to infectious diseases, and access to medical care. It has also been suggested that a key role is represented by improved education that might help the individual to make healthier lifestyle choices (Oeppen and Vaupel, 2002; Christensen et al., 2009; Lutz et al., 2019; Accardi et al., 2019). A comprehensive analysis of all the demographic aspects of the aging of the world population is covered in detail in Chapter 2, including theories concerning both epidemiological and demographic transition. However, the rapid increase of older people is nonetheless accompanied by an increase in the number of people with chronic age-related diseases. The mechanisms responsible for many age-related diseases are discussed in Chapters 3 and 5, whereas cellular senescence, as a contributing factor, is treated in Chapter 4. Senescent cells can, indeed, accumulate at etiological sites of the most common age-related diseases throughout the lifespan, and probably there is a lag between senescent cell accumulation and development of functional effects.

Older people are less resistant to environmental insults and pathological stimuli, including infections. In our society, the perception of the phenomenon of aging

Human Aging
https://doi.org/10.1016/B978-0-12-822569-1.00001-9

comprises a decreased ability to survive due to chronic diseases and the combined loss of mobility, sensory and cognitive functions with an exponential increase in healthcare costs, linked to the greater number of older people in the Western world. Preventive therapeutic interventions are therefore urgently needed to slow down aging (Longo et al., 2015). Thus the aging of the population is a challenge for many societies, because it is responsible for the increase in public spending in pension, health, and social security programs for older people. In many countries, the already high public spending limits the possibility of an increase in spending linked to aging. That could damage the economic growth and the general quality of life of the remaining population if governments must divert public spending from investments in education and infrastructure to finance programs for older people. In this context, Chapter 17 discusses how relevant and timely political solutions are needed to ensure economic sustainability for the health and well-being of citizens of all ages. Effective planning and fair policies should be based on a deep awareness of the priorities to be assigned, of the vulnerabilities to be safeguarded, but also of the prejudices that could affect our choices. Therefore an extensive ethical assessment on the issues related to aging will be essential. Hence, the time has come to seriously address how to slow down human aging, in other words to study how to age successfully. It has been calculated that if the age-related risk of death, frailty, and disability could be reduced by about half at all ages, in the future, people who will reach the age of 60 may look like current 53-year-olds, and so on. On the other hand, if one ages in good health, there would be a net gain for society in social, economic, and health terms (Olshansky et al., 2007; Farrelly, 2008).

Frailty is a multisystem impairment which makes an individual more vulnerable to external or internal stressors. Sarcopenia, the age-dependent loss of muscle mass and function, is considered as the biological substrate whereby the consequences of physical frailty develop. Sarcopenia severely impairs the individual quality of life, increasing the occurrence of falls, disability, institutionalization, and leading to increased mortality. Chapter 10 focuses on sarcopenia as a multifactorial syndrome, discussing the major contributors to its development, and the possible management of sarcopenia as well as the new advances in treatments.

Understanding mechanisms of aging (Chapter 1) as well as of age-related diseases (Chapters 3, 4, and 5) must play a leading role in the search for new strategies to increase the health of an older population highly susceptible to the diseases of aging. As discussed in this book, studying the pathophysiology of aging can provide important clues as to how to develop drugs or lifestyle changes that can slow or delay aging or at least prevent age-related diseases. One of the greatest deficiencies in our current knowledge is a poor understanding of the relative importance of the different genetic, epigenetic, and environmental variables, as well as serendipity, in favoring the predisposition to diseases and conditioning the quality of aging. Careful phenotyping of different aging patterns, the collation of genetic data, and the current "explosion" of molecular genetics techniques and big data may soon add important missing pieces to the aging puzzle. Several factors, processes, and mechanisms seem capable of determining the aging process, but the reductionist nature of experimental techniques is still focused on the study of individual mechanisms. This is very limiting since the synergy and interaction between the mechanisms are probably important. Consequently, the research objectives should no longer tend to identify a single cause (e.g., a single gene or the decline of a system crucial for the survival of an organism). It is believed that different causes and processes are likely to interact simultaneously and operate at different levels of functional organization. Alterations of molecular events associated with aging can in turn lead to cellular changes, and the latter in turn can contribute to changes in organs and systems. This has allowed, in multicellular and complex organisms, the study of the interaction between intrinsic (genetic), extrinsic (environmental), and stochastic causes (random damage in vital molecules), providing a useful approach to understanding the aging process and longevity. Hence, a systems-biology approach becomes essential. Understanding human aging also requires an integrated approach able to combine, beyond purely biological studies, studies concerning various social sciences, including demographic, historical, anthropological disciplines. Researches conducted on long-lived individuals have in fact shown that, in comparison with the less old people, they have a higher level of education, better economic conditions, better family support, and fewer negative individual experiences (Caruso, 2019). Furthermore, as discussed in the next paragraph, also the role of gender/sex clearly has to be considered.

18.2 Aging and gender medicine

Gender medicine is a new science that represents an important prospect for the third millennium. Over the past 20 years, several studies have unequivocally documented the existence of the effect of gender in biomedicine, i.e., differences in pathophysiology, clinical manifestation of the disease, response to treatments, in relation to gender. This highlights the need for medicine to consider women as entities different from men, not underestimating the biological-hormonal and anatomical differences. Gender medicine takes into account sexual dimorphism and, therefore, studies the influence of sex, in the biological meaning (linked to hormones and sex chromosomes), and of gender, in the social meaning (linked to education and social conventions), on the differences between men and women, in terms of prevention of diseases, manifestation of clinical signs, therapeutic approach, prognosis, psychological and social impact. It is now known that there are physiological differences between men and women in the response to the etiological agents of diseases, although it has been argued that, without absolute social equity, it is not possible to establish the real difference between men and women (Baggio et al., 2013; Machluf et al., 2020).

The most obvious difference in medicine between men and women is the one reported in Chapter 2, i.e., that women live longer than men. In the Western world, women live 5–6 years longer than men and, therefore, 85% of centenarians are female. Although this difference is considered the norm, it is a relatively recent demographic phenomenon due to the reduction of infections and the increase in the death rate linked to cancer and cardiovascular disease. In particular, cardiovascular disease is the main condition associated with excess male mortality, especially in the birth cohorts of 1900–35 (Beltrán-Sánchez et al., 2015). Despite the fact that women live longer than men, the healthy life expectancy is the same in both sexes, so the 5-year advantage for women are mostly years of disability, mainly due to the consequences of cardiovascular, osteoarticular, and neurological diseases (Freedman et al., 2016).

It is debated whether women live longer than men for reasons of sex or gender, that is, for biological or cultural differences. However, given that even in other animal species, females often live longer than males, this phenomenon cannot be attributed solely to cultural behavior, but should be due, at least in part, to biological causes, such as the presence of the two X chromosomes or the production of different hormones. For instance, testosterone increases low-density lipoprotein levels and reduces high-density lipoprotein levels, while estrogen has the opposite effect with beneficial effects on the risk of cardiovascular disease. Another possibility is that women, due to their menstrual cycle, have a relative iron deficiency for about 30–40 years compared to men. Iron is a crucial catalyst for the production of reactive oxygen species (ROS), as an effect of metabolism, by the mitochondria. A reduction in available iron would result in a lower production of ROS. Indeed, diets rich in iron have been associated with a significantly higher risk of heart disease. It follows that women suffer from cardiovascular diseases such as heart attack or stroke, about 10 years later than men. Moreover, relative iron deficiency is known to protect against bacterial infections (Candore et al., 2010; Caruso et al., 2013).

Testosterone could be responsible for aggressive behaviors and a risky lifestyle such as dangerous driving, alcohol abuse, the use of weapons, all behaviors that can increase the mortality rate. Hence, cultural factors also contribute to aging and longevity. Women tend to have a quieter activity within the family or at work, with less risky lifestyles for their health. They smoke less, eat less, cope with stress better, take better care of their body and health. These are gender differences, that is, the characteristics that society delineates as male or female. They are valid in older generations today, but not in future ones, as girls today tend to copy many deleterious aspects of male culture (aggression, alcoholism, and smoking) (Caruso et al., 2013).

Thus gender differences can advantage women. However, gender differences may also disadvantage women. Both financial and cultural factors may be responsible for reduced consumption of food, since women are often more prone to renounce food in favor of their relatives, and hence females are more frequently subjected to malnutrition. Food intake and composition can negatively modulate the immune response through the lack of micronutrients such as zinc and copper and vitamins, essential for function and survival of immune cells (Caruso et al., 2013).

Steroid hormones, linking to specific receptors, differentially modulate the immune system. In general, while estrogens increase the immune response, progesterone and androgens have immune-suppressive actions. As an example, estrogens activate the mitogen–activated protein kinase (MAPK) pathway, leading to the downstream activation of nuclear factor kappa B (NF-κB) signaling pathway. Both MAPK and NF-κB pathways are involved in the expression of genes involved in immunological responses and in genes encoding antioxidant enzymes (Candore et al., 2010; Caruso et al., 2013; Caruso and Vasto, 2016).

In addition to hormones, the X chromosome is the other most intuitive biological factor potentially responsible for the differences in the immune response between men and women, because some genes involved in immunity map to this chromosome. The presence of a single X chromosome in men allows the phenotypic display of the corresponding recessive mutations. In addition, sex also modulates other immune response genes located on autosomes (Ellegren and Parsch, 2007). Another relevant biological process different between men and women is the rate at which telomeres shorten, faster in men than in women. However, telomere shortening may be either a cause for immunosenescence or its consequence (Barrett and Richardson, 2011).

Therefore the strength and types of immune responses are different between males and females. Estrogens promote while androgens suppress immune responses during infections, after vaccination or in case of autoimmunity. On these grounds, men are more susceptible to many infections, while women suffer more from infectious diseases with enhanced immunopathological impact. This sexual dimorphism in the immune response thus means that women are more resistant to infections, but they have higher incidence of autoimmune diseases compared to men. However, the relevance of autoimmune disorders for lifespan appears to be negligible (Candore et al., 2010; Caruso et al., 2013; Giefing-Kröll et al., 2015; Caruso and Vasto, 2016). It is noteworthy that it has been suggested that a more rapid aging of the immune system occurs in men than women (Hirokawa et al., 2013; Caruso et al., 2013). Finally, estrogens mediate an antiinflammatory effect and a cytoprotective effect against ROS, enhancing the levels of various antioxidant enzymes at the mitochondrial level. Accordingly, the mitochondria of the cells of female individuals produce less ROS than those of the cells of male ones (Candore et al., 2010; Vina et al., 2011). (The role of ROS in aging and age-related diseases is treated in Chapter 3.)

Despite their biomedical relevance, gender differences seem to be still poorly considered and inadequately investigated in aging studies (Ostan et al., 2016).

18.3 The role of immune-inflammatory responses in aging and age-related diseases, and therapeutic interventions

In 1969 Roy Walford published his landmark book, "The Immunologic Theory of Aging" (Walford, 1969). He hypothesized that faulty immune processes play a relevant role in aging of humans and of all mammals. Therefore he was the first to note and promote the power of modern immunological approaches as tools for the analysis of aging. Research has repeatedly confirmed the insightful predictions made by him regarding the role of the immune system in various pathologies associated with aging. Indeed, in accord with his original hypothesis on the role of immunosenescence in human aging, there is evidence that the previously identified diseases of aging are closely linked to dysregulation of the immune function and excessive inflammation (Effros, 2005; Caruso and Vasto, 2016).

In older people, a variety of changes of innate and acquired immunity have been described and viewed as deleterious, hence the term immunosenescence. Indeed, immunosenescence is a complex process involving multiple reorganizational and developmentally regulated changes, rather than simple unidirectional decline of the whole immune functionality (Caruso and Vasto, 2016). Chapter 7 summarizes recent data on the dynamic reassessment of immune changes with aging. Most of the attention of the scientific community has been devoted to changes of the T and B cell compartment, which are characterized by a more pronounced decline relative to innate immunity, and that are responsible of most of the immune defects that we can observe in older people, i.e., the increased incidence of infections, the reduced response to vaccination, and the higher frequency of autoimmune diseases and, maybe, cancer. Thus successful aging appears to be inextricably linked with optimal functioning of the immune system.

A further aspect of immunosenescence is examined in Chapter 16, exploring the role of Cytomegalovirus (CMV), one of the most ubiquitous human latent persistent viruses, and immune aging. Following the concept whereby CMV is responsible for substantial phenomenology associated with aging of the immune system, particularly of T cells (Pawelec et al., 2004, 2005), intense research followed. One, particularly incisive, performed in monozygotic twins demonstrated that twins discordant for CMV seropositivity showed significant differences in 58% of immune parameters (Goldeck et al., 2016). Thus these results clearly imply that CMV is a strong manipulator of the immune system. However, further studies (reviewed in Nikolich-Žugich et al., 2020) suggest that interaction of CMV with immunocompetent host abounds in nuances, with some costs and some benefits that on balance appear compatible with long and healthy coexistence, although stratifying CMV-seropositive and CMW-seropositive individuals has revealed a significant difference in lifespan (Savva et al., 2013).

Vaccination is one of the most effective medical interventions ever introduced, preventing millions of cases of infections worldwide every year. However, vaccines are commonly believed to be less effective in providing protection in older adults, due to the decline seen in immunity in this population. Chapter 8 summarizes data regarding the immunogenicity and efficacy of influenza, pneumococcal, and herpes zoster vaccines currently in use for older people, and provides a perspective on possible new methodological approaches for future vaccine development specifically for this age group. Concerning strategies for achieving optimal functioning of the immune system, and an increased response to vaccines in older people, available options may include immunomodulation with interleukin-7 as a growth factor for naïve T cells, monoclonal antibodies that affect immune checkpoints, and drugs that inhibit MAPKs and their interaction with nutrient sensing (signaling) pathways (NSPs, treated in Chapter 3) represent a promising therapeutic approach. Finally, the inclusion of appropriate combinations of Toll-like receptor agonists may enhance the efficacy of vaccination-mediated immunity in older adults (Aiello et al., 2019a).

Closely linked to immunosenescence, being one of its hallmarks, is inflammaging (Aiello et al., 2019a), that is the development and progression of a chronic, low-grade, and, therefore, for a long time subclinical, inflammatory state. Although the progression of inflammaging is currently recognized as one of the main driving forces of aging and one of the main risk factors for morbidity and mortality in older subjects, current knowledge on causative agents of inflammaging itself and of chronic, aging-related diseases is still incomplete. In Chapter 5, the authors argue that these chronic inflammatory diseases of older people in fact develop throughout life as an excessive immune reaction to specific stress factors. Therefore inflammaging might not be the cause of these diseases; however, it can be the trigger, leading to manifestation of aging-related diseases. In this context, the best intervention should aim to regulate the balance between pro- and anti–inflammatory signals and the most appropriate reaction to chronic stimulations to avoid/delay the appearance of these diseases. In their opinion, this goal can be achieved by polyvaccination, the use of appropriate antiviral drugs or antibiotics, personalized use of anti-inflammatory treatment, preventing and curing microbiome dysbiosis (see also Chapter 6), preventing accumulation of senescent cells, a source of pro–inflammatory mediators, and disrupting these already present with the use of "senolytic" drugs that specifically target senescent cells inducing apoptosis.

As discussed in Chapter 4, increasing efforts have been devoted to identify natural or synthetic compounds with senolytic activity, i.e., the ability to selectively remove senescent cells by inducing apoptosis, and immunotherapies using immune cell-mediated clearance of senescent cells that are currently the most promising strategies to fight aging and age-related diseases.

Concerning dysbiosis, Chapter 6 points out that age-related gut microbiota changes are well known and are associated with systemic inflammation. In turn, inflammation is responsible for high amount of ROS in gut that cause pathogen boom in lumen.

Coming back to the counteracting of inflammatory status, other interventions may be considered, e.g., physical exercise and diet. Physical exercise can decrease inflammation as suggested by the fact that exercise deprivation induces insulin resistance, impaired glucose uptake and hyperlipidemia, i.e., a cluster of physiological abnormalities similar to pro-inflammatory metabolic syndrome (see Chapter 14); an antiinflammatory diet, such as the "Mediterranean" diet, i.e., a diet rich in fruit and vegetables and poor in meat and refined sugars (Leonardi et al., 2018) may also be beneficial.

18.4 Slowing aging and fighting age-related diseases

As argued in Chapter 1, aging results from a reduced ability to adapt to the environment due to an exhaustion of self-organizing systems. Aging is a natural phenomenon, the product of the interaction between genetic, environmental, and lifestyle factors that manifests itself as a progressive decrease in the ability to adapt to stimuli and insults from the environment. This event inexorably leads the organism to a condition of greater susceptibility and vulnerability to diseases, with a consequent increase in mortality in an age-dependent manner. In fact, the reality for which every human being ages is inscribed within the normal biological cycle (birth/life/death) to which all living beings, including humans, are subjected. This concept, concerning the inevitability of the phenomenon of which as humans we are all aware, is underlined by the famous aphorism of Charles Augustin De Sainte Beuve, "Aging is the only system that has been found to live long" (https://fr.wikipedia.org/wiki/Charles-Augustin_Sainte-Beuve). Therefore the different antiaging strategies can only improve and/or delay the deterioration and duration of life, but certainly not abrogate a phenomenon that is independent of the will of the individual.

In the past year hope has come from hormonal therapy as an antiaging strategy because endocrine function plays a pivotal role in many phases of human life and development, from intrauterine development to the transition through puberty and adulthood. However, as discussed in Chapter 11, evidence concerning the effects of endocrine changes in successful and unsuccessful aging is still confusing. While some benefits have been clearly proven for treatment of endocrine deficiencies, such as testosterone replacement therapy in male hypogonadism, there is still no evidence concerning the benefits of supplementation for other hormones, such as growth hormone or dehydro-epiandrosterone, in older age. There is some reason to suspect that these changes, while seemingly hazardous for human health, might be beneficial; therefore, future studies should be aimed at investigating to which extent treatment should be suggested.

Lifestyle represents one relevant aspect in aging research since it is the most important modifiable factor that affects the aging process. A healthy lifestyle can limit the damage caused by environmental challenges that we face each day. Among the lifestyle factors that may influence successful aging and longevity, physical activity and healthy dietary habits have a great impact, as already alluded to. Both can have systemic antiaging effects, playing a significant role in determining the well-being of older people, and in delaying and reducing the risk of onset of diseases. In Chapter 14 the authors summarize the impact of nutritional strategies and exercise or physical activity on the hallmarks of aging, the nine molecular and cellular features that characterize aging phenomenon. There are two key points related to nutritional strategies and physical activity. i.e., to clarify the dose of "treatment" and the age at which it is administered. The issue of the dose needs to take hormetic effects into account. The term "hormesis" comes from the Greek verb meaning "to stimulate." By translating the concept to human system, it is described as an adaptive response of a cell or an organism characterized by a biphasic dose-dependent response to a variety of stressors. The systemic or cellular response will be beneficial with a low to modest stimulation, detrimental in consequence of high doses (Calabrese et al., 2007; Rattan, 2012). The intensity and the type of physical exercise and of dietary strategies could differentially act on aging, counteracting the hallmarks or reinforcing their harmful effects.

The matter of diet has been also exhaustively treated in Chapter 12. The authors point out that the optimal diet model for promoting successful aging and longevity and health is the one capable of satisfying at least the points reported considering data from centenarian studies, and all accompanied by an optimal modulation of the light-dark cycle and adequate quality and quantity of sleep. Epidemiological studies on Italian populations with a high incidence of centenarians have concluded that these subjects had the following common denominators: high intake of fruit, legumes, and vegetables; outdoor movement; fewer calorific intake than current recommendations suggest; prevalence of local products and gastronomy; and correct circadian distribution of meals, which includes a hearty breakfast, light snacks, and frugal dinner (Vasto et al., 2012; Franceschi et al., 2018). The diet must guarantee over time the quality of the food introduced, inspired by the food choices typical of the Mediterranean model with oriental influences. In order to best accompany successful aging, it is advisable to adopt a calorie intake that, depending on the proposed diet, provides a quantity of calories that is between a range of 15% and 20% around the basal metabolic rate in subjects with a low level of physical activity. From data published in literature, it appears that a sustainable calorie restriction regimen for humans may start from a calorie intake 15%–20% lower than the basal metabolic rate, since this calorie intake has been shown to be safe in human studies. Furthermore, significant effects of calorie restriction (10%–15% lower than the basal metabolism) are demonstrated even in humans, and adopting a balanced diet with that caloric intake is considered effective and safe but must provide a protein content of around 0.95–1 g/kg body weight. For older people slightly lower amounts are indicated and animal proteins should be limited. Both a low-calorie diet and the regular use of functional and mimetic compounds (compounds that produce the effects of calorie restriction without the need to restrict energy intake) could be the nutritional basis for successful aging and longevity, obviously only together with all the other measure that characterize an appropriate lifestyle (Rickman et al., 2011; Aiello et al., 2019b). Chapters 12 and 14 discuss the interactions of diet with NSPs, also extensively treated in Chapter 3.

A particular aspect of diet concerns some compounds principally contained in fruit and vegetables, i.e., nutraceuticals. This term is a portmanteau word, a combination of "nutrition" and "pharmaceutical," and refers to "naturally derived bioactive compounds that are found in foods, dietary supplements and herbal products, and have health promoting, disease preventing, and/or medicinal properties." Several nutraceuticals exhibit antiaging features by acting on the inflammatory status and on the prevention of oxidative reaction. That has been suggested to result in a significant reduction of all risk factors for age-related diseases, enhancing the attainment of successful aging. In this context, Chapter 13 summarizes the available clinical evidence supporting the use of selected botanicals and phytochemicals with confirmed activity on the human central nervous system and demonstrated effects on modulating cognitive decline as an example of age-related disease. As in many such studies, however, discrepant results have been reported, but as discussed by the authors, reasons for this may depend on the different dosages tested, the different levels of quality of the extracts, and the different study designs. The authors conclude that studies addressing whether long-term dietary intake of single or combined botanicals can reduce the severity and incidence of neurodegenerative and other age-related diseases appear critical and crucial for the future.

Another aspect of the effects of plant compounds is treated in Chapter 9. The authors provide a brief overview of biological mechanisms underlying stress resilience and explore how resilience changes throughout the lifespan. Emerging research has focused on biological resilience elicited by consumption of phytonutrients, particularly vitamins, plant polyphenols, known to be antiinflammatory and antioxidant. Activation of stress-responsive mechanisms following moderate and chronic consumption of low doses of plant polyphenols induces "vitagenes," activating brain resilience mechanisms effective in the prevention of neuroinflammation and aging-associated cognitive decline, as well as neuropsychiatric disorder pathogenesis, thereby improving brain health life and longevity in animals and humans. Identifying personalized biomarker signatures of resilience can help researchers to characterize biologically vulnerable individuals and, at the other end of the scale, resilient individuals, such as centenarians, that represent the best model of longevity and successful aging (Caruso, 2019).

Thus there is a compelling body of evidence indicating that adherence to healthy dietary patterns and intake of plant-derived bioactive compounds may serve as potential strategy to preserve body function, slowing the progression of aging and reducing the incidence of age-related disease, i.e., an approach called nutrigerontology (Aiello et al., 2016). The aim of Chapter 15 is, therefore, to discuss recent advances in nutritional science, with an emphasis on biomarkers that might predict successful aging. The authors summarize current knowledge concerning potential biomarkers for assessing the casual relationship between aging and nutrition.

Several other strategies under development have shown significant promise for slowing aging or delaying the onset of age-related diseases. Multiple preclinical studies suggest that in addition to elimination of senescent cells by senolytics, upregulation of autophagy (treated in Chapter 3) through autophagy stimulators and juvenile plasma transfusion (as molecules circulating in the young blood can rejuvenate the aging cells and tissues in models) are seemingly promising approaches to support normal health during aging and also to postpone age-related diseases. However, these treatments need to be critically evaluated in clinical trials to determine both their long-term efficacy and lack of adverse effects on the function of various tissues and organs. On the other hand, other approaches such as different types of intermittent fasting, regular exercise, the most promising antioxidants resveratrol and curcumin treatments are ready for large-scale clinical trials, as they are noninvasive and appear to have minimal side effects (Shetty et al., 2018).

A recent review analyzed the mechanisms of action on hallmarks of aging of metformin, a biguanide used in the treatment of Type II Diabetes (Kulkarni et al., 2020). It is the first drug to be tested for its age-targeting effects in the large clinical trial, i.e., TAME (targeting aging using metformin). It improves NSPs, enhances autophagy and intercellular communication, protects against macromolecular damage, delays stem cell aging, modulates mitochondrial function, regulates transcription, and lowers telomere attrition and senescence (Chapter 3). However, as recently pointed out (Soukas et al., 2019), not all individuals prescribed metformin have the same benefit and some develop side effects. Therefore a better understanding of the effects of the drug in humans is needed.

A growing body of research has highlighted mTOR inhibitors, i.e., immunosuppressive drugs rapamycin and everolimus (mTOR precisely means mechanistic target of rapamycin), as promising treatments for several age-related pathologies, prolonging lifespan, especially in all four major animal models of aging, yeast, worms, flies, and mice (Aiello et al., 2019a, b). In a recent review (Selvarani et al., 2020), authors focused on areas where, in mice, there is strong evidence for positive effects of the drug in aging and age-related diseases, i.e., lifespan, cardiovascular system, central nervous system, immune system, and cell senescence. The authors conclude that it is time that preclinical studies be focused on taking rapamycin to the clinic. The role of mTOR in aging is discussed in Chapter 3. Finally, further possible future strategies are represented by cellular and genetic therapies, including genetic reprogramming. In particular, genetically reprogramming cells into induced pluripotent stem cells might rejuvenate any tissue type (Stahl and Brown, 2015).

18.5 Conclusions and future perspectives

A long life in a healthy, vigorous, youthful body has always been one of humanity's greatest dreams. Recent progress in genetic manipulations and calorie-restricted diets in laboratory animals hold forth the promise that someday science should enable us to exert total control over own biological aging (Jirillo et al., 2008). In the present book, antiaging strategies aimed not to rejuvenate but to slow aging and to delay or avoid the onset of age-related diseases have been discussed, hence we will able to substantially slow down the aging process, extending our productive, youthful lives. In fact, as stated 20 years ago by Hayflick (2000), if the main goal of biomedical research enterprises is to resolve and negate causes of death, then every older person becomes a testimony to its successes. Researchers have an obligation to emphasize that the goal of research on aging is not to increase human longevity regardless of the consequences but to increase active longevity free from disability and functional dependence.

References

Accardi G, Aiello A, Vasto S, Caruso C. Chance and causality in ageing and longevity. In: Caruso C, editor. Centenarians. Cham: Springer; 2019. p. 1–21. https://doi.org/10.1007/978-3-030-20762-5_1.

Aiello A, Accardi G, Candore G, Carruba G, Davinelli S, Passarino G, Scapagnini G, Vasto S, Caruso C. Nutrigerontology: a key for achieving successful ageing and longevity. Immun Ageing 2016;13:17. https://doi.org/10.1186/s12979-016-0071-2. PMID: 27213002; PMCID: PMC4875663.

Aiello A, Farzaneh F, Candore G, Caruso C, Davinelli S, Gambino CM, Ligotti ME, Zareian N, Accardi G. Immunosenescence and its hallmarks: how to oppose aging strategically? A review of potential options for therapeutic intervention. Front Immunol 2019a;10. https://doi.org/10.3389/fimmu.2019.02247. 2247.

Aiello A, Caruso C, Accardi G. Slow-ageing diets. In: Gu D, Dupre M, editors. Encyclopedia of gerontology and population aging. Cham: Springer; 2019b. https://doi.org/10.1007/978-3-319-69892-2_134-1.

Baggio G, Corsini A, Floreani A, Giannini S, Zagonel V. Gender medicine: a task for the third millennium. Clin Chem Lab Med 2013;51:713–27. https://doi.org/10.1515/cclm-2012-0849.

Barrett ELB, Richardson DS. Sex differences in telomeres and lifespan. Aging Cell 2011;10(6):913–21. https://doi.org/10.1111/j.1474-9726.2011.00741.x.

Beltrán-Sánchez H, Finch CE, Crimmins EM. Twentieth century surge of excess adult male mortality. Proc Natl Acad Sci U S A 2015;112:8993–8. https://doi.org/10.1073/pnas.1421942112.

Calabrese EJ, Bachmann KA, Bailer AJ, Bolger PM, Borak J, Cai L, Cedergreen N, Cherian MG, Chiueh CC, Clarkson TW, Cook RR, Diamond DM, Doolittle DJ, Dorato MA, Duke SO, Feinendegen L, Gardner DE, Hart RW, Hastings KL, Hayes AW, Hoffmann GR, Ives JA, Jaworowski Z, Johnson TE, Jonas WB, Kaminski NE, Keller JG, Klaunig JE, Knudsen TB, Kozumbo WJ, Lettieri T, Liu SZ, Maisseu A, Maynard KI, Masoro EJ, McClellan RO, Mehendale HM, Mothersill C, Newlin DB, Nigg HN, Oehme FW, Phalen RF, Philbert MA, Rattan SI, Riviere JE, Rodricks J, Sapolsky RM, Scott BR, Seymour C, Sinclair DA, Smith-Sonneborn J, Snow ET, Spear L, Stevenson DE, Thomas Y, Tubiana M, Williams GM, Mattson MP. Biological stress response terminology: integrating the concepts of adaptive response and preconditioning stress within a hormetic dose-response framework. Toxicol Appl Pharmacol 2007;222:122–8. https://doi.org/10.1016/j.taap.2007.02.015.

Candore G, Balistreri CR, Colonna-Romano G, Lio D, Listì F, Vasto S, Caruso C. Gender-related immune-inflammatory factors, age-related diseases, and longevity. Rejuvenation Res 2010;13:292–7. https://doi.org/10.1089/rej.2009.0942.

Caruso C, editor. Centenarians. Cham: Springer; 2019. p. 1–179. https://doi.org/10.1007/978-3-030-20762-5.

Caruso C, Vasto S. Immunity and aging. In: Ratcliffe MJH, editor. Encyclopedia of immunobiology. vol. 5. Oxford: Academic Press; 2016. p. 127–32.

Caruso C, Accardi G, Virruso C, Candore G. Sex, gender and immunosenescence: a key to understand the different lifespan between men and women? Immun Ageing 2013;10:20. https://doi.org/10.1186/1742-4933-10-20.

Christensen K, Doblhammer G, Rau R, Vaupel JW. Ageing populations: the challenges ahead. Lancet 2009;374:1196–208. https://doi.org/10.1016/S0140-6736(09)61460-4.

Effros RB. Roy Walford and the immunologic theory of aging. Immun Ageing 2005;2:7. https://doi.org/10.1186/1742-4933-2-7.

Ellegren H, Parsch J. The evolution of sex-biased genes and sex-biased gene expression. Nat Rev Genet 2007;8(9):689–98. https://doi.org/10.1038/nrg2167.

Farrelly C. Has the time come to take on time itself? BMJ 2008;337:a414. https://doi.org/10.1136/bmj.a414.

Franceschi C, Ostan R, Santoro A. Nutrition and inflammation: are centenarians similar to individuals on calorie-restricted diets? Annu Rev Nutr 2018;38:329–56. https://doi.org/10.1146/annurev-nutr-082117-05163.

Freedman VA, Wolf DA, Spillman BC. Disability-free life expectancy over 30 years: a growing female disadvantage in the US population. Am J Public Health 2016;106:1079–85. https://doi.org/10.2105/AJPH.2016.303089.

Giefing-Kröll C, Berger P, Lepperdinger G, Grubeck-Loebenstein B. How sex and age affect immune responses, susceptibility to infections, and response to vaccination. Aging Cell 2015;14:309–21. https://doi.org/10.1111/acel.12326.

Goldeck D, Larsen LA, Christiansen L, Christensen K, Hamprecht K, Pawelec G, Derhovanessian E. J Gerontol A Biol Sci Med Sci 2016;71(12):1537–43. https://doi.org/10.1093/gerona/glv230. Epub 2016. PMID: 26755680.

Hayflick L. The future of ageing. Nature 2000;408:267–9. https://doi.org/10.1038/35041709.

Hirokawa K, Utsuyama M, Hayashi Y, Kitagawa M, Makinodan T, Fulop T. Slower immune system aging in women versus men in the Japanese population. Immun Ageing 2013;10:19. https://doi.org/10.1186/1742-4933-10-19.

Jirillo E, Candore G, Magrone T, Caruso C. A scientific approach to anti-ageing therapies: state of the art. Curr Pharm Des 2008;14(26):2637–42. https://doi.org/10.2174/138161208786264070.

Kulkarni AS, Gubbi S, Barzilai N. Benefits of metformin in attenuating the hallmarks of aging. Cell Metab 2020;32(1):15–30. https://doi.org/10.1016/j.cmet.2020.04.001. Epub 2020. PMID: 32333835; PMCID: PMC7347426.

Leonardi GC, Accardi G, Monastero R, Nicoletti F, Libra M. Ageing: from inflammation to cancer. Immun Ageing 2018;15:1. https://doi.org/10.1186/s12979-017-0112-5.

Longo VD, Antebi A, Bartke A, Barzilai N, Brown-Borg HM, Caruso C, Curiel TJ, de Cabo R, Franceschi C, Gems D, Ingram DK, Johnson TE, Kennedy BK, Kenyon C, Klein S, Kopchick JJ, Lepperdinger G, Madeo F, Mirisola MG, Mitchell JR, Passarino G, Rudolph KL, Sedivy JM, Shadel GS, Sinclair DA, Spindler SR, Suh Y, Vijg J, Vinciguerra M, Fontana L. Interventions to slow aging in humans: are we ready? Aging Cell 2015;14:497–510. https://doi.org/10.1111/acel.12338.

Lutz W, Crespo Cuaresma J, Kebede E, Prskawetz A, Sanderson WC, Striessnig E. Education rather than age structure brings demographic dividend. Proc Natl Acad Sci U S A 2019;116:12798–803. https://doi.org/10.1073/pnas.1820362116.

Machluf Y, Chaiter Y, Tal O. Gender medicine: lessons from COVID-19 and other medical conditions for designing health policy. World J Clin Cases 2020;8:3645–68. https://doi.org/10.12998/wjcc.v8.i17.3645.

Nikolich-Žugich J, Čicin-Šain L, Collins-McMillen D, Jackson S, Oxenius A, Sinclair J, Snyder C, Wills M, Lemmermann N. Advances in cytomegalovirus (CMV) biology and its relationship to health, diseases, and aging. GeroScience 2020;42(2):495–504. https://doi.org/10.1007/s11357-020-00170-8.

Oeppen J, Vaupel JW. Demography. Broken limits to life expectancy. Science 2002;296:1029–31. https://doi.org/10.1126/science.1069675.

Olshansky SJ, Perry D, Miller RA, Butler RN. Pursuing the longevity dividend: scientific goals for an aging world. Ann N Y Acad Sci 2007;1114:11–3. https://doi.org/10.1196/annals.1396.050.

Ostan R, Monti D, Gueresi P, Bussolotto M, Franceschi C, Baggio G. Gender, aging and longevity in humans: an update of an intriguing/neglected scenario paving the way to a gender-specific medicine. Clin Sci (Lond) 2016;130:1711–25. https://doi.org/10.1042/CS20160004.

Pawelec G, Akbar A, Caruso C, Effros R, Grubeck-Loebenstein B, Wikby A. Is immunosenescence infectious? Trends Immunol 2004;25(8):406–10. https://doi.org/10.1016/j.it.2004.05.006. PMID: 15275638.

Pawelec G, Akbar A, Caruso C, Solana R, Grubeck-Loebenstein B, Wikby A. Human immunosenescence: is it infectious? Immunol Rev 2005 Jun;205:257–68. https://doi.org/10.1111/j.0105-2896.2005.00271.x. PMID: 15882359.

Rattan SI. Rationale and methods of discovering hormetins as drugs for healthy ageing. Expert Opin Drug Discovery 2012;7:439–48. https://doi.org/10.1517/17460441.2012.677430.

Rickman AD, Williamson DA, Martin CK, Gilhooly CH, Stein RI, Bales CW, Roberts S, Das SK. The CALERIE study: design and methods of an innovative 25% caloric restriction intervention. Contemp Clin Trials 2011;32(6):874–81. https://doi.org/10.1016/j.cct.2011.07.002.

Savva GM, Pachnio A, Kaul B, Morgan K, Huppert FA, Brayne C, Moss PA, Medical Research Council Cognitive Function and Ageing Study. Cytomegalovirus infection is associated with increased mortality in the older population. Aging Cell 2013;12(3):381–7. https://doi.org/10.1111/acel.12059. Epub 2013. PMID: 23442093.

Selvarani R, Mohammed S, Richardson A. Effect of rapamycin on aging and age-related diseases-past and future. GeroScience 2020. https://doi.org/10.1007/s11357-020-00274-1. Epub ahead of print. PMID: 33037985.

Shetty AK, Kodali M, Upadhya R, Madhu LN. Emerging anti-aging strategies—scientific basis and efficacy. Aging Dis 2018;9(6):1165–84. https://doi.org/10.14336/AD.2018.1026.

Soukas AA, Hao H, Wu L. Metformin as anti-aging therapy: is it for everyone? Trends Endocrinol Metab 2019;30(10):745–55. https://doi.org/10.1016/j.tem.2019.07.015. Epub 2019. PMID: 31405774; PMCID: PMC6779524.

Stahl EC, Brown BN. Cell therapy strategies to combat immunosenescence. Organogenesis 2015;11(4):159–72. https://doi.org/10.1080/15476278.2015.1120046.

Vasto S, Scapagnini G, Rizzo C, Monastero R, Marchese A, Caruso C. Mediterranean diet and longevity in sicily: survey in a sicani mountains population. Rejuvenation Res 2012;15(2):184–8. https://doi.org/10.1089/rej.2011.1280.

Vina J, Gambini J, Lopez-Grueso R, Abdelaziz KM, Jove M, Borras C. Females live longer than males: role of oxidative stress. Curr Pharm Des 2011;17:3959–65. https://doi.org/10.2174/138161211798764942.

Walford RL. The immunologic theory of aging. Copenhagen: Munksgaard; 1969.

Index

Note: Page numbers followed by *f* indicate figures and *t* indicate tables.

A

Ability to survive, 10
Adaptive immunity
  B cells, 116
  T cells, 114–115
Adjuvants
  alum/aluminum salts, 132–133
  AS03, 133
  complete Freund's adjuvant (CFA), 130
  dendritic cells (DCs), 132
  diphtheria, 130
  incomplete Freund's adjuvant (IFA), 130
  MF59, 133
Adrenal androgens, 211–212
Advanced Glycation Endproducts
  (AGEs), 94–95
Ageism, 330–331
Age structural transition, 20
Aging-related diseases (ARDs), 342–343.
    See also Pathobiology
  allo-priming, 98–99
  chronic inflammation
    Alzheimer's disease (AD), 96–97
    atherosclerosis, 95–96
    chronic inflammatory characteristics,
      93–94
    chronic obstructive pulmonary disease
      (COPD), 95
    frailty syndrome, 96
    Parkinson's disease (PD), 97
    T2DM, 94–95
  COVID-19, 97–98
  future perspective, 99
  incidence, 89
  interventions, 98–99, 99*f*
Aging theories, 4–6
Allo-priming, 98–99
Altered intercellular communication, 63
Alzheimer's disease (AD), 96–97
Anabolic resistance, 182–183
Androgen deficiency, 209–210
Andropause, 209–210
Animal machine, 3, 9–10

B

Bacopa monnieri, 264–265
B cells, 116
Bioethics, 332
Biological age, 39–40
Biomarkers. *See* Nutritional biomarkers
Biorhythms
  AMP/ATP ratio, 227–228
  biological clocks, 220
  and cell cycle, 227
  circadian biorhythms, 220
  Circadian Locomotor Output Cycles Kaput
    (CLOCK) and BMAL1, 226
  classification, 219
  duration of the cycle, 220
  external light, 220–221
  light-dark cycle, 219
  permanent/rotating night shifts, 221

Anti-Müllerian hormone
  (AMH), 208
Antioxidants, 57–59
Antioxidant vitamins
  chronic disease, 299–300
  vitamin A, 300–301
  vitamin C, 300
  vitamin E, 300
Apoptosis, 181–182
AS03, 133
AS04, 140
Asiatic diets, 273–274
Atherosclerosis, 95–96
Autoimmune thyroid disease, 213
Autophagy
  ATG genes expression, 52–53
  autophagosome, 52
  caloric restriction (CR), 53
  metabolic demands, 52
  metformin, 54
  neurodegeneration, 53–54
  rapamycin, 54
  regulation, 53
  types, 46

Biorhythms *(Continued)*
  set, 219
  sleep duration, 221–222
  TTFLs, 226
  zeitgebers, 220

C

Caloric restriction (CR)
  longevity, 238
  micronutrients, 238
  mimetics, 239
  risk, 238–239
Camelia sinensis (tea plant), 262–263
Cancer and aging
  convergent mechanisms, 64, 65*f*
  divergent mechanisms, 64–65, 66*f*
  mortality, 63–64
  senescence response, 65–66
Carotenoids
  α- and β-carotene, 304–305
  invasive and noninvasive approaches, 303–304
  lutein, 305
  lycopene, 304–305
  phytonutrients, 303–304
  zeaxanthin, 305
Causality and chance
  epigenetics, 8–9
  gene expression and alternative splicing, 9
  genetic control, 8
  stochastic processes, 7–8
Causes of death, 25
Cellular resilience, 155
Cellular senescence, 60, 65–66, 284.
        See also Senescence-associated secretory
        phenotype (SASP)
  and aging, 82–83
  biomarkers, 75
  characteristics, 75, 89–90
  future perspectives, 83–84
  mechanism, 89–90
  molecular biomarkers, 310
  triggers, 82–83
Central oscillator, 223–224
Chaperone-mediated autophagy, 46
Characteristics of aging, 1, 2*t*, 41
Chronic inflammation
  Alzheimer's disease (AD), 96–97
  atherosclerosis, 95–96

chronic inflammatory characteristics,
        93–94
chronic obstructive pulmonary disease
        (COPD), 95
  frailty syndrome, 96
  Parkinson's disease (PD), 97
  T2DM, 94–95
Chronic obstructive pulmonary disease
        (COPD), 95
Chronobiology
  biorhythms (*see* Biorhythms)
  central oscillator, 223–224
  chronodisruption
    caffeine, 232
    clock gene expression, 233–234
    darkness circadian effect, 231
    dietary regimens, 233
    exercise, 233
    factors, 235*f*
    fats, 231
    genetic variations, 234
    glucocorticoids, 231
    Heat Shock Proteins (HSP), 231
    intensity and frequency of exposure, 230
    light stimuli, 230
    MAPkinase pathway, 230
    mealtimes, 232–233
    melatonin blood levels, 230–231
    night work and jet lag, 229–230
    resveratrol, 232
    salt intake, 232
  clock-controlled genes (*see* Clock-controlled
        genes)
  future perspectives, 241–242
  peripheral oscillator, 224–225
Chrononutrition
  caloric restriction (CR), 237–239
  future perspectives, 242
  macronutrients, 234–236
  meals composition, 239–241
  meals frequency, 236–237
Circadian oscillators, 222–223
Clock-controlled genes
  adenylate cyclase activity, 227–228
  and cell cycle, 227
  CLOCK and BMAL1, 226
  epiphysis, 228
  epithelia, 229
  gene transcription, 226

immune system, 229
leptin modulation, 228
liver oscillator, 229
melatonin, 228
PER-CRY dimer, 226
thyroid hormone, 228
TTFLs, 226
Clock CpGs, 44
Cocoa, 263–264
Cognitive decline
    Bacopa monnieri, 264–265
    Camelia sinensis (tea plant), 262–263
    Crocus sativus L., 265–266
    Curcuma longa, 266
    Ginkgo biloba (GB), 257–259
    Theobroma cacao, 263–264
    Vitis vinifera, 259–261
Complexity
    biological age, 39–40
    component of complexity, 38–39
    concept, 37–38, 38f
    degree of complexity, 37–38
    entropy, 40
    healthy physiological processes, 40
    loss of complexity, 39
    presbycusis, 39
    residual energy, 40–41
COVID-19, 97–98
Crocus sativus L., 265–266
Cultural trend, 330
Curcuma longa, 266
Curcumin, 266
Cytomegalovirus (CMV)
    all-cause mortality, 323
    cardiovascular mortality, 323
    congenital infection, 319
    future perspectives, 324–325
    host immune response, 319–320
        multisite latency, 322
        persistent tissue-localized
            replication, 320–322
        primary acute infection, 320
    and immune aging, 324
    immunocompetent hosts, 319–320
    incidence, 319
    inflammation, 322–323
    longevity and chronic diseases, 323
    seroprevalence, 319
    smoldering infection, 319–320

D
Definition of aging, 1–2
Dehydroepiandrosterone (DHEA), 183–184,
    211–212
Dementia, 255–256
Demographic transition theory, 21
Dendritic cells (DCs), 118
Disposable soma theory, 5–6
DNA and chromosomes, 308
Dysbiosis, 108–109

E
Economic old-age dependency ratio (EOADR),
    30–31
Endocrine factors, 183–184
Endocrine function, 207, 214. *See also* Hormones
Energy supply, 180
Epidemiological transition theory, 21–22
Epigenetics, 278–280
    biological mechanisms, 43
    CpG sites, 43–44
    DNA methylation, 44
    epigenetic modifications, 45
    histone modifications, 43
    nutrition and dietary compounds, 45
    phenoAge, 44–45
    phenotype, 43
    senescent phenotypes, 76–77
Error/stochastic theories, 4
Ethics of aging, 341–342
    age and aging, 332–333
    bioethics, 332
    discrimination and abuse, 337
    field of investigation, 333
    future perspectives, 337–338
    medical resource allocation
        approaches, 334–335
        dramatically scarce medical resources, 336–337
        egalitarian approach, 334–335
        good aging, 333
        healthcare costs, 335–336
        individualistic liberalism/conservative
            liberalism, 334
        issue of justice, 333
        needs-based approach, 335
        social determinants of health, 333–334
        utilitarian approach, 336–337
    moral and social attitudes

Ethics of aging *(Continued)*
    ageism, 331
    cultural background, 330
    vulnerability, 331–332
  policies, 338
  population aging, 329–330
Eudaimonic well-being, 156–157
Evidence-based medicine (EBM), 255–256
Evolutionary theories, 5

F

Field of investigation, 333
Frailty, 96, 342
  assessment, 174, 175–176t
  definition
    conceptual definitions, 173–174
    general definition, 173
    operational definitions, 174
  domains, 173–174, 174f
  frailty index, 174
  future perspectives, 192–193
  physical frailty, 176 *(see also* Sarcopenia)
Free radical theory, 4, 105

G

Gender differences, 16, 22–23, 23f, 24t
Gender medicine
  cardiovascular disease, 344
  food intake and composition, 344–345
  immune responses, 345
  iron deficiency, 344
  representation, 343
  steroid hormones, 345
  X chromosome, 345
Genetically programmed theories, 4
Genomic instability, 274–276
  chromosomal aberrations, 42
  histone deacetylase enzymes, 42
  nuclear DNA damage, 41–42
  progeroid syndromes, 42
  sirtuin function, 42
Ginkgo biloba (GB), 257–259
Global Activity Limitation Indicator (GALI), 27–29
Glucocorticoids, 211
Glutathione disulfide (GSSG), 160–166
Green tea, 262–263
Growth hormone deficiency (GHD), 210–211
Gut homeostasis, 108–109

H

Hallmarks of aging
  antagonistic hallmarks
    cellular senescence, 284
    deregulated nutrient sensing, 281–283
    mitochondrial dysfunction, 283–284
  definition, 273–274
  epigenetics, 278–280
  future perspectives, 286–287
  genomic instability, 274–276
  integrative hallmarks
    altered intercellular communication, 285–286
    definition, 284
    stem cell exhaustion, 285
  physical activity, 275f
  proteostasis, 280–281
  slow-aging diets, 273–274, 275f
  telomere attrition, 277–278
Health transition theory, 22
Hedonic well-being, 156–157
Hormesis, 160–166, 286–287
Hormones
  adrenal function
    adrenal androgens, 211–212
    glucocorticoids, 211
    mineralocorticoids, 212
  endocrine physiology, 207
  future perspective, 214
  gonadal function
    andropause, 209–210
    menopause, 208–209
  growth hormone (GH), 210–211
  hypothalamus, 207–208
  pituitary gland, 207–208
  thyroid function, 212–213
Hyperthyroidism and hypothyroidism, 213

I

Idiopathic pulmonary fibrosis (IPF), 95
Immune-inflammatory responses
  cytomegalovirus (CMV), 346
  faulty immune processes, 346
  gut microbiota, 347
  immunosenescence, 346
  inflammaging, 347
  physical exercise and diet, 348
  senolytic drugs, 347
  vaccination, 346–347

Immune system
  adaptive immunity
      B cells, 116
      T cells, 114–115
  future perspectives, 121–122
  immunosenescence, 113
  inflammaging
      causes and mechanisms, 119
      CMV infection, 119–120
      complement system, 120
      gut microbiota, 120
      mitochondria, 120
      pro- and anti-inflammatory
          molecules, 121
      risk factor, 119
  innate immunity
      dendritic cells (DCs), 118
      mast cells, eosinophils, and basophils, 118–119
      monocytes and macrophages, 117
      neutrophils, 117
  life expectancy, 113
  -omics approach, 121–122
Immunosenescence, 82–83, 90–92, 113, 129–130,
      346. *See also* Immune system
Inflammaging, 58, 82–83
  causes and mechanisms, 119
  cellular senescence, 89–92
  chronic inflammation
      Alzheimer's disease (AD), 96–97
      atherosclerosis, 95–96
      chronic inflammatory characteristics, 93–94
      chronic obstructive pulmonary disease
          (COPD), 95
      frailty syndrome, 96
      Parkinson's disease (PD), 97
      T2DM, 94–95
  CMV infection, 119–120
  complement system, 120
  cytokines, 90–92
  definition, 90–92
  gut microbiome, 90–92
  gut microbiota, 120
  immunosenescence, 90–92
  innate immunity, 90–92
  interventions, 98–99, 99f
  mitochondria, 120
  molecular inflammation, 92
  pro- and anti-inflammatory molecules, 121
  risk factor, 119

senoinflammation, 92–93
Innate immunity
  dendritic cells (DCs), 118
  mast cells, eosinophils, and
      basophils, 118–119
  monocytes and macrophages, 117
  neutrophils, 117
Intercellular communication, 285–286
Interleukin-7, 143–144

L
Labor-intensive long-term care (LTC), 32–33
Langerhans cells (LC), 118
Life expectancy, 16–20, 18–19f, 341
Lifestyle
  altered intercellular communication,
      285–286
  cellular senescence, 284
  deregulated nutrient sensing, 281–283
  epigenetics, 278–280
  genomic instability, 274–276
  Mediterranean/Asiatic diets, 273–274
  mitochondrial dysfunction, 283–284
  nongenetic factors, 273–274
  physical activity and diet, 273–274
  proteostasis, 280–281
  stem cell exhaustion, 285
  telomere attrition, 277–278
Longevity assurance processes, 166–167

M
Macroautophagy, 46. *See also* Autophagy
Macronutrients, 234–236
Male late-onset hypogonadism, 209
Malnutrition-Sarcopenia Syndrome (MSS),
      188–189
Mast cells, eosinophils, and basophils, 118–119
Meal composition
  afternoon snack, 240–241
  breakfast, 240
  carbohydrates, 239–240
  lunch, 240
  midmorning snack, 240
  plasma glucose, 240
  protein, 241
  recommendations, 239
  season vegetables, 241
Meal frequency, 236–237

Mechanisms of aging, 342–343
  DNA damage, 6
  energy dispersal, 7
  entropy, 6–7, 7t
  proteins damage, 6
Mediterranean diet (MD), 234–235, 255, 273–274
Menopause, 208–209
Mesenchymal stem/stromal cells (MSCs), 80–82
Metabolism, 309
MF59, 133
Microautophagy, 46
MicroRNA (miRNA), 79–80
Mineralocorticoids, 212
Mitochondrial dysfunction, 283–284, 309
  antioxidants, 57–59
  energy homeostasis, 55–56
  inflammaging, 58
  oxidative stress, 58–59
  protein glycation, 59–60, 59f
  reactive species, 57, 57t
  ROS production, 56–57, 57t
  ROS regulation, 56, 56f
Mitochondrial Free Radical Theory of Aging
      (MFRTA), 107–108
Molecular inflammation concept, 92
Monocytes and macrophages, 117
Moral and social attitudes
  ageism, 331
  cultural background, 330
  vulnerability, 331–332
Mortality rates, 1, 2f
mRNA-1273 vaccine, 147
Muscle fibers, 179
Myeloid DCs (mDC), 118
Myokines, 184–185

N

Neutrophils, 117
Nutraceutical approach
  Bacopa monnieri, 264–265
  botanicals/phytochemicals and their effects, 258t
  Camelia sinensis (tea plant), 262–263
  chronic age-related diseases and cognitive
      impairment, 255–256, 256f
  Crocus sativus L., 265–266
  Curcuma longa, 266
  data selection, 256–257
  dementia, 255–256
  future perspectives, 266–267

  Ginkgo biloba (GB), 257–259
  mechanisms of action, 256f
  Mediterranean diet (MD), 255
  resveratrol, 259–261
  Theobroma cacao, 263–264
  Vitis vinifera, 259–261
Nutrient sensing pathways (NSPs), 281–283
  activation, 46–47
  autophagy, 46, 52–54
  FOXO3, 47–48
  insulin/IGF-1 pathway, 48–49
  mTOR pathway, 49–50
  nutrient-sensing longevity genes, 309
  representation, 46, 46f
  sirtuin pathway, 50–51
Nutrigerontology, 350
Nutritional biomarkers
  biochemical measures, 297
  carotenoids
    α- and β-carotene, 304–305
    invasive and noninvasive approaches, 303–304
    lutein, 305
    lycopene, 304–305
    phytonutrients, 303–304
    zeaxanthin, 305
  definition, 295–297
  future perspectives, 310
  lifestyle factors, 295–297, 296f
  measurement, 295–297
  minerals
    selenium, 298–299
    zinc, 298
  molecular biomarkers
    cell senescence, 310
    DNA and chromosomes, 308
    metabolism, 309
    mitochondria, 309
    RNA and transcriptome, 308
  nutritional status, 295–297
  omics techniques, 295–297, 296f
  polyphenols
    biochemical assessment, 307
    classification and chemical structure, 305, 306f
    flavonoids, 305–307
  polyunsaturated fatty acids (PUFA)
    AA/EPA ratio, 303
    dietary sources, 301–302
    importance, 303
    n-3 index, 302–303, 302f

vitamins
  analytical considerations, 299
  antioxidant vitamins, 299–301
  vitamin D, 301
Nutritional status, 295–297

O

Occurrence of aging
  aging due to damage, 3
  disposable soma theory, 5–6
  free radical theory, 4
  genetically programmed theory, 4
  mutation accumulation theory, 5
  natural environment, 3
  natural selection and chronological age, 5
  programmed aging, 3
Old-age dependency ratio (OADR), 29
Oscillations
  central oscillator, 223–224
  feedback mechanisms, 222–223
  origin, 223
  peripheral oscillator, 224–225
  suprachiasmatic nucleus (SCN), 223
Oxidative stress, 105–108
Oxy-inflammaging. *See* Molecular inflammation
    concept

P

Parkinson's disease (PD), 97
Pathobiology
  aging as disease, 36
  aging *vs.* age-related diseases, 35–36
  altered intercellular communication, 63
  cancer and aging
    convergent mechanisms, 64, 65*f*
    divergent mechanisms, 64–65, 66*f*
    mortality, 63–64
    senescence response, 65–66
  cellular senescence, 60
  complexity
    biological age, 39–40
    component of complexity, 38–39
    concept, 37–38, 38*f*
    degree of complexity, 37–38
    entropy, 40
    healthy physiological processes, 40
    loss of complexity, 39
    presbycusis, 39
    residual energy, 40–41

epigenetics
  biological mechanisms, 43
  CpG sites, 43–44
  DNA methylation, 44
  epigenetic modifications, 45
  histone modifications, 43
  nutrition and dietary compounds, 45
  phenoAge, 44–45
  phenotype, 43
future perspectives, 66
genomic instability
  chromosomal aberrations, 42
  histone deacetylase enzymes, 42
  nuclear DNA damage, 41–42
  progeroid syndromes, 42
  sirtuin function, 42
hallmarks of aging, 41
markers, 37
mitochondrial dysfunction
  antioxidants, 57–59
  energy homeostasis, 55–56
  inflammaging, 58
  oxidative stress, 58–59
  protein glycation, 59–60, 59*f*
  reactive species, 57, 57*t*
  ROS production, 56–57, 57*t*
  ROS regulation, 56, 56*f*
multiple chronic diseases, 36
normality (health) *vs.* disease, 36–37
nutrient sensing pathways (NSPs)
  activation, 46–47
  autophagy, 46, 52–54
  FOXO3, 47–48
  insulin/IGF-1 pathway, 48–49
  mTOR pathway, 49–50
  representation, 46, 46*f*
  sirtuin pathway, 50–51
proteostasis, 54–55
stem cell exhaustion, 61–62
telomere attrition, 60
Peripheral oscillators
  Food Entrainable Oscillator (FEO), 225
  metabolic processes, 224
  suprachiasmatic nucleus
    (SCN) activity, 224–225
Physical activity, 275*f*
Physical frailty, 176
Physical inactivity/sedentary behavior, 189
Physiological syndrome, 173–174. *See also* Frailty

Plasmacytoid DCs (pDC), 118
Pneumonia
    antibody titers, 136
    community-acquired pneumonia (CAP), 137
    conjugate vaccine, 137
    incidence, 136
    mortality rates, 136
    polysaccharide vaccine, 137
    serotype replacement, 137
Polyunsaturated fatty acids (PUFA)
    AA/EPA ratio, 303
    dietary sources, 301–302
    importance, 303
    n-3 index, 302–303, 302f
Population aging, 329–330
    age structural transition, 20
    causes of death, 25
    centenarians, 15, 15t
    chronological age, 13
    demographic transition theory, 21
    economics
        economic old-age dependency ratio
            (EOADR), 30–31
        measures, 30t
        old-age dependency ratio
            (OADR), 29
        prospective old-age dependency ratio
            (POADR), 29
    education, 25
    epidemiological transition theory, 21–22
    gender, 16, 22–23, 23f, 24t
    Global Activity Limitation Indicator
        (GALI), 27–29
    health transition theory, 22
    healthy life years/disability-free life expectancy,
        26
    immigrants, 16
    income, 13–15, 14t
    labor-intensive long-term care
        (LTC), 32–33
    life expectancy
        at birth, 16–19, 18f
        child mortality, 20
        quality of life, 19–20, 19f
    mental health, 29
    policies, 31–32
    poor/fair health, 26, 27f
    population pyramids, 16, 17f, 31
    socioeconomic status, 26

    Years of Good Life (YoGL), 27, 28f
Population pyramids, 16, 17f, 31
Presbycusis, 39
Preventive/therapeutic strategies, 98–99, 99f
Progeroid syndromes, 42
Prokaryotic kingdom and aging, 2–3
Prospective old-age dependency ratio
    (POADR), 29
Protein glycation, 59–60, 59f
Proteostasis, 54–55, 280–281

R
Reactive oxygen species (ROS)
    biological function, 107
    dysbiosis, 108–109
    enzymes, 105–107
    free radical damage theory, 105
    future perspective, 109
    generation, 105–107
    gut homeostasis, 108–109
    inflammation, 108–109
    mitochondrial DNA, 105–107
    oxidative stress theory, 107–108
    production, 181–182
    signaling, 108–109
Resilience signaling
    brain modulation, 156–157
    brain regional specificity
        cognitive and affective processes, 158
        executive control and emotional arousal
            function, 158–159, 159f
        glucocorticoid receptor stimulation, 159
        neurogenesis and mental health, 159–160
        stress/brain trauma, 158–159
    description, 155
    endophenotypes, 156–157
    future perspectives, 169–170
    gasotransmitter, 160–166
    glutathione disulfide (GSSG), 160–166
    hydrogen sulfide, 160–166
    longevity assurance processes, 166–168
    neuronal plasticity, 160–166
    NLRP3 inflammasome, 160–166
    plant polyphenols/vitagenes, 166–169
    positive affect, 156–157
    preconditioning and postconditioning methods,
        156–157
    redox-resilient molecules, 160–166
    resilience processes, 155, 156f

sulfhydration, 160–166
thioredoxin (Trx) pathway, 160–166
Resveratrol, 259–261
Reverse vaccinology, 143
RNA and transcriptome, 308

S

Saffron, 265–266
Sarcopenia, 342
  anabolic resistance, 182–183
  characteristics, 176–177
  comorbidities, 177
  development and progression, 179*f*
  diagnostic criteria and cutoff values, 177, 178*t*
  endocrine factors, 183–184
  genetic components, 185–188, 186–187*t*
  inflammation, 184–185
  lifestyle risk factors
    malnutrition, 188–189
    physical inactivity/sedentary behavior, 189
  management
    anabolic medications, 191–192
    nutrition, 190–191
    pharmacological treatments, 191–192
    physical activity, 190–191
  mitochondrial dysfunction
    apoptosis, 181–182
    energy supply, 180
    reactive oxygen species (ROS) production, 181–182
    skeletal muscle atrophy, 180–181
  muscle structure and function changes
    muscle fibers, 179
    satellite cells, 179–180
  prevalence, 177
  weight loss, 188–189
SARS-CoV-2 vaccines, 147
Satellite cells, 179–180
Senescence-associated secretory phenotype (SASP)
  acquisition, 75
  and aging, 82–83
  appearance, 76–77, 77*t*
  characteristics, 75
  components, 78
  epithelial cells (ECs), 80–82
  extracellular vesicles (EVs), 79–80
  future perspectives, 83–84
  mesenchymal stem/stromal cells (MSCs), 80–82
  miRNA and extracellular vesicles (EVs), 79–80

senolytics, 83–84
signaling pathway, 76–77, 77*t*
Senile dementia, 255–256
Senoinflammation, 92–93
Senolytics, 83–84
Skeletal muscle atrophy, 180–181
Sleep restriction, 222
Slowing aging
  diet, 273–274, 349–350
  future perspectives, 351–352
  hormonal therapy, 348
  lifestyle, 348–349
  metformin, 351
  mTOR inhibitors, 351
  nutraceuticals, 350
  nutrigerontology, 350
  plant compounds, 350
  vitagenes, 350
Socioeconomic status, 26
Somatopause, 210
Stem cell exhaustion, 61–62, 285
Systemic factors, 3

T

T2DM, 94–95
T cells
  adipocytes, 115
  chronic CMV infection, 115
  phenotype, 114
  thymic involution, 114
Telomere attrition, 60, 277–278
Testosterone, 183–184
Theobroma cacao, 263–264
Thyroid function, 212–213
Thyroid stimulating hormone (TSH), 207
Toxic AGEs (TAGEs), 94–95

V

Vaccination
  adjuvants
    alum/aluminum salts, 132–133
    AS03, 133
    complete Freund's adjuvant (CFA), 130
    dendritic cells (DCs), 132
    diphtheria, 130
    incomplete Freund's adjuvant (IFA), 130
    MF59, 133

Vaccination *(Continued)*
  efficiency, 129
  future perspectives, 146–147
  immunosenescence, 129–130
  influenza
    acute respiratory viral infections, 134
    adjuvanted vaccine, 135–136
    humoral immunity, 135
    laboratory-confirmed hospitalized
      influenza, 136
    license, 135
    public health, 136
    split-virus/subunit formulations, 134
  interleukin-7, 143–144
  mitogen-activated protein and adenosine
      monophosphate-activated protein kinases
    bioactive compounds, 145–146
    hydroxytyrosol, 146
    oleuropein, 146
    pathway, 144–145
    sestrins, 144–145
    small-molecule kinase inhibitors (SMKIs), 145
  mRNA-1273 vaccine, 147
  reverse vaccinology, 143
  SARS-CoV-2 vaccines, 147
  Streptococcus pneumoniae
    antibody titers, 136
    community-acquired pneumonia (CAP), 137
    conjugate vaccine, 137
    incidence, 136
    mortality rates, 136
    polysaccharide vaccine, 137
    serotype replacement, 137
  TLR agonists
    AS04, 140
    combined adjuvants, 141
    efficiencies, 139–140
    pathogen-associated molecular patterns
      (PAMPs), 139
  varicella zoster virus (VZV)
    incidence, 138

    live-attenuated vaccine, 138
    occurrence, 138
    recombinant vaccine, 138–139
  viral vectors, 142–143
  virosomes, 141–142
Variables of aging, 1–2
Varicella zoster virus (VZV)
  incidence, 138
  live-attenuated vaccine, 138
  occurrence, 138
  recombinant vaccine, 138–139
Viral vectors, 142–143
Virosomes, 141–142
Vitagenes, 350
  adaptive responses, 169
  Aβ deposition, 167–168
  cellular redox homeostasis, 167–168
  3,4-dihydroxyphenylacetic acid (DOPAC),
    167–168
  epigenetic memory, 166–167
  HO-1 activity, 167–168
  inflammatory response and oxidative stress,
    168–169
  longevity assurance processes, 166–167
Vitamin D, 183–184
Vitamins
  analytical considerations, 299
  antioxidant vitamins, 299–301
  vitamin D, 301
Vitis vinifera, 259–261
Vulnerability, 331–332

W

Weight loss, 188–189
Western diet, 234–235

Y

Years of Good Life (YoGL), 27, 28*f*

Z

Zeitgebers, 220

9780128225691